Accession no.
36189309

KU-572-473

System Analysis Design UML Version 2.0

AN OBJECT-ORIENTED APPROACH

Fourth Edition

International Student Version

David Tegarden
Virginia Tech

Alan Dennis
Indiana University

Barbara Haley Wixom
University of Virginia

LIS - LIBRARY
Date 13-12-13
Fund C-che
Order No. 244981X
University of Chester

WILEY

John Wiley & Sons, Inc.

Copyright © 2013 John Wiley & Sons Singapore Pte. Ltd.

Cover image from © DrHitch/Shutterstock

Contributing Subject Matter Expert: Robert Dollinger, University of Wisconsin-Stevens Point

Founded in 1807, John Wiley & Sons, Inc. has been a valued source of knowledge and understanding for more than 200 years, helping people around the world meet their needs and fulfill their aspirations. Our company is built on a foundation of principles that include responsibility to the communities we serve and where we live and work. In 2008, we launched a Corporate Citizenship Initiative, a global effort to address the environmental, social, economic, and ethical challenges we face in our business. Among the issues we are addressing are carbon impact, paper specifications and procurement, ethical conduct within our business and among our vendors, and community and charitable support. For more information, please visit our website: www.wiley.com/go/citizenship.

All rights reserved. **This book is authorized for sale in Europe, Asia, Africa and the Middle East only and may not be exported outside of these territories.** Exportation from or importation of this book to another region without the Publisher's authorization is illegal and is a violation of the Publisher's rights. The Publisher may take legal action to enforce its rights. The Publisher may recover damages and costs, including but not limited to lost profits and attorney's fees, in the event legal action is required.

No part of this publication may be reproduced, stored in a retrieval system, or transmitted in any form or by any means, electronic, mechanical, photocopying, recording, scanning, or otherwise, except as permitted under Section 107 or 108 of the 1976 United States Copyright Act, without either the prior written permission of the Publisher or authorization through payment of the appropriate per-copy fee to the Copyright Clearance Center, Inc., 222 Rosewood Drive, Danvers, MA 01923, website www.copyright.com. Requests to the Publisher for permission should be addressed to the Permissions Department, John Wiley & Sons, Inc., 111 River Street, Hoboken, NJ 07030, (201) 748-6011, fax (201) 748-6008, website http://www.wiley.com/go/permissions.

ISBN: 978-1-118-09236-1

Printed in Asia

10 9 8 7 6 5 4 3 2 1

CONTENTS

Chapter 6
Behavioral Modeling   236

■ PART TWO
DESIGN   271

Chapter 7
System Design   273

Chapter 8
Class and Method Design   317

Chapter 9
Data Base Design   367

Chapter 10
User Interface Design   412

Chapter 13
Installation 545

Chapter 11
Architecture 473

PART THREE
IMPLEMENTATION 513

Chapter 12
Development 515

PREFACE

PURPOSE OF THIS BOOK

Systems Analysis and Design (SAD) is an exciting, active field in which analysts continually learn new techniques and approaches to develop systems more effectively and efficiently. However there is a core set of skills that all analysts need to know—no matter what approach or methodology is used. All information systems projects move through the four phases of planning, analysis, design, and implementation; all projects require analysts to gather requirements, model the business needs, and create blueprints for how the system should be built; and all projects require an understanding of organizational behavior concepts like change management and team building. Today, the cost of developing modern software is composed primarily of the cost associated with the developers themselves and not the computers. As such, object-oriented approaches to developing information systems hold much promise in controlling these costs.

Today, the most exciting change to systems analysis and design is the move to object-oriented techniques, which view a system as a collection of self-contained objects that have both data and processes. This change has been accelerated through the creation of the Unified Modeling Language (UML). UML provides a common vocabulary of object-oriented terms and diagramming techniques that is rich enough to model any systems development project from analysis through implementation.

This book captures the dynamic aspects of the field by keeping students focused on doing SAD while presenting the core set of skills that we feel every systems analyst needs to know today and in the future. This book builds on our professional experience as systems analysts and on our experience in teaching SAD in the classroom.

This book will be of particular interest to instructors who have students do a major project as part of their course. Each chapter describes one part of the process, provides clear explanations on how to do it, gives a detailed example, and then has exercises for the students to practice. In this way, students can leave the course with experience that will form a rich foundation for further work as a systems analyst.

OUTSTANDING FEATURES

A Focus on Doing SAD

The goal of this book is to enable students to do SAD—not just read about it, but understand the issues so they can actually analyze and design systems. The book introduces each major technique, explains what it is, explains how to do it, presents an example, and provides opportunities for students to practice before they do it for real in a project. After reading each chapter, the student will be able to perform that step in the system development process.

Rich Examples of Success and Failure

The book includes a running case about a fictitious company called CD Selections. Each chapter shows how the concepts are applied in situations at CD Selections. Unlike running cases in other books, we have tried to focus these examples on planning, managing, and executing the activities described in the chapter, rather than on detailed dialogue between fictitious actors. In this way, the running case serves as a template that students can apply to their own work. Each chapter also includes numerous Concepts in Action boxes, many of which were written by Dr. Bruce White from Quinnipiac University, that describe how real companies succeeded—and failed—in performing the activities in the chapter.

Real World Focus

The skills that students learn in a systems analysis and design course should mirror the work that they ultimately will do in real organizations. We have tried to make this book as "real" as possible by building extensively on our experience as professional systems analysts for organizations such as Arthur Andersen, IBM, the U.S. Department of Defense, and the Australian Army. We have also worked with a diverse industry advisory board of IS professionals and consultants in developing the book and have incorporated their stories, feedback, and advice throughout. Many students who use this book will eventually use the skills on the job in a business environment, and we believe they will have a competitive edge in understanding what successful practitioners feel is relevant in the real world.

Project Approach

We have presented the topics in this book in the order in which an analyst encounters them in a typical project. Although the presentation is necessarily linear (because students have to learn concepts in the way in which they build on each other), we emphasize the iterative, complex nature of SAD as the book unfolds. The presentation of the material should align well with courses that encourage students to work on projects because it presents topics as students need to apply them.

WHAT'S NEW IN THIS EDITION

In this edition, we have increased the coverage of and better organized the text around the enhanced Unified Process; provided a greater focus on nonfunctional requirements; provided a greater emphasis on the iterative and incremental development associated with object-oriented analysis and design; added figures and examples, along with additional explanatory text that addresses some of the more difficult concepts to learn; better aligned the CD selections case material; and did some major reorganization. Details of the major changes are as follows:

1. Given the lack of object-oriented programming experience of the typical student and the importance of understanding basic object-oriented concepts to perform object-oriented systems analysis and design, the appendix entitled "Basic Characteristics of Object-Oriented Systems" has been incorporated in Chapter 1.

2. Due to the popularity of the so-called agile approaches to systems development, we have greatly increased their coverage throughout the text. In Chapter 1, we have expanded the coverage of both XP and SCRUM. In Chapter 2, we have added a section regarding "Jelled" teams and their importance when considering staffing requirements of projects. In Chapter 3, we have added story cards and

task lists as additional approaches for gathering and documenting requirements. We have also greatly increased the focus on testing throughout the text. For example, the verification and validation material in the Moving On to Design chapter has been distributed over the analysis modeling chapters and the Moving On to Design chapter.

3. Given the differences between traditional project management and object-oriented project management, the project management material has been rewritten to reflect more of an object-oriented flavor. However, since much of the traditional project management material is still useful within an object-oriented context, we still cover it, e.g., net present value and return on investment, break-even point, work breakdown structures, Gantt charts, network diagrams and PERT analysis. The reorganization and rewriting of project management material allowed us to apply better the iterative and incremental development characteristics of object-oriented systems development to project management. Finally, we replaced the project size estimation section with an expansion of the use case points material that was in the functional modeling chapter in the previous edition.

4. To increase the focus on business processes, we have reorganized and expanded the functional modeling material. In this edition, minimize the potential overload of different notations used for business process modeling, e.g., the business process modeling notation or data flow diagrams, we have aligned the use case construct with the idea of a business process. Consequently, a use case diagram can be used to provide an overview of the different business processes and how they interrelate. Each use case can then be decomposed by creating an activity diagram to represent the details of each use case. Furthermore, each use case can be described with a use case description.

5. As in the third edition, the material included within the analysis modeling chapters has been more tightly coupled. This is especially true with regard to the idea of iterative and incremental development. The text now emphasizes that systems must be incrementally built by iterating over each of the models and over the intersection of the models. For example, the normal flow of events contained within a use-case description is associated with the activities on an activity diagram, the operations on a class diagram, the behaviors on the CRC cards, the messages on sequence and communication diagrams, and transitions on behavioral state machines. As such, any change to any one of these most likely will force changes in the others. Furthermore, we have extended CRUD analysis to CRUDE analysis that includes the idea of simply executing a method associated with another object.

6. With regards to the requirements determination, we have expanded the coverage of non-functional requirements throughout the design modeling chapters.

7. We have expanded the material that addresses global concerns. For example, we have created a new section that addresses international and cultural issues with regard to user interface design and we have expanded the coverage of cultural issues with regards to construction and the installation and operations of information systems.

8. Given all of the technological changes that have taken place since the third edition, we have also included material that addresses NoSQL data stores, mobile computing, social media, cloud computing, and green IT in the design modeling chapters.

9. To decrease the cognitive load required for much of the material in the text, additional figures and explanatory material have been added.

Finally, to provide a more complete version of the CD Selection case, we have moved the case to an online format. However, at the end of each chapter in the text, a very short synopsis of the case is provided.

ORGANIZATION OF THIS BOOK

This edition of the book is loosely organized around the phases and workflows of the enhanced Unified Process. Each chapter has been written to teach students specific tasks that analysts need to accomplish over the course of a project, and the deliverables that will be produced from the tasks. As students complete the chapters, they will realize the iterative and incremental nature of the tasks in object-oriented systems development.

Chapter 1 introduces the SDLC, systems development methodologies, roles and skills needed for a systems analyst, the basic characteristics of object-oriented systems, object-oriented systems analysis, the Unified Process, and the UML. Chapter 2 presents topics related to the project management workflow of the Unified Process, including project identification, system request, feasibility analysis, project selection, traditional project management tools (including work breakdown structures, network diagrams, and PERT analysis), project effort estimation using use case points, evolutionary work breakdown structures, iterative workplans, scope management, timeboxing, risk management, and staffing the project. Chapter 2 also addresses issues related to the Environment and Infrastructure management workflows of the Unified Process.

Part One focuses on creating analysis models. Chapter 3 introduces students to an assortment of analysis techniques to help with business automation, business improvement, and business process reengineering, a variety of requirements-gathering techniques that are used to determine the functional and nonfunctional requirements of the system, and to a system proposal. Chapter 4 focuses on constructing business process and functional models using use case diagrams, activity diagrams, and use case descriptions. Chapter 5 addresses producing structural models using CRC cards, class diagrams, and object diagrams. Chapter 6 tackles creating behavioral models using sequence diagrams, communication diagrams, behavioral state machines, and CRUDE analysis and matrices. Chapters 4 through 6 also cover the verification and validation of the models described in each chapter.

Part Two addresses design modeling. In Chapter 7, students learn how to verify and validate the analysis models created during analysis modeling and to evolve the analysis models into design models via the use of factoring, partitions, and layers. The students also learn to create an alternative matrix that can be used to compare custom, packaged, and outsourcing alternatives. Chapter 8 concentrates on designing the individual classes and their respective methods through the use of contracts and method specifications. Chapter 9 presents the issues involved in designing persistence for objects. These issues include the different storage formats that can be used for object persistence, how to map an object-oriented design into the chosen storage format, and how to design a set of data access and manipulation classes that act as a translator between the classes in the application and the object persistence. This chapter also focuses on the nonfunctional requirements that impact the data management layer. Chapter 10 presents the design of the human–computer interaction layer, where students learn how to design user interfaces using use scenarios, windows navigation diagrams, storyboards, windows layout diagrams, HTML prototypes, language prototypes, real use cases, interface standards, and user interface templates, to perform user interface evaluations using heuristic evaluation, walkthrough evaluation, interactive evaluation, and formal usability testing, and to address nonfunctional requirements such as user interface layout, content awareness, aesthetics, user experience, and consistency. This chapter also addresses issues related to mobile computing, social media, and international and cultural issues with regards to user interface design. Chapter 11 focuses on the physical architecture and infrastructure design, which includes deployment diagrams and hardware/software specification. In today's world, this also includes issues related to cloud computing and green IT. This chapter, like the previous design chapters, covers the impact that nonfunctional requirements can have on the physical architecture layer.

Part Three provides material that is related to the construction, installation, and operations of the system. Chapter 12 focuses on system construction, where students learn how to build, test, and document the system. Installation and operations are covered in Chapter 13, where students learn about the conversion plan, change management plan, support plan, and project assessment. Additionally, these chapters address the issues related to developing systems in a flat world, where developers and users are distributed throughout the world.

ACKNOWLEDGMENTS

For the fourth edition, we would like to thank the students of the ACIS 3515: Information Systems Development I and ACIS 3516: Information Systems Development II classes at Virginia Tech for giving many suggestions that drove most of the changes from the third edition to the fourth edition. We would like to especially thank Ashley, Ben, Daniel, Jason, Jason, Jason (yes, there were three of them), Kyle, Lucy, and Omar. Their suggestions were invaluable in improving the text and examples.

We would like to thank the following reviewers for their helpful and insightful comments on the fourth edition: David Champion, DeVry University, Columbus, OH campus; Jeff Cummings, Indiana University; Junhua Ding, East Carolina University; Robert Dollinger, University of Wisconsin-Stevens Point; Abhijit Dutt, Carnegie Mellon University; Yujong Hwang, DePaul University; Zongliang Jiang, North Carolina A&T State University; Raymond Kirsch, La Salle University; Gillian Lee, Lander University; Steve Machon, DeVry University; Makoto Nakayama, College of CDM, DePaul University; Parasuraman Nurani, Devry University; Selwyn Piramuthu, University of Florida; Iftikhar Sikder, Cleveland State University; Fan Zhao, Florida Gulf Coast University; and Dan Zhu, Iowa State University.

For the third edition, we would like to thank the students of the ACIS 3515: Information Systems Development I and ACIS 3516: Information Systems Development II classes at Virginia Tech for giving many suggestions that drove most of the changes from the second edition to the third edition. Their feedback was invaluable in improving the text and examples.

We would also like to thank the following reviewers for their helpful and insightful comments on the first, second, and third editions: Evans Adams, Fort Lewis College; Murugan Anandarajon, Drexel University; Ron Anson, Boise State University; Noushin Ashrafi, University of Massachusetts, Boston; Dirk Baldwin, University of Wisconsin-Parkside; Robert Barker, University of Louisville; Qing Cao, University of Missouri–Kansas City; Terry Fox, Baylor University; Ahmad Ghafarian, North Georgia College & State University; Donald Golden, Cleve-land State University; Cleotilde Gonzalez, Carnegie Melon University; Daniel V. Goulet, University of Wisconsin–Stevens Point; Harvey Hayashi, Loyalist College of Applied Arts and Technology; Scott James, Saginaw Valley State University; Rajiv Kishore, State University of New York–Buffalo; Ravindra Krovi, University of Akron; Jean-Piere Kuilboer, University of Massachusetts, Boston; Leo

Legorreta, California State University Sacramento; Diane Lending, James Madison University; Major Fernando Maymi, West Point University; Daniel Mittleman, DePaul University; Fred Niederman, Saint Louis University; H. Robert Pajkowski, DeVry Institute of Technology, Scarborough, Ontario; June S. Park, University of Iowa; Graham Peace, West Virginia University; Tom Pettay, DeVry Institute of Technology, Columbus, Ohio; J. Drew Procaccino, Rider University; Neil Ramiller, Portland State University; Eliot Rich, University at Albany, State University of New York; Marcus Rothenberger, University of Wisconsin–Milwaukee; Carl Scott, University of Houston; Keng Siau, University of Nebraska–Lincoln; Jonathan Trower, Baylor University; June Verner, Drexel University; Anna Wachholz, Sheridan College; Bill Watson, Indiana University–Purdue University Indianapolis; Randy S.Weinberg, Carnegie Mellon University; Eli J.Weissman, DeVry Institute of Technology, Long Island City, NY; Heinz Roland Weistroffer, Virginia Commonwealth University; Amy Wilson, DeVry Institute of Technology, Decatur, GA; Amy Woszczynski, Kennesaw State University; and Vincent C.Yen, Wright State University.

SUPPLEMENTS http://www.wiley.com/go/global/dennis

Instructor's Resources Web Site

- PowerPoint slides, which instructors can tailor to their classroom needs and that students can use to guide their reading and studying activities
- Test Bank, that includes a variety of questions ranging from multiple choice to essay style questions. A computerized version of the Test Bank will also be available.

Online Instructor's Manual

The Instructor's Manual provides resources to support the instructor both inside and out of the classroom:

- Short experiential exercises that instructors can use to help students experience and understand key topics in each chapter.
- Short stories have been provided by people working in both corporate and consulting environments for instructors to insert into lectures to make concepts more colorful and real
- Additional minicases for every chapter allow students to perform some of the key concepts that were learned in the chapter.
- Solutions to end of chapter questions and exercises are provided.

Student Website

- Relevant Web links, including career resources Web site.
- Web quizzes help students prepare for class tests.

Software Tools

Three Software Tools can be purchased with the text in special packages:

1. Visible Systems Corporation's Visible Analyst Student Edition.
2. Microsoft's Visio.
3. Microsoft's Project.

A 60-day trial edition of Microsoft Project can be purchased with the textbook. Note that Microsoft has changed their policy and no longer offers the 120-day trial previously available.

Another option now available to education institutions adopting this Wiley textbook is a free 3-year membership to the MSDN **Academic Alliance.** The MSDN AA is designed to provide the easiest and most inexpensive way for academic departments to make the latest Microsoft software available in labs, classrooms, and on student and instructor PCs.

Microsoft Project 2007 software is available through this Wiley and Microsoft publishing partnership, free of charge with the adoption of any qualified Wiley textbook. Each copy of Microsoft Project is the full version of the software, with no time limitations, and can be used indefinitely for educational purposes. For more information about the MSDN AA program, go to http://msdn.microsoft.com/academic/.

Contact your local Wiley sales representative for details, including pricing and ordering information.

CHAPTER 1

INTRODUCTION TO SYSTEMS ANALYSIS AND DESIGN

Chapter 1 introduces the systems development life cycle (SDLC), the fundamental four-phase model (planning, analysis, design, and implementation) common to all information systems development projects. It describes the evolution of system development methodologies and discusses the roles and skills required of a systems analyst. The chapter then overviews the basic characteristics of object-oriented systems and the fundamentals of object-oriented systems analysis and design, and closes with a description of the Unified Process and its extensions and the Unified Modeling Language.

OBJECTIVES

- Understand the fundamental systems development life cycle and its four phases
- Understand the evolution of systems development methodologies
- Be familiar with the different roles played by and the skills of a systems analyst
- Be familiar with the basic characteristics of object-oriented systems
- Be familiar with the fundamental principles of object-oriented systems analysis and design
- Be familiar with the Unified Process, its extensions, and the Unified Modeling Language

CHAPTER OUTLINE

Introduction
The Systems Development Life Cycle
   Planning
   Analysis
   Design
   Implementation
Systems Development Methodologies
   Structured Design
   Rapid Application Development (RAD)
   Agile Development
   Selecting the Appropriate Development
     Methodology
Typical Systems Analyst Roles and Skills
   Business Analyst
   Systems Analyst
   Infrastructure Analyst

   Change Management Analyst
   Project Manager
Basic Characteristics of Object-Oriented Systems
   Classes and Objects
   Methods and Messages
   Encapsulation and Information Hiding
   Inheritance
   Polymorphism and Dynamic Binding
Object-Oriented Systems Analysis and Design (OOSAD)
   Use-Case Driven
   Architecture-Centric
   Iterative and Incremental
   Benefits of Object-Oriented Systems
     Analysis and Design

INTRODUCTION

The *systems development life cycle (SDLC)* is the process of understanding how an information system (IS) can support business needs by designing a system, building it, and delivering it to users. If you have taken a programming class or have programmed on your own, this probably sounds pretty simple. Unfortunately, it is not. A 1996 survey by the Standish Group found that 42 percent of all corporate IS projects were abandoned before completion. A similar study done in 1996 by the General Accounting Office found 53 percent of all U.S. government IS projects were abandoned. Unfortunately, many of the systems that are not abandoned are delivered to the users significantly late, cost far more than planned, and have fewer features than originally planned. Most of us would like to think that these problems only happen to "other" people or "other" organizations, but they happen in most companies. Even Microsoft has a history of failures and overdue projects (e.g., Windows 1.0, Windows 95).[1] Although we would like to promote this book as a silver bullet that will keep you from IS failures, we readily admit that a silver bullet that guarantees IS development success simply does not exist. Instead, this book provides you with several fundamental concepts and many practical techniques that you can use to improve the probability of success.

The key person in the SDLC is the systems analyst, who analyzes the business situation, identifies opportunities for improvements, and designs an information system to implement them. Being a systems analyst is one of the most interesting, exciting, and challenging jobs around. Systems analysts work with a variety of people and learn how they conduct business. Specifically, they work with a team of systems analysts, programmers, and others on a common mission. Systems analysts feel the satisfaction of seeing systems that they designed and developed make a significant business impact, knowing that they contributed unique skills to make that happen.

However, the primary objective of a systems analyst is not to create a wonderful system; instead, it is to create value for the organization, which for most companies means increasing profits (government agencies and not-for-profit organizations measure value differently). Many failed systems have been abandoned because the analysts tried to build a wonderful system without clearly understanding how the system would fit with an organization's goals, current business processes, and other information systems to provide value. An investment in an information system is like any other investment, such as a new machine tool. The goal is not to acquire the tool, because the tool is simply a means to an end; the goal is to enable the organization to perform work better so it can earn greater profits or serve its constituents more effectively.

This book introduces the fundamental skills a systems analyst needs. This pragmatic book discusses best practices in systems development; it does not present a general survey of systems development that covers everything about the topic. By definition, systems analysts *do things* and challenge the current way that organizations work. To get the most

---

[1] For more information on the problem, see Capers Jones, *Patterns of Software System Failure and Success* (London: International Thompson Computer Press, 1996); Capers Jones, *Assessment and Control of Software Project Risks* (Englewood Cliffs, NJ: Yourdon Press, 1994); Julia King, "IS Reins in Runaway Projects," *Computer world* (February 24, 1997).

out of this book, you will need to actively apply to your own systems development project the ideas and concepts in the examples and in the "Your Turn" exercises that are presented throughout. This book guides you through all the steps for delivering a successful information system. Also, it illustrates how one organization (called CD Selections) applies the steps in one project (developing a Web-based CD sales system). By the time you finish the book, you won't be an expert analyst, but you will be ready to start building systems for real.

This chapter first introduces the basic SDLC that IS projects follow. This life cycle is common to all projects, although the focus and approach to each phase of the life cycle may differ. The next section describes three fundamentally different types of systems development methodologies: structured design, rapid application development, and agile development. The third section describes the roles played by and the skills necessary for a systems analyst. The final four sections introduce the fundamental characteristics of object-oriented systems, object-oriented systems analysis and design, a specific object-oriented systems development methodology (the Unified Process), and a specific object-oriented systems development graphical notation (the Unified Modeling Language).

CONCEPTS 1–A An Expensive False Start

IN ACTION

A real-estate group in the federal government cosponsored a data warehouse with the information technology (IT) department. In the formal proposal written by IT, costs were estimated at $800,000, the project's duration was estimated to be eight months, and the responsibility for funding was defined as the business unit's. The IT department proceeded with the project before it even knew if the project had been accepted.

The project actually lasted two years because requirements gathering took nine months instead of one and a half, the planned user base grew from 200 to 2,500, and the approval process to buy technology for the project took a year. Three weeks before technical delivery, the IT director canceled the project. This failed endeavor cost the organization and taxpayers $2.5 million.

Source: Hugh J. Watson et al., "Data Warehousing Failure: Case Studies and Findings," *The Journal of Data Warehousing* 4, (no. 1) (1999): 44–54.

Questions

1. Why did this system fail?
2. Why would a company spend money and time on a project and then cancel it?
3. What could have been done to prevent this?

THE SYSTEMS DEVELOPMENT LIFE CYCLE

In many ways, building an information system is similar to building a house. First, the house (or the information system) starts with a basic idea. Second, this idea is transformed into a simple drawing that is shown to the customer and refined (often through several drawings, each improving on the last) until the customer agrees that the picture depicts what he or she wants. Third, a set of blueprints is designed that presents much more detailed information about the house (e.g., the type of water faucets, where the telephone jacks will be placed). Finally, the house is built following the blueprints, often with some changes directed by the customer as the house is erected.

The SDLC has a similar set of four fundamental *phases*: planning, analysis, design, and implementation. Different projects might emphasize different parts of the SDLC or approach the SDLC phases in different ways, but all projects have elements of these four phases. Each *phase* is itself composed of a series of *steps*, which rely upon *techniques* that produce *deliverables* (specific documents and files that provide understanding about the project).

For example, in applying for admission to a university, all students go through the same phases: information gathering, applying, and accepting. Each of these phases has steps; for example, information gathering includes steps such as searching for schools, requesting information, and reading brochures. Students then use techniques (e.g., Internet searching) that can be applied to steps (e.g., requesting information) to create *deliverables* (e.g., evaluations of different aspects of universities).

In many projects, the SDLC phases and steps proceed in a logical path from start to finish. In other projects, the project teams move through the steps consecutively, incrementally, iteratively, or in other patterns. In this section, we describe the phases, the actions, and some of the techniques that are used to accomplish the steps at a very high level. Not all organizations follow the SDLC in exactly the same way. As we shall shortly see, there are many variations on the overall SDLC.

For now, there are two important points to understand about the SDLC. First, you should get a general sense of the phases and steps through which IS projects move and some of the techniques that produce certain deliverables. Second, it is important to understand that the SDLC is a process of *gradual refinement*. The deliverables produced in the analysis phase provide a general idea of the shape of the new system. These deliverables are used as input to the design phase, which then refines them to produce a set of deliverables that describes in much more detailed terms exactly how the system will be built. These deliverables, in turn, are used in the implementation phase to produce the actual system. Each phase refines and elaborates on the work done previously.

Planning

The *planning phase* is the fundamental process of understanding *why* an information system should be built and determining how the project team will go about building it. It has two steps:

1. During *project initiation,* the system's business value to the organization is identified: How will it lower costs or increase revenues? Most ideas for new systems come from outside the IS area (e.g., from the marketing department, accounting department) in the form of a *system request.* A system request presents a brief summary of a business need, and it explains how a system that supports the need will create business value. The IS department works together with the person or department that generated the request (called the *project sponsor*) to conduct a *feasibility analysis.*

   The feasibility analysis examines key aspects of the proposed project:

   - The idea's technical feasibility (Can we build it?)
   - The economic feasibility (Will it provide business value?)
   - The organizational feasibility (If we build it, will it be used?)

   The system request and feasibility analysis are presented to an information systems *approval committee* (sometimes called a steering committee), which decides whether the project should be undertaken.

2. Once the project is approved, it enters *project management.* During project management, the *project manager* creates a *workplan,* staffs the project, and puts techniques in place to help the project team control and direct the project through the entire SDLC. The deliverable for project management is a *project plan,* which describes how the project team will go about developing the system.

Analysis

The *analysis phase* answers the questions of *who* will use the system, *what* the system will do, and *where* and *when* it will be used. During this phase, the project team investigates any current system(s), identifies opportunities for improvement, and develops a concept for the new system.

This phase has three steps:

1. An *analysis strategy* is developed to guide the project team's efforts. Such a strategy usually includes an analysis of the current system (called the *as-is system*) and its problems and then ways to design a new system (called the *to-be system*).

2. The next step is *requirements gathering* (e.g., through interviews or questionnaires). The analysis of this information—in conjunction with input from the project sponsor and many other people—leads to the development of a concept for a new system. The system concept is then used as a basis to develop a set of business *analysis models,* which describe how the business will operate if the new system is developed. The set of models typically includes models that represent the data and processes necessary to support the underlying business process.

3. The analyses, system concept, and models are combined into a document called the *system proposal,* which is presented to the project sponsor and other key decision makers (e.g., members of the approval committee) who decide whether the project should continue to move forward.

The system proposal is the initial deliverable that describes what business requirements the new system should meet. Because it is really the first step in the design of the new system, some experts argue that it is inappropriate to use the term "analysis" as the name for this phase; some argue a better name would be "analysis and initial design." Most organizations continue to use the name *analysis* for this phase, however, so we use it in this book as well. Just keep in mind that the deliverable from the analysis phase is both an analysis and a high-level initial design for the new system.

Design

The *design phase* decides *how* the system will operate, in terms of the hardware, software, and network infrastructure; the user interface, forms, and reports; and the specific programs, databases, and files that will be needed. Although most of the strategic decisions about the system were made in the development of the system concept during the analysis phase, the steps in the design phase determine exactly how the system will operate. The design phase has four steps:

1. The *design strategy* is first developed. It clarifies whether the system will be developed by the company's own programmers, whether the system will be outsourced to another firm (usually a consulting firm), or whether the company will buy an existing software package.

2. This leads to the development of the basic *architecture design* for the system, which describes the hardware, software, and network infrastructure to be used. In most cases, the system will add or change the infrastructure that already exists in the organization. The *interface design* specifies how the users will move through the system (e.g., navigation methods such as menus and on-screen buttons) and the forms and reports that the system will use.

3. The *database and file specifications* are developed. These define exactly what data will be stored and where they will be stored.

4. The analyst team develops the *program design,* which defines the programs that need to be written and exactly what each program will do.

This collection of deliverables (architecture design, interface design, database and file specifications, and program design) is the *system specification* that is handed to the programming team for implementation. At the end of the design phase, the feasibility analysis and project plan are reexamined and revised, and another decision is made by the project sponsor and approval committee about whether to terminate the project or continue.

Implementation

The final phase in the SDLC is the *implementation phase,* during which the system is actually built (or purchased, in the case of a packaged software design). This is the phase that usually gets the most attention, because for most systems it is the longest and most expensive single part of the development process. This phase has three steps:

1. System *construction* is the first step. The system is built and tested to ensure it performs as designed. Because the cost of bugs can be immense, testing is one of the most critical steps in implementation. Most organizations give more time and attention to testing than to writing the programs in the first place.

2. The system is installed. *Installation* is the process by which the old system is turned off and the new one is turned on. It may include a direct cutover approach (in which the new system immediately replaces the old system), a parallel conversion approach (in which both the old and new systems are operated for a month or two until it is clear that there are no bugs in the new system), or a phased conversion strategy (in which the new system is installed in one part of the organization as an initial trial and then gradually installed in others). One of the most important aspects of conversion is the development of a *training plan* to teach users how to use the new system and help manage the changes caused by the new system.

3. The analyst team establishes a *support plan* for the system. This plan usually includes a formal or informal post-implementation review as well as a systematic way for identifying major and minor changes needed for the system.

CONCEPTS

IN ACTION

1–B Keeping Up with Consumer Electronics

Consumer electronics is a very competitive business. What might be the success story of the year one year is a forgotten item two years later. Rapid product commoditization makes the consumer electronics marketplace very competitive. Getting the right products to market at the right time with the right components is an ongoing challenge for telecommunications and consumer electronics goods companies.

Questions

1. What external data analysis should a consumer electronics company use to determine marketplace needs and its abilities to compete effectively in a marketplace?

2. Staying one step ahead of competitors requires a corporate strategy and the support of information systems. How can information systems and systems analysts contribute to an aggressive corporate strategy?

SYSTEMS DEVELOPMENT METHODOLOGIES

A *methodology* is a formalized approach to implementing the SDLC (i.e., it is a list of steps and deliverables). There are many different systems development methodologies, and each one is unique, based on the order and focus it places on each SDLC phase. Some methodologies are formal standards used by government agencies, whereas others have been developed by consulting firms to sell to clients. Many organizations have internal methodologies that have been honed over the years, and they explain exactly how each phase of the SDLC is to be performed in that company.

There are many ways to categorize methodologies. One way is by looking at whether they focus on business processes or the data that support the business. A *process-centered*

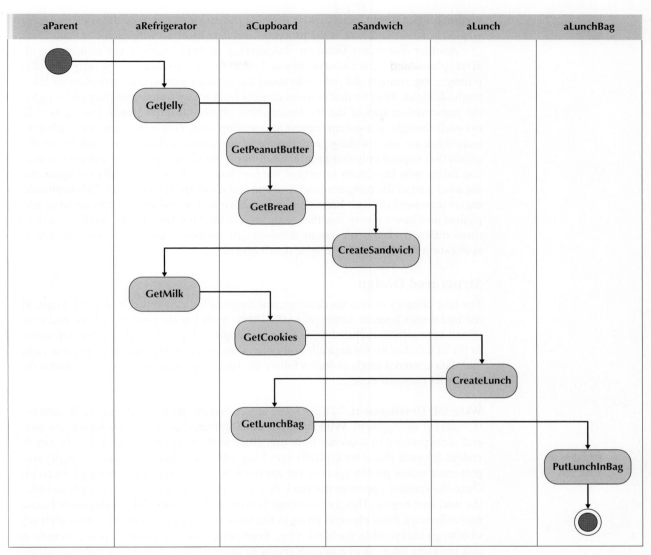

| aParent | aRefrigerator | aCupboard | aSandwich | aLunch | aLunchBag |

FIGURE 1-1 A Simple Behavioral Model for Making a Simple Lunch

methodology emphasizes process models as the core of the system concept. In Figure 1-1, for example, process-centered methodologies would focus first on defining the processes (e.g., assemble sandwich ingredients). *Data-centered* methodologies emphasize data models as the core of the system concept. In Figure 1-1, data-centered methodologies would focus first on defining the contents of the storage areas (e.g., refrigerator) and how the contents were organized.[2] By contrast, *object-oriented methodologies* attempt to balance the focus between process and data by incorporating both into one model. In Figure 1-1, these

[2] The classic modern process-centered methodology is that by Edward Yourdon, *Modern Structured Analysis* (Englewood Cliffs, NJ: Yourdon Press, 1989). An example of a data-centered methodology is information engineering; see James Martin, *Information Engineering*, vols. 1–3 (Englewood Cliffs, NJ: Prentice Hall, 1989). A widely accepted standardized non–object-oriented methodology that balances processes and data is IDEF; see FIPS 183, *Integration Definition for Function Modeling*, Federal Information Processing Standards Publications, U.S. Department of Commerce, 1993.

methodologies would focus first on defining the major elements of the system (e.g., sandwiches, lunches) and look at the processes and data involved with each element.

Another important factor in categorizing methodologies is the sequencing of the SDLC phases and the amount of time and effort devoted to each.[3] In the early days of computing, programmers did not understand the need for formal and well-planned life-cycle methodologies. They tended to move directly from a very simple planning phase right into the construction step of the implementation phase—in other words, from a very fuzzy, not-well-thought-out system request into writing code. This is the same approach that you sometimes use when writing programs for a programming class. It can work for small programs that require only one programmer, but if the requirements are complex or unclear, you might miss important aspects of the problem and have to start all over again, throwing away part of the program (and the time and effort spent writing it). This approach also makes teamwork difficult because members have little idea about what needs to be accomplished and how to work together to produce a final product. In this section, we describe three different classes of system development methodologies: structured design, rapid application development, and agile development.

Structured Design

The first category of systems development methodologies is called *structured design*. These methodologies became dominant in the 1980s, replacing the previous ad hoc and undisciplined approach. Structured design methodologies adopt a formal step-by-step approach to the SDLC that moves logically from one phase to the next. Numerous process-centered and data-centered methodologies follow the basic approach of the two structured design categories outlined next.

Waterfall Development   The original structured design methodology (still used today) is *waterfall development*. With waterfall development–based methodologies, the analysts and users proceed in sequence from one phase to the next (see Figure 1-2). The key deliverables for each phase are typically very long (often hundreds of pages in length) and are presented to the project sponsor for approval as the project moves from phase to phase. Once the sponsor approves the work that was conducted for a phase, the phase ends and the next one begins. This methodology is referred to as waterfall development because it moves forward from phase to phase in the same manner as a waterfall. Although it is possible to go backward in the SDLC (e.g., from design back to analysis), it is extremely difficult (imagine yourself as a salmon trying to swim upstream against a waterfall, as shown in Figure 1-2).

Structured design also introduced the use of formal modeling or diagramming techniques to describe the basic business processes and the data that support them. Traditional structured design uses one set of diagrams to represent the processes and a separate set of diagrams to represent data. Because two sets of diagrams are used, the systems analyst must decide which set to develop first and use as the core of the system: process-model diagrams or data-model diagrams. There is much debate over which should come first, the processes or the data, because both are important to the system. As a result, several different structured design methodologies have evolved that follow the basic steps of the waterfall model but use different modeling approaches at different times. Those that attempt to emphasize process-model diagrams as the core of the system are process centered, whereas those that emphasize data-model diagrams as the core of the system concept are data centered.

---

[3] A good reference for comparing systems development methodologies is Steve McConnell, *Rapid Development* (Redmond, WA: Microsoft Press, 1996).

FIGURE 1-2
A Waterfall
Development–based
Methodology

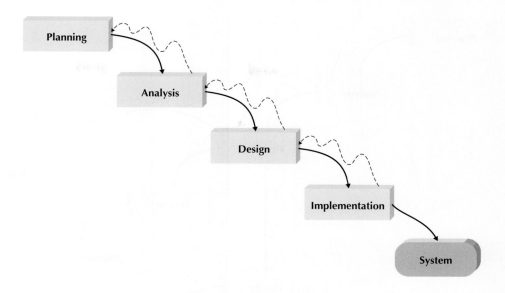

The two key advantages of the structured design waterfall approach are that it identifies system requirements long before programming begins and it minimizes changes to the requirements as the project proceeds. The two key disadvantages are that the design must be completely specified before programming begins and that a long time elapses between the completion of the system proposal in the analysis phase and the delivery of the system (usually many months or years). Lengthy deliverables often result in poor communication; the result is that important requirements can be overlooked in the voluminous documentation. Users are rarely prepared for their introduction to the new system, which occurs long after the initial idea for the system was introduced. If the project team misses important requirements, expensive post-implementation programming may be needed (imagine yourself trying to design a car on paper; how likely would you be to remember interior lights that come on when the doors open or to specify the right number of valves on the engine?).

A system can also require significant rework because the business environment has changed from the time that the analysis phase occurred. When changes do occur, it means going back to the initial phases and following the change through each of the subsequent phases in turn.

Parallel Development   *Parallel development* methodology attempts to address the problem of long delays between the analysis phase and the delivery of the system. Instead of doing design and implementation in sequence, it performs a general design for the whole system and then divides the project into a series of distinct subprojects that can be designed and implemented in parallel. Once all subprojects are complete, the separate pieces are integrated and the system is delivered (see Figure 1-3).

The primary advantage of this methodology is that it can reduce the time to deliver a system; thus, there is less chance of changes in the business environment causing rework. However, the approach still suffers from problems caused by paper documents. It also adds a new problem: Sometimes the subprojects are not completely independent; design decisions made in one subproject can affect another, and the end of the project can require significant integration efforts.

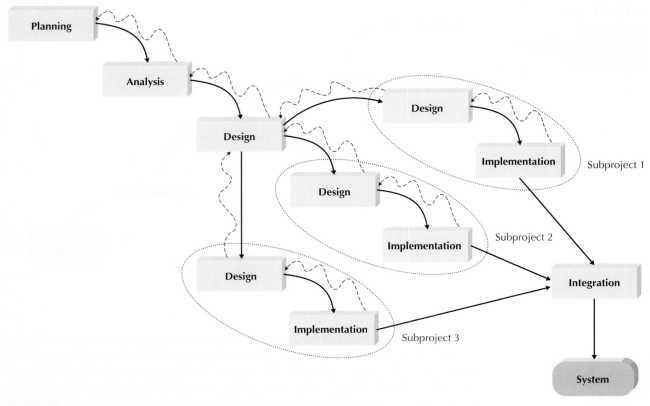

FIGURE 1-3   A Parallel Development-based Methodology

Rapid Application Development (RAD)

A second category of methodologies includes *rapid application development (RAD)*-based methodologies. These are a newer class of systems development methodologies that emerged in the 1990s. RAD-based methodologies attempt to address both weaknesses of structured design methodologies by adjusting the SDLC phases to get some part of the system developed quickly and into the hands of the users. In this way, the users can better understand the system and suggest revisions that bring the system closer to what is needed.[4]

Most RAD-based methodologies recommend that analysts use special techniques and computer tools to speed up the analysis, design, and implementation phases, such as computer-aided software engineering (CASE) tools, joint application design (JAD) sessions, fourth-generation or visual programming languages that simplify and speed up programming (e.g., Visual Basic), and code generators that automatically produce programs from design specifications. The combination of the changed SDLC phases and the use of these tools and techniques improves the speed and quality of systems development. However, there is one possible subtle problem with RAD-based methodologies: managing user expectations. Owing to the use of the tools and techniques that can improve the speed and quality of systems development, user expectations of what is possible can change dramatically. As a

---

[4] One of the best RAD books is Steve McConnell, *Rapid Development* (Redmond, WA: Microsoft Press, 1996).

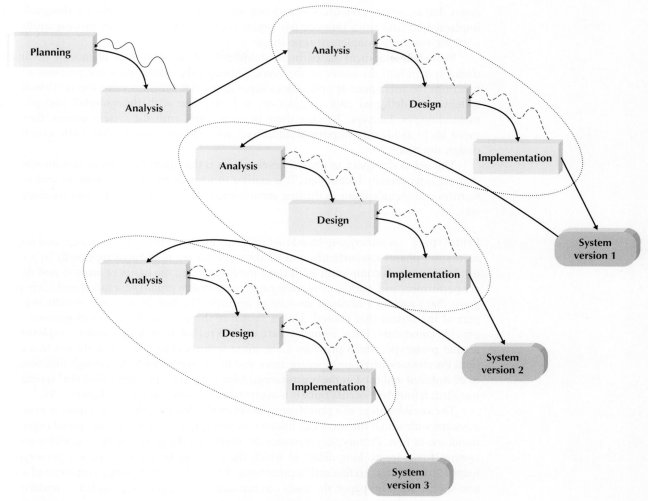

FIGURE 1-4   A Phased Development-based Methodology

user better understands the information technology (IT), the systems requirements tend to expand. This was less of a problem when using methodologies that spent a lot of time thoroughly documenting requirements. Process-centered, data-centered, and object-oriented methodologies that follow the basic approaches of the three RAD categories are described in the following sections.

Phased Development   A *phased development*-based methodology breaks an overall system into a series of *versions* that are developed sequentially. The analysis phase identifies the overall system concept, and the project team, users, and system sponsor then categorize the requirements into a series of versions. The most important and fundamental requirements are bundled into the first version of the system. The analysis phase then leads into design and implementation—but only with the set of requirements identified for version 1 (see Figure 1-4).

Once version 1 is implemented, work begins on version 2. Additional analysis is performed based on the previously identified requirements and combined with new ideas and issues that arose from the users' experience with version 1. Version 2 then is designed and implemented, and work immediately begins on the next version. This process continues until the system is complete or is no longer in use.

Phased development–based methodologies have the advantage of quickly getting a useful system into the hands of the users. Although the system does not perform all the functions the users need at first, it does begin to provide business value sooner than if the system were delivered after completion, as is the case with the waterfall and parallel methodologies. Likewise, because users begin to work with the system sooner, they are more likely to identify important additional requirements sooner than with structured design situations.

The major drawback to phased development is that users begin to work with systems that are intentionally incomplete. It is critical to identify the most important and useful features and include them in the first version and to manage users' expectations along the way.

Prototyping   A *prototyping*-based methodology performs the analysis, design, and implementation phases concurrently, and all three phases are performed repeatedly in a cycle until the system is completed. With these methodologies, the basics of analysis and design are performed, and work immediately begins on a *system prototype,* a quick-and-dirty program that provides a minimal amount of features. The first prototype is usually the first part of the system that is used. This is shown to the users and the project sponsor, who provide comments. These comments are used to reanalyze, redesign, and re-implement a second prototype, which provides a few more features. This process continues in a cycle until the analysts, users, and sponsor agree that the prototype provides enough functionality to be installed and used in the organization. After the prototype (now called the "system") is installed, refinement occurs until it is accepted as the new system (see Figure 1-5).

The key advantage of a prototyping-based methodology is that it *very* quickly provides a system with which the users can interact, even if it is not ready for widespread organizational use at first. Prototyping reassures the users that the project team is working on the system (there are no long delays in which the users see little progress), and prototyping helps to more quickly refine real requirements. Rather than attempting to understand a system specification on paper, the users can interact with the prototype to better understand what it can and cannot do.

The major problem with prototyping is that its fast-paced system releases challenge attempts to conduct careful, methodical analysis. Often the prototype undergoes such significant changes that many initial design decisions become poor ones. This can cause

FIGURE 1-5

A Prototyping-based Methodology

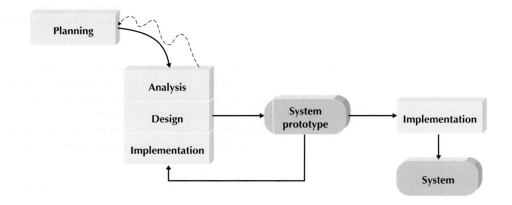

problems in the development of complex systems because fundamental issues and problems are not recognized until well into the development process. Imagine building a car and discovering late in the prototyping process that you have to take the whole engine out to change the oil (because no one thought about the need to change the oil until after it had been driven 10,000 miles).

Throwaway Prototyping   *Throwaway prototyping*-based methodologies are similar to prototyping-based methodologies in that they include the development of prototypes; however, throwaway prototypes are done at a different point in the SDLC. These prototypes are used for a very different purpose than those previously discussed, and they have a very different appearance (see Figure 1-6).

The throwaway prototyping–based methodologies have a relatively thorough analysis phase that is used to gather information and to develop ideas for the system concept. However, users might not completely understand many of the features they suggest, and there may be challenging technical issues to be solved. Each of these issues is examined by analyzing, designing, and building a *design prototype*. A design prototype is not a working system; it is a product that represents a part of the system that needs additional refinement, and it contains only enough detail to enable users to understand the issues under consideration. For example, suppose users are not completely clear on how an order-entry system should work. The analyst team might build a series of HTML pages viewed using a Web browser to help the users visualize such a system. In this case, a series of mock-up screens *appear* to be a system, but they really do nothing. Or suppose that the project team needs to develop a sophisticated graphics program in Java. The team could write a portion of the program with pretend data to ensure that they could do a full-blown program successfully.

A system developed using this type of methodology probably relies on several design prototypes during the analysis and design phases. Each of the prototypes is used to minimize the risk associated with the system by confirming that important issues are understood before the real system is built. Once the issues are resolved, the project moves into design and implementation. At this point, the design prototypes are thrown away, which is an important difference between these methodologies and prototyping methodologies, in which the prototypes evolve into the final system.

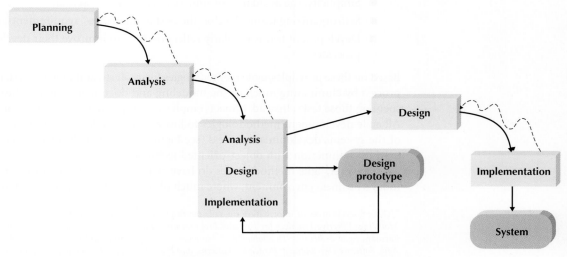

FIGURE 1-6   A Throwaway Prototyping–based Methodology

Throwaway prototyping-based methodologies balance the benefits of well-thought-out analysis and design phases with the advantages of using prototypes to refine key issues before a system is built. It can take longer to deliver the final system as compared to prototyping-based methodologies (because the prototypes do not become the final system), but this type of methodology usually produces more stable and reliable systems.

Agile Development[5]

A third category of systems development methodologies is still emerging today: *agile development*. All agile development methodologies are based on the agile manifesto and a set of twelve principles. The emphasis of the manifesto is to focus the developers on the working conditions of the developers, the working software, the customers, and addressing changing requirements instead of focusing on detailed systems development processes, tools, all-inclusive documentation, legal contracts, and detailed plans. These programming-centric methodologies have few rules and practices, all of which are fairly easy to follow. These methodologies are typically based only on the twelve principles of agile software. These principles include the following:

- Software is delivered early and continuously through the development process, satisfying the customer.
- Changing requirements are embraced regardless of when they occur in the development process.
- Working software is delivered frequently to the customer.
- Customers and developers work together to solve the business problem.
- Motivated individuals create solutions; provide them the tools and environment they need and trust them to deliver.
- Face-to-face communication within the development team is the most efficient and effective method of gathering requirements.
- The primary measure of progress is working, executing software.
- Both customers and developers should work at a pace that is sustainable. That is, the level of work could be maintained indefinitely without any worker burnout.
- Agility is heightened through attention to both technical excellence and good design.
- Simplicity, the avoidance of unnecessary work, is essential.
- Self-organizing teams develop the best architectures, requirements, and designs.
- Development teams regularly reflect on how to improve their development processes.

Based on these principles, agile methodologies focus on streamlining the system-development process by eliminating much of the modeling and documentation overhead and the time spent on those tasks. Instead, projects emphasize simple, iterative application development.[6] All agile development methodologies follow a simple cycle through the traditional phases of the systems development process (see Figure 1-7). Virtually all agile methodologies are used in conjunction with object-oriented technologies.

However, agile methodologies do have critics. One of the major criticisms deals with today's business environment, where much of the actual information systems development

---

[5] Three good sources of information on agile development and object-oriented systems are S. W. Ambler, *Agile Modeling: Effective Practices for Extreme Programming and The Unified Process* (New York: Wiley, 2002); C. Larman, *Agile & Iterative Development: A Manager's Guide* (Boston: Addison-Wesley, 2004); and R. C. Martin, *Agile Software Development: Principles, Patterns, and Practices* (Upper Saddle River, NJ: Prentice Hall, 2003).

[6] See www.agilealliance.com.

FIGURE 1-7
Typical Agile
Development
Methodology

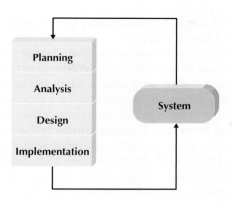

is offshored, outsourced, and/or subcontracted. Given agile development methodologies requiring co-location of the development team, this seems to be a very unrealistic assumption. A second major criticism is that if agile development is not carefully managed, and by definition it is not, the development process can devolve into a prototyping approach that essentially becomes a "programmers gone wild" environment where programmers attempt to hack together solutions. A third major criticism, based on the lack of actual documentation created during the development of the software, raises issues regarding the auditability of the systems being created. Without sufficient documentation, neither the system nor the systems-development process can be assured. A fourth major criticism is based on whether agile approaches can deliver large mission-critical systems.

Even with these criticisms, given the potential for agile approaches to address the application backlog and to provide timely solutions to many business problems, agile approaches should be considered in some circumstances. Furthermore, many of the techniques encouraged by attending to the underlying purpose of the agile manifesto and the set of twelve agile principles are very useful in object-oriented systems development. Two of the more popular examples of agile development methodologies are extreme programming (XP) and Scrum. We describe both of these methodologies in the next two sections.

Extreme Programming[7]   *Extreme programming* (*XP*) is founded on four core values: communication, simplicity, feedback, and courage. These four values provide a foundation that XP developers use to create any system. First, the developers must provide rapid feedback to the end users on a continuous basis. Second, XP requires developers to follow the KISS principle.[8] Third, developers must make incremental changes to grow the system, and they must not only accept change, they must embrace change. Fourth, developers must have a quality-first mentality. XP also supports team members in developing their own skills. Three of the key principles that XP uses to create successful systems are continuous testing, simple coding performed by pairs of developers, and close interactions with end users to build systems very quickly.

Testing and efficient coding practices are the core of XP. Code is tested each day and is placed into an integrative testing environment. If bugs exist, the code is backed out until

---

[7] For more information, see K. Beck, *eXtreme Programming Explained: Embrace Change* (Reading, MA: Addison-Wesley, 2000), C. Larman, *Agile & Iterative Development: A Manager's Guide* (Boston: Addison-Wesley, 2004), M. Lippert, S. Roock, and H. Wolf, *eXtreme Programming in Action: Practical Experiences from Real World Projects* (New York: Wiley, 2002), or www.extremeprogramming.com.

[8] Keep it simple, stupid.

it is completely free of errors. XP relies heavily on refactoring, which is a disciplined way to restructure code to keep it simple.

An XP project begins with user stories that describe what the system needs to do. Then, programmers code in small, simple modules and test to meet those needs. Users are required to be available to clear up questions and issues as they arise. Standards are very important to minimize confusion, so XP teams use a common set of names, descriptions, and coding practices. XP projects deliver results sooner than even the RAD approaches, and they rarely get bogged down in gathering requirements for the system.

XP adherents claim many strengths associated with developing software using XP. Programmers work closely with all stakeholders, and communication among all stakeholders is improved. Continuous testing of the evolving system is encouraged. The system is developed in an evolutionary and incremental manner, which allows the requirements to evolve as the stakeholders understand the potential that the technology has in providing a solution to their problem. Estimation is task driven and is performed by the programmer who will implement the solution for the task under consideration. Because all programming is done in pairs, a shared responsibility for each software component develops among the programmers. Finally, the quality of the final product increases during each iteration.

For small projects with highly motivated, cohesive, stable, and experienced teams, XP should work just fine. However, if the project is not small or the teams aren't jelled,[9] the success of an XP development effort is doubtful. This tends to throw into doubt the whole idea of bringing outside contractors into an existing team environment using XP.[10] The chance of outsiders jelling with insiders might simply be too optimistic. XP requires a great deal of discipline; otherwise projects will become unfocused and chaotic. XP is recommended only for small groups of developers—no more than ten developers—and it is not advised for large mission-critical applications. Owing to the lack of analysis and design documentation, there is only code documentation associated with XP, so maintaining large systems built with XP may be impossible. And because mission-critical business information systems tend to exist for a long time, the utility of XP as a business information system development methodology is in doubt. Finally, the methodology needs a lot of on-site user input, something to which many business units cannot commit.[11] However, some of the techniques associated with XP are useful in object-oriented systems development. For example, user stories, pair programming, and continuous testing are invaluable tools from which object-oriented systems development could benefit.

Scrum[12]   *Scrum* is a term that is well known to rugby fans. In rugby, a scrum is used to restart a game (see Figure 1-8). In a nutshell, the creators of the Scrum method believe that no matter how much you plan, as soon as the software begins to be developed, chaos breaks

---

[9] A *jelled team* is one that has low turnover, a strong sense of identity, a sense of eliteness, a feeling that they jointly own the product being developed, and enjoyment in working together. For more information regarding jelled teams, see T. DeMarco and T. Lister, *Peopleware: Productive Projects and Teams* (New York: Dorset/House, 1987).

[10] Considering the tendency for offshore outsourcing, this is a major obstacle for XP to overcome. For more information on offshore outsourcing, see P. Thibodeau, "ITAA Panel Debates Outsourcing Pros, Cons," *Computerworld Morning Update* (September 25, 2003), and S. W. Ambler, "Chicken Little Was Right," *Software Development* (October 2003).

[11] Many of the observations on the utility of XP as a development approach were based on conversations with Brian Henderson-Sellers.

[12] For more information, see C. Larman, *Agile & Iterative Development: A Manager's Guide* (Boston: Addison-Wesley, 2004); K. Schwaber and M. Beedle, *Agile Software Development with Scrum* (Upper Saddle River, NJ: Prentice Hall, 2001); and R. Wysocki, *Effective Project Management: Traditional, Agile, Extreme,* 5th Ed. (Indianapolis, IN: Wiley Publishing, 2009).

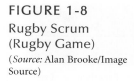

FIGURE 1-8
Rugby Scrum
(Rugby Game)
(*Source:* Alan Brooke/Image
Source)

out and the plans go out the window.[13] The best you can do is to react to where the prover-bial rugby ball squirts out. You then sprint with the ball until the next scrum. In the case of the Scrum methodology, a sprint lasts thirty working days. At the end of the sprint, a system is delivered to the customer.

Of all systems development approaches, on the surface, Scrum is the most chaotic. To control some of the innate chaos, Scrum development focuses on a few key practices. Teams are self-organized and self-directed. Unlike other approaches, Scrum teams do not have a designated team leader. Instead, teams organize themselves in a symbiotic manner and set their own goals for each sprint (iteration). Once a sprint has begun, Scrum teams do not consider any additional requirements. Any new requirements that are uncovered are placed on a backlog of requirements that still need to be addressed. At the beginning of every work-day, a Scrum meeting takes place. At the end of each sprint, the team demonstrates the soft-ware to the client. Based on the results of the sprint, a new plan is begun for the next sprint.

Scrum meetings are one of the most interesting aspects of the Scrum development process. The team members attend the meetings, but anyone can attend. However, with very few exceptions, only team members may speak. One prominent exception is manage-ment providing feedback on the business relevance of the work being performed by the specific team. In this meeting, all team members stand in a circle and report on what they accomplished during the previous day, state what they plan to do today, and describe any-thing that blocked progress the previous day. To enable continuous progress, any block identified is dealt with within one hour. From a Scrum point of view, it is better to make a "bad" decision about a block at this point in development than to not make a decision. Because the meetings take place each day, a bad decision can easily be undone. Larman[14] suggests that each team member should report any additional requirements that have been uncovered during the sprint and anything that the team member learned that could be use-ful for other team members to know.

[13] Scrum developers are not the first to question the use of plans. One of President Eisenhower's favorite maxims was, "In preparing for battle I have always found that plans are useless, but planning is indispensable." M. Dobson, *Streetwise Project Management: How to Manage People, Processes, and Time to Achieve the Results You Need* (Avon, MA: F+W Publications, 2003) p. 43.

[14] C. Larman, *Agile & Iterative Development: A Manager's Guide* (Boston: Addison-Wesley, 2004).

Ability to Develop Systems	Structured Methodologies		RAD Methodologies			Agile Methodologies	
	Waterfall	Parallel	Phased	Prototyping	Throwaway Prototyping	XP	SCRUM
With Unclear User Requirements	Poor	Poor	Good	Excellent	Excellent	Excellent	Excellent
With Unfamiliar Technology	Poor	Poor	Good	Poor	Excellent	Good	Good
That Are Complex	Good	Good	Good	Poor	Excellent	Good	Good
That Are Reliable	Good	Good	Good	Poor	Excellent	Excellent	Excellent
With a Short Time Schedule	Poor	Good	Excellent	Excellent	Good	Excellent	Excellent
With Schedule Visibility	Poor	Poor	Excellent	Excellent	Good	Excellent	Excellent

FIGURE 1-9   Criteria for Selecting a Methodology

One of the major criticisms of Scrum, as with all agile methodologies, is that it is questionable whether Scrum can scale up to develop very large, mission-critical systems. A typical Scrum team size is no more than seven members. The only organizing principle put forth by Scrum followers to address this criticism is to organize a scrum of scrums. Each team meets every day, and after the team meeting takes place, a representative (not leader) of each team attends a scrum-of-scrums meeting. This continues until the progress of entire system has been determined. Depending on the number of teams involved, this approach to managing a large project is doubtful. However, as in XP and other agile development approaches, many of the ideas and techniques associated with Scrum development are useful in object-oriented systems development, such as the focus of a Scrum meeting, the evolutionary and incremental approach to identifying requirements, and the incremental and iterative approach to the development of the system.

Selecting the Appropriate Development Methodology

Because there are many methodologies, the first challenge faced by analysts is selecting which methodology to use. Choosing a methodology is not simple, because no one methodology is always best. (If it were, we'd simply use it everywhere!) Many organizations have standards and policies to guide the choice of methodology. You will find that organizations range from having one "approved" methodology to having several methodology options to having no formal policies at all.

Figure 1-9 summarizes some important criteria for selecting a methodology. One important item not discussed in this figure is the degree of experience of the analyst team. Many of the RAD-based methodologies require the use of new tools and techniques that have a significant learning curve. Often these tools and techniques increase the complexity of the project and require extra time for learning. However, once they are adopted and the team becomes experienced, the tools and techniques can significantly increase the speed at which the methodology can deliver a final system.

Clarity of User Requirements   When the user requirements for a system are unclear, it is difficult to understand them by talking about them and explaining them with written reports. Users normally need to interact with technology to really understand what a new system can do and how to best apply it to their needs. RAD and agile methodologies are usually more appropriate when user requirements are unclear.

Familiarity with Technology   When the system will use new technology with which the analysts and programmers are not familiar (e.g., the first Web development project with Java), early application of the new technology in the methodology will improve the chance of success. If the system is designed without some familiarity with the base technology, risks increase because the tools might not be capable of doing what is needed. Throwaway prototyping–based methodologies are particularly appropriate if users lack familiarity with technology because they explicitly encourage the developers to develop design prototypes for areas with high risks. Phased development–based methodologies are good as well, because they create opportunities to investigate the technology in some depth before the design is complete. Also, owing to the programming-centric nature of agile methodologies, both XP and Scrum are appropriate. Although you might think prototyping-based methodologies are also appropriate, they are much less so because the early prototypes that are built usually only scratch the surface of the new technology. It is generally only after several prototypes and several months that the developers discover weaknesses or problems in the new technology.

System Complexity   Complex systems require careful and detailed analysis and design. Throwaway prototyping–based methodologies are particularly well suited to such detailed analysis and design, but prototyping-based methodologies are not. The traditional structured design–based methodologies can handle complex systems, but without the ability to get the system or prototypes into the users' hands early on, some key issues may be overlooked. Although phased development–based methodologies enable users to interact with the system early in the process, we have observed that project teams who follow these tend to devote less attention to the analysis of the complete problem domain than they might using other methodologies. Finally, agile methodologies are a mixed bag when it comes to system complexity. If the system is going to be a large one, then, owing to the lack of formal project management techniques used, agile methodologies will perform poorly. However, if the system is small to medium size, then agile approaches will be excellent. We rate them good on these criteria.

System Reliability   System reliability is usually an important factor in system development; after all, who wants an unreliable system? However, reliability is just one factor among several. For some applications, reliability is truly critical (e.g., medical equipment, missile-control systems), whereas for other applications (e.g., games, Internet video) it is merely important. Because throwaway prototyping methodologies combine detailed analysis and design phases with the ability for the project team to test many different approaches through design prototypes before completing the design, they are appropriate when system reliability is a high priority. Prototyping methodologies are generally not a good choice when reliability is critical because it lacks the careful analysis and design phases that are essential for dependable systems. However, owing to the heavy focus on testing, evolutionary and incremental identification of requirements, and iterative and incremental development, agile methods may be the best overall approach.

Short Time Schedules   Projects that have short time schedules are well suited for RAD-based and agile methodologies because these methodologies are designed to increase the speed of development. RAD-based and agile methodologies are excellent choices when timelines are short because they best enable the project team to adjust the functionality in the system based on a specific delivery date, and if the project schedule starts to slip, it can be readjusted by removing functionality from the version or prototype under development. Waterfall-based methodologies are the worst choice when time is at a premium because they do not allow easy schedule changes.

Schedule Visibility   One of the greatest challenges in systems development is determining whether a project is on schedule. This is particularly true of the structured design methodologies because design and implementation occur at the end of the project. The RAD-based methodologies move many of the critical design decisions earlier in the project to help project managers recognize and address risk factors and keep expectations in check. However, given the daily progress meetings associated with Agile approaches, schedule visibility is always on the proverbial front burner.

YOUR   **1-1 Selecting a Methodology**

TURN

*S*uppose you are an analyst for the Roanoke Software Consulting Company (RSCC), a large consulting firm with offices around the world. The company wants to build a new knowledge management system that can identify and track the expertise of individual consultants anywhere in the world based on their education and the various consulting projects on which they have worked. Assume that this is a new idea that has never before been attempted in RSCC or elsewhere. RSCC has an international network, but the offices in each country may use somewhat different hardware and software. RSCC management wants the system up and running within a year.

Question

What type of methodology would you recommend that RSCC use? Why?

TYPICAL SYSTEMS ANALYST ROLES AND SKILLS

It is clear from the various phases and steps performed during the SDLC that the project team needs a variety of skills. Project members are *change agents* who identify ways to improve an organization, build an information system to support them, and train and motivate others to use the system. Leading a successful organizational change effort is one of the most difficult jobs that someone can do. Understanding what to change and how to change it—and convincing others of the need for change—requires a wide range of skills. These skills can be broken down into six major categories: technical, business, analytical, interpersonal, management, and ethical.

Analysts must have the technical skills to understand the organization's existing technical environment, the technology that will make up the new system, and the way both can fit into an integrated technical solution. Business skills are required to understand how IT can be applied to business situations and to ensure that the IT delivers real business value. Analysts are continuous problem solvers at both the project and the organizational level, and they put their analytical skills to the test regularly.

Analysts often need to communicate effectively one-on-one with users and business managers (who often have little experience with technology) and with programmers (who often have more technical expertise than the analyst). They must be able to give presentations to large and small groups and write reports. Not only do they need to have strong interpersonal abilities, but they also need to manage people with whom they work and they need to manage the pressure and risks associated with unclear situations.

Finally, analysts must deal fairly, honestly, and ethically with other project team members, managers, and system users. Analysts often deal with confidential information or information that, if shared with others, could cause harm (e.g., dissent among employees); it is important to maintain confidence and trust with all people.

Role	Responsibilities
Business analyst	Analyzing the key business aspects of the system Identifying how the system will provide business value Designing the new business processes and policies
Systems analyst	Identifying how technology can improve business processes Designing the new business processes Designing the information system Ensuring that the system conforms to information systems standards
Infrastructure analyst	Ensuring the system conforms to infrastructure standards Identifying infrastructure changes needed to support the system
Change management analyst	Developing and executing a change management plan Developing and executing a user training plan
Project manager	Managing the team of analysts, programmers, technical writers, and other specialists Developing and monitoring the project plan Assigning resources Serving as the primary point of contact for the project

FIGURE 1-10
Project Team Roles

In addition to these six general skill sets, analysts require many specific skills associated with roles performed on a project. In the early days of systems development, most organizations expected one person, the analyst, to have all the specific skills needed to conduct a systems development project. Some small organizations still expect one person to perform many roles, but because organizations and technology have become more complex, most large organizations now build project teams containing several individuals with clearly defined responsibilities. Different organizations divide the roles differently, but Figure 1-10 presents one commonly used set of project team roles. Most IS teams include many other individuals, such as the *programmers,* who actually write the programs that make up the system, and *technical writers,* who prepare the help screens and other documentation (e.g., users manuals and systems manuals).

Business Analyst

A *business analyst* focuses on the business issues surrounding the system. These issues include identifying the business value that the system will create, developing ideas and suggestions for how the business processes can be improved, and designing the new processes and policies in conjunction with the systems analyst. This individual likely has business experience and some type of professional training (e.g., the business analyst for accounting systems is likely a CPA [in the United States] or a CA [in Canada]). He or she represents the interests of the project sponsor and the ultimate users of the system. A business analyst assists in the planning and design phases but is most active in the analysis phase.

Systems Analyst

A *systems analyst* focuses on the IS issues surrounding the system. This person develops ideas and suggestions for how information technology can improve business processes, designs the new business processes with help from the business analyst, designs the new information system, and ensures that all IS standards are maintained. A systems analyst likely has significant training and experience in analysis and design, programming, and even areas of the business. He or she represents the interests of the IS department and works intensively through the project but perhaps less so during the implementation phase.

Infrastructure Analyst

An *infrastructure analyst* focuses on the technical issues surrounding how the system will interact with the organization's technical infrastructure (e.g., hardware, software, networks, and databases). An infrastructure analyst's tasks include ensuring that the new information system conforms to organizational standards and identifying infrastructure changes needed to support the system. This individual probably has significant training and experience in networking, database administration, and various hardware and software products. He or she represents the interests of the organization and IS group that will ultimately have to operate and support the new system once it has been installed. An infrastructure analyst works throughout the project but perhaps less so during planning and analysis phases.

Change Management Analyst

A *change management analyst* focuses on the people and management issues surrounding the system installation. The roles of this person include ensuring that the adequate documentation and support are available to users, providing user training on the new system, and developing strategies to overcome resistance to change. This individual should have significant training and experience in organizational behavior in general and change management in particular. He or she represents the interests of the project sponsor and users for whom the system is being designed. A change management analyst works most actively during the implementation phase but begins laying the groundwork for change during the analysis and design phases.

Project Manager

A project manager is responsible for ensuring that the project is completed on time and within budget and that the system delivers all benefits intended by the project sponsor. The role of the project manager includes managing the team members, developing the project plan, assigning resources, and being the primary point of contact when people outside the team have questions about the project. This individual likely has significant experience in project management and has probably worked for many years as a systems analyst beforehand. He or she represents the interests of the IS department and the project sponsor. The project manager works intensely during all phases of the project.

YOUR TURN   1-2 **Being an Analyst**

*S*uppose you decide to become an analyst after you graduate. Decide what type of analyst you would prefer to be and what types of courses you should take before you graduate. Then decide the type of summer job or internship you should seek.

Question

Develop a short plan that describes how you will prepare for your career as an analyst.

BASIC CHARACTERISTICS OF OBJECT-ORIENTED SYSTEMS

Object-oriented systems focus on capturing the structure and behavior of information systems in little modules that encompass both data and process. These little modules are known as *objects*. In this section, we describe the basic characteristics of object-oriented systems, which include classes, objects, methods, messages, encapsulation, information hiding, inheritance, polymorphism, and dynamic binding.[15]

Classes and Objects

A *class* is the general template we use to define and create specific instances, or objects. Every object is associated with a class. For example, all the objects that capture information about patients could fall into a class called Patient, because there are attributes (e.g., name, address, birth date, phone, and insurance carrier) and methods (e.g., make appointment, calculate last visit, change status, and provide medical history) that all patients share (see Figure 1-11).

An *object* is an instantiation of a class. In other words, an object is a person, place, or thing about which we want to capture information. If we were building an appointment system for a doctor's office, classes might include Doctor, Patient, and Appointment. The specific patients, such as Jim Maloney, Mary Wilson, and Theresa Marks, are considered *instances*, or objects, of the patient class (see Figure 1-11).

Each object has *attributes* that describe information about the object, such as a patient's name, birth date, address, and phone number. Attributes are also used to represent relationships between objects; for example, there could be a department attribute in an employee object with a value of a department object that captures in which department the employee object works. The *state* of an object is defined by the value of its attributes and its relationships with other objects at a particular point in time. For example, a patient might have a state of new or current or former.

Each object also has *behaviors*. The behaviors specify what the object can do. For example, an appointment object can probably schedule a new appointment, delete an appointment, and locate the next available appointment. In object-oriented programming, behaviors are implemented as methods (see the next section).

FIGURE 1-11
Classes and Objects

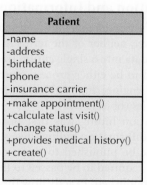

---

[15] In Chapter 8, we review the basic characteristics of object-oriented systems in more detail.

FIGURE 1-12

Messages and
Methods

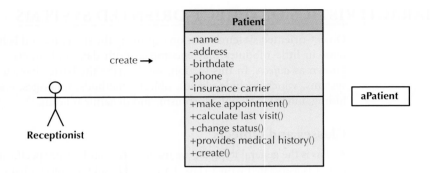

One of the more confusing aspects of object-oriented systems development is the fact that in most object-oriented programming languages, both classes and instances of classes can have attributes and methods. Class attributes and methods tend to be used to model attributes (or methods) that deal with issues related to all instances of the class. For example, to create a new patient object, a message is sent to the Patient class to create a new instance of itself. However, in this book, we focus primarily on attributes and methods of objects and not of classes.

Methods and Messages

Methods implement an object's behavior. A method is nothing more than an action that an object can perform. As such, a method is analogous to a function or procedure in a traditional programming language such as C, COBOL, or Pascal. *Messages* are information sent to objects to trigger methods. A message is essentially a function or procedure call from one object to another object. For example, if a patient is new to the doctor's office, the receptionist sends a create message to the application. The patient class receives the create message and executes its create() method (see Figure 1-12), which then creates a new object: a Patient (see Figure 1-12).

Encapsulation and Information Hiding

The ideas of *encapsulation* and *information hiding* are interrelated in object-oriented systems. However, neither of the terms is new. Encapsulation is simply the combination of process and data into a single entity. Traditional approaches to information systems development tend to be either process-centric (e.g., structured systems) or data-centric (e.g., information engineering). Object-oriented approaches combine process and data into holistic entities (objects).

Information hiding was first promoted in structured systems development. The principle of information hiding suggests that only the information required to use a software module be published to the user of the module. Typically, this implies that the information required to be passed to the module and the information returned from the module are published. Exactly how the module implements the required functionality is not relevant. We really do not care how the object performs its functions, as long as the functions occur.

In object-oriented systems, combining encapsulation with the information-hiding principle suggests that the information-hiding principle be applied to objects instead of merely applying it to functions or processes. Thus, objects are treated like black boxes.

The fact that we can use an object by calling methods is the key to reusability because it shields the internal workings of the object from changes in the outside system, and it keeps the system from being affected when changes are made to an object. In Figure 1-12, notice how a message (create) is sent to an object, yet the internal algorithms needed to respond to the message are hidden from other parts of the system. The only information that an object needs to know is the set of operations, or methods, that other objects can perform and what messages need to be sent to trigger them.

YOUR TURN   **1-3 Encapsulation and Information Hiding**

Come up with a set of examples of using encapsulation and information hiding in everyday life. For example, is there any information about yourself that you would not mind if everyone knew? How would someone retrieve this information? What about personal information that you would prefer to be private? How would you prevent someone from retrieving it?

Inheritance

Inheritance, as an information systems development characteristic, was proposed in data modeling in the late 1970s and the early 1980s. The data modeling literature suggests using inheritance to identify higher-level, or more general, classes of objects. Common sets of attributes and methods can be organized into *superclasses.* Typically, classes are arranged in a hierarchy whereby the superclasses, or general classes, are at the top and the *subclasses,* or specific classes, are at the bottom. In Figure 1-13, Person is a superclass to the classes Doctor and Patient. Doctor, in turn, is a superclass to General Practitioner and Specialist. Notice how a class (e.g., Doctor) can serve as a superclass and subclass concurrently. The relationship between the class and its superclass is known as the *a-kind-of* relationship. For example in Figure 1-13, a General Practitioner is a-kind-of Doctor, which is a-kind-of Person.

FIGURE 1-13

Class Hierarchy with Abstract and Concrete Classes

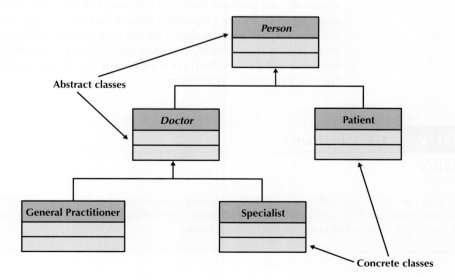

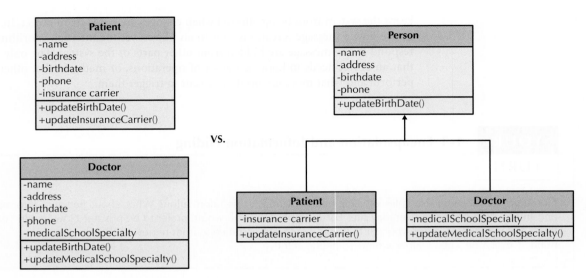

FIGURE 1-14   Inheritance Advantage?

Subclasses *inherit* the appropriate attributes and methods from the superclasses above them. That is, each subclass contains attributes and methods from its parent superclass. For example, Figure 1-13 shows that both Doctor and Patient are subclasses of Person and therefore inherit the attributes and methods of the Person class. Inheritance makes it simpler to define classes. Instead of repeating the attributes and methods in the Doctor and Patient classes separately, the attributes and methods that are common to both are placed in the Person class and inherited by the classes below it. Notice how much more efficient inheritance hierarchies of object classes are than the same objects without an inheritance hierarchy (see Figure 1-14).

Most classes throughout a hierarchy lead to instances; any class that has instances is called a *concrete class*. For example, if Mary Wilson and Jim Maloney are instances of the Patient class, Patient would be considered a concrete class (see Figure 1-11). Some classes do not produce instances because they are used merely as templates for other, more-specific classes (especially classes located high up in a hierarchy). The classes are referred to as *abstract classes*. Person is an example of an abstract class. Instead of creating objects from Person, we create instances representing the more-specific classes of Specialist and Patient, both types of Person (see Figure 1-13). What kind of class is the General Practitioner class? Why?

YOUR   **1-4 Inheritance**

TURN

*S*ee if you can come up with at least three different classes that you might find in a typical business situation. Select one of the classes and create at least a three-level   inheritance hierarchy using the class. Which of the classes are abstract, if any, and which ones are concrete?

Polymorphism and Dynamic Binding

Polymorphism means that the same message can be interpreted differently by different classes of objects. For example, inserting a patient means something different than inserting an appointment. Therefore, different pieces of information need to be collected and stored. Luckily, we do not have to be concerned with *how* something is done when using objects. We can simply send a message to an object, and that object will be responsible for interpreting the message appropriately. For example, if an artist sent the message Draw yourself to a square object, a circle object, and a triangle object, the results would be very different, even though the message is the same. Notice in Figure 1-15 how each object responds appropriately (and differently) even though the messages are identical.

Polymorphism is made possible through *dynamic binding*. Dynamic, or late, binding is a technique that delays typing the object until run-time. The specific method that is actually called is not chosen by the object-oriented system until the system is running. This is in contrast to *static binding*. In a statically bound system, the type of object is determined at compile-time. Therefore, the developer has to choose which method should be called instead of allowing the system to do it. This is why most traditional programming languages have complicated decision logic based on the different types of objects in a system. For example, in a traditional programming language, instead of sending the message Draw yourself to the different types of graphical objects in Figure 1-15, we would have to write decision logic using a case statement or a set of if statements to determine what kind of graphical object we wanted to draw, and we would have to name each draw function differently (e.g., draw square, draw circle, or draw triangle). This obviously makes the system much more complicated and difficult to understand.

FIGURE 1-15
Polymorphism

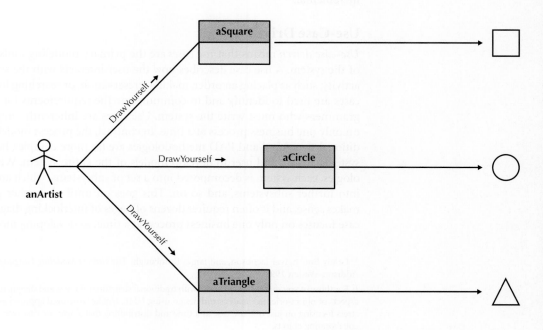

YOUR
TURN

1-5 Polymorphism and Dynamic Binding

Can you think of any way you use polymorphism and/or dynamic binding in your everyday life? For example, when you are told to do some task, do you always perform the task the same way everyone else you know does?

Do you always perform the task the same way or does the method of performance depend on where you are when you perform the task?

OBJECT-ORIENTED SYSTEMS ANALYSIS AND DESIGN (OOSAD)

Object-oriented approaches to developing information systems, technically speaking, can use any of the traditional methodologies. However, the object-oriented approaches are most associated with a phased development RAD or agile methodology. The primary difference between a traditional approach like structured design and an object-oriented approach is how a problem is decomposed. In traditional approaches, the problem-decomposition process is either process-centric or data-centric. However, processes and data are so closely related that it is difficult to pick one or the other as the primary focus. Based on this lack of congruence with the real world, new *object-oriented methodologies* have emerged that use the RAD-based sequence of SDLC phases but attempt to balance the emphasis between process and data by focusing the decomposition of problems on objects that contain both data and processes.

According to the creators of the Unified Modeling Language (UML), Grady Booch, Ivar Jacobson, and James Rumbaugh,[16] any modern object-oriented approach to developing information systems must be use-case driven, architecture-centric, and iterative and incremental.

Use-Case Driven

Use-case driven means that *use cases* are the primary modeling tools defining the behavior of the system. A use case describes how the user interacts with the system to perform some activity, such as placing an order, making a reservation, or searching for information. The use cases are used to identify and to communicate the requirements for the system to the programmers who must write the system. Use cases are inherently simple because they focus on only one business process at a time. In contrast, the process model diagrams used by traditional structured and RAD methodologies are far more complex because they require the systems analyst and user to develop models of the entire system. With traditional methodologies, each system is decomposed into a set of subsystems, which are, in turn, decomposed into further subsystems, and so on. This goes on until no further process decomposition makes sense, and it often requires dozens of pages of interlocking diagrams. In contrast, a use case focuses on only one business process at a time, so developing models is much simpler.[17]

---

[16] Grady Booch, Ivar Jacobson, and James Rumbaugh, *The Unified Modeling Language User Guide* (Reading, MA: Addison-Wesley, 1999).

[17] For those of you that have experience with traditional structured analysis and design, this is one of the most unusual aspects of object-oriented analysis and design using UML. Unlike structured approaches, object-oriented approaches stress focusing on just one use case at a time and distributing that single use case over a set of communicating and collaborating objects.

Architecture-centric

Any modern approach to systems analysis and design should be architecture-centric. *Architecture-centric* means that the underlying software architecture of the evolving system specification drives the specification, construction, and documentation of the system. Modern object-oriented systems analysis and design approaches should support at least three separate but interrelated architectural views of a system: functional, static, and dynamic. The *functional,* or *external, view* describes the behavior of the system from the perspective of the user. The *structural,* or *static, view* describes the system in terms of attributes, methods, classes, and relationships. The *behavioral,* or *dynamic, view* describes the behavior of the system in terms of messages passed among objects and state changes within an object.

Iterative and Incremental

Modern object-oriented systems analysis and design approaches emphasize *iterative* and *incremental* development that undergoes continuous testing and refinement throughout the life of the project. This implies that the systems analysts develop their understanding of a user's problem by building up the three architectural views little by little. The systems analyst does this by working with the user to create a functional representation of the system under study. Next, the analyst attempts to build a structural representation of the evolving system. Using the structural representation of the system, the analyst distributes the functionality of the system over the evolving structure to create a behavioral representation of the evolving system. As an analyst works with the user in developing the three architectural views of the evolving system, the analyst iterates over each of, and among, the views. That is, as the analyst better understands the structural and behavioral views, the analyst uncovers missing requirements or misrepresentations in the functional view. This, in turn, can cause changes to be cascaded back through the structural and behavioral views. All three architectural views of the system are interlinked and dependent on each other (see Figure 1-16). As each increment and iteration is completed, a more complete representation of the user's real functional requirements is uncovered.

Benefits of Object-Oriented Systems Analysis and Design

Concepts in the object-oriented approach enable analysts to break a complex system into smaller, more manageable modules, work on the modules individually, and easily piece the modules back together to form an information system. This modularity makes systems development easier to grasp, easier to share among members of a project team, and easier to communicate to users, who are needed to provide requirements and confirm how well the system meets the requirements throughout the systems development process. By modularizing

FIGURE 1-16
Iterative and
Incremental
Development

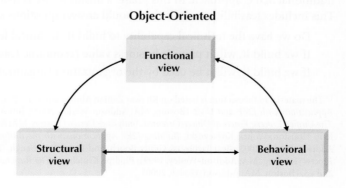

systems development, the project team actually is creating reusable pieces that can be plugged into other systems efforts or used as starting points for other projects. Ultimately, this can save time because new projects don't have to start completely from scratch.

Many people argue that "object-think" is a much more realistic way to think about the real world. Users typically do not think in terms of data or process; instead, they see their business as a collection of logical units that contain both, so communicating in terms of objects improves the interaction between a user and an analyst or developer.

THE UNIFIED PROCESS

The Unified Process is a specific methodology that maps out when and how to use the various Unified Modeling Language (UML) techniques for object-oriented analysis and design. The primary contributors were Grady Booch, Ivar Jacobsen, and James Rumbaugh of Rational. Whereas the UML provides structural support for developing the structure and behavior of an information system, the Unified Process provides the behavioral support. The Unified Process, of course, is use-case driven, architecture-centric, and iterative and incremental. Furthermore, the Unified Process is a two-dimensional systems development process described by a set of phases and workflows. The phases are inception, elaboration, construction, and transition. The workflows include business modeling, requirements, analysis, design, implementation, test, deployment, configuration and change management, project management, and environment. In the remainder of this section, we describe the phases and workflows of the Unified Process.[18] Figure 1-17 depicts the Unified Process.

Phases

The *phases* of the Unified Process support an analyst in developing information systems in an iterative and incremental manner. The phases describe how an information system evolves through time. Depending on which development phase the evolving system is currently in, the level of activity varies over the *workflows*. The curve in Figure 1-17 associated with each workflow approximates the amount of activity that takes place during the specific phase. For example, the inception phase primarily involves the business modeling and requirements workflows, while practically ignoring the test and deployment workflows. Each phase contains a set of iterations, and each iteration uses the various workflows to create an incremental version of the evolving information system. As the system evolves through the phases, it improves and becomes more complete. Each phase has objectives, a focus of activity over the workflows, and incremental deliverables. Each of the phases is described next.

Inception   In many ways, the *inception phase* is very similar to the planning phase of a traditional SDLC approach. In this phase, a business case is made for the proposed system. This includes feasibility analysis that should answer questions such as the following:

Do we have the technical capability to build it (technical feasibility)?

If we build it, will it provide business value (economic feasibility)?

If we build it, will it be used by the organization (organizational feasibility)?

---

[18] The material in this section is based on Khawar Zaman Ahmed and Cary E. Umrysh, *Developing Enterprise Java Applications with J2EE and UML* (Boston, MA: Addison-Wesley, 2002); Jim Arlow and Ila Neustadt, *UML and The Unified Process: Practical Object-Oriented Analysis & Design* (Boston, MA: Addison-Wesley, 2002); Peter Eeles, Kelli Houston, Wojtek Kozacynski, *Building J2EE Applications with the Rational Unified Process*, (Boston, MA: Addison-Wesley, 2003); Ivar Jacobson, Grady Booch, and James Rumbaugh, *The Unified Software Development Process* (Reading, MA: Addison-Wesley, 1999); Phillipe Krutchten, *The Rational Unified Process: An Introduction*, 2nd ed. (Boston, MA: Addison-Wesley, 2000).

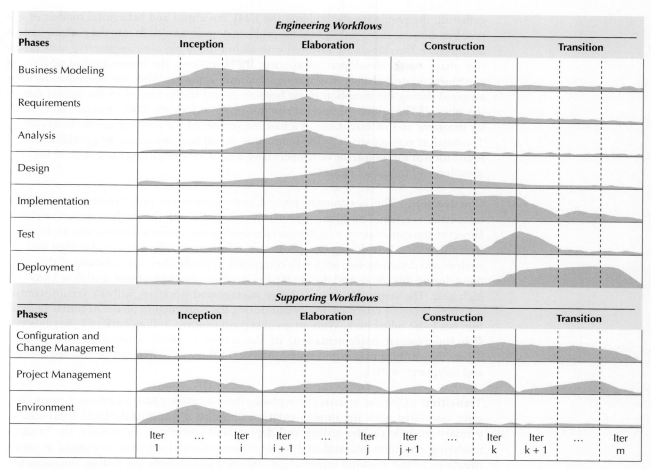

FIGURE 1-17 The Unified Process

To answer these questions, the development team performs work related primarily to the business modeling, requirements, and analysis workflows. In some cases, depending on the technical difficulties that could be encountered during the development of the system, a throwaway prototype is developed. This implies that the design, implementation, and test workflows could also be involved. The project management and environment supporting workflows are very relevant to this phase. The primary deliverables from the inception phase are a vision document that sets the scope of the project, identifies the primary requirements and constraints, sets up an initial project plan, and describes the feasibility of and risks associated with the project, the adoption of the necessary environment to develop the system, and some aspects of the problem domain classes being implemented and tested.

Elaboration When we typically think about object-oriented systems analysis and design, the activities related to the *elaboration phase* of the Unified Process are the most relevant. The *analysis* and *design workflows* are the primary focus during this phase. The elaboration phase continues with developing the vision document, including finalizing the business case, revising the risk assessment, and completing a project plan in sufficient detail to allow the stakeholders to be able to agree with constructing the actual final system. It deals with gathering the requirements, building the UML structural and behavioral models of the problem domain, and detailing how the problem domain models fit into the evolving system architecture. Developers are involved with all but the *deployment engineering workflow* in this phase. As the developers iterate over the workflows, the importance of addressing configuration and change management becomes apparent. Also, the development tools acquired during the inception phase become critical to the success of the project during this phase.[19] The primary deliverables of this phase include the UML structure and behavior diagrams and an executable of a baseline version of the evolving information system. The baseline version serves as the foundation for all later iterations. By providing a solid foundation at this point, the developers have a basis for completing the system in the construction and transition phases.

Construction   The *construction phase* focuses heavily on programming the evolving information system. This phase is primarily concerned with the *implementation workflow*. However, the *requirements workflow* and the *analysis* and *design* workflows also are involved with this phase. It is during this phase that missing requirements are identified and the analysis and design models are finally completed. Typically, there are iterations of the workflows during this phase, and during the last iteration, the deployment workflow kicks into high gear. The *configuration and change management workflow*, with its version-control activities, becomes extremely important during the construction phase. At times, an iteration has to be rolled back. Without good version controls, rolling back to a previous version (incremental implementation) of the system is nearly impossible. The primary deliverable of this phase is an implementation of the system that can be released for beta and acceptance testing.

Transition   Like the construction phase, the *transition phase* addresses aspects typically associated with the implementation phase of a traditional SDLC approach. Its primary focus is on the testing and deployment workflows. Essentially, the business modeling, requirements, and analysis workflows should have been completed in earlier iterations of the evolving information system. Furthermore, the testing workflow will have been executing during the earlier phases of the evolving system. Depending on the results from the testing workflow, some redesign and programming activities on the design and implementation workflows could be necessary, but they should be minimal at this point. From a managerial perspective, the project management, configuration and change management, and environment are involved. Some of the activities that take place are beta and acceptance testing, fine-tuning the design and implementation, user training, and rolling out the final product onto a production platform. Obviously, the primary deliverable is the actual executable information system. The other deliverables include user manuals, a plan to support the users, and a plan for upgrading the information system in the future.

Workflows

The workflows describe the tasks or activities that a developer performs to evolve an information system over time. The workflows of the Unified Process are grouped into two broad categories: engineering and supporting.

---

[19] With UML comprising fourteen different, related diagramming techniques, keeping the diagrams coordinated and the different versions of the evolving system synchronized is typically beyond the capabilities of a mere mortal systems developer. These tools typically include project management and CASE tools. We describe the use of these tools in Chapter 2.

Engineering Workflows   Engineering workflows include business-modeling, requirements, analysis, design, implementation, test, and deployment workflows. The engineering workflows deal with the activities that produce the technical product (i.e., the information system).

Business Modeling Workflow   The *business-modeling workflow* uncovers problems and identifies potential projects within a user organization. This workflow aids management in understanding the scope of the projects that can improve the efficiency and effectiveness of a user organization. The primary purpose of business modeling is to ensure that both developer and user organizations understand where and how the to-be-developed information system fits into the business processes of the user organization. This workflow is primarily executed during the inception phase to ensure that we develop information systems that make business sense. The activities that take place on this workflow are most closely associated with the planning phase of the traditional SDLC; however, requirements gathering, and use-case and business process modeling techniques also help us to understand the business situation.

Requirements Workflow   In the Unified Process, the requirements workflow includes eliciting both functional and nonfunctional requirements. Typically, requirements are gathered from project stakeholders, such as end users, managers within the end user organization, and even customers. There are many different ways to capture requirements, including interviews, observation techniques, joint application development, document analysis, and questionnaires. The requirements workflow is used the most during the inception and elaboration phases. The identified requirements are very helpful for developing the vision document and the use cases used throughout the development process. Additional requirements tend to be discovered throughout the development process. In fact, only the transition phase tends to have few, if any, additional requirements identified.

Analysis Workflow   The analysis workflow primarily addresses the creation of an analysis model of the problem domain. In the Unified Process, the analyst begins designing the architecture associated with the problem domain; using the UML, the analyst creates structural and behavior diagrams that depict a description of the problem domain classes and their interactions. The primary purpose of the analysis workflow is to ensure that both the developer and user organizations understand the underlying problem and its domain without overanalyzing. If they are not careful, analysts can create *analysis paralysis,* which occurs when the project becomes so bogged down with analysis that the system is never actually designed or implemented. A second purpose of the analysis workflow is to identify useful reusable classes for class libraries. By reusing predefined classes, the analyst can avoid reinventing the wheel when creating the structural and behavior diagrams. The analysis workflow is predominantly associated with the elaboration phase, but like the requirements workflow, it is possible that additional analysis will be required throughout the development process.

Design Workflow   The design workflow transitions the analysis model into a form that can be used to implement the system: the *design model.* Whereas the analysis workflow concentrated on understanding the problem domain, the design workflow focuses on developing a solution that will execute in a specific environment. Basically, the design workflow simply enhances the description of the evolving information system by adding classes that address the environment of the information system to the evolving analysis model. The design workflow uses activities such as detailed problem domain class design, optimization of the evolving information system, database design, user-interface design, and physical architecture design. The design workflow is associated primarily with the elaboration and construction phases of the Unified Process.

Implementation Workflow   The primary purpose of the implementation workflow is to create an executable solution based on the design model (i.e., programming). This includes not only writing new classes but also incorporating reusable classes from executable class libraries into the evolving solution. As with any programming activity, the new classes and their interactions with the incorporated reusable classes must be tested. Finally, in the case of multiple groups performing the implementation of the information system, the implementers also must integrate the separate, individually tested modules to create an executable version of the system. The implementation workflow is associated primarily with the elaboration and construction phases.

Testing Workflow   The primary purpose of the *testing workflow* is to increase the quality of the evolving system. Testing goes beyond the simple unit testing associated with the implementation workflow. In this case, testing also includes testing the integration of all modules used to implement the system, user acceptance testing, and the actual alpha testing of the software. Practically speaking, testing should go on throughout the development of the system; testing of the analysis and design models occurs during the elaboration and construction phases, whereas implementation testing is performed primarily during the construction and, to some degree, transition phases. Basically, at the end of each iteration during the development of the information system, some type of test should be performed.

Deployment Workflow   The deployment workflow is most associated with the transition phase of the Unified Process. The deployment workflow includes activities such as software packaging, distribution, installation, and beta testing. When actually deploying the new information system into a user organization, the developers might have to convert the current data, interface the new software with the existing software, and train the end user to use the new system.

Supporting Workflows   The supporting workflows include the project management, configuration and change management, and environment workflows. The supporting workflows focus on the managerial aspects of information systems development.

Project Management Workflow   Whereas the other workflows associated with the Unified Process are technically active during all four phases, the *project management workflow* is the only truly cross-phase workflow. The development process supports incremental and iterative development, so information systems tend to grow or evolve over time. At the end of each iteration, a new incremental version of the system is ready for delivery. The project management workflow is quite important owing to the complexity of the two-dimensional development model of the Unified Process (workflows and phases). This workflow's activities include risk identification and management, scope management, estimating the time to complete each iteration and the entire project, estimating the cost of the individual iteration and the whole project, and tracking the progress being made toward the final version of the evolving information system.

Configuration and Change Management Workflow   The primary purpose of the configuration and change management workflow is to keep track of the state of the evolving system. In a nutshell, the evolving information system comprises a set of artifacts, including, for example, diagrams, source code, and executables. During the development process, these artifacts are modified. A substantial amount of work—and, hence, money—is involved in developing the artifacts. The artifacts themselves should be handled as any expensive asset would be handled—access controls must be put into place to safeguard the artifacts from being stolen or destroyed. Furthermore, because the artifacts are modified on a regular, if not continuous, basis, good version control mechanisms should be established. Finally, a good deal of project management information needs to be captured (e.g., author, time, and location of each modification). The configuration and change management workflow is associated mostly with the construction and transition phases.

Environment Workflow   During the development of an information system, the development team needs to use different tools and processes. The *environment workflow* addresses these needs. For example, a CASE tool that supports the development of an object-oriented information system via the UML could be required. Other tools necessary include programming environments, project management tools, and configuration management tools. The environment workflow involves acquiring and installing these tools. Even though this workflow can be active during all of the phases of the Unified Process, it should be involved primarily with the inception phase.

Extensions to the Unified Process

As large and as complex as the Unified Process is, many authors have pointed out a set of critical weaknesses. First, the Unified Process does not address staffing, budgeting, or contract management issues. These activities were explicitly left out of the Unified Process. Second, the Unified Process does not address issues relating to maintenance, operations, or support of the product once it has been delivered. Thus it is not a complete software process; it is only a development process. Third, the Unified Process does not address cross- or inter-project issues. Considering the importance of reuse in object-oriented systems development and the fact that in many organizations employees work on many different projects at the same time, leaving out inter-project issues is a major omission.

To address these omissions, Ambler and Constantine suggest adding a production phase and two workflows: the operations and support workflow and the infrastructure management workflow (see Figure 1-18).[20] In addition to these new workflows, the test, deployment, and environment workflows are modified, and the project management and the configuration and change management workflows are extended into the production phase. These extensions are based on alternative object-oriented software processes: the OPEN process (Object-oriented Process, Environment, and Notation) and the Object-Oriented Software Process.[21] The new phase, the new workflows, and the modifications and extensions to the existing workflows are described next.

[20] S. W. Ambler and L. L. Constantine, *The Unified Process Inception Phase: Best Practices in Implementing the UP* (Lawrence, KS: CMP Books, 2000); S. W. Ambler and L. L. Constantine, *The Unified Process Elaboration Phase: Best Practices in Implementing the UP* (Lawrence, KS: CMP Books, 2000); S. W. Ambler and L. L. Constantine, *The Unified Process Construction Phase: Best Practices in Implementing the UP* (Lawrence, KS: CMP Books, 2000); S. W. Ambler and L. L. Constantine, *The Unified Process Transition and Production Phases: Best Practices in Implementing the UP* (Lawrence, KS: CMP Books, 2002).

[21] S. W. Ambler, *Process Patterns—Building Large-Scale Systems Using Object Technology* (Cambridge, UK: SIGS Books/Cambridge University Press, 1998); S. W. Ambler, *More Process Patterns—Delivering Large-Scale Systems Using Object Technology* (Cambridge, UK: SIGS Books/Cambridge University Press, 1999); I. Graham, B. Henderson-Sellers, and H. Younessi, *The OPEN Process Specification* (Harlow, UK: Addison-Wesley, 1997); B. Henderson-Sellers and B. Unhelkar, *OPEN Modeling with UML* (Harlow, UK: Addison-Wesley, 2000).

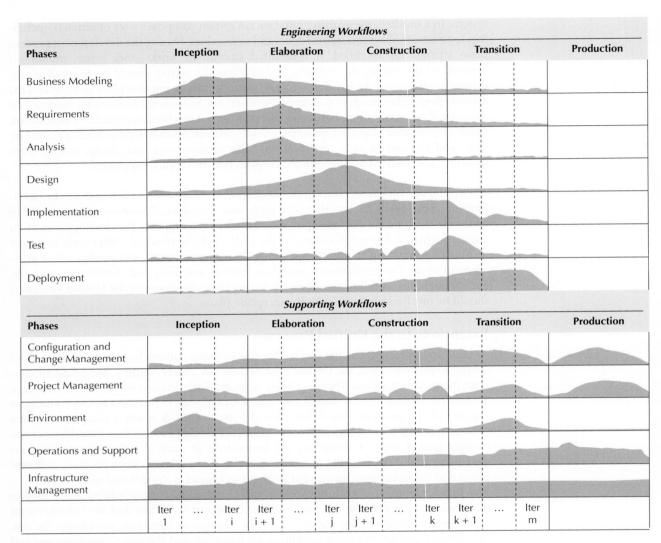

FIGURE 1-18   The Enhanced Unified Process

Production Phase   The *production phase* is concerned primarily with issues related to the software product after it has been successfully deployed. This phase focuses on issues related to updating, maintaining, and operating the software. Unlike the previous phases, there are no iterations or incremental deliverables. If a new release of the software is to be developed, then the developers must begin a new run through the first four phases. Based on the activities that take place during this phase, no engineering workflows are relevant. The supporting workflows that are active during this phase include the configuration and change management workflow, the project management workflow, the new operations and support workflow, and the infrastructure management workflow.

Operations and Support Workflow   The *operations and support workflow*, as you might guess, addresses issues related to supporting the current version of the software and operating the software on a daily basis. Activities include creating plans for the operation and support of the software product once it has been deployed, creating training and user documentation, putting into place necessary backup procedures, monitoring and optimizing the performance of the software, and performing corrective maintenance on the software. This workflow becomes active during the construction phase; its level of activity increases throughout the transition and, finally, the production phase. The workflow finally drops off when the current version of the software is replaced by a new version. Many developers are under the false impression that once the software has been delivered to the customer, their work is finished. In most cases, the work of supporting the software product is much more costly and time consuming than the original development. At that point, the developer's work may have just begun.

Infrastructure Management Workflow The *infrastructure management workflow's* primary purpose is to support the development of the infrastructure necessary to develop object-oriented systems. Activities such as development and modification of libraries, standards, and enterprise models are very important. When the development and maintenance of a problem-domain architecture model goes beyond the scope of a single project and reuse is going to occur, the infrastructure management workflow is essential. Another very important set of cross-project activities is the improvement of the software development process. Because the activities on this workflow tend to affect many projects and the Unified Process focuses only on a specific project, the Unified Process tends to ignore these activities (i.e., they are simply beyond the scope and purpose of the Unified Process).

Existing Workflow Modifications and Extensions In addition to the workflows that were added to address deficiencies contained in the Unified Process, existing workflows had to be modified and/or extended into the production phase. These workflows include the test, deployment, environment, project management, and configuration and change management workflows.

Test Workflow For high-quality information systems to be developed, testing should be done on every deliverable, including those created during the inception phase. Otherwise, less than high-quality systems will be delivered to the customer.

Deployment Workflow Legacy systems exist in most corporations today, and these systems have databases associated with them that must be converted to interact with the new systems. Owing to the complexity of deploying new systems, the conversion requires significant planning. Therefore, the activities on the deployment workflow need to begin in the inception phase instead of waiting until the end of the construction phase, as suggested by the Unified Process.

Environment Workflow The environment workflow needs to be modified to include activities related to setting up the operations and production environment. The actual work performed is similar to the work related to setting up the development environment that was performed during the inception phase. In this case, the additional work is performed during the transition phase.

Project Management Workflow Even though the project management workflow does not include staffing the project, managing the contracts among the customers and vendors, and managing the project's budget, these activities are crucial to the success of any software development project. We suggest extending project management to include these activities. This workflow should additionally occur in the production phase to address issues such as training, staff management, and client relationship management.

Enhanced UP Phases	Chapters
Inception	2–4
Elaboration	3–11
Construction	8, 12
Transition	12–13
Production	13

Enhanced UP Engineering Workflows	Chapters
Business Modeling	2–5
Requirements	3–5, 10
Analysis	3–7
Design	7–11
Implementation	9, 12
Test	4–7, 12
Deployment	13

Enhanced UP Supporting Workflows	Chapters
Project Management	2, 13
Configuration and Change Management	13
Environment	2
Operations and Support	13
Infrastructure Management	2

FIGURE 1-19   The Enhanced Unified Process and the Textbook Organization

Configuration and Change Management Workflow   The configuration and change management workflow is extended into the new production phase. Activities performed during the production phase include identifying potential improvements to the operational system and assessing the potential impact of the proposed changes. Once developers have identified these changes and understood their impact, they can schedule the changes to be made and deployed with future releases.

Figure 1-19 shows the chapters in which the Enhanced Unified Process's phases and workflows are covered. Given the offshore outsourcing and automation of information technology,[22] in this textbook, we focus primarily on the elaboration phase and the business modeling, requirements, analysis, design, and project management workflows of the Enhanced Unified Process. However, as Figure 1-12 shows, the other phases and workflows

[22] See Thomas L. Friedman, *The World Is Flat: A Brief History of the Twenty-First Century, Updated and Expanded Edition* (New York: Farrar, Straus, and Giroux, 2006); and Daniel H. Pink, *A Whole New Mind: Why Right-Brainers Will Rule the Future* (New York: Riverhead Books, 2006).

are covered. In many object-oriented systems development environments today, code generation is supported. Thus, from a business perspective, we believe the activities associated with these workflows are the most important.

YOUR
TURN

1-6 Object-Oriented Systems Analysis and Design Methodology

*R*eview Figures 1-17, 1-18, and 1-19. Based on your understanding of the Unified Process and the Enhanced Unified Process, suggest a set of steps for an alternative object-oriented systems development method. Be sure that the steps are capable of delivering an executable and maintainable system.

THE UNIFIED MODELING LANGUAGE

Until 1995, object concepts were popular but implemented in many different ways by different developers. Each developer had his or her own methodology and notation (e.g., Booch, Coad, Moses, OMT, OOSE, SOMA.)[23] Then in 1995, Rational Software brought three industry leaders together to create a single approach to object-oriented systems development. Grady Booch, Ivar Jacobson, and James Rumbaugh worked with others to create a standard set of diagramming techniques known as the *Unified Modeling Language (UML)*. The objective of UML was to provide a common vocabulary of object-oriented terms and diagramming techniques rich enough to model any systems development project from analysis through implementation. In November 1997, the *Object Management Group (OMG)* formally accepted UML as the standard for all object developers. During the following years, the UML has gone through multiple minor revisions. The current version of UML, Version 2.4, was released by the OMG in January 2011.

Version 2.4 of the UML defines a set of fourteen diagramming techniques used to model a system. The diagrams are broken into two major groupings: one for modeling the structure of a system and one for modeling behavior. *Structure diagrams* provide a way to represent the data and static relationships in an information system. The structure diagrams include class, object, package, deployment, component, and composite structure diagrams. *Behavior diagrams* provide the analyst with a way to depict the dynamic relationships among the instances or objects that represent the business information system. They also allow modeling of the dynamic behavior of individual objects throughout their lifetime. The behavior diagrams support the analyst in modeling the functional requirements of an evolving information system. The behavior modeling diagrams include

---

[23] See Grady Booch, *Object-Oriented Analysis and Design with Applications,* 2nd ed. (Redwood City, CA: Benjamin/Cummings, 1994); Peter Coad and Edward Yourdon, *Object-Oriented Analysis,* 2nd ed. (Englewood Cliffs, NJ: Yourdon Press, 1991); Peter Coad and Edward Yourdon, *Object-Oriented Design* (Englewood Cliffs, NJ: Yourdon Press, 1991); Brian Henderson-Sellers and Julian Edwards, *Book Two of Object-Oriented Knowledge: The Working Object* (Sydney, Australia: Prentice Hall, 1994); James Rumbaugh, Michael Blaha, William Premerlani, Frederick Eddy, and William Lorensen, *Object-Oriented Modeling and Design* (Englewood Cliffs, NJ: Prentice Hall, 1991); Ivar Jacobson, Magnus Christerson, Patrik Jonsson, and Gunnar Overgaard, *Object-Oriented Software Engineering: A Use Case Approach* (Wokingham, England: Addison-Wesley, 1992); Ian Graham, *Migrating to Object Technology* (Wokingham, England: Addison-Wesley, 1994).

Diagram Name	Used to...	Primary Phase
Structure Diagrams		
Class	Illustrate the relationships between classes modeled in the system	Analysis, Design
Object	Illustrate the relationships between objects modeled in the system; used when actual instances of the classes will better communicate the model	Analysis, Design
Package	Group other UML elements together to form higher-level constructs; implementation	Analysis, Design
Deployment	Show the physical architecture of the system; can also be used to show software components being deployed onto the physical architecture	Physical Design, Implementation
Component	Illustrate the physical relationships among the software components; implementation	Physical Design
Composite Structure Design	Illustrate the internal structure of a class, i.e., the relationships among the parts of a class	Analysis
Behavioral Diagrams		
Activity	Illustrate business workflows independent of classes, the flow of activities in a use case, or detailed design of a method	Analysis, Design
Sequence	Model the behavior of objects within a use case; focuses on the time-based ordering of an activity	Analysis, Design
Communication	Model the behavior of objects within a use case; focus on the communication among a set of collaborating objects of an activity	Analysis, Design
Interaction Overview	Illustrate an overview of the flow of control of a process	Analysis, Design
Timing	Illustrate the interaction among a set of objects and the state changes they go through along a time axis	Analysis, Design
Behavioral State Machine	Examine the behavior of one class	Analysis, Design
Protocol State Machine	Illustrate the dependencies among the different interfaces of a class	Analysis, Design
Use-Case	Capture business requirements for the system and illustrate the interaction between the system and its environment.	Analysis

FIGURE 1-20   UML 2.3 Diagram Summary

activity, sequence, communication, interaction overview, timing, behavior state machine, protocol state machine, and use-case diagrams.[24] Figure 1-20 provides an overview of these diagrams.

Depending on where in the development process the system is, different diagrams play a more important role. In some cases, the same diagramming technique is used throughout the development process. In that case, the diagrams start off very conceptual and

[24] The material contained in this section is based on the *Unified Modeling Language: Superstructure Version 2.4, ptc/2010-11-14* (www.uml.org). Additional useful references include Michael Jesse Chonoles and James A. Schardt, *UML 2 for Dummies* (Indianapolis, IN: Wiley, 2003); Hans-Erik Eriksson, Magnus Penker, Brian Lyons, and David Fado, *UML 2 Toolkit* (Indianapolis, IN: Wiley, 2004); and Kendall Scott, *Fast Track UML 2.0* (Berkeley, CA: Apress, 2004). For a complete description of all diagrams, see www.uml.org.

abstract. As the system is developed, the diagrams evolve to include details that ultimately lead to generating and developing code. In other words, the diagrams move from documenting the requirements to laying out the design. Overall, the consistent notation, integration among the diagramming techniques, and application of the diagrams across the entire development process makes the UML a powerful and flexible language for analysts and developers. Later chapters provide more detail on using a subset of the UML in object-oriented systems analysis and design. In particular, these chapters describe activity, use-case, class, object, sequence, communication, package, and deployment diagrams and the behavior state machines. We also introduce an optional UML diagram, the windows navigation diagram, that is an extension to the behavioral state machine that is used to design user navigation through an information system's user interfaces.

APPLYING THE CONCEPTS AT CD SELECTIONS

Throughout this book, many new concepts about object-oriented systems analysis and design are introduced. As a way to make these new concepts more relevant, we apply them to a fictitious company called CD Selections. CD Selections is a chain of 50 music stores in California, with headquarters in Los Angeles. Annual sales last year were $50 million, and they have been growing at about 3 to 5 percent per year for the past few years. The firm has been interested in expanding their presence beyond California. Margaret Mooney, Vice President of Marketing, has become excited by and concerned about the rise of Internet sites selling CDs and sites such as iTunes that sell digital music. She believes that the Internet has great potential, but she wants to use it in the right way. Rushing into e-commerce without considering its effect on existing brick-and-mortar stores and the implications on existing systems at CD Selections could cause more harm than good.

Currently, CD Selections has a website that provides basic information about the company and about each of its stores (e.g., map, operating hours, phone number). The website was developed by an Internet consulting firm and is hosted by a prominent local Internet service provider (ISP) in Los Angeles. The IT department at CD Selections has become experienced with Internet technology as it has worked with the ISP to maintain the site; however, it still has a lot to learn when it comes to conducting business over the web. Margaret is interested in investigating the possibility of creating an e-commerce site that will work with the current systems used by CD Selections. In future chapters, we revisit CD Selections to see how the concepts introduced in the individual chapters affect Margaret, and the team developing a web-based solution for CD Selections.

SUMMARY

The Systems Development Life Cycle

All systems development projects follow essentially the same fundamental process, called the system development life cycle (SDLC). The SDLC starts with a planning phase in which the project team identifies the business value of the system, conducts a feasibility analysis, and plans the project. The second phase is the analysis phase, in which the team develops an analysis strategy, gathers information, and builds a set of analysis models. In the next phase, the design phase, the team develops the physical design, architecture design, interface design, database and file specifications, and program design. In the final phase, implementation, the system is built, installed, and maintained.

The Evolution of Systems Development Methodologies

System development methodologies are formalized approaches to implementing an SDLC. System development methodologies have evolved over the decades. Structured design methodologies, such as waterfall and parallel development, emphasize decomposition of a problem by focusing on either process decomposition (process-centric methodologies) or data decomposition. They produce a solid, well-thought-out system but can overlook requirements because users must specify them early in the design process before seeing the actual system. RAD-based methodologies attempt to speed up development and make it easier for users to specify requirements by having parts of the system developed sooner, either by producing different versions (phased development) or by using prototypes (prototyping, throwaway prototyping) through the use of CASE tools and fourth-generation or visual programming languages. However, RAD-based methodologies still tend to be either process-centric or data-centric. Agile development methodologies, such as XP and Scrum, focus on streamlining the SDLC by eliminating many of the tasks and time associated with defining and documenting requirements. Several factors influence the choice of a methodology: clarity of the user requirements, familiarity with the base technology, system complexity, need for system reliability, time pressures, and the need to see progress on the time schedule.

Systems Analysts' Roles and Skills

The project team needs a variety of skills. All analysts need to have general skills, such as change management, ethics, communications, and technical skills. However, different kinds of analysts require specific skills in addition to these. Business analysts usually have business skills that help them to understand the business issues surrounding the system, and systems analysts also have significant experience in analysis and design and programming. The infrastructure analyst focuses on technical issues surrounding how the system will interact with the organization's technical infrastructure, and the change management analyst focuses on people and management issues surrounding the system installation. In addition to analysts, project teams include a project manager, programmers, technical writers, and other specialists.

Basic Characteristics of Object-Oriented Systems

A class is a template on which objects can be instantiated. An object is a person, place, or thing about which we want to capture information. Each object has attributes and methods. The methods are executed by objects sending messages that trigger them. Encapsulation and information hiding allows an object to conceal its inner processes and data from the other objects. Polymorphism and dynamic binding allow a message to be interpreted differently by different kinds of objects. However, if polymorphism is not used in a semantically consistent manner, it can make an object design incomprehensible. Classes can be arranged in a hierarchical fashion in which subclasses inherit attributes and methods from superclasses to reduce the redundancy in development.

Object-Oriented Systems Analysis and Design

Object-oriented systems analysis and design (OOSAD) is most associated with phased-development RAD- and agile-based methodologies, where the time spent in each phase is very short. OOSAD uses a use-case-driven, architecture-centric, iterative, and incremental information systems development approach. It supports three different views of the evolving system: functional, static, and dynamic. OOSAD allows the analyst to decompose complex problems into smaller, more manageable components using a commonly accepted set of notations. Also, many people believe that users do not think in terms of data or processes but, instead, think in terms of a collection of collaborating objects. OOSAD allows the analyst to interact with the user with objects from the user's environment instead of a set of separate processes and data.

The Unified Process

One of the most popular approaches to object-oriented systems analysis and design is the Unified Process. The Unified Process is a two-dimensional systems development process described with a set of phases and workflows. The phases consist of the inception, elaboration, construction, and transition phases. The workflows are organized into two subcategories: engineering and supporting. The engineering workflows include business modeling, requirements, analysis, design, implementation, test, and deployment workflows, and the supporting workflows comprise the project management, configuration and change management, and environment workflows. Depending on which development phase the evolving system is currently in, the level of activity will vary over the workflows.

The Unified Modeling Language

The Unified Modeling Language (UML) is a standard set of diagramming techniques that provide a graphical representation rich enough to model any systems development project from analysis through implementation. Today most object-oriented systems analysis and design approaches use the UML to depict an evolving system. The UML uses a set of different diagrams to portray the various views of the evolving system. The diagrams are grouped into two broad classifications: structure and behavior. The structure diagrams include class, object, package, deployment, component, and composite structure diagrams. The behavior diagrams include activity, sequence, communication, interaction overview, timing, behavior state machine, protocol state machine, and use-case diagrams.

KEY TERMS

Abstract classes, 26
Agile development, 14
A-kind-of, 25
Analysis model, 5
Analysis paralysis, 33
Analysis phase, 4
Analysis strategy, 5
Analysis workflow, 33
Approval committee, 4
Architecture-centric, 29
Architecture design, 5
As-is system, 5
Attribute, 23
Behavior, 23
Behavior diagrams, 39
Behavioral view, 29
Business analyst, 21
Business modeling workflow, 33

Change agent, 20
Change management analyst, 22
Class, 23
Concrete classes, 25
Configuration and change
   management workflow, 32
Construction, 6
Construction phase, 32
Database and file specification, 5
Data-centered methodology, 7
Deliverable, 3
Deployment workflow, 34
Design model, 33
Design phase, 5
Design prototype, 13
Design strategy, 5
Design workflow, 31
Dynamic binding, 27

Dynamic view, 29
Elaboration phase, 31
Encapsulation, 24
Engineering workflow, 32
Environment workflow, 35
External view, 29
Extreme programming (XP), 15
Feasibility analysis, 4
Functional view, 29
Gradual refinement, 4
Implementation phase, 6
Implementation workflow, 32
Inception phase, 30
Incremental, 29
Information hiding, 24
Infrastructure analyst, 22
Infrastructure management
   workflow, 37

QUESTIONS

1. Compare and contrast phases, steps, techniques, and deliverables.
2. Describe the major phases in the SDLC.
3. Describe the principal steps in the planning phase. What are the major deliverables?
4. Describe the principal steps in the analysis phase. What are the major deliverables?
5. Describe the principal steps in the design phase. What are the major deliverables?
6. Describe the principal steps in the implementation phase. What are the major deliverables?
7. What are the roles of a project sponsor and the approval committee?
8. What does *gradual refinement* mean in the context of SDLC?
9. Compare and contrast process-centered methodologies with data-centered methodologies.
10. Compare and contrast structured design-based methodologies in general to RAD-based methodologies in general.
11. Compare and contrast extreme programming and throwaway prototyping.
12. Describe the major elements in, and issues with, waterfall development.

13. Describe the major elements in, and issues with, parallel development.
14. Describe the major elements in, and issues with, phased development.
15. Describe the major elements in, and issues with, prototyping.
16. Describe the major elements in, and issues with, throwaway prototyping.
17. Describe the major elements in, and issues with, XP.
18. Describe the major elements in, and issues with, Scrum.
19. What are the key factors in selecting a methodology?
20. What are the major roles played by a systems analyst on a project team?
21. Compare and contrast the role of a systems analyst, business analyst, and infrastructure analyst.
22. What is the difference between classes and objects?
23. What are methods and messages?
24. Why are encapsulation and information hiding important characteristics of object-oriented systems?
25. What is meant by polymorphism when applied to object-oriented systems?
26. Compare and contrast dynamic and static binding.
27. What is a use case?
28. What is meant by use-case driven?

29. What is the Unified Modeling Language?
30. Who is the Object Management Group?
31. What is the primary purpose of structure diagrams? Give some examples of structure diagrams.
32. For what are behavior diagrams used? Give some examples of behavior diagrams.
33. Why is it important for an OOSAD approach to be architecture-centric?
34. What does it mean for an OOSAD approach to be incremental and iterative?

35. What are the phases and workflows of the Unified Process?
36. Compare the phases of the Unified Process with the phases of the waterfall model.
37. Which phase in the SDLC is most important? Why?
38. Describe the major elements and issues with an object-oriented approach to developing information systems.

EXERCISES

A. Suppose you are a project manager using a waterfall development–based methodology on a large and complex project. Your manager has just read the latest article in *Computerworld* that advocates replacing this methodology with prototyping and comes to you requesting that you switch. What would you say?

B. The basic types of methodologies discussed in this chapter can be combined and integrated to form new hybrid methodologies. Suppose you were to combine throwaway prototyping with the use of waterfall development. What would the methodology look like? Draw a picture (similar to those in Figures 1–2 through 1–7). How would this new methodology compare to the others?

C. Look on the web for different kinds of job opportunities that are available for people who want analyst positions. Compare and contrast the skills that the ads ask for to the skills that we presented in this chapter.

D. Think about your ideal analyst position. Write an ad to hire someone for that position. What requirements would the job have? What skills and experience would be required? How would an applicant be able to demonstrate having the appropriate skills and experience?

E. Using your favorite web search engine, find alternative descriptions of the basic characteristics of object-oriented systems.

F. Look up object-oriented programming in Wikipedia. Write a short report based on its entry.

G. Choose an object-oriented programming language, such as C++, Java, Objective-C, Smalltalk, or VB.Net, and use the web to find out how the language supports the basic characteristics of object-oriented systems.

H. Assume that you have been assigned the task of creating an object-oriented system that could be used to support students in finding an appropriate apartment to live in next semester. What are the different types of objects (i.e., classes) you would want to include in your system? What attributes or methods would you want to include in their definition? Is it possible to arrange them into an inheritance hierarchy? If so, do it. If not, why not?

I. Create an inheritance hierarchy that could be used to represent the following classes: accountant, customer, department, employee, manager, organization, and salesperson.

J. Investigate IBM's Rational Unified Process (RUP) on the web. RUP is a commercial version that extends aspects of the Unified Process. Write a brief memo describing how it is related to the Unified Process as described in this chapter. (*Hint:* A good website with which to begin is www.ibm.com/software/awdtools/rup/.)

K. Suppose you are a project manager who typically has been using a waterfall development–based methodology on a large and complex project. Your manager has just read the latest article in *Computerworld* that advocates replacing this methodology with the Unified Process and comes to you requesting you to switch. What do you say?

L. Suppose you are an analyst working for a small company to develop an accounting system. Would you use the Unified Process to develop the system, or would you prefer one of the other approaches? Why?

M. Suppose you are an analyst developing a new information system to automate the sales transactions and manage inventory for each retail store in a large chain. The system would be installed at each store and exchange data with a mainframe computer at the company's head office. Would you use the Unified Process to develop the system or would you prefer one of the other approaches? Why?

LIBRARY, UNIVERSITY OF CHESTER

N. Suppose you are an analyst working for a small company to develop an accounting system. What type of methodology would you use? Why?

O. Suppose you are an analyst developing a new executive information system intended to provide key strategic information from existing corporate databases to senior executives to help in their decision making. What type of methodology would you use? Why?

P. Investigate the Unified Modeling Language on the web. Write a paragraph news brief describing the current state of the UML. (*Hint:* A good website with which to begin is www.uml.org.)

Q. Investigate the Object Management Group (OMG) on the web. Write a report describing the purpose of the OMG and what it is involved with besides the UML. (*Hint:* A good website with which to begin is www.omg.org.)

R. Using the web, find a set of CASE tools that support the UML. A couple of examples include Poseidon, Rational Rose, and Visual Paradigm. Find at least two more. Write a short report describing how well they support the UML, and make a recommendation as to which one you believe would be best for a project team to use in developing an object-oriented information system using the UML.

MINICASES

1. Maria Carrasquillo is the logistics manager for El Corte de Mejias, a Spanish company that makes and distributes sporting goods throughout Europe. El Corte de Mejias started as a small specialty company, but has grown rapidly over the past three years. However, its information systems (IS) have not kept pace. Señora Carrasquillo requested that the IS department develop a logistics system that would ensure that the right products are in stock in El Corte de Mejias stores. Due to the massive backlog of work faced by the IS department, her request was given low priority. After six months of inaction by the department, Maria decided to take matters into her own hands. Based on her friends' advice, Maria purchased a simple database software package and created a logistics system on her own.

   Although Maria's system has been installed for about six weeks, it still does not work correctly and is error prone. Maria's assistant is so mistrustful of the system that she has secretly gone back to using her old paper-based system, since it is much more reliable.

   Over dinner one evening, Maria complained to a systems analyst friend, "I don't know what went wrong with this project. It seemed pretty simple to me. Those IS guys wanted me to follow their elaborate set of steps and tasks, but I didn't think all that would really apply to a PC-based system. I just thought I could build this system and tweak it around without all the fuss and bother of the methodology the IS guys were pushing for. I mean, that applies only to big, expensive systems, right?"

   Assuming you are Maria's systems analyst friend, how would you respond to her complaint?

2. Mike is a computer specialist with good experience in programming and application development. Mike's friend, George, decided to start a new DVD rental business and asked for his assistance in creating a computer application to automate the activities in the new store. Mike found the challenge quite interesting, and he was quite excited about the perspective of helping his good friend, so, although busy, he decided to commit to the project. Both Mike and George really had only a vague idea about how the application should look like, so they decided to meet every evening for one hour at George's house in order to figure things out. Enthusiastically, Mike and George went on with their project. Soon enough, it turned out that this project was subjected to some inherent constraints: George did not really have a budget for the development of the application, which meant that Mike was going to be the only developer on the team. On the other hand, George was willing to spend as much time as needed with Mike to make the project successful. His only concern was to have a functional application within 8 weeks, which is when the grand opening of his DVD rental shop was scheduled. Since Mike really did not want to let his friend down, he started thinking seriously about what would be the best way to approach this project. What methodology would you advise Mike to use? Why? How should Mike proceed to avoid any possible pitfalls with his approach?

3. Star Software Inc. was a small software company that specialized in the development of custom applications for the businesses in the region. When the company received an offer from the local Piggly Wiggly grocery store to develop a new computerized system for their

stock management, James, the CEO of Star Software Inc., was very happy to accept it. Star Software Inc. had recently delivered a similar application to a little store in Nekoosa, and since things had gone very well, James was confident that this time, he hit the jackpot with this deal. He even offered Piggly Wiggly discounts and incentives they could not refuse. James quickly created a small team of two, consisting one project manager and one change management analyst, to deploy and implement the existing project at Piggly Wiggly. Everything seemed to be fine at first, and the Star Software Inc. team did its job as requested. But a couple of weeks after the application was in place, Star Software Inc. started receiving complaints from Piggly Wiggly. First of all, the application was too slow and seemed to get slower by the day; second, most of the product codes Piggly Wiggly used to work with were not in the system, and the trouble did not stop here. What went wrong with the Piggly Wiggly project? What mistakes did James make in handling this project?

CHAPTER 2

PROJECT MANAGEMENT

This chapter primarily describes the project management workflow of the Unified Process. The first step in the process is to identify a project that will deliver value to the business and to create a system request that provides basic information about the proposed system. Second, the analysts perform a feasibility analysis to determine the technical, economic, and organizational feasibility of the system; if appropriate, the system is selected and the development project begins. Third, the project manager estimates the functionality of the project and identifies the tasks that need to be performed. Fourth, the manager staffs the project. Finally, the manager identifies the tools, standards, and process to be used; identifies opportunities for reuse; determines how the current project fits into the portfolio of projects currently under development; and identifies opportunities to update the overall structure of the firm's portfolio of systems current in use.

OBJECTIVES

- Understand the importance of linking the information system to business needs.
- Be able to create a system request.
- Understand how to assess technical, economic, and organizational feasibility.
- Be able to perform a feasibility analysis.
- Understand how projects are selected in some organizations.
- Become familiar with work breakdown structures, Gantt charts, and network diagrams.
- Become familiar with use-case–driven effort estimation.
- Be able to create an iterative project workplan.
- Understand how to manage the scope, refine the estimates, and manage the risk of a project.
- Become familiar with how to staff a project.
- Understand how the environment and infrastructure workflows interact with the project management workflow.

CHAPTER OUTLINE

Introduction
Project Identification
  System Request
Feasibility Analysis
  Technical Feasibility
  Economic Feasibility
  Organizational Feasibility
Project Selection

Traditional Project Management Tools
  Work Breakdown Structures
  Gantt Chart
  Network Diagram
Project Effort Estimation
Creating and Managing the Workplan
  Evolutionary Work Breakdown
    Structures and Iterative Workplans

INTRODUCTION

Think about major projects that occur in the lives of people, such as throwing a big party, a wedding, or a graduation celebration. Months are spent in advance identifying and performing all the tasks that need to get done, such as sending out invitations and selecting a menu, and time and money are carefully allocated among them. Along the way, decisions are recorded, problems are addressed, and changes are made. The increasing popularity of the party planner, a person whose sole job is to coordinate a party, suggests how tough this job can be. In the end, the success of any party has a lot to do with the effort that went into planning along the way. System development projects can be much more complicated than the projects we encounter in our personal lives—usually, more people are involved (e.g., the organization), the costs are higher, and more tasks need to be completed. Owing to the complexity of software and software development, it is virtually impossible to "know" all of the possible things that could happen during system development projects. Therefore, it is not surprising that "party planners" exist for information systems projects: They are called *project managers*.

Project management is the process of planning and controlling the development of a system within a specified time frame at a minimum cost with the right functionality.[1] In general, a *project* is a set of activities with a starting point and an ending point meant to create a system that brings value to the business. A project manager has the primary responsibility for managing the hundreds of tasks and roles that need to be carefully coordinated. Today, project management is an actual profession, and analysts spend years working on projects before tackling the management of them. However, in many cases, unreasonable demands set by project sponsors and business managers can make project management very difficult. Too often, the approach of the holiday season, the chance at winning a proposal with a low bid, or a funding opportunity pressures project managers to promise systems long before they are able to deliver them. These overly optimistic timetables are thought to be one of the biggest problems that projects face; instead of pushing a project forward faster, they result in delays. Another source is the changing

---

[1] For a very good comprehensive description of project management for information systems, see R.K. Wysocki, *Effective Project Management: Traditional, Agile, Extreme,* 5th Ed. (Indianapolis, IN: Wiley Publishing, 2009). Also, the Project Management Institute (www.pmi.org) and the Information Systems Special Interest Group of the Project Management Institute (www.pmi-issig.org) have valuable resources on information systems project management. Finally, the following are good books on project management for object-oriented projects: G. Booch, *Object Solutions: Managing the Object-Oriented Project* (Menlo Park, CA: Addison-Wesley, 1996); M. R. Cantor, *Object-Oriented Project Management with UML* (New York: Wiley, 1998); A. Cockburn, *Surviving Object-Oriented Projects: A Manager's Guide* (Reading, MA: Addison-Wesley, 1998); I. Jacobson, G. Booch, and J. Rumbaugh, *The Unified Software Development Process* (Reading, MA: Addison-Wesley, 1999); and W. Royce, *Software Project Management: A Unified Framework* (Reading, MA: Addison-Wesley, 1998).

nature of the information technology on which information systems are deployed. The promise of new information technology innovations can appear so attractive that organizations begin projects even if they are not sure what value these technologies offer, because they believe that the technologies are somehow important in their own right. Problems can usually be traced back to the very beginning of the development of the system, where too little attention was given to identifying the business value and understanding the risks associated with the project.

During the inception phase of the Unified Process of a new systems development project, someone—a manager, staff member, sales representative, or systems analyst—typically identifies some business value that can be gained from using information technology. New systems development projects should start from a business need or opportunity. Many ideas for new systems or improvements to existing ones arise from the application of a new technology, but an understanding of technology is usually secondary to a solid understanding of the business and its objectives. This does not mean that technical people should not recommend new systems projects. In fact, the ideal situation is for both IT people (i.e., the experts in systems) and the business people (i.e., the experts in business) to work closely to find ways for technology to support business needs. In this way, organizations can leverage the exciting innovative technologies that are available while ensuring that projects are based upon real business objectives, such as increasing sales, improving customer service, and decreasing operating expenses. Ultimately, information systems need to affect the organization's bottom line (in a positive way!). To ensure that a real business need is being addressed, the affected business organization (called the *project sponsor*), proposes the new systems development project using a *system request*. The system request effectively kicks off the inception phase for the new systems development project. The request is forwarded to an *approval committee* for consideration. The approval committee reviews the request and makes an initial determination of whether to investigate the proposal or not. If the committee initially approves the request, the systems development team gathers more information to determine the feasibility of the project.

A *feasibility analysis* plays an important role in deciding whether to proceed with an information systems development project. It examines the technical, economic, and organizational pros and cons of developing the system, and it gives the organization a slightly more detailed picture of the advantages of investing in the system as well as any obstacles that could arise. In most cases, the project sponsor works closely with the development team to develop the feasibility analysis. Once the feasibility analysis has been completed, it is submitted to the approval committee, along with a revised system request. The committee then decides whether to approve the project, decline the project, or table it until additional information is available. Projects are selected by weighing risks and return, and by making trade-offs at the organizational level.

Once the committee has approved a project, the development team must carefully plan for the actual development of the system. Because we are following a Unified Process-based approach, the systems development workplan will evolve throughout the development process. Given this evolutionary approach, one critical success factor for project management is to start with a realistic assessment of the work that needs to be accomplished and then manage the project according to that assessment. This can be achieved by carefully creating and managing the workplan, estimating the effort to develop the system, staffing the project, and coordinating project activities.

In addition to covering the above material, this chapter also covers three traditional project management tools that are very useful to manage object-oriented systems development projects: work breakdown structures, Gantt charts, and network diagrams.

CONCEPTS

IN ACTION

2-A Interview with Lyn McDermid, CIO, Dominion Virginia Power

A CIO needs to have a global view when identifying and selecting projects for her organization. I would get lost in the trees if I were to manage on a project-by-project basis. Given this, I categorize my projects according to my three roles as a CIO, and the mix of my project portfolio changes depending on the current business environment.

My primary role is to **keep the business running**. That means every day when each person comes to work, he or she can perform his or her job efficiently. I measure this using various service-level, cost, and productivity measures. Projects that keep the business running could have a high priority if the business were in the middle of a merger or a low priority if things were running smoothly and it were "business as usual."

My second role is to push **innovation that creates value for the business**. I manage this by looking at our lines of business and asking which lines of business create the most value for the company. These are the areas for which I should be providing the most value. For example, if we had a highly innovative marketing strategy, I would push for innovation there. If operations were running smoothly, I would push less for innovation in that area.

My third role is strategic, to look beyond today and find **new opportunities** for both IT and the business of providing energy. This may include investigating process systems, such as automated meter reading, or looking into the possibilities of wireless technologies.

—Lyn McDermid

PROJECT IDENTIFICATION

A project is identified when someone in the organization identifies a *business need* to build a system. This could occur within a business unit or IT, come from a steering committee charged with identifying business opportunities, or evolve from a recommendation made by external consultants. Examples of business needs include supporting a new marketing campaign, reaching out to a new type of customer, or improving interactions with suppliers. Sometimes, needs arise from some kind of "pain" within the organization, such as a drop in market share, poor customer service levels, or increased competition. Other times, new business initiatives and strategies are created, and a system is required to enable them.

Business needs also can surface when the organization identifies unique and competitive ways of using IT. Many organizations keep an eye on *emerging technology*, which is technology that is still being developed and is not yet viable for widespread business use. For example, if companies stay abreast of technology such as the Internet, smart cards, and mobile devices in their earliest stages, they can develop business strategies that leverage the capabilities of these technologies and introduce them into the marketplace as a *first mover*. Ideally, they can take advantage of this first-mover advantage by making money and continuing to innovate while competitors trail behind.

The *project sponsor* is someone who recognizes the strong business need for a system and has an interest in seeing the system succeed. He or she will work throughout the development process to make sure that the project is moving in the right direction from the perspective of the business. The project sponsor serves as the primary point of contact for the system. Usually, the sponsor of the project is from a business function, such as marketing, accounting, or finance; however, members of the IT area also can sponsor or cosponsor a project.

The size or scope of a project determines the kind of sponsor needed. A small departmental system might require sponsorship from only a single manager, whereas a large organizational initiative might need support from the entire senior management team and even the CEO. If a project is purely technical in nature (e.g., improvements to the existing IT infrastructure or research into the viability of an emerging technology), then sponsorship from IT is appropriate. When projects have great importance to the business yet are technically complex, joint sponsorship by both the business and IT may be necessary.

The business need drives the high-level *business requirements* for the system. Requirements are what the information system will do, or the *functionality* it will contain. They need to be explained at a high level so that the approval committee and, ultimately, the project team understand what the business expects from the final product. Business requirements are the features and capabilities the information system will have to include, such as the ability to collect customer orders online or the ability for suppliers to receive inventory information as orders are placed and sales are made.

The project sponsor also should have an idea of the *business value* to be gained from the system, both in tangible and intangible ways. *Tangible value* can be quantified and measured easily (e.g., 2 percent reduction in operating costs). An *intangible value* results from an intuitive belief that the system provides important, but hard-to-measure, benefits to the organization (e.g., improved customer service or a better competitive position).

Once the project sponsor identifies a project that meets an important business need and he or she can identify the system's business requirements and value, it is time to formally initiate the project. In most organizations, project initiation begins with a technique called a *system request*.

YOUR TURN

2-1 Identify Tangible and Intangible Value

*D*ominion Virginia Power is one of the nation's ten largest investor-owned electric utilities. The company delivers power to more than two million homes and businesses in Virginia and North Carolina. In 1997, the company overhauled some of its core processes and technology. The goal was to improve customer service and cut operations costs by developing a new workflow and geographic information system. When the project was finished, service engineers who had sifted through thousands of paper maps could use computerized searches to pinpoint the locations of electricity poles. The project helped the utility improve management of all its facilities, records, maps, scheduling, and human resources. That, in turn, helped increase employee productivity, improve customer response times, and reduce the costs of operating crews.

Source: Computerworld (November 11, 1997).

Questions

1. What kinds of things does Dominion Virginia Power do that require it to know power pole locations? How often does it do these things? Who benefits if the company can locate power poles faster?

2. Based on your answers to question 1, describe three tangible benefits that the company can receive from its new computer system. How can these be quantified?

3. Based on your answers to question 1, describe three intangible benefits that the company can receive from its new computer system. How can these be quantified?

System Request

A system request is a document that describes the business reasons for building a system and the value that the system is expected to provide. The project sponsor usually completes this form as part of a formal system project selection process within the organization. Most system requests include five elements: project sponsor, business need, business requirements, business value, and *special issues* (see Figure 2-1). The sponsor describes the person who will serve as the primary contact for the project, and the business need presents the reasons prompting the project. The business requirements of the project refer to

Element	Description	Examples
Project Sponsor	The person who initiates the project and who serves as the primary point of contact for the project on the business side.	Several members of the Finance department Vice President of Marketing project IT Manager Steering committee CIO CEO
Business Need	The business-related reason for initiating the system.	Increase sales Improve market share Improve access to information Improve customer service Decrease product defects Streamline supply acquisition Processes
Business Requirements	The business capabilities that the system will provide.	Provide online access to information Capture customer demographic information Include product search capabilities Produce management reports Include online user support
Business Value	The benefits that the system will create for the organization.	A 3 percent increase in sales A 1 percent increase in market share Reduction in headcount by 5 FTEs* $200,000 cost savings from decreased supply costs $150,000 savings from removal of existing system
Special Issues or Constraints	Issues that are relevant to the implementation of the system and decisions made by the committee about the project.	Government-mandated deadline for May 30 System needed in time for the Christmas holiday season Top-level security clearance needed by project team to work with data

*Full-time equivalent

FIGURE 2-1
Elements of the
System Request Form

the business capabilities that the system will need to have, and the business value describes the benefits that the organization should expect from the system. Special issues are included on the document as a catch-all for other information that should be considered in assessing the project. For example, the project may need to be completed by a specific deadline. Project teams need to be aware of any special circumstances that could affect the outcome of the system. Figure 2-2 shows a template for a system request.

System Request—Name of Project	
Project Sponsor:	Name of project sponsor
Business Need:	Short description of business need
Business Requirements:	Description of business requirements
Business Value:	Expected value that the system will provide
Special Issues or Constraints:	Any additional information that may be relevant to the stakeholders

FIGURE 2-2
System Request
Template

CONCEPTS

IN ACTION

2-B   Interview with Don Hallacy, President, Technology Services, Sprint Corporation

At Sprint, network projects originate from two vantage points—IT and the business units. IT projects usually address infrastructure and support needs. The business unit projects typically begin after a business need is identified locally, and a business group informally collaborates with IT regarding how a solution can be delivered to meet customer expectations.

Once an idea is developed, a more formal request process begins, and an analysis team is assigned to investigate and validate the opportunity. This team includes members from the user community and IT, and they scope out at a high level what the project will do; create estimates for technology, training, and business development costs; and create a business case. This contains the economic value-add and the net present value of the project.

Of course, not all projects undergo this rigorous process. The larger the project, the more time is allocated to the analysis team. It is important to remain flexible and not let the process consume the organization. At the beginning of each budgetary year, specific capital expenditures are allocated for operational improvements and maintenance. Moreover, this money is set aside to fund quick projects that deliver immediate value without going through the traditional approval process.

—*Don Hallacy*

The completed system request is submitted to the approval committee for consideration. This approval committee could be a company steering committee that meets regularly to make information systems decisions, a senior executive who has control of organizational resources, or any other decision-making body that governs the use of business investments. The committee reviews the system request and makes an initial determination, based on the information provided, of whether to investigate the proposal or not. If so, the next step is to conduct a feasibility analysis.

YOUR

TURN

2–2   Create a System Request

Think about your own university or college, and choose an idea that could improve student satisfaction with the course enrollment process. Currently, can students enroll for classes from anywhere? How long does it take? Are directions simple to follow? Is online help available?

Next, think about how technology can help support your idea. Would you need completely new technology? Can the current system be changed?

Question

Create a system request that you could give to the administration that explains the sponsor, business need, business requirements, and potential value of the project. Include any constraints or issues that should be considered.

FEASIBILITY ANALYSIS

Once the need for the system and its business requirements have been defined, it is time to create a more detailed business case to better understand the opportunities and limitations associated with the proposed project. Feasibility analysis guides the organization in determining whether or not to proceed with a project. Feasibility analysis also identifies the

---

Technical Feasibility: Can We Build It?
- Familiarity with Functional area: Less familiarity generates more risk
- Familiarity with Technology: Less familiarity generates more risk
- Project Size: Large projects have more risk
- Compatibility: The harder it is to integrate the system with the company's existing technology, the higher the risk

Economic Feasibility: Should We Build It?
- Development costs
- Annual operating costs
- Annual benefits (cost savings and revenues)
- Intangible costs and benefits

Organizational Feasibility: If We Build It, Will They Come?
- Project champion(s)
- Senior management
- Users
- Other stakeholders
- Is the project strategically aligned with the business?

FIGURE 2-3
Feasibility Analysis Assessment Factors

important *risks* associated with the project that must be addressed if the project is approved. As with the system request, each organization has its own process and format for the feasibility analysis, but most include three types: technical feasibility, economic feasibility, and organizational feasibility. The results of these analyses are combined into a *feasibility study*, which is given to the approval committee at the end of project initiation (see Figure 2-3).

Although we now discuss feasibility analysis within the context of initiating a project, most project teams will revise their feasibility study throughout the development process and revisit its contents at various checkpoints during the project. If at any point the project's risks and limitations outweigh its benefits, the project team may decide to cancel the project or make necessary improvements.

Technical Feasibility

The first type of feasibility analysis addresses the *technical feasibility* of the project: the extent to which the system can be successfully designed, developed, and installed by the IT group. Technical feasibility analysis is, in essence, a *technical risk analysis* that strives to answer this question: *Can* we build it?[2]

Many risks can endanger the successful completion of a project. First is the users' *and* analysts' lack of *familiarity with the functional area*. When analysts are unfamiliar with the business functional area, they have a greater chance of misunderstanding the users or of missing opportunities for improvement. The risk increases dramatically when the users themselves are less familiar with an application, such as with the development of a system to support a business innovation (e.g., Microsoft starting up a new Internet dating service). In general, developing new systems is riskier than producing extensions to an existing system because existing systems tend to be better understood.

---

[2] We use *build it* in the broadest sense. Organizations can also choose to buy a commercial software package and install it, in which case, the question might be, Can we select the right package and successfully install it?

Familiarity with the technology is another important source of technical risk. When a system uses technology that has not been used before *within the organization*, there is a greater chance that problems will occur and delays will be incurred because of the need to learn how to use the technology. Risk increases dramatically when the technology itself is new (e.g., Android, iPad).

Project size is an important consideration, whether measured as the number of people on the development team, the length of time it will take to complete the project, or the number of distinct features in the system. Larger projects present more risk, both because they are more complicated to manage and because there is a greater chance that important system requirements will be overlooked or misunderstood. The extent to which the project is highly integrated with other systems (which is typical of large systems) can cause problems because complexity increases when many systems must work together.

Finally, project teams need to consider the *compatibility* of the new system with the technology that already exists in the organization. Systems are rarely built in a vacuum—they are built in organizations that already have numerous systems in place. New technology and applications need to be able to be integrated with the existing environment for many reasons. They might rely on data from existing systems, they might produce data that feed other applications, and they might have to use the company's existing communications infrastructure. A new customer relationship management (CRM) system, for example, has little value if it does not use customer data found across the organization in existing sales systems, marketing applications, and customer service systems.

The assessment of a project's technical feasibility is not cut and dried because in many cases, some interpretation of the underlying conditions is needed (e.g., how large a project needs to grow before it becomes less feasible). One approach is to compare the project under consideration with prior projects undertaken by the organization. Another option is to consult with experienced IT professionals in the organization or external IT consultants; often, they are able to judge whether a project is feasible from a technical perspective.

CONCEPTS
IN ACTION

2-C Caring for Grandpa and Grandma

Health care is a big industry in the United States, and with the baby boomers born in the late 1940s and 1950s (after World War II) starting to retire, there will be huge demands for senior health care. The desire is for better technologies to allow grandpa and grandma to live independently in their own homes or apartments longer— and not to use the more expensive options of nursing homes and assisted-living centers. Some technologies include vital-sign monitoring and reporting; motion detectors that sense if somebody has fallen; sensors to turn off the stove

that might have been left on; and Internet portals so that family members can check on the health of their loved ones.

Questions

1. How can technology assist with keeping retirees healthy?
2. How can technology help keep retirees out of expensive nursing homes and centers?

Economic Feasibility

The second element of a feasibility analysis is to perform an *economic feasibility* analysis (also called a *cost–benefit analysis*), which identifies the financial risk associated with the project. It attempts to answer the question, *Should* we build the system? Economic feasibility is determined by identifying costs and benefits associated with the system, assigning values to

1. Identifying Costs and Benefits	List the tangible costs and benefits for the project. Include both one-time and recurring costs.
2. Assigning Values to Costs and Benefits	Work with business users and IT professionals to create numbers for each of the costs and benefits. Even intangibles should be valued if at all possible.
3. Determining Cash Flow	Project what the costs and benefits will be over a period of time, usually three to five years. Apply a growth rate to the numbers, if necessary.
4. Determining Net Present Value (NPV)	Calculate what the value of future costs and benefits are if measured by today's standards. You will need to select a rate of growth to apply the NPV formula.
5. Determining Return on Investment (ROI)	Calculate how much money the organization will receive in return for the investment it will make using the ROI formula.
6. Determining the Break-Even Point	Find the first year in which the system has greater benefits than costs. Apply the break-even formula using figures from that year. This will help you understand how long it will take before the system creates real value for the organization.
7. Graphing the Break-Even Point	Plot the yearly costs and benefits on a line graph. The point at which the lines cross is the break-even point.

FIGURE 2-4
Steps for Conducting
Economic Feasibility

them, and then calculating the cash flow and return on investment for the project. The more expensive the project, the more rigorous and detailed the analysis should be. Figure 2-4 lists the steps in performing a cost–benefit analysis; each step is described in the following sections.

Identifying Costs and Benefits   The first task when developing an economic feasibility analysis is to identify the kinds of costs and benefits the system will have and list them along the left-hand column of a spreadsheet. Figure 2-5 lists examples of costs and benefits that may be included.

Costs and benefits can be broken down into four categories: development costs, operational costs, tangible benefits, and intangibles. *Development costs* are tangible expenses incurred during the construction of the system, such as salaries for the project team, hardware and software expenses, consultant fees, training, and office space and equipment. Development costs are usually thought of as one-time costs. *Operational costs* are tangible costs required to operate the system, such as the salaries for operations staff, software licensing fees, equipment upgrades, and communications charges. Operational costs are usually thought of as ongoing costs.

Revenues and cost savings are the *tangible benefits* the system enables the organization to collect or the tangible expenses the system enables the organization to avoid. Tangible benefits could include increased sales, reductions in staff, and reductions in inventory. Of course, a project also can affect the organization's bottom line by reaping *intangible benefits* or incurring *intangible costs*. Intangible costs and benefits are more difficult to incorporate into the economic feasibility because they are based on intuition and belief rather than "hard numbers." Nonetheless, they should be listed in the spreadsheet along with the tangible items.

Development Costs	Operational Costs
Development Team Salaries	Software Upgrades
Consultant Fees	Software Licensing Fees
Development Training	Hardware Repairs
Hardware and Software	Hardware Upgrades
Vendor Installation	Operational Team Salaries
Office Space and Equipment	Communications Charges
Data Conversion Costs	User Training
Tangible Benefits	**Intangible Benefits**
Increased Sales	Increased Market Share
Reductions in Staff	Increased Brand Recognition
Reductions in Inventory	Higher Quality Products
Reductions in IT Costs	Improved Customer Service
Better Supplier Prices	Better Supplier Relations

FIGURE 2-5

Example Costs and Benefits for Economic Feasibility

Assigning Values to Costs and Benefits   Once the types of costs and benefits have been identified, analysts assign specific dollar values to them. This might seem impossible; how can someone quantify costs and benefits that haven't happened yet? And how can those predictions be realistic? Although this task is very difficult, analysts have to do the best they can to come up with reasonable numbers for all the costs and benefits. Only then can the approval committee make an educated decision about whether or not to move ahead with the project.

The best strategy for estimating costs and benefits is to rely on the people who have the clearest understanding of them. For example, costs and benefits related to the technology or the project itself can be provided by the company's IT group or external consultants, and business users can develop the numbers associated with the business (e.g., sales projections, order levels). Analysts can also consider past projects, industry reports, and vendor information, although these approaches probably will be a bit less accurate. All the estimates will probably be revised as the project proceeds.

Sometimes it is acceptable for analysts to list intangible benefits, such as improved customer service, without assigning a dollar value, whereas other times they have to make estimates regarding the value of an intangible benefit. If at all possible, they should quantify intangible costs or benefits. Otherwise, it will not be apparent whether the costs and benefits have been realized. Consider a system that is supposed to improve customer service. This is intangible, but assume that the greater customer service will decrease the number of customer complaints by 10 percent each year over three years, and that $200,000 is spent on phone charges and phone operators who handle complaint calls. Suddenly, there are some very tangible numbers with which to set goals and measure the original intangible benefit.

Figure 2-6 shows costs and benefits along with assigned dollar values. Notice that the customer service intangible benefit has been quantified based on fewer customer complaint phone calls. The intangible benefit of being able to offer services that competitors currently offer was not quantified, but it was listed so that the approval committee will consider the benefit when assessing the system's economic feasibility.

Benefits[a]	
Increased sales	500,000
Improved customer service[b]	70,000
Reduced inventory costs	68,000
Total benefits	**638,000**
Development costs	
2 servers @ $125,000	250,000
Printer	100,000
Software licenses	34,825
Server software	10,945
Development labor	1,236,525
Total development costs	**1,632,295**
Operational costs	
Hardware	54,000
Software	20,000
Operational labor	111,788
Total operational costs	**185,788**
Total costs	**1,818,083**

[a] An important yet intangible benefit will be the ability to offer services that our competitors currently offer.

[b] Customer service numbers have been based on reduced costs for customer complaint phone calls.

FIGURE 2-6
Assigning Values to
Costs and Benefits

CONCEPTS

IN ACTION

2-D Intangible Value at Carlson Hospitality

I conducted a case study at Carlson Hospitality, a global leader in hospitality services, encompassing more than 1,300 hotel, resort, restaurant, and cruise ship operations in seventy-nine countries. One of its brands, Radisson Hotels & Resorts, researched guest stay information and guest satisfaction surveys. The company was able to quantify how much of a guest's lifetime value can be attributed to his or her perception of the stay experience. As a result, Radisson knows how much of the collective future value of the enterprise is at stake given the perceived quality of the stay experience. Using this model, Radisson can confidently show that a 10 percent increase in customer satisfaction among the 10 percent of highest-quality customers will capture a one-point market share for the brand. Each point in market share for the Radisson brand is worth $20 million in additional revenue.

—*Barbara Wixom*

Question

How can a project team use this information to help determine the economic feasibility of a system?

Determining Cash Flow A formal cost–benefit analysis usually contains costs and benefits over a selected number of years (usually three to five years) to show cash flow over time (see Figure 2-7). When using this *cash-flow method,* the years are listed across the top of the spreadsheet to represent the time period for analysis, and numeric values are entered in the appropriate cells within the spreadsheet's body. Sometimes fixed amounts are entered into the columns. For example, Figure 2-7 lists the same amount for customer complaint calls and inventory costs for all five years. Usually, amounts are augmented by

	2011	2012	2013	2014	2015	Total
Increased sales	500,000	530,000	561,800	595,508	631,238	
Reduction in customer complaint calls	70,000	70,000	70,000	70,000	70,000	
Reduced inventory costs	68,000	68,000	68,000	68,000	68,000	
TOTAL BENEFITS:	638,000	668,000	699,800	733,508	769,238	
PV OF BENEFITS:	619,417	629,654	640,416	651,712	663,552	3,204,752
PV OF ALL BENEFITS:	619,417	1,249,072	1,889,488	2,541,200	3,204,752	
2 Servers @ $125,000	250,000	0	0	0	0	
Printer	100,000	0	0	0	0	
Software licenses	34,825	0	0	0	0	
Server software	10,945	0	0	0	0	
Development labor	1,236,525	0	0	0	0	
TOTAL DEVELOPMENT COSTS:	1,632,295	0	0	0	0	
Hardware	54,000	81,261	81,261	81,261	81,261	
Software	20,000	20,000	20,000	20,000	20,000	
Operational labor	111,788	116,260	120,910	125,746	130,776	
TOTAL OPERATIONAL COSTS:	185,788	217,521	222,171	227,007	232,037	
TOTAL COSTS:	1,818,083	217,521	222,171	227,007	232,037	
PV OF COSTS:	1,765,129	205,034	203,318	201,693	200,157	2,575,331
PV OF ALL COSTS:	1,765,129	1,970,163	2,173,481	2,375,174	2,575,331	
TOTAL PROJECT BENEFITS COSTS:	(1,180,083)	450,479	477,629	506,501	537,201	
YEARLY NPV:	(1,145,712)	424,620	437,098	450,019	463,395	629,421
CUMULATIVE NPV:	(1,145,712)	(721,091)	(283,993)	166,026	629,421	
RETURN ON INVESTMENT:	24.44%	(629,421/2,575,331)				
BREAK-EVEN POINT:	3.63 years	[break-even occurs in year 4; (450,019 − 166,026)/450,019 = 0.63]				
INTANGIBLE BENEFITS:	This service is currently provided by competitors Improved customer satisfaction					

FIGURE 2-7   Cost–Benefit Analysis

some rate of growth to adjust for inflation or business improvements, as shown by the 6 percent increase that is added to the sales numbers in the sample spreadsheet. Finally, totals are added to determine what the overall benefits will be; the higher the overall total, the greater the economic feasibility of the solution.

Determining Net Present Value and Return on Investment   There are several problems with the cash-flow method because it does not consider the time value of money (i.e., a dollar today is *not* worth a dollar tomorrow), and it does not show the overall "bang for the buck" that the organization is receiving from its investment. Therefore, some project teams add additional calculations to the spreadsheet to provide the approval committee with a more accurate picture of the project's worth.

Net present value (NPV) is used to compare the present value of future cash flows with the investment outlay required to implement the project. Consider the table in Figure 2-8, which shows the future worth of a dollar investment today, given different numbers of years and different rates of change. If you have a friend who owes you a dollar today but instead gives you a dollar three years from now, you've been had! Given a 10 percent increase in value, you'll be receiving the equivalent of 75 cents in today's terms.

Number of years	6%	10%	15%
1	0.943	0.909	0.870
2	0.890	0.826	0.756
3	0.840	0.751	0.572
4	0.792	0.683	0.497

This table shows how much a dollar today is worth one to four years from now in today's terms using different interest rates.

FIGURE 2-8
The Value of a Future Dollar Today

NPV can be calculated in many different ways, some of which are extremely complex. Figure 2-9 shows a basic calculation that can be used in your cash flow analysis to get more relevant values. In Figure 2-7, the present value of the costs and benefits are calculated first (i.e., they are shown at a discounted rate). Then, net present value is calculated, and it shows the discounted rate of the combined costs and benefits.

The *return on investment* (ROI) is a calculation listed somewhere on the spreadsheet that measures the amount of money an organization receives in return for the money it spends. A high ROI results when benefits far outweigh costs. ROI is determined by finding the total benefits less the costs of the system and dividing that number by the total costs of the system (see Figure 2-9). ROI can be determined per year or for the entire project over a period of time. One drawback of ROI is that it considers only the end points of the investment, not the cash flow in between, so it should not be used as the sole indicator of a project's worth. The spreadsheet in Figure 2-7 shows an ROI figure.

Determining the Break-Even Point   If the project team needs to perform a rigorous cost–benefit analysis, it might need to include information about the length of time before the project will break even, or when the returns will match the amount invested in the project. The greater the time it takes to break even, the riskier the project. The *break-even point* is determined by looking at the cash flow over time and identifying the year in which the

Calculation	Definition	Formula
Present Value (PV)	The amount of an investment today compared to that same amount in the future, taking into account inflation and time.	$\dfrac{\text{Amount}}{(1 + \text{interest rate})^n}$ n = number of years in future
Net Present Value (NPV)	The present value of benefit less the present value of costs.	PV Benefits − PV Costs
Return on Investment (ROI)	The amount of revenues or cost savings results from a given investment.	$\dfrac{\text{Total benefits} - \text{Total costs}}{\text{Total costs}}$
Break-Even Point	The point in time at which the costs of the project equal the value it has delivered.	$\dfrac{\text{Yearly NPV* } - \text{ Cumulative NPV}}{\text{Yearly NPV*}}$

*Use the Yearly NPV amount from the first year in which the project has a positive cash flow.

Add the above amount to the year in which the project has a positive cash flow.

FIGURE 2-9   Financial Calculations Used For Cost–Benefit Analysis

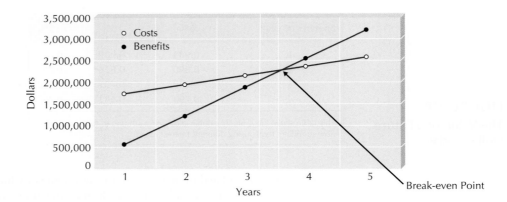

FIGURE 2-10
Break-even Graph

benefits are larger than the costs (see Figure 2-7). Then, the difference between the yearly and cumulative NPV for that year is divided by the yearly NPV to determine how far into the year the break-even point will occur. See Figure 2-9 for the break-even calculation. The break-even point also can be depicted graphically, as shown in Figure 2-10. The cumulative present value of the costs and benefits for each year are plotted on a line graph; the point at which the lines cross is the break-even point.

CONCEPTS
IN ACTION

2-E The FBI Pulls the Plug on a Project

*T*he FBI's failure to roll out an expanded computer system that would help agents investigate criminals and terrorists is the latest in a series of costly technology blunders by the government over more than a decade. Experts blame poor planning, rapid industry advances, and the massive scope of some complex projects, whose price tags can run into billions of dollars at U.S. agencies with tens of thousands of employees. "There are very few success stories," said Paul Brubaker, former deputy chief information officer at the Pentagon. "Failures are very common, and they've been common for a long time." The FBI said earlier this month it might shelve its custom-built, $170 million "Virtual Case File" project because it is inadequate and outdated. The system was intended to help agents, analysts, and others around the world share information without using paper or time-consuming scanning of documents. Officials said commercial software might accomplish some of what the FBI needs. The bureau's

mess—the subject of an investigation by the Justice Department and an upcoming congressional hearing—was the latest black eye among ambitious technology upgrades by the government since the 1990s.

Question

Some systems like this are very complex. They must have security, and they must interface among the FBI, CIA, and other government agencies as well as state and local law enforcement groups. Such complexity can take years to build and is almost guaranteed to fail because of newer technologies that come along during the wait. How might you keep a complex project on track? What commercial software might work in this case (as mentioned in the case?)

Source: www.securityfocus.com/news/10383

Alternatives to Traditional Cost–Benefit Analysis   Concerns have been raised about the appropriateness of using traditional cost–benefit analysis with NPV and ROI to determine economic feasibility of an IT project. One of the major problems of using traditional cost–benefit analysis to determine the economic feasibility of an IT investment is that traditional cost–benefit analysis is based on the assumption that the investor must either invest now or not invest at all. However, in most IT investment decisions, the decision to invest is not a now-or-never decision. In most situations, an information system is already in place, and the decision to replace or upgrade the current information system can usually be delayed. Different proposals have been made to overcome some of the weaknesses in traditional cost–benefit analysis. For example, economic production models, activity-based costing, and balanced score cards have been suggested.[3]

In this section, we describe the primary alternative that has been proposed for object-oriented systems: *option pricing models (OPMs)*.[4] At this point in time, OPMs have had limited use in economic feasibility analysis for IT investment decisions in industry. In fact, there is some controversy as to whether an instrument created for a traded asset (stock) can be used in evaluating IT investment opportunities. However, the preliminary research results demonstrate that their use in IT investment evaluations may be warranted. OPMs have shown promise in evaluating the potential future value of an investment in IT. In many cases in which traditional cost–benefit analysis of investments in IT has predicted that the investment would be a failure, OPMs have shown that it might indeed be feasible.

With object-oriented systems, where classes are designed not only for the current application but also for use in future development efforts, an investment in developing a class or a set of classes can pay dividends well beyond the original system development effort. Furthermore, with the iterative and incremental development emphasis in object-oriented systems development approaches, an object-oriented project can be viewed as a sequence of smaller projects. Thus, you might treat investments in an object-oriented project much as you would an investment in a call option in finance. A *call option* is essentially a contract that gives the right to purchase an amount of stock for a given price for a specified period of time to the purchaser of the call option. However, a call option does not create an obligation to buy the stock.

Treating an IT investment as a call option allows management, at a relevant point in the future, to determine whether additional investment into the evolving system is reasonable. This gives management the flexibility to determine the economic feasibility of a project to decide whether to continue with the development as planned, to abandon the project, to expand the scope of the project, to defer future development, or to shrink the current development effort. In many ways, treating IT investments as call options simply allows management to delay investment decisions until more information is available. Once the decision is made to invest (i.e., the call option is exercised), the decision is consid-

---

[3] See, for example, Q. Hu, R. Plant, and D. Hertz, "Software Cost Estimation Using Economic Production Models," *Journal of MIS* 15, no. 1 (Summer 1998): 143–163; G. Ooi and C. Soh, "Developing an Activity-based Approach for System Development and Implementation," *ACM Data Base for Advances in Information Systems* 34, no. 3 (Summer 2003): 54–71; and K. Milis and R. Mercken, "The Use of the Balanced Scorecard for the Evaluation of Information and Communication Technology Projects," *International Journal of Project Management* 22 (2004): 87–97.

[4] For more information regarding the use of option pricing models in evaluating economic feasibility of information systems, see M. Benaroch and R. Kauffman, "A Case for Using Real Options Pricing Analysis to Evaluate Information Technology Project Investments," *Information Systems Research* 10, no. 1 (March 1999): 70–86; M. Benaroch and R. Kauffman, "Justifying Electronic Banking Network Expansion Using Real Options Analysis," *MIS Quarterly* 24, no. 2 (June 2000): 197–225; Q. Dai, R. Kauffman, and S. March, "Analyzing Investments in Object-Oriented Middleware," *Ninth Workshop on Information Technologies and Systems* (December 1999): 45–50; A. Kambil, J. Henderson, and H. Mohsenzadeh, Strategic Management of Information Technology Investments: An Options Perspective, in R. D. Banker, R. J. Kauffman, and M. A. Mahmood (eds.), *Strategic Information Technology Management: Perspectives on Organizational Growth and Competitive Advantage* (Harrisburg, PA: Idea Group, 1993); A. Taudes, "Software Growth Options," *Journal of Management Information Systems* 15, no. 1 (Summer 1998): 165–185; A. Taudes, M. Feurstein, and A. Mild, "Options Analysis of Software Platform Decisions: A Case Study," *MIS Quarterly* 24, no. 2 (June 2000): pp. 227–243.

ered irreversible. The idea of irreversible decisions is one of the fundamental assumptions on which OPMs are based. This assumption fits quite well with modern object-oriented systems development approaches, in which, once an iteration has begun, an increment is completed before another investment decision is made.

Researchers have studied many different OPMs in terms of their applicability to IT investment.[5] However, all OPMs share a common thread: Both the direct benefit of the proposed project and the indirect value (option value) must be computed to determine economic feasibility of an IT investment using an OPM. The direct benefit can be computed using the traditional NPV, whereas the value of the option can be computed using one of the OPMs in the literature. Given that the minimum expected value of an option is always zero, the minimum estimated value for investing using an OPM will be the same as the value given by the traditional approach. However, when the expected value of the option (e.g., future iterations or projects) exceeds zero, an OPM will give an estimated value greater than the traditional approach. The actual calculation of the value of an option is quite complex and is beyond the scope of this book. However, given how well the OPMs fit the object-oriented systems development approaches, it seems reasonable that OPMs should be considered as alternatives to evaluating IT investments in object-oriented systems.

Organizational Feasibility

The final type of feasibility analysis is to assess the *organizational feasibility* of the system, how well the system ultimately will be accepted by its users and incorporated into the ongoing operations of the organization. There are many organizational factors that can have an effect on the project, and seasoned developers know that organizational feasibility can be the most difficult feasibility dimension to assess. In essence, an organizational feasibility analysis attempts to answer the question, If we build it, will they come?

One way to assess the organizational feasibility of the project is to understand how well the goals of the project align with business objectives. *Strategic alignment* is the fit between the project and business strategy—the greater the alignment, the less risky the project will be from an organizational feasibility perspective. For example, if the marketing department has decided to become more customer focused, then a CRM project that produces integrated customer information would have strong strategic alignment with marketing's goal. Many IT projects fail when the IT department initiates them, because there is little or no alignment with business unit or organizational strategies.

A second way to assess organizational feasibility is to conduct a *stakeholder analysis*.[6] A stakeholder is a person, group, or organization that can affect (or will be affected by) a new system. In general, the most important stakeholders in the introduction of a new system are the project champion, system users, and organizational management (see Figure 2-11), but systems sometimes affect other stakeholders as well. For example, the IS department can be a stakeholder of a system because IS jobs or roles may be changed significantly after its implementation. One key stakeholder outside of the champion, users, and management in Microsoft's project that embedded Internet Explorer as a standard part of Windows was the U.S. Department of Justice.

---

[5] Two of the more important OPMs used for evaluating IT investments are the binomial OPM and the Black-Scholes OPM. For more information on these models, see J. C. Hull, *Options, Futures, and Other Derivative Securities* (Englewood Cliffs, NJ: Prentice-Hall, 1993).

[6] A good book that presents a series of stakeholder analysis techniques is R. O. Mason and I. I. Mittroff, *Challenging Strategic Planning Assumptions: Theory, Cases, and Techniques* (New York: Wiley, 1981).

	Role	Techniques for improvement
Champion	A champion: • Initiates the project • Promotes the project • Allocates his or her time to project • Provides resources	• Make a presentation about the objectives of the project and the proposed benefits to those executives who will benefit directly from the system • Create a prototype of the system to demonstrate its potential value
Organizational Management	Organizational managers: • Know about the project • Budget enough money for the project • Encourage users to accept and use the system	• Make a presentation to management about the objectives of the project and the proposed benefits • Market the benefits of the system using memos and organizational newsletters • Encourage the champion to talk about the project with his or her peers
System Users	Users: • Make decisions that influence the project • Perform hands-on activities for the project • Ultimately determine whether the project is successful by using or not using the system	• Assign users official roles on the project team • Assign users specific tasks to perform with clear deadlines • Ask for regular feedback from users (e.g., at weekly meetings)

FIGURE 2-11   Some Important Stakeholders for Organizational Feasibility

The *champion* is a high-level, non–information-systems executive who is usually, but not always, the project sponsor who created the system request. The champion supports the project with time, resources (e.g., money), and political support within the organization by communicating the importance of the system to other organizational decision makers. More than one champion is preferable because if the champion leaves the organization, the support could leave as well.

Whereas champions provide day-to-day support for the system, *organizational management* also needs to support the project. Such management support conveys to the rest of the organization the belief that the system will make a valuable contribution and that necessary resources will be made available. Ideally, management should encourage people in the organization to use the system and to accept the many changes that the system will likely create.

A third important group of stakeholders are the *system users* who will ultimately use the system once it has been installed in the organization. Too often, the project team meets with users at the beginning of a project and then disappears until after the system is created. In this situation, rarely does the final product meet the expectations and needs of those who are supposed to use it because needs change and users become savvier as the project progresses. User participation should be promoted throughout the development process to make sure that the final system will be accepted and used by getting users actively involved in the development of the system (e.g., performing tasks, providing feedback, making decisions).

Finally, the feasibility study helps organizations make wiser investments regarding information systems because it forces project teams to consider technical, economic, and organizational factors that can affect their projects. It protects IT professionals from criticism by keeping the business units educated about decisions and positioned as the leaders in the decision-making process. Remember, the feasibility study should be revised several times during the project at points where the project team makes critical decisions about the system (e.g., before the design begins). It can be used to support and explain the critical choices that are made throughout the development process.

YOUR
TURN

2-3 Create a Feasibility Analysis

*T*hink about the idea that you developed in Your Turn 2-2 to improve your university or college course enrollment.

Questions

1. List three things that influence the technical feasibility of the system.

2. List three things that influence the economic feasibility of the system.

3. List three things that influence the organizational feasibility of the system.

4. How can you learn more about the issues that affect the three kinds of feasibility?

PROJECT SELECTION

Once the feasibility analysis has been completed, it is submitted to the approval committee, along with a revised system request. The committee then decides whether to approve the project, decline the project, or table it until additional information is available. At the project level, the committee considers the value of the project by examining the business need (found in the system request) and the risks of building the system (presented in the feasibility analysis).

Before approving the project, however, the committee also considers the project from an organizational perspective; it has to keep in mind the company's entire portfolio of projects. This way of managing projects is called *portfolio management*. Portfolio management takes into consideration the different kinds of projects that exist in an organization—large and small, high risk and low risk, strategic and tactical. (See Figure 2-12 for the different ways of classifying projects.) A good project portfolio has the most appropriate mix of projects for the organization's needs. The committee acts as portfolio manager with the goal of maximizing the cost–benefit performance and other important factors of the projects in their portfolio. For example, an organization might want to keep high-risk projects to less than 20 percent of its total project portfolio.

Size	What is the size? How many people are needed to work on the project?
Cost	How much will the project cost the organization?
Purpose	What is the purpose of the project? Is it meant to improve the technical infrastructure? Support a current business strategy? Improve operations? Demonstrate a new innovation?
Length	How long will the project take before completion? How much time will go by before value is delivered to the business?
Risk	How likely is it that the project will succeed or fail?
Scope	How much of the organization is affected by the system? A department? A division? The entire corporation?
Return on investment	How much money does the organization expect to receive in return for the amount the project costs?

FIGURE 2-12
Ways to Classify
Projects

The approval committee must be selective about where to allocate resources, because the organization has limited funds. This involves *trade-offs*, in which the organization must give up something in return for something else to keep its portfolio well balanced. If there are three potentially high-payoff projects, yet all have very high risk, then perhaps only one of the projects will be selected. Also, there are times when a system at the project level makes good business sense, but it does not make sense at the organization level. Thus, a project may show a very strong ROI and support important business needs for a part of the company, but it is not selected. This could happen for many reasons—because there is no money in the budget for another system, the organization is about to go through some kind of change (e.g., a merger or an implementation of a company-wide system like an enterprise resource plan [ERP]), projects that meet the same business requirements already are under way, or the system does not align well with current or future corporate strategy.

YOUR TURN   2-4 To Select or Not to Select

It seems hard to believe that an approval committee would not select a project that meets real business needs, has a high potential ROI, and has a positive feasibility analysis. Think of a company you have worked for or know about. Describe a scenario in which a project might be very attractive at the project level but not at the organization level.

CONCEPTS IN ACTION   2-F Interview with Carl Wilson, CIO, Marriott Corporation

At Marriott, we don't have IT projects—we have business initiatives and strategies that are enabled by IT. As a result, the only time a traditional "IT project" occurs is when we have an infrastructure upgrade that will lower costs or leverage better-functioning technology. In this case, IT has to make a business case for the upgrade and prove its value to the company.

The way IT is involved in business projects in the organization is twofold. First, senior IT positions are filled by people with good business understanding. Second, these people are placed on key business committees and forums where the real business happens, such as finding ways to satisfy guests. Because IT has a seat at the table, we are able to spot opportunities to support business strategy. We look for ways in which IT can enable or better support business initiatives as they arise.

Therefore, business projects are proposed, and IT is one component of them. These projects are then evaluated the same as any other business proposal, such as a new resort—by examining the return on investment and other financial measures.

At the organizational level, I think of projects as must-do's, should-do's, and nice-to-do's. The must-do's are required to achieve core business strategy, such as guest preference. The should-do's help grow the business and enhance the functionality of the enterprise. These can be somewhat untested, but good drivers of growth. The nice-to-do's are more experimental and look farther out into the future.

The organization's project portfolio should have a mix of all three kinds of projects, with a much greater proportion devoted to the must-do's.

—*Carl Wilson*

YOUR TURN 2-5 Project Selection

*I*n April 1999, one of Capital Blue Cross's health-care insurance plans had been in the field for three years but hadn't performed as well as expected. The ratio of premiums to claims payments wasn't meeting historic norms. To revamp the product features or pricing to boost performance, the company needed to understand why it was underperforming. The stakeholders came to the discussion already knowing they needed better extraction and analysis of usage data to understand product shortcomings and recommend improvements.

After listening to input from the user teams, the stakeholders proposed three options. One was to persevere with the current manual method of pulling data from flat files via ad hoc reports and retyping it into spreadsheets.

The second option was to write a program to dynamically mine the needed data from Capital's customer information control system (CICS). While the system was processing claims, for instance, the program would pull out up-to-the-minute data at a given point in time for users to analyze.

The third alternative was to develop a decision support system to allow users to make relational queries from a data mart containing a replication of the relevant claims and customer data. Each of these alternatives was evaluated on cost, benefits, risks, and intangibles.

Questions

1. What are three costs, benefits, risks, and intangibles associated with each project?
2. Based on your answer to question 1, which project would you choose?

Source: Richard Pastore, "Capital Blue Cross," *CIO Magazine* (February 15, 2000).

CONCEPTS IN ACTION 2-G A Project That Does Not Get Selected

*H*ygeia Travel Health is a Toronto-based health insurance company whose clients are the insurers of foreign tourists to the United States and Canada. Its project selection process is relatively straightforward. The project evaluation committee, consisting of six senior executives, splits into two groups. One group includes the CIO, along with the heads of operations and research and development, and it analyzes the costs of every project. The other group consists of the two chief marketing officers and the head of business development, and they analyze the expected benefits. The groups are permanent, and to stay objective, they don't discuss a project until both sides have evaluated it. The results are then shared, both on a spreadsheet and in conversation. Projects are then approved, passed over, or tabled for future consideration.

Last year, the marketing department proposed purchasing a claims database filled with detailed information on the costs of treating different conditions at different facilities. Hygeia was to use this information to estimate how much money insurance providers were likely to owe on a given claim if a patient was treated at a certain hospital as opposed to any other. For example, a 45-year-old man suffering a heart attack might accrue $5,000 in treatment costs at hospital A but only $4,000 at hospital B. This information would allow Hygeia to recommend the cheaper hospital to its customer. That would save the customer money and help differentiate Hygeia from its competitors.

The benefits team used the same three-meeting process to discuss all the possible benefits of implementing the claims database. Members of the team talked to customers and made a projection using Hygeia's past experience and expectations about future business trends. The verdict: The benefits team projected a revenue increase of $210,000. Client retention would rise by 2 percent. Overall, profits would increase by 0.25 percent.

The costs team, meanwhile, came up with large estimates: $250,000 annually to purchase the database and an additional $71,000 worth of internal time to make the information usable. Putting it all together, it was a financial loss of $111,000 in the first year.

The project still could have been good for marketing—maybe even good enough to make the loss acceptable. But some of Hygeia's clients were also in the claims information business and therefore were potential competitors. This, combined with the financial loss, was enough to make the company reject the project.

Source: Ben Worthen, "Two Teams are Better than One" *CIO Magazine,* (July 15, 2001).

CONCEPTS   **2-H Trade-offs**

IN ACTION

I was once on a project to develop a system that should have taken a year to build. Instead, the business need demanded that the system be ready within five months—impossible!

On the first day of the project, the project manager drew a triangle on a white board to illustrate some trade-offs that he expected to occur over the course of the project. The corners of the triangle were labeled Functionality, Time, and Money. The manager explained, "We have too little time. We have an unlimited budget. We will not be measured by the bells and whistles that this system contains. So over the next several weeks, I want you as developers to keep this triangle in mind and do everything it takes to meet this five-month deadline."

At the end of the five months, the project was delivered on time; however, the project was incredibly over budget, and the final product was "thrown away" after it was used because it was unfit for regular use. Remarkably, the business users felt that the project was very successful because it met the very specific business needs for which it was built. They believed that the trade-offs that were made were worthwhile.

—Barbara Wixom

Questions

1. What are the risks in stressing only one corner of the triangle?
2. How would you have managed this project? Can you think of another approach that might have been more effective?

TRADITIONAL PROJECT MANAGEMENT TOOLS

Before we get to actually creating a workplan that is suitable to manage and control an object-oriented systems development project, we need to introduce a set of project management tools that have been used to successfully manage traditional software development projects (and many other types of projects): a work-breakdown structure, a Gantt chart, and a network diagram. However, before we can begin covering these tools, we must first understand what a task is. A *task* is a unit of work that will be performed by a member or members of the development team, such as feasibility analysis. Each task is described by information such as its name, start and completion dates, person assigned to complete the task, deliverables, completion status, priority, resources needed, estimated time to complete the task, and the actual time it took to complete the task (see Figure 2-13). The first thing a project manager must do is to identify the tasks that need to be accomplished and determine how long each task will take. Tasks and their identification and documentation are the basis of all three of these tools. Once the tasks have been identified and documented,

Workplan Information	Example
Name of the task	Perform economic feasibility
Start date	Jan 05, 2010
Completion date	Jan 19, 2010
Person assigned to the task	Project sponsor: Mary Smith
Deliverable(s)	Cost–benefit analysis
Completion status	Open
Priority	High
Resources that are needed	Spreadsheet software
Estimated time	16 hours
Actual time	14.5 hours

FIGURE 2-13
Task Information

they are organized within a work breakdown structure that is used to drive the creation of Gantt charts and network diagrams that can be used to graphically portray a traditional workplan. These techniques help a project manager understand and manage the project's progress over time.

Work Breakdown Structures

If a project manager prefers to begin from scratch, he or she can use a structured, top-down approach whereby high-level tasks are first defined and then broken down into subtasks. For example, Figure 2-14 shows a list of high-level tasks needed to implement a new IT training class. Some of the main steps in the process include identifying vendors, creating and administering a survey, and building new classrooms. Each step is then broken down in turn and numbered in a hierarchical fashion. There are eight subtasks (i.e., 7.1–7.8) for creating and administering a survey, and there are three subtasks (7.2.1–7.2.3) that make up the review initial survey task. A list of tasks hierarchically numbered in this way is called a *work breakdown structure (WBS)*. The number of tasks and level of detail depend on the complexity and size of the project. At a minimum, the WBS must include the duration of the task, the current statuses of the tasks (i.e., open, complete), and the *task dependencies*, which occur when one task cannot be performed until another task is completed. For example, Figure 2-14 shows that incorporating changes to the survey (task 7.4) takes a week to perform, but it cannot occur until after the survey is reviewed (task 7.2) and pilot tested (task 7.3). Key *milestones*, or important dates, are also identified on the workplan.

There are two basic approaches to organizing a traditional WBS: by development phase or by product. For example, if a firm decided that it needed to develop a website, the firm could create a WBS based on the inception, elaboration, construction, and

Task Number	Task Name	Duration (in weeks)	Dependency	Status
1	Identify vendors	2		Complete
2	Review training materials	6	1	Complete
3	Compare vendors	2	2	In Progress
4	Negotiate with vendors	3	3	Open
5	Develop communications information	4	1	In Progress
6	Disseminate information	2	5	Open
7	Create and administer survey	4	6	Open
7.1	Create initial survey	1		Open
7.2	Review initial survey	1	7.1	Open
7.2.1	Review by Director of IT Training	1		Open
7.2.2	Review by Project Sponsor	1		Open
7.2.3	Review by Representative Trainee	1		Open
7.3	Pilot test initial survey	1	7.1	Open
7.4	Incorporate survey changes	1	7.2, 7.3	Open
7.5	Create distribution list	0.5		Open
7.6	Send survey to distribution list	0.5	7.4, 7.5	Open
7.7	Send follow-up message	0.5	7.6	Open
7.8	Collect completed surveys	1	7.6	Open
8	Analyze results and choose vendor	2	4, 7	Open
9	Build new classrooms	11	1	In Progress
10	Develop course options	3	8, 9	Open

FIGURE 2-14
Work Breakdown
Structure

transition phases of the Unified Process. In this case, a typical task that would take place during inception would be feasibility analysis. This task would be broken down into the different types of feasibility analysis: technical, economic, and organizational. Each of these would be further broken down into a set of subtasks. Alternatively, the firm could organize the workplan along the lines of the different products to be developed. For example, in the case of a website, the products could include applets, application servers, database servers, the various sets of web pages to be designed, a site map, and so on. Then these would be further decomposed into the different tasks associated with the phases of the development process. Either way, once the overall structure is determined, tasks are identified and included in the WBS. We return to the topic of WBSs and their use in iterative planning later in this chapter.

Gantt Chart

A *Gantt chart* is a horizontal bar chart that shows the same task information as the project WBS but in a graphical way. Sometimes a picture really is worth a thousand words, and the Gantt chart can communicate the high-level status of a project much faster and easier than the WBS. Creating a Gantt chart is simple and can be done using a spreadsheet package, graphics software (e.g., Microsoft Visio), or a project management package.

First, tasks are listed as rows in the chart, and time is listed across the top in increments based on the needs of the projects (see Figure 2-15). A short project may be divided into hours or days, whereas a medium-sized project may be represented using weeks or months. Horizontal bars are drawn to represent the duration of each task; the bar's beginning and end mark exactly when the task will begin and end. As people work on tasks, the appropriate bars are filled in proportionately to how much of the task is finished. Too many tasks on a Gantt chart can become confusing, so it's best to limit the number of tasks to around twenty or thirty. If there are more tasks, break them down into subtasks and create Gantt charts for each level of detail.

There are many things a project manager can see quickly by looking at a Gantt chart. In addition to seeing how long tasks are and how far along they are, the project manager also can tell which tasks are sequential, which tasks occur at the same time, and which tasks overlap in some way. He or she can get a quick view of tasks that are ahead of schedule and behind schedule by drawing a vertical line on today's date. If a bar is not filled in and is to the left of the line, that task is behind schedule.

There are a few special notations that can be placed on a Gantt chart. Project milestones are shown using upside-down triangles or diamonds. Arrows are drawn between the task bars to show task dependencies. Sometimes, the names of people assigned to each task are listed next to the task bars to show what human resources have been allocated to the tasks.

Network Diagram

A second graphical way to look at project workplan information is the *network diagram* that lays out the project tasks in a flowchart (see Figure 2-16). *Program Evaluation and Review Technique (PERT)* is a network analysis technique that can be used when the individual task time estimates are fairly uncertain. Instead of simply putting a point estimate for the duration estimate, PERT uses three time estimates: optimistic, most likely, and a pessimistic. It then combines the three estimates into a single weighted average estimate using the following formula:

$$\text{PERT weighted average} = \frac{\text{optimistic estimate} + (4 * \text{most likely estimate}) + \text{pessimistic estimate}}{6}$$

ID	Task Name	Duration	Start	Finish	Prede	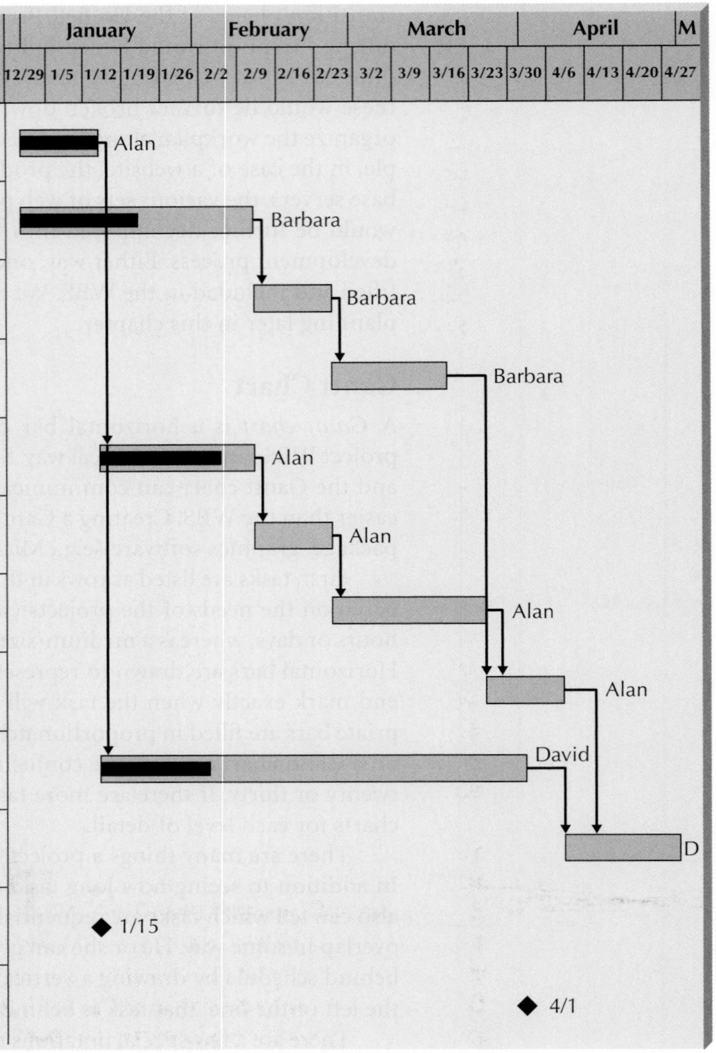 January					February			March				April			M
						12/29	1/5	1/12	1/19	1/26	2/2	2/9	2/16	2/23	3/2	3/9	3/16	3/23	3/30	4/6	4/13	4/20	4/27
1	Identify vendors	2 wks	Wed 1/1/12	Tue 1/14/12		Alan
2	Review training materials	6 wks	Wed 1/1/12	Tue 2/11/12		Barbara
3	Compare vendors	2 wks	Wed 2/12/12	Tue 2/25/12	2	Barbara
4	Negotiate with vendors	3 wks	Wed 2/26/12	Tue 3/8/12	3	Barbara
5	Develop communications information	4 wks	Wed 1/15/12	Tue 2/11/12	1	Alan
6	Disseminate information	2 wks	Wed 2/12/12	Tue 2/25/12	5	Alan
7	Create and administer survey	4 wks	Wed 2/26/12	Tue 3/25/12	6	Alan
8	Analyze results and choose	2 wks	Wed 3/26/12	Tue 4/8/12	4, 7	Alan
9	Build new classroom	11 wks	Wed 1/15/12	Tue 4/1/12	1	David
10	Develop course options	3 wks	Wed 4/9/12	Tue 4/29/12	8, 9	D
11	Budget Meeting	1 day	Wed 1/15/12	Wed 1/15/12		◆ 1/15
12	Software Installation	1 day	Tue 4/1/12	Tue 4/1/12		◆ 4/1

FIGURE 2-15 Gantt Chart

The network diagram is drawn as a node-and-arc type of graph that shows time estimates in the nodes and task dependencies on the arcs. Each *node* represents an individual task, and a line connecting two nodes represents the dependency between two tasks. Partially completed tasks are usually displayed with a diagonal line through the node, and completed tasks contain crossed lines.

Network diagrams are the best way to communicate task dependencies because they lay out the tasks in the order in which they need to be completed. The *critical path method* (CPM) simply allows the identification of the critical path in the network. The critical path is the longest path from the project inception to completion. The critical path shows all the tasks that must be completed on schedule for a project as a whole to finish on schedule. If any tasks on the critical path take longer than expected, the entire project will fall behind. Each task on the critical path is a *critical task*, and they are usually depicted in a unique way;

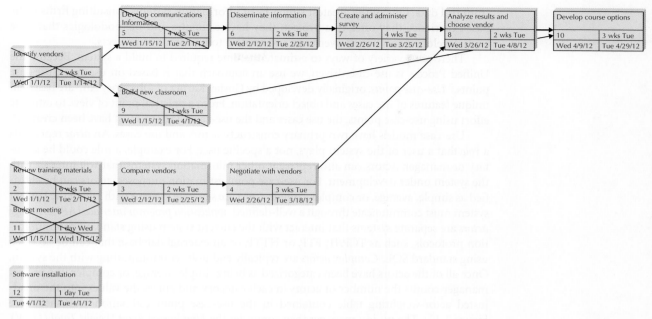

FIGURE 2-16   Network Diagram

in Figure 2-16 they are shown with double borders (see tasks 5, 6, 7, 8, and 10). CPM can be used with or without PERT.

PROJECT EFFORT ESTIMATION

The science (or art) of project management is in making *trade-offs* among three important concepts: the functionality of the system, the time to complete the project (when the project will be finished), and the cost of the project. Think of these three things as interdependent levers that the project manager controls throughout the development of the system. Whenever one lever is pulled, the other two levers are affected in some way. For example, if a project manager needs to readjust a deadline to an earlier date, then the only solutions are to decrease the functionality of the system or to increase costs by adding more people or having them work overtime. Often, a project manager has to work with the project sponsor to change the goals of the project, such as developing a system with less functionality or extending the deadline for the final system, so that the project has reasonable goals that can be met. In the beginning of the project, the manager needs to estimate each of these levers and then continuously assess how to roll out the project in a way that meets the organization's needs. *Estimation* is the process of assigning projected values for time and effort. The estimates developed at the start of a project are usually based on a range of possible values and gradually become more specific as the project moves forward. That is, the range of values for the inception phase will be much greater than for the transition phase.

The numbers used to calculate these estimates can come from several sources. For example, they can be taken from projects with similar tasks and technologies or provided by experienced developers. Generally speaking, the numbers should be conservative. A good practice is to keep track of the actual values for time and effort during the development process so that numbers can be refined along the way and the next project can benefit from real data. One of the greatest strengths of information systems consulting firms is the past experience they offer to a project; they have estimates and methodologies that have been developed and honed over time and applied to hundreds of projects.

There are a variety of ways to estimate the time required to build a system. Because the Unified Process is use-case driven, we use an approach that is based on use cases: use-case points.[7] *Use-case points*, originally developed by Gustav Karner of Objectory AB,[8] are based on unique features of use cases and object orientation. From a practical point of view, to estimate effort using use-case points, the use cases and the use-case diagram must have been created.[9]

Use-case models have two primary constructs: actors and use cases. An *actor* represents a *role* that a user of the system plays, not a specific user. For example, a role could be secretary or manager. Actors can also represent other information systems that will interact with the system under development. For use-case point estimation purposes, actors can be classified as simple, average, or complex. *Simple actors* are separate systems with which the current system must communicate through a well-defined *application program interface (API)*. *Average actors* are separate systems that interact with the current system using standard communication protocols, such as TCP/IP, FTP, or HTTP, or an external database that can be accessed using standard SQL. *Complex actors* are typically end users communicating with the system. Once all of the actors have been categorized as being simple, average, or complex, the project manager counts the number of actors in each category and enters the values into the unadjusted actor-weighting table contained in the use-case point estimation worksheet (see Figure 2-17). The project manager then computes the *Unadjusted Actor Weight Total (UAW)*. This is computed by summing the individual results that were computed by multiplying the weighting factor by the number of actors of each type. For example, if we assume that the use-case diagram has zero simple, zero average, and four complex actors that interact with the system being developed, the UAW will equal 12 (see Figure 2-18).

A *use case* represents a major business process that the system will perform that benefits the actor(s) in some manner. Depending on the number of unique transactions that the use case must address, such as actors, a use case can be categorized as being simple, average, or complex. A use case is classified as a *simple use case* if it supports one to three transactions, as an *average use case* if it supports four to seven transactions, or as a *complex use case* if it supports more than seven transactions. Once all of the use cases have been successfully categorized, the project manager enters the number of each type of use case into the unadjusted use-case weighting table contained in the use-case point estimation worksheet (see Figure 2-17). By multiplying by the appropriate weights and summing the results, we get the value for the *unadjusted use-case weight total (UUCW)*. For example, if we assume that we have three simple use cases, four average use cases, and one complex use case, the value for the unadjusted use-case weight total is 70 (see Figure 2-18). Next, the project manager computes the value of the *unadjusted use-case points (UUCP)* by simply summing the unadjusted actor weight total and the unadjusted use-case weight total. In this case, the value of the UUCP equals 82 (see Figure 2-18).

Use-case point based estimation also has a set of factors that are used to adjust the use-case point value. In this case, there are two sets of factors: *technical complexity factors (TCFs)* and *environmental factors (EFs)*. There are thirteen separate technical factors and eight separate environmental factors. The purpose of these factors is to allow the project as a whole to

---

[7] The material in this section is based on descriptions of use-case points contained in Raul R. Reed, Jr., *Developing Applications with Java and UML* (Reading, MA: Addison-Wesley, 2002); Geri Schneider and Jason P. Winters, *Applying Use Cases: A Practical Guide* (Reading, MA: Addison-Wesley, 1998); and Kirsten Ribu, "Estimating Object-Oriented Software Projects with Use Cases" (Master's thesis, University of Oslo, 2001).

[8] Objectory AB was acquired by Rational in 1995 and Rational is now part of IBM.

[9] We cover the details of use-case modeling in Chapter 4.

Unadjusted Actor Weighting Table:

Actor Type	Description	Weighting Factor	Number	Result
Simple	External System with well-defined API	1		
Average	External System using a protocol-based interface, e.g., HTTP, TCT/IP, or a database	2		
Complex	Human	3		
		Unadjusted Actor Weight Total (UAW)		

Unadjusted Use-Case Weighting Table:

Use-Case Type	Description	Weighting Factor	Number	Result
Simple	1–3 transactions	5		
Average	4–7 transactions	10		
Complex	>7 transactions	15		
		Unadjusted Use-Case Weight Total (UUCW)		

Unadjusted Use-Case Points (UUCP) = UAW + UUCW

Technical Complexity Factors:

Factor Number	Description	Weight	Assigned Value (0 – 5)	Weighted Value	Notes
T1	Distributed system	2.0			
T2	Response time or throughput performance objectives	1.0			
T3	End user online efficiency	1.0			
T4	Complex internal processing	1.0			
T5	Reusability of code	1.0			
T6	Ease of installation	0.5			
T7	Ease of use	0.5			
T8	Portability	2.0			
T9	Ease of change	1.0			
T10	Concurrency	1.0			
T11	Special security objectives included	1.0			
T12	Direct access for third parties	1.0			
T13	Special user training required	1.0			
			Technical Factor Value (TFactor)		

*Technical Complexity Factor (TCF) = 0.6 + (0.01 * TFactor)*

Environmental Factors:

Factor Number	Description	Weight	Assigned Value (0 – 5)	Weighted Value	Notes
E1	Familiarity with system development process being used	1.5			
E2	Application experience	0.5			
E3	Object-oriented experience	1.0			
E4	Lead analyst capability	0.5			
E5	Motivation	1.0			
E6	Requirements stability	2.0			
E7	Part-time staff	−1.0			
E8	Difficulty of programming language	−1.0			
			Environmental Factor Value (EFactor)		

*Environmental Factor (EF) = 1.4 + (−0.03 * EFactor)*
*Adjusted Use-Case Points (UCP) = UUCP * TCF * ECF*
*Effort in Person-Hours = UCP * PHM*

TEMPLATE can be found at www.wiley.com /go/global/ tegarden

FIGURE 2-17   Use-Case Point Estimation Worksheet

Unadjusted Actor Weighting Table:

Actor Type	Description	Weighting Factor	Number	Result
Simple	External system with well-defined API	1	0	0
Average	External system using a protocol-based interface, e.g., HTTP, TCT/IP, or a database	2	0	0
Complex	Human	3	4	12
		Unadjusted Actor Weight Total (UAW)		12

Unadjusted Use-Case Weighting Table:

Use-Case Type	Description	Weighting Factor	Number	Result
Simple	1–3 transactions	5	3	15
Average	4–7 transactions	10	4	40
Complex	>7 transactions	15	1	15
		Unadjusted Use-Case Weight Total (UUCW)		70

Unadjusted Use-Case Points (UUCP) = UAW + UUCW $82 = 12 + 70$

Technical Complexity Factors:

Factor Number	Description	Weight	Assigned Value (0 – 5)	Weighted Value	Notes
T1	Distributed system	2.0	0	0	
T2	Response time or throughput performance objectives	1.0	5	5	
T3	End user online efficiency	1.0	3	3	
T4	Complex internal processing	1.0	1	1	
T5	Reusability of code	1.0	1	1	
T6	Ease of installation	0.5	2	1	
T7	Ease of use	0.5	4	2	
T8	Portability	2.0	0	0	
T9	Ease of change	1.0	2	2	
T10	Concurrency	1.0	0	0	
T11	Special security objectives included	1.0	0	0	
T12	Direct access for third parties	1.0	0	0	
T13	Special user training required	1.0	0	0	
		Technical Factor Value (TFactor)		15	

Technical Complexity Factor (TCF) = 0.6 + (0.01 * TFactor) $0.75 = 0.6 + (0.01 * 15)$

Environmental Factors:

Factor Number	Description	Weight	Assigned Value (0 – 5)	Weighted Value	Notes
E1	Familiarity with system development process being used	1.5	4	6	
E2	Application experience	0.5	4	2	
E3	Object-oriented experience	1.0	4	4	
E4	Lead analyst capability	0.5	5	2.5	
E5	Motivation	1.0	5	5	
E6	Requirements stability	2.0	5	10	
E7	Part-time staff	−1.0	0	0	
E8	Difficulty of programming language	−1.0	4	−4.0	
		Environmental Factor Value (EFactor)		25.5	

Environmental Factor (EF) = 1.4 + (−0.03 * EFactor) $0.635 = 1.4 + (-0.03 * 25.5)$
Adjusted Use-Case Points (UCP) = UUCP * TCF * ECF $33.3375 = 70 * 0.75 * 0.635$
Effort in Person-Hours = UCP * PHM $666.75 = 20 * 33.3375$

FIGURE 2-18   Use-Case Point Estimation for the Appointment System

be evaluated for the complexity of the system being developed and the experience levels of the development staff, respectively. Obviously, these types of factors can affect the effort that a team requires to develop a system. Each of these factors is assigned a value between 0 and 5, 0 indicating that the factor is irrelevant to the system under consideration and 5 indicating that the factor is essential for the system to be successful. The assigned values are then multiplied by their respective weights. These weighted values are then summed up to create a *technical factor value (TFactor)* and an *environmental factor value (EFactor)* (see Figure 2-17).

The technical factors include the following (see Figure 2-17):

- Whether the system is going to be a distributed system
- The importance of response time
- The efficiency level of the end user using the system
- The complexity of the internal processing of the system
- The importance of code reuse
- How easy the installation process has to be
- The importance of the ease of using the system
- How important it is for the system to be able to be ported to another platform
- Whether system maintenance is important
- Whether the system is going to have to handle parallel and concurrent processing
- The level of special security required
- The level of system access by third parties
- Whether special end user training is to be required.

Assuming the values for the technical factors are T1 (0), T2 (5), T3 (3), T4 (1), T5 (1), T6 (2), T7 (4), T8 (0), T9 (2), T10 (0), T11 (0), T12 (0), and T13 (0), respectively, the *technical factor value (TFactor)* is computed as the weighted sum of the individual technical factors. In this case, TFactor equals 15 (see Figure 2-18). Plugging this value into the *technical complexity factor (TCF)* equation $(0.6 + (.01 * TFactor)$ of the use-case point worksheet gives a value of .75 for the TCF of the system (see Figures 2-17 and 2-18).

The environmental factors include the following (see Figure 2-17):

- The level of experience the development staff has with the development process being used
- The application being developed
- The level of object-oriented experience
- The level of capability of the lead analyst
- The level of motivation of the development team to deliver the system
- The stability of the requirements
- Whether part-time staff have to be included as part of the development team
- The difficulty of the programming language being used to implement the system

Assuming the values for the environmental factors were E1 (4), E2 (4), E3 (4), E4 (5), E5 (5), E6 (5), E7 (0), and E8 (4) gives an *environmental factor value (EFactor)* of 25.5 (See Figure 2-14). Like the TFactor, Efactor is simply the sum of the weighted values. Using the *environmental factor (EF)* equation $(1.4 + (-0.03 * EFactor)$ of the use-case point worksheet produces a value of .635 for the EF of the system (see Figures 2-17 and 2-18). Plugging the TCF and EF values, along with the UUCP value computed earlier, into the adjusted use-case points equation $(UUCP * TCF * EF)$ of the worksheet yields a value of 33.3375 *adjusted use-case points (UCP)* (see Figure 2-18).

Now that we know the estimated size of the system by means of the value of the adjusted use-case points, we are ready to estimate the *effort* required to build the system. In Karner's original work, he suggested simply multiplying the number of use-case points by 20 to estimate the number of person-hours required to build the system. However, based on additional experiences using use-case points, a decision rule to determine the value of the *person-hours multiplier (PHM)* has been created that suggests using either 20 or 28, based on the values assigned to the individual environmental factors. The decision rule is:

If the sum of (number of Efactors E1 through E6 assigned value < 3) and
(number of Efactors E7 and E8 assigned value > 3)
$\leqslant 2$
PHM = 20
Else, if the sum of (number of Efactors E1 through E6 assigned value < 3) and
(number of Efactors E7 and E8 assigned value > 3)
= 3 or 4
PHM = 28
Else
Rethink project; it has too high of a risk for failure

Based on these rules, because none of Efactors E1 through E6 have a value less than 3 and only Efactors E8 has a value greater than 3, the sum of the number EFactors is 1. Thus, the system should use a PHM of 20. Plugging the values for UCP (33.3375) and PHM (20) into the effort equation (UCP * PHM) gives an estimated number of person-hours of 666.75 hours (see Figures 2-17 and 2-18).

YOUR TURN

2-6 Project Estimation

*I*magine that job hunting has been going so well that you need to develop a system to support your efforts. The system should allow you to input information about the companies with which you interview, the interviews and office visits that you have scheduled, and the offers you receive. It should be able to produce reports, such as a company contact list, an interview schedule, and an office visit schedule, as well as produce Thank You letters to be brought into a word processor to customize. You also need the system to answer queries, such as the number of interviews by city and your average offer amount.

Questions

1. Determine the number and type (simple, average, and complex) of actors there are for this system. Compute the value for the Unadjusted Actor Weight Total.

2. Determine the number and type (simple, average, and complex) of uses cases there are for this system.

Compute the value for the Unadjusted Use-Case Weight Total.

3. Compute the value for the Unadjusted Use-Case Points.

4. Assume values for the technical complexity factors are T1(0), T2(1), T3(2), T4(2), T5(0), T6(1), T7(2), T8(0), T9(0), T10(0), T11(0), T12(0), and T13(0). Compute the Technical Factor Value.

5. Compute the value for the Technical Complexity Factor.

6. Assume values for the environmental factors are E1(4), E2(3), E3(3), E4(3), E5(4), E6(3), E7(0), and E8(3). Compute the Environmental Factor Value.

7. Compute the value for the Environmental Factor.

8. Compute the value for the Adjusted Use-Case Points.

9. Compute the estimated effort in person-hours.

CREATING AND MANAGING THE WORKPLAN

Once a project manager has a general idea of the functionality and effort for the project, he or she creates a *workplan*, which is a dynamic schedule that records and keeps track of all the tasks that need to be accomplished over the course of the project. The workplan lists each task, along with important information about it, such as when it needs to be completed, the person assigned to do the work, and any deliverables that will result. The level of detail and the amount of information captured by the workplan depend on the needs of the project, and the detail usually increases as the project progresses.

The overall objectives for the system should be listed on the system request, and it is the project manager's job to identify all the tasks that need to be accomplished to meet those objectives. This sounds like a daunting task. How can someone know everything that needs to be done to build a system that has never been built before?

One approach for identifying tasks is to get a list of tasks that has already been developed and to modify it. There are standard lists of tasks, or methodologies, that are available for use as a starting point. As we stated in Chapter 1, a *methodology* is a formalized approach to implementing a systems development process (i.e., it is a list of steps and deliverables). A project manager can take an existing methodology, select the steps and deliverables that apply to the current project, and add them to the workplan. If an existing methodology is not available within the organization, methodologies can be purchased from consultants or vendors, or books such as this textbook can serve as a guide. Because most organizations have a methodology they use for projects, using an existing methodology is the most popular way to create a workplan. In our case, because we are using a Unified Process–based methodology, we can use the phases, workflows, and iterations as a starting point to create an evolutionary work breakdown structure and an iterative workplan.

Evolutionary Work Breakdown Structures and Iterative Workplans

Because object-oriented systems approaches to systems analysis and design support incremental and iterative development, any project planning approach for object-oriented systems development also requires an incremental and iterative process. In the description of the enhanced Unified Process in Chapter 1, the development process was organized around iterations, phases, and workflows. In many ways, a workplan for an incremental and iterative development process is organized in a similar manner. For each iteration, there are different tasks executed on each workflow. This section describes an incremental and iterative process using evolutionary WBSs for project planning that can be used with object-oriented systems development.

According to Royce,[10] most approaches to developing conventional WBSs tend to have three underlying problems:

- *They tend to be focused on the design of the information system being developed.* The creation of the WBS forces the premature decomposition of the system design and the tasks associated with creating the design of the system. Where the problem domain is well understood, tying the structure of the workplan to the product to be created makes sense. However, in cases where the problem domain is not well understood, the analyst must commit to the architecture of the system being developed before the requirements of the system are fully understood.

---

[10] Walker Royce, *Software Project Management: A Unified Framework* (Reading, MA: Addison-Wesley, 1998).

■ *They tend to force too many levels of detail very early on in the systems development process for large projects or they tend to allow too few levels of detail for small projects.* Because the primary purpose of a WBS is to allow cost estimation and scheduling to take place, in conventional approaches to planning, the WBS must be done correctly and completely at the beginning of the development process. To say the least, this is a very difficult task to accomplish with any degree of validity. In such cases, it is no wonder that cost and schedule estimation for many information systems development projects tend to be wildly inaccurate.

■ *Because they are project specific, they are very difficult to compare across projects.* This leads to ineffective learning across the organization. Without some standard approach to create WBSs, it is difficult for project managers to learn from previous projects managed by others. This tends to encourage the reinventing of the wheel and allows managers to make the same mistakes that previous managers have made.

Evolutionary WBSs allow the analyst to address all three problems by allowing the development of an *iterative workplan*. First, evolutionary WBSs are organized in a standard manner across all projects: by workflows, phases, and then the specific tasks that are accomplished during an individual iteration. This decouples the structure of an evolutionary WBS from the structure of the design of the product and prevents prematurely committing to a specific architecture of a new system. Second, evolutionary WBSs are created in an incremental and iterative manner. This encourages a more realistic view of both cost and schedule estimation. Third, because the structure of an evolutionary WBS is not tied to any specific project, evolutionary WBSs enable the comparison of the current project to earlier projects. This supports learning from past successes and failures.

In the case of the enhanced Unified Process, the workflows are the major points listed in the WBS. Next, each workflow is decomposed along the phases of the enhanced Unified Process. After that, each phase is decomposed along the tasks that are to be completed to create the deliverables associated with an individual iteration contained in each phase (see Figure 1-18). The template for the first two levels of an evolutionary WBS for the enhanced Unified Process would look like Figure 2-19.

As each iteration through the development process is completed, additional iterations and tasks are added to the WBS (i.e., the WBS evolves along with the evolving information system).[11] For example, typical activities for the inception phase of the project management workflow would include identifying the project, performing the feasibility analysis, selecting the project, and estimating the effort. The inception phase of the requirements workflow would include determining the requirements gathering and analysis techniques, identifying functional and nonfunctional requirements, interviewing stakeholders, developing a vision document, and developing use cases. Probably no tasks are associated with the inception phase of the operations and support workflow. A sample evolutionary WBS for planning the inception phase of the enhanced Unified Process, based on Figures 1-18 and 2-19, is shown in Figure 2-20. Notice the last two tasks for the project management workflow are "create workplan for first iteration of the elaboration phase" and "assess the inception phase"; the last two things to do are to plan for the next iteration in the development of the evolving system and to assess the current iteration. As the project moves through later phases, each workflow

---

[11] Good sources that help explain this approach are Phillippe Krutchen, "Planning an Iterative Project," *The Rational Edge* (October 2002); and Eric Lopes Cordoza and D. J. de Villiers,"Project Planning Best Practices," *The Rational Edge* (August 2003).

I. Business Modeling	V. Implementation	IX. Project Management
a. Inception	**a.** Inception	**a.** Inception
b. Elaboration	**b.** Elaboration	**b.** Elaboration
c. Construction	**c.** Construction	**c.** Construction
d. Transition	**d.** Transition	**d.** Transition
e. Production	**e.** Production	**e.** Production
II. Requirements	VI. Test	X. Environment
a. Inception	**a.** Inception	**a.** Inception
b. Elaboration	**b.** Elaboration	**b.** Elaboration
c. Construction	**c.** Construction	**c.** Construction
d. Transition	**d.** Transition	**d.** Transition
e. Production	**e.** Production	**e.** Production
III. Analysis	VII. Deployment	XI. Operations and Support
a. Inception	**a.** Inception	**a.** Inception
b. Elaboration	**b.** Elaboration	**b.** Elaboration
c. Construction	**c.** Construction	**c.** Construction
d. Transition	**d.** Transition	**d.** Transition
e. Production	**e.** Production	**e.** Production
IV. Design	VIII. Configuration and Change Management	XII. Infrastructure Management
a. Inception	**a.** Inception	**a.** Inception
b. Elaboration	**b.** Elaboration	**b.** Elaboration
c. Construction	**c.** Construction	**c.** Construction
d. Transition	**d.** Transition	**d.** Transition
e. Production	**e.** Production	**e.** Production

FIGURE 2-19
Evolutionary WBS Template for the Enhanced Unified Process

	Duration	Dependency
I. Business Modeling		
a. Inception		
1. Understand current business situation	0.50 days	
2. Uncover business process problems	0.25 days	
3. Identify potential projects	0.25 days	
b. Elaboration		
c. Construction		
d. Transition		
e. Production		
II. Requirements		
a. Inception		
1. Identify appropriate requirements-analysis technique	0.25 days	
2. Identify appropriate requirements-gathering techniques	0.25 days	
3. Identify functional and nonfunctional requirements		II.a.1, II.a.2
A. Perform JAD sessions	3 days	
B. Perform document analysis	5 days	II.a.3.A
C. Conduct interviews		II.a.3.A
1. Interview project sponsor	0.5 days	
2. Interview inventory system contact	0.5 days	
3. Interview special order system contact	0.5 days	
4. Interview ISP contact	0.5 days	
5. Interview CD Selection web contact	0.5 days	
6. Interview other personnel	1 day	
D. Observe retail store processes	0.5 days	II.a.3.A

FIGURE 2-20
Evolutionary WBS for a Single–Iteration–based Inception Phase

	Duration	Dependency
4. Analyze current systems	4 days	II.a.1, II.a.2
5. Create requirements definition		II.a.3, II.a.4
A. Determine requirements to track	1 day	
B. Compile requirements as they are elicited	5 days	II.a.5.A
C. Review requirements with sponsor	2 days	II.a.5.B
b. Elaboration		
c. Construction		
d. Transition		
e. Production		
III. Analysis		
a. Inception		
1. Identify business processes	3 days	
2. Identify use cases	3 days	III.a.1
b. Elaboration		
c. Construction		
d. Transition		
e. Production		
IV. Design		
a. Inception		
1. Identify potential classes	3 days	III.a
b. Elaboration		
c. Construction		
d. Transition		
e. Production		
V. Implementation		
a. Inception		
b. Elaboration		
c. Construction		
d. Transition		
e. Production		
VI. Test		
a. Inception		
b. Elaboration		
c. Construction		
d. Transition		
e. Production		
VII. Deployment		
a. Inception		
b. Elaboration		
c. Construction		
d. Transition		
e. Production		
VIII. Configuration and Change Management		
a. Inception		
1. Identify necessary access controls for developed artifacts	0.25 days	
2. Identify version control mechanisms for developed artifacts	0.25 days	
b. Elaboration		
c. Construction		
d. Transition		
e. Production		

FIGURE 2-20
Continued

	Duration	Dependency
IX. Project Management		
a. Inception		
1. Create workplan for the inception phase	1 day	
2. Create system request	1 day	
3. Perform feasibility analysis		IX.a.2
A. Perform technical feasibility analysis	1 day	
B. Perform economic feasibility analysis	2 days	
C. Perform organizational feasibility analysis	2 days	
4. Identify project effort	0.50 days	IX.a.3
5. Identify staffing requirements	0.50 days	IX.a.4
6. Compute cost estimate	0.50 days	IX.a.5
7. Create workplan for first iteration of the elaboration phase	1 day	IX.a.1
8. Assess inception phase	1 day	I.a, II.a, III.a IV.a, V.a, VI.a VII.a, VIII.a, IX.a, X.a, XI.a XII.a
b. Elaboration		
c. Construction		
d. Transition		
e. Production		
X. Environment		
a. Inception		
1. Acquire and install CASE tool	0.25 days	
2. Acquire and install programming environment	0.25 days	
3. Acquire and install configuration and change management tools	0.25 days	
4. Acquire and install project management tools	0.25 days	
b. Elaboration		
c. Construction		
d. Transition		
e. Production		
XI. Operations and Support		
a. Inception		
b. Elaboration		
c. Construction		
d. Transition		
e. Production		
XII. Infrastructure Management		
a. Inception		
1. Identify appropriate standards and enterprise models	0.25 days	
2. Identify reuse opportunities, such as patterns, frameworks, and libraries	0.50 days	
3. Identify similar past projects	0.25 days	
b. Elaboration		
c. Construction		
d. Transition		
e. Production		

FIGURE 2-20

Continued

has tasks added to its iterations. For example, the analysis workflow will have the creation of the functional, structural, and behavioral models during the elaboration phase. Finally, when an iteration includes a lot of complex tasks, traditional tools, such as Gantt charts and network diagrams, can be used to detail the workplan for that specific iteration.

Managing Scope

An analyst may assume that a project will be safe from scheduling problems because he or she carefully estimated and planned the project up front. However, the most common reason for schedule and cost overruns—*scope creep*—occurs after the project is under way. Scope creep happens when new requirements are added to the project after the original project scope was defined and frozen. It can happen for many reasons: Users might suddenly understand the potential of the new system and realize new functionality that would be useful; developers might discover interesting capabilities to which they become very attached; a senior manager might decide to let this system support a new strategy that was developed at a recent board meeting.

Fortunately, using an iterative and incremental development process allows the team to deal with changing requirements in an effective way. However, the more extensive the change becomes, the greater the impact on cost and schedule. Therefore, the project manager plays a critical role in managing this change to keep scope creep to a reasonable level.

The keys are to identify the requirements as well as possible in the beginning of the project and to apply analysis techniques effectively. For example, if needs are fuzzy at the project's onset, a combination of intensive meetings with the users and prototyping would allow users to "experience" the requirements and better visualize how the system could support their needs. In fact, the use of meetings and prototyping has been found to reduce scope creep to less than 5 percent on a typical project.

Of course, some requirements may be missed no matter what precautions are taken, but several practices can help control additions to the task list. First, the project manager should allow only absolutely necessary requirements to be added after the project begins. Even at that point, members of the project team should carefully assess the ramifications of the addition and present the assessment to the users. For example, it may require two more person-months of work to create a newly defined report, which would throw off the entire project deadline by several weeks. Any change that is implemented should be carefully tracked so that an audit trail exists to measure the change's impact.

Sometimes changes cannot be incorporated into the present system even though they truly would be beneficial. In this case, these additions to scope should be recorded as future enhancements to the system. The project manager can offer to provide functionality in future releases of the system, thus getting around telling someone "no."

Timeboxing

Another approach to scope management is a technique called *timeboxing*. Up until now, we have described task-oriented projects. In other words, we have described projects that have a schedule driven by the tasks that need to be accomplished, so the greater number of tasks and requirements, the longer the project will take. Some companies have little patience for development projects that take a long time, and these companies take a time-oriented approach that places meeting a deadline above delivering functionality.

Think about the use of word processing software. For 80 percent of the time, only 20 percent of the features, such as the spelling checker, boldfacing, and cutting and pasting, are used. Other features, such as document merging and creating mailing labels, may be nice to have, but they are not a part of day-to-day needs. The same goes for other software applications; most users rely on only a small subset of their capabilities. Ironically, most

FIGURE 2-21
Steps for Timeboxing

1. Set the date for system delivery.
2. Prioritize the functionality that needs to be included in the system.
3. Build the core of the system (the functionality ranked as most important).
4. Postpone functionality that cannot be provided within the time frame.
5. Deliver the system with core functionality.
6. Repeat steps 3 through 5 to add refinements and enhancements.

developers agree that typically 75 percent of a system can be provided relatively quickly, with the remaining 25 percent of the functionality demanding most of the time.

To resolve this incongruency, the technique of timeboxing has become quite popular, especially when using RAD and agile methodologies. This technique sets a fixed deadline for a project and delivers the system by that deadline no matter what, even if functionality needs to be reduced. Timeboxing ensures that project teams don't get hung up on the final finishing touches that can drag out indefinitely, and it satisfies the business by providing a product within a relatively short time frame.

Several steps are involved in implementing timeboxing on a project (see Figure 2-21). First, set the date of delivery for the proposed goals. The deadline should not be impossible to meet, so it is best to let the project team determine a realistic due date. If you recall from Chapter 1, the Scrum agile methodology sets all of its timeboxes to 30 working days. Next, build the core of the system to be delivered; you will find that timeboxing helps create a sense of urgency and helps keep the focus on the most important features. Because the schedule is absolutely fixed, functionality that cannot be completed needs to be postponed. It helps if the team prioritizes a list of features beforehand to keep track of what functionality the users absolutely need. Quality cannot be compromised, regardless of other constraints, so it is important that the time allocated to activities is not shortened unless the requirements are changed (e.g., don't reduce the time allocated to testing without reducing features). At the end of the time period, a high-quality system is delivered, but it is likely that future iterations will be needed to make changes and enhancements. In that case, the timeboxing approach can be used once again.

CONCEPTS

IN ACTION

2-1 Faster Products to Market—with IT

*T*ravelers Insurance Company of Hartford, Connecticut, has adopted agile development methodologies. The insurance field can be competitive, and Travelers wanted to have the shortest "time to implement" in the field. Travelers set up development teams of six people: two systems analysts, two representatives from the user group (such as claim services), a project manager, and a clerical support person. In the agile approach, the users are physically assigned to the development team for the project. Although at first it might seem that the users might just be sitting around drinking coffee and watching the developers come up with appropriate software solutions, this is not the case. The rapport that is developed within the team allows instant communication. The interaction is very profound. The resulting software product is delivered quickly—and generally with all the features and nuances that the users wanted.

Questions

1. Could this be done differently, such as having the users review the program on a weekly basis rather than taking the users away from their real job to work on development?

2. What mindset does an analyst need to work on such an approach?

Refining Estimates

The estimates that are produced during inception need to be refined as the project progresses. This does not mean that estimates were poorly done at the start of the project; rather, it is virtually impossible to develop an exact assessment of the project's schedule at the beginning of the development process. A project manager should expect to be satisfied with broad ranges of estimates that become more and more specific as the project's product becomes better defined.

In many respects, estimating what an IS development project will cost, how long it will take, and what the final system will actually do follows a *hurricane model*. When storms and hurricanes first appear in the Atlantic or Pacific, forecasters watch their behavior and, on the basis of minimal information about them (but armed with lots of data on previous storms), attempt to predict when and where the storms will hit and what damage they will do when they arrive. As storms move closer to North America, forecasters refine their tracks and develop better predictions about where and when they are most likely to hit and their force when they do. The predictions become more and more accurate as the storms approach a coast, until they finally arrive.

In planning, when a system is first requested, the project sponsor and project manager attempt to predict how long the development process will take, how much it will cost, and what it will ultimately do when it is delivered (i.e., its functionality). However, the estimates are based on very little knowledge of the system. As the system moves into the elaboration, more information is gathered, the system concept is developed, and the estimates become even more accurate and precise. As the system moves closer to completion, the accuracy and precision increase, until the final system is delivered (see Figure 2-22).

According to one of the leading experts in software development,[12] a well-done project plan (prepared at the end of inception) has a 100 percent margin of error for project

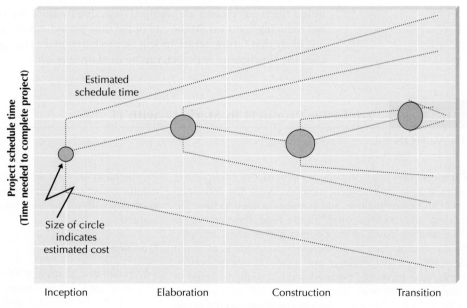

FIGURE 2-22
Hurricane Model

[12] Barry W. Boehm et al., "Cost Models for Future Software Life Cycle Processes: COCOMO 2.0," in J. D. Arthur and S. M. Henry (eds.), *Annals of Software Engineering: Special Volume on Software Process and Product Measurement* (Amsterdam: J. C. Baltzer AG Science Publishers, 1995).

Assumptions	Actions	Level of Risk
If you assume the rest of the project is simpler than the part that was late and is also simpler than believed when the original schedule estimates were made, you can make up lost time.	Do not change schedule.	High risk
If you assume the rest of the project is simpler than the part that was late and is no more complex than the original estimate assumed, you can't make up the lost time, but you will not lose time on the rest of the project.	Increase the entire schedule by the total amount of time that you are behind (e.g., if you missed the scheduled date by two weeks, move the rest of the schedule dates to two weeks later). If you included padded time at the end of the project in the original schedule, you might not have to change the promised system delivery date; you'll just use up the padded time.	Moderate risk
If you assume that the rest of the project is as complex as the part that was late (your original estimates were too optimistic), then all the scheduled dates in the future underestimate the real time required by the same percentage as the part that was late.	Increase the entire schedule by the percentage of weeks that you are behind (e.g., if you are two weeks late on part of the project that was supposed to take eight weeks, you need to increase all remaining time estimates by 25 percent). If this moves the new delivery date beyond what is acceptable to the project sponsor, the scope of the project must be reduced.	Low risk

FIGURE 2-23
Possible Actions When a Schedule Date Is Missed

cost and a 25 percent margin of error for schedule time. In other words, if a carefully done project plan estimates that a project will cost $100,000 and take twenty weeks, the project will actually cost between $0 and $200,000 and take between fifteen and twenty-five weeks.

What happens if you overshoot an estimate (e.g., analysis ends up lasting two weeks longer than expected)? There are a number of ways to adjust future estimates. If the project team finishes a step ahead of schedule, most project managers shift the deadlines sooner by the same amount but do not adjust the promised completion date. The challenge, however, occurs when the project team is late in meeting a scheduled date. Three possible responses to missed schedule dates are presented in Figure 2-23. If an estimate proves too optimistic early in the project, planners should not expect to make up for lost time—very few projects end up doing this. Instead, they should change future estimates to include an increase similar to the one that was experienced. For example, if the first phase was completed 10 percent over schedule, planners should increase the rest of their estimates by 10 percent.

Managing Risk

One final facet of project management is *risk management*, the process of assessing and addressing the risks that are associated with developing a project. Many things can cause risks: weak personnel, scope creep, poor design, and overly optimistic estimates. The project team must be aware of potential risks so that problems can be avoided or controlled well ahead of time.

Typically, project teams create a *risk assessment*, or a document that tracks potential risks along with an evaluation of the likelihood of each risk and its potential impact on the

Risk Assessment	
RISK 1:	The development of this system likely will be slowed considerably because project team members have not programmed in Java prior to this project.
Likelihood of risk:	High probability of risk.
Potential impact on the project:	This risk will probably increase the time to complete programming tasks by 50 percent.
Ways to address this risk:	

It is very important that time and resources are allocated to up-front training in Java for the programmers who are used for this project. Adequate training will reduce the initial learning curve for Java when programming begins. Additionally, outside Java expertise should be brought in for at least some part of the early programming tasks. This person should be used to provide experiential knowledge to the project team so that Java-related issues (of which novice Java programmers would be unaware) are overcome.

| **RISK 2:** | ... |

FIGURE 2-24
Sample Risk
Assessment

project (Figure 2-24). A paragraph or two is also included to explain potential ways that the risk can be addressed. There are many options: the risk could be publicized, avoided, or even eliminated by dealing with its root cause. For example, imagine that a project team plans to use new technology but its members have identified a risk in the fact that its members do not have the right technical skills. They believe that tasks may take much longer to perform because of a high learning curve. One plan of attack could be to eliminate the root cause of the risk—the lack of technical experience by team members—by finding the time and resources needed to provide proper training to the team.

Most project managers keep abreast of potential risks, even prioritizing them according to their magnitude and importance. Over time, the list of risks will change as some items are removed and others surface. The best project managers, however, work hard to keep risks from having an impact on the schedule and costs associated with the project.

STAFFING THE PROJECT

Staffing the project includes determining how many people should be assigned to the project, matching people's skills with the needs of the project, motivating them to meet the project's objectives, and minimizing the conflict that will occur over time. The deliverables for this part of project management are a staffing plan, which describes the number and kinds of people who will work on the project, the overall reporting structure, and the project charter, which describes the project's objectives and rules. However, before describing the development of a staffing plan, how to motivate people, and how to handle conflict, we describe a set of characteristics of jelled teams.

Characteristics of a Jelled Team[13]

The idea of a jelled team has existed for a long time. To begin with, most (if not all) student groups are *not* representative of the idea of a jelled team, and you may have never had

---

[13] The material in the section is based on T. DeMarco and T. Lister, *Peopleware: Productive Projects and Teams,* 2nd Ed. (New York: Dorset House, 1999); and P. Lencioni, *The Five Dysfunctions of a Team: A Leadership Fable* (San Francisco: Jossey-Bass, 2002).

the opportunity to appreciate the effectiveness of a true team and not simply a group. In fact, DeMarco and Lister point out that teams are not created; they are grown. And, given typical class projects, the ability to grow a team, compared to being assigned to or forming a group, is very limited. However, growing development teams is crucial in information systems development. The whole set of agile software development approaches hinges on growing jelled teams. Otherwise, agile development approaches would totally fail.

In this section, we describe some of the characteristics of jelled teams. But before we do this, we should define the phrase *jelled team.* According to DeMarco and Lister,[14] "[a] jelled team is a group of people so strongly knit that the whole is greater than the sum of the parts. The production of such a team is greater than that of the same people working in unjelled form." They go on to state that a jelled "team can become almost unstoppable, a juggernaut for success." When is the last time that you worked with a group on a class project that could be described "a juggernaut for success"? Demarco and Lister identify five characteristics of a jelled team.

First, jelled teams have a very low turnover during a project. Typically, members of a jelled team feel a responsibility to the other team members. This responsibility is felt so intensely that for a member to leave the team, the member would feel that they were letting the team down and that they were breaking a bond of trust. Lencioni also identifies the "absence of trust" as the primary cause for dysfunctional teams.

Second, jelled teams have a strong sense of identity. In many classes, when you are part of a group, the group chooses some cute name to identify the group and differentiate it from the other groups. However, in this case, it is not simply the choosing of a name. It is instead evolving every member into something that only exists within the team. This can be seen when members of the team tend to do non–work-related activities together, e.g., do lunch together as a team or form a basketball team composed of only members of the development team.

Third, the strong sense of identity tends to lead the team into feeling a sense of eliteness. The members of a jelled development team almost have a swagger about the way they relate to nonteam employees. That is, if you are not a member of the team, then what are you? Good examples that come to mind that possess this sense of eliteness outside of the scope of information systems development teams are certain sports teams, U.S. Navy Seal teams, or big city police force SWAT teams. In all three examples, each team member is highly competent in their specialty area and each other team member knows (not thinks) that they can depend on the team members performing their individual jobs with a very high-level of skill.

Fourth, during the development process, jelled teams feel that the team owns the information system being developed and not any one individual member. In many ways, you could almost say that jelled teams are a little communistic in nature. By this we mean that the individual contributions to the effort are not important to a true team. The only things that matter are the output of the team. However, this is not to imply that a member who does not deliver their fair share will not go unpunished. In a jelled team, any member who is not producing is actually breaking their bond of trust with the other team members (see the first characteristic).

The final characteristic of a jelled team is that they really enjoy (have fun) doing their work. The members actually like to go to work and be with their team members. Much of this can be attributed to the level of challenge they receive. If the project is challenging and the members of the team are going to learn something from completing the project, the members of a jelled team will enjoy tackling the project.

---

[14] T. DeMarco and T. Lister, *Peopleware: Productive Projects and Teams,* 2nd Ed., p. 123.

When a team jells, they will avoid Lencioni's dysfunctions. The lack of trust is the primary cause of a team's becoming dysfunctional. Lencioni describes four other causes of a team's becoming dysfunctional that can come from the lack of trust. First, dysfunctional teams fear conflict, whereas members of a jelled team never fear conflict.[15] Going to a member of a jelled team and admitting that you do not know how to do something is no big deal. In fact, it provides a method for the team member to help out, which would increase the level of trust between the two members. Second, dysfunctional teams do not have a commitment to the team from the individual members. Instead, they tend to focus on their individual performance instead of the team's performance. This can even be to the detriment of the development team. Obviously, this is not an issue for jelled teams. Third, dysfunctional teams try to avoid accountability. With jelled teams, accountability is not an issue. Members of a jelled team feel a high level of responsibility to the other team members. No team member ever wants to let down the team. Furthermore, owing to the bond that holds jelled teams together, no member has any problem with holding other members accountable for their performance (or lack of performance). Fourth, dysfunctional teams do not pay attention to the team's results. Again, in this case, the cause of this dysfunction is that the individual members only focus on their individual goals. From a team management perspective, the team leader should focus on getting the goals of the team aligned; a jelled team will attain the goals.

Staffing Plan

The first step to staffing is determining the average number of staff needed for the project. To calculate this figure, divide the total person-months of effort by the optimal schedule. So to complete a forty-person-month project in ten months, a team should have an average of four full-time staff members, although this may change over time as different specialists enter and leave the team (e.g., business analysts, programmers, technical writers).

Many times, the temptation is to assign more staff to a project to shorten the project's length, but this is not a wise move. Adding staff resources does not translate into increased productivity; staff size and productivity share a disproportionate relationship, mainly because it is more difficult to coordinate a large number of staff members. The more a team grows, the more difficult it becomes to manage. Imagine how easy it is to work on a two-person project team: the team members share a single line of communication. But adding two people increases the number of communication lines to six, and greater increases lead to more dramatic gains in communication complexity. Figure 2-25 illustrates the impact of adding team members to a project team.

One way to reduce efficiency losses on teams is to understand the complexity that is created in numbers and to build in a *reporting structure* that tempers its effects. The general rule is to keep team sizes to fewer than eight to ten people; therefore, if more people are needed, create sub-teams. In this way, the project manager can keep the communication effective within small teams, which, in turn, communicate to a contact at a higher level in the project.

After the project manager understands how many people are needed for the project, he or she creates a *staffing plan* that lists the roles and the proposed reporting structure that are required for the project. Typically, a project has one project manager, who oversees the overall progress of the development effort, with the core of the team comprising the various types of analysts described in Chapter 1. A *functional lead* is usually assigned to manage

---

[15] When conflict occurs, it is necessary to address it in an effective manner. We discuss how to handle conflict later in the chapter.

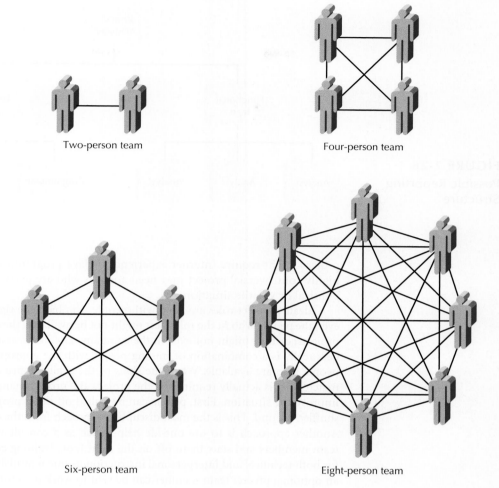

FIGURE 2-25
Increasing Complexity with Larger Teams

Two-person team

Four-person team

Six-person team

Eight-person team

a group of analysts, and a *technical lead* oversees the progress of a group of programmers and more technical staff members.

There are many structures for project teams; Figure 2-26 illustrates one possible configuration of a project team. After the roles are defined and the structure is in place, the project manager needs to think about which people can fill each role. Often, one person fills more than one role on a project team.

When you make assignments, remember that people have *technical skills* and *interpersonal skills*, and both are important on a project. Technical skills are useful when working with technical tasks (e.g., programming in Java) and in trying to understand the various roles that technology plays in the particular project (e.g., how a web server should be configured on the basis of a projected number of hits from customers). Interpersonal skills, on the other hand, include interpersonal and communication abilities that are used when dealing with business users, senior management executives, and other members of the project team. They are particularly critical when performing the requirements-gathering activities and when addressing organizational feasibility issues. Each project requires unique technical and interpersonal skills. For example, a web-based

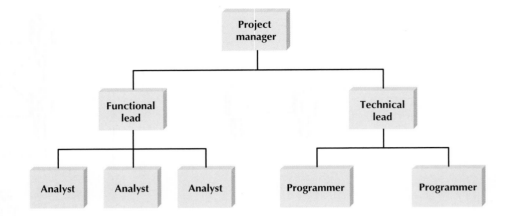

FIGURE 2-26
Possible Reporting
Structure

project might require Internet experience or Java programming knowledge, whereas a highly controversial project may need analysts who are particularly adept at managing political or volatile situations.

Ideally, project roles are filled with people who have the right skills for the job. However, the people who fit the roles best might not be available; they may be working on other projects, or they might not exist in the company. Therefore, assigning project team members really is a combination of finding people with the appropriate skill sets and finding people who are available. When the skills of the available project team members do not match what is actually required by the project, the project manager has several options to improve the situation. First, people can be pulled off other projects, and resources can be shuffled around. This is the most disruptive approach from the organization's perspective. Another approach is to use outside help—such as a consultant or contractor—to train team members and start them off on the right foot. Training classes are usually available for both technical and interpersonal instruction if time is available. Mentoring may also be an option; a project team member can be sent to work on another similar project so that he or she can return with skills to apply to the current job.

YOUR TURN   2-7 **Staffing Plan**

Now it is time to staff the project that was described in Your Turn 2-6. On the basis of the effort required for the project, how many people will be needed on the project? Given this number, select classmates who will work with you on your project.

Questions

1. What roles will be needed to develop the project? List them and write short descriptions for each of

these roles, almost as if you had to advertise for the positions.

2. Which roles will each classmate perform? Will some people perform multiple roles?

3. What will the reporting structure be for the project?

Motivation

Assigning people to tasks isn't enough; project managers need to motivate the people to make the project a success. *Motivation* has been found to be the number one influence on people's performance,[16] but determining how to motivate the team can be quite difficult. You might think that good project managers motivate their staff by rewarding them with money and bonuses, but most project managers agree that this is the last thing that should be done. The more often managers reward team members with money, the more they expect it—and most times monetary motivation won't work. Pink[17] has suggested a set of principles to follow to motivate individuals in twenty-first century firms. In this section, we adapt his suggestions to information systems development teams.

Pink suggests that to motivate individuals we should consider some form of the 20 percent time rule. This rule suggests that 20 percent of an employee's time should be spent on some idea in which they believe. The project does not have to be related in any way to the project at hand. On the surface, this sounds like a colossal waste of time, but you should not throw this idea away. Have you used Gmail? This is how Google's Gmail and Google News became software products. If 20 percent sounds too high, Pink suggests that you consider 10 percent to begin with.

He recommends that firms should be willing to fund small "Now That" awards. These awards are given as small signs of appreciation for doing a great job. However, these awards are not given by a manager to an employee but by an employee to a peer of the employee. The awards are monetary, but they are very small, typically $50. As such, they really are not relevant from a monetary perspective. However, they are very relevant because they are given by one of the employee's colleagues to show that some action that the employee did was appreciated.

He endorses the idea of applying Robert Reich's (President's Clinton's Secretary of Labor) pronoun test. If an employee (or team member) refers to the firm (the team) as "they," then there is the real possibility that the employee feels disengaged or possibly alienated. On the other hand, when employees refer to the firm as "we," they obviously feel like they are part of the organization. From a team perspective, this could be an indication that the team has begun to jell.

Pink suggests that management should periodically consider giving each employee a day on which they can work on anything they want. In some ways, this is related to the 20 percent rule. It does not necessarily require one day a week (20 percent), but it does require some deliverable. The deliverable can be a new utility program that could be used by lots of different projects, it could be a new prototype of a new software product, or it could be an improvement for a business process that is used internally. The major point here is to provide team members with the ability to focus on interesting and challenging problems that might (or might not) provide results to the firm's bottom line. Regardless, it demonstrates an amount of trust and respect that the firm has for its employees.

He recommends that managers remove the issue of compensation from the motivation equation. By this, he means that all employees should be paid a sufficient amount so that compensation awards are not an issue. Technical employees on project teams are much more motivated by recognition, achievement, the work itself, responsibility, advancement, and the chance to learn new skills.[18] Simplistic financial awards, such as raises that are perceived as being unjust, can actually demotivate the overall team and lower overall performance.

[16] Barry W. Boehm, *Software Engineering Economics* (Englewood Cliffs, NJ: Prentice Hall, 1981). One of the best books on managing project teams is that by Tom DeMarco and Timothy Lister, *Peopleware: Productive Projects and Teams* (New York: Dorset House, 1987).

[17] D. H. Pink, *Drive: The Surprising Truth About What Motivates Us* (New York, NY: Riverhead Books, 2009).

[18] F. H. Hertzberg, "One More Time: How Do You Motivate Employees?" *Harvard Business Review* (January–February 1968).

He advocates that twenty-first century bosses (team leaders) need to be willing to give up control. Many of the agile development approaches make similar suggestions. Appelo[19] suggests that an open door policy that is supported by a team leader actually can be self-defeating. In the case of software development teams, an open door policy implies that the team leader has a door that can be left open, whereas the poor individual team member does not have an office with a door. In this case, Appelo suggests that the team leader move from the office with a door to the same shared space in which the team resides. One of Pink's other ideas is for the team leader to not use controlling language, such as telling the team member that they "must" do something. Instead, the team leader should ask the team member to "consider" or "think about" the idea instead. In some ways, a true team leader should never receive credit for any ideas associated with the team. Instead, a team leader should make suggestions and encourage the team members to consider ideas and, most importantly, let the team member and the team receive the credit.

Pink provides evidence that intrinsic motivation is very important for twenty-first century knowledge workers. In this case, he states that to intrinsically motivate individuals, you must provide them with a degree of autonomy, support them in such a way that they can master their area of expertise, and encourage them to pursue projects with a purpose. Providing team members with autonomy relates to the jelled team concept of trust. Team leaders need to trust the team members to deliver the software for which they are responsible. Supporting team members so that they can master their area of expertise can be as simple as providing support to attend conferences, seminars, and training sessions that deal with the member's area of expertise. It also could imply providing the team member with a high-end development environment. For example, when building information visualization and virtual reality applications, special hardware and software environments can make it much easier to master the technology to develop the application. Finally, today it is very important for team members to feel that what they are doing can make a difference. A team leader should encourage the team members to tackle problems that can make a difference in people's lives. This can easily be accomplished through the use of the 20 percent rule. Figure 2-27 lists some motivational don'ts to avoid demotivating team members.

Handling Conflict

The third component of staffing is organizing the project to minimize conflict among group members. *Group cohesiveness* (the attraction that members feel to the group and to other members) contributes more to productivity than do project members' individual capabilities or experiences.[20] Clearly defining the roles on the project and holding team members accountable for their tasks is a good way to begin mitigating potential conflict on a project. Some project managers develop a *project charter,* which lists the project's norms and ground rules. For example, the charter may describe when the project team should be at work, when staff meetings will be held, how the group will communicate with each other, and the procedures for updating the workplan as tasks are completed. Figure 2-28 lists additional techniques that can be used at the start of a project to keep conflict to a minimum.

[19] J. Appelo, *Management 3.0: Leading Agile Developers, Developing Agile Leaders* (Upper Saddle River, NJ: Addison-Wesley, 2011).

[20] B. Lakhanpal, "Understanding the Factors Influencing the Performance of Software Development Groups: An Exploratory Group-Level Analysis," *Information and Software Technology* 35, no. 8 (1993): 468–473.

Don'ts	Reasons
Assign unrealistic deadlines	Few people will work hard if they realize that a deadline is impossible to meet.
Ignore good efforts	People will work harder if they feel like their work is appreciated. Often, all it takes is public praise for a job well done.
Create a low-quality product	Few people can be proud of working on a project that is of low quality.
Give everyone on the project a raise	If everyone is given the same reward, then high-quality people will believe that mediocrity is rewarded—and they will resent it.
Make an important decision without the team's input	Buy-in is very important. If the project manager needs to make a decision that greatly affects the members of her team, she should involve them in the decision-making process.
Maintain poor working conditions	A project team needs a good working environment or motivation will go down the tubes. This includes lighting, desk space, technology, privacy from interruptions, and reference resources.

FIGURE 2-27
Motivational Don'ts

Source: Steve McConnell, *Adapted Rapid Development* (Redmond, WA: Microsoft Press, 1996).

CONCEPTS

IN ACTION

2-J RFID—Promising Technology?

*S*ome animals are extremely valuable. For centuries, horse thieves have stolen horses, so now most horses have tattoos in their mouths. Likewise, purebred pets, such as dog show winners, are valuable animals. What if there were a better way to positively identify valuable animals?

Radio frequency identification (or RFID) has been used for many years in airplanes and on toll roads (like E-ZPass and FAST LANE in the United States) as well as in libraries so that books and materials are not taken out of the library without being checked out. With RFID, a low-frequency radio transmitter, when bombarded with a radio wave, replies with a unique signal. Some animal owners have inserted RFID chips into their pets' shoulders so that they can be identified. The code is unique and cannot be changed. It would be possible to track a stolen race horse

if the horse came into range of an RFID device. Likewise, a pet shop or a veterinarian could identify a valuable pet.

Questions

1. If you were working for a state consumer-protection agency, what requirements might you place on pet shops to ensure that animals for sale have not been stolen?
2. What technological requirements might be needed in the system proposal?
3. What ethical issues might be involved?
4. If your system project team did not have the correct technical background, what might you do?

- Clearly define plans for the project.
- Make sure the team understands how the project is important to the organization.
- Develop detailed operating procedures and communicate these to the team members.
- Develop a project charter.
- Develop schedule commitments ahead of time.
- Forecast other priorities and their possible impact on the project.

FIGURE 2-28
Conflict-Avoidance Strategies

Source: H. J. Thamhain and D. L. Wilemon, "Conflict Management in Project Life Cycles," *Sloan Management Review* (Spring 1975).

YOUR

TURN

2-8 Project Charter

*G*et together with several of your classmates and pretend that you are all staff on the project described in Your Turn 2-6. Discuss what would most motivate each of you to perform well on the project. List three potential sources of conflict that could surface as you work together.

Question

Develop a project charter that lists five rules that all team members need to follow. How might these rules help avoid potential team conflict?

ENVIRONMENT AND INFRASTRUCTURE MANAGEMENT

The environment and infrastructure management workflows support the development team throughout the development process. The environment workflow primarily deals with choosing the correct set of tools that will be used throughout the development process and identifying the appropriate set of standards to be followed during the development process. Infrastructure management workflow deals with choosing the appropriate level and type of documentation that will be created during the development process. Other activities associated with the infrastructure management workflow include developing, modifying, and reusing predefined components, frameworks, libraries, and patterns. The topic of reuse is discussed in later chapters (see Chapters 5 and 8).

CASE Tools

Computer-aided software engineering (CASE) is a category of software that automates all or part of the development process. Some CASE software packages are used primarily to support the analysis workflow to create integrated diagrams of the system and to store information regarding the system components (often called *upper CASE*), whereas others support the design workflow that can be used to generate code for database tables and system functionality (often called *lower CASE*). *Integrated CASE*, or I-CASE, contains functionality found in both upper CASE and lower CASE tools in that it supports tasks that happen throughout the system-development process. CASE comes in a wide assortment of flavors in terms of complexity and functionality, and many good tools are available in the marketplace to support object-oriented systems development (e.g., ArgoUml, Enterprise Architect, Metamill, Poseidon, Visual Paradigm, and Rational Rose).

The benefits of using CASE are numerous. With CASE tools, tasks can be completed and altered much faster, development information is centralized, and information is illustrated through diagrams, which are typically easier to understand. Potentially, CASE can reduce maintenance costs, improve software quality, and enforce discipline. Some project teams even use CASE to assess the magnitude of changes to the project. Many modern CASE tools that support object-oriented systems development support a development technique known as *round-trip engineering*. Round-trip engineering supports not only code generation but also the reverse engineering of UML diagrams from code. For example, Poseidon for UML can generate Java that can be modified by the programmers, at which point the UML diagrams will be out of date and no longer accurately represent the code. However, Poseidon for UML also supports the generation of UML diagrams from code. In this way, the system can evolve via diagrams and via code in a round-trip manner.

Of course, like anything else, CASE should not be considered a silver bullet for project development. The advanced CASE tools are complex applications that require significant training and experience to achieve real benefits. Our experience has shown that CASE is a helpful way to support the communication and sharing of project diagrams and technical specifications so long as it is used by trained developers who have applied CASE on past projects.

The central component of any CASE tool is the *CASE repository*, otherwise known as the information repository or data dictionary. The CASE repository stores the diagrams and other project information, such as screen and report designs, and it keeps track of how the diagrams fit together. For example, most CASE tools warn you if you place a field on a screen design that doesn't exist in your structural model. As the project evolves, project team members perform their tasks using CASE.

Standards

Members of a project team need to work together, and most project management software and CASE tools provide access privileges to everyone working on the system. When people work together, however, things can get pretty confusing. To make matters worse, people sometimes are reassigned in the middle of a project. It is important that their project knowledge does not leave with them and that their replacements can get up to speed quickly.

One way to make certain that everyone is performing tasks in the same way and following the same procedures is to create *standards* that the project team must follow. Standards can include formal rules for naming files, forms that must be completed when goals are reached, and programming guidelines. Figure 2-29 shows some examples of the types

Types of Standards	Examples
Documentation standards	The date and project name should appear as a header on all documentation.
	All margins should be set to 1 inch.
	All deliverables should be added to the project binder and recorded in its table of contents.
Coding standards	All modules of code should include a header that lists the programmer, last date of update, and a short description of the purpose of the code.
	Indentation should be used to indicate loops, if-then-else statements, and case statements.
	On average, every program should include one line of comments for every five lines of code.
Procedural standards	Record actual task progress in the work plan every Monday morning by 10 AM.
	Report to project update meeting on Fridays at 3:30 PM.
	All changes to a requirements document must be approved by the project manager.
Specification requirement standards	Name of program to be created
	Description of the program's purpose
	Special calculations that need to be computed
	Business rules that must be incorporated into the program
	Pseudocode
	Due date
User interface design standards	Labels will appear in boldface text, left-justified, and followed by a colon.
	The tab order of the screen will move from top left to bottom right.
	Accelerator keys will be provided for all updatable fields.

FIGURE 2-29

A Sampling of Project Standards

YOUR

TURN

2-9 Computer-Aided Software Engineering Tool Analysis

*S*elect a CASE tool—one that you will use for class, a program that you own, or a tool that you can examine over the web. Create a list of the capabilities that are offered by the CASE tool.

Question

Would you classify the CASE tool as upper CASE, lower CASE, or I-CASE? Why?

of standards that a project can create. When a team forms standards and then follows them, the project can be completed faster because task coordination becomes less complex.

Standards work best when they are created at the beginning of each major phase of the project and communicated clearly to the entire project team. As the team moves forward, new standards are added when necessary. Some standards (e.g., file naming conventions, status reporting) are applied during the entire development process, whereas others (e.g., programming guidelines) are appropriate only for certain tasks.

Documentation

Finally, during the inception phase of the infrastructure workflow, project teams put into place good *documentation* standards that include detailed information about the tasks of the Unified Process. Typically, the standards for the required documentation are set by the development organization. The development team only needs to ascertain which documentation standards are appropriate for the current systems development project. Often, the documentation is stored in a *project binder(s)* that contain all the deliverables and all the internal communication that takes place—the history of the project. The good news is that Unified Process has a set of standard documentation that is expected. The documentation typically includes the system request, the feasibility analysis, the original and later versions of the effort estimation, the evolving workplan, and UML diagrams for the functional, structural, and behavioral models.

A poor project management practice is waiting until the last minute to create documentation; this typically leads to an undocumented system that no one understands. In fact, many problems that companies had updating their systems to handle the year 2000 crisis were the result of the lack of documentation. Good project teams learn to document a system's history as it evolves while the details are still fresh in their memory. In most case tools that support object-oriented systems development, some of the documentation can be automated. For example, if the programming language chosen to implement the system in is Java, then it is possible to automatically create HTML manual pages that will describe the classes being implemented. This is accomplished through the javadoc[21] tool that is part of the Java development environment. Other tools enable the developer to automatically generate HTML documentation for the UML diagrams, e.g., umldoc, which is part of the Poseidon for UML CASE tool.[22] Even though virtually all developers hate creating documentation and documentation takes valuable time, it is a good investment that will pay off in the long run.

[21] See www.oracle.com/technetwork/java/javase/documentation/index-jsp-135444.html.

[22] See www.gentleware.com/fileadmin/media/viewlets/text/UMLdoc.viewlet/UMLdoc_viewlet_swf.html.

CONCEPTS IN ACTION

2-K Poor Naming Standards

I once started on a small project (four people) in which the original members of the project team had not set up any standards for naming electronic files. Two weeks into the project, I was asked to write a piece of code that would be referenced by other files that had already been written. When I finished my piece, I had to go back to the other files and make changes to reflect my new work. The only problem was that the lead programmer decided to name the files using his initials (e.g., GG1.prg, GG2.prg, GG3.prg)—and there were more than 200 files! I spent two days opening every one of those files because there was no way to tell what their contents were.

Needless to say, from then on, the team created a code for file names that provided basic information regarding the file's contents, and they kept a log that recorded the file name, its purpose, the date of the last update, and the programmer for every file on the project.

—*Barbara Wixom*

Question

Think about a program that you have written in the past. Would another programmer be easily able to make changes to it? Why or why not?

PRACTICAL TIP

Avoiding Classic Planning Mistakes

*A*s Seattle University's David Umphress has pointed out, watching most organizations develop systems is like watching reruns of *Gilligan's Island*. At the beginning of each episode, someone comes up with a cockamamie scheme to get off the island, and it seems to work for a while, but something goes wrong and the castaways find themselves right back where they started—stuck on the island. Similarly, most companies start new projects with grand ideas that seem to work, only to make a classic mistake and deliver the project behind schedule, over budget, or both. Here we summarize four classic mistakes in the planning and project management aspects of the project and discuss how to avoid them:

1. *Overly optimistic schedule:* Wishful thinking can lead to an overly optimistic schedule that causes analysis and design to be cut short (missing key requirements) and puts intense pressure on the programmers, who produce poor code (full of bugs).

   Solution: Don't inflate time estimates; instead, explicitly schedule slack time at the end of each phase to account for the variability in estimates.

2. *Failing to monitor the schedule:* If the team does not regularly report progress, no one knows if the project is on schedule.

Solution: Require team members to report progress (or the lack of progress) honestly every week. There is no penalty for reporting a lack of progress, but there are immediate sanctions for a misleading report.

3. *Failing to update the schedule:* When a part of the schedule falls behind (e.g., information gathering uses all the slack in item 1 plus two weeks), a project team often thinks it can make up the time later by working faster. It can't. This is an early warning that the entire schedule is too optimistic.

   Solution: Immediately revise the schedule and inform the project sponsor of the new end date or use timeboxing to reduce functionality or move it into future versions.

4. *Adding people to a late project:* When a project misses a schedule, the temptation is to add more people to speed it up. This makes the project take longer because it increases coordination problems and requires staff to take time to explain what has already been done.

   Solution: Revise the schedule, use timeboxing, throw away bug-filled code, and add people only to work on an isolated part of the project.

Source: Adapted from Steve McConnell, *Rapid Development* (Redmond, WA: Microsoft Press, 1996), pp. 29–50.

APPLYING THE CONCEPTS AT CD SELECTIONS

In this chapter, we introduced how object-oriented systems development projects were managed. Specifically, we described how projects were identified and how the identification led to a system request. Next, we presented the three different types of feasibility analyses and how their results helped in selecting a project. After that, we reviewed a set of traditional project management tools that can be applied to planning and managing of an object-oriented systems development project and demonstrated employing use-case points as a method that can be used to estimate the effort it will take to develop an object-oriented system. We next discussed the use of evolutionary work breakdown structures and iterative workplans in conjunction with the Unified Process. We then covered the issues related to assigning the right people to the development team. Finally, we described topics associated with the environment and infrastructure management workflows of the Unified Process. In this installment of the CD Selections case, we see how Margaret and the development team work through all of these topics with regard to the web-based solution that they hope to create.

SUMMARY

Project Identification

Potential projects can be identified by a member of an organization that has identified a business need that can be addressed through the application of information technology. The first step in the process is to identify the business value for the system by developing a system request that provides basic information about the proposed system. Next, the analysts perform a feasibility analysis to determine the technical, economic, and organizational feasibility of the system; if appropriate, the system is approved and the development project begins.

Feasibility Analysis

A feasibility analysis is used to provide more detail about the risks associated with the proposed system, and it includes technical, economic, and organizational feasibilities. The technical feasibility focuses on whether the system *can* be built by examining the risks associated with the users' and analysts' familiarity with the functional area, familiarity with the technology, and the project's size. The economic feasibility addresses whether the system *should* be built. It includes a cost–benefit analysis of development costs, operational costs, tangible benefits, and intangible costs and benefits. Finally, the organizational feasibility assesses how well the system will be accepted by its users and incorporated into the ongoing operations of the organization. The strategic alignment of the project and a stakeholder analysis can be used to assess this feasibility dimension.

Project Selection

Once the feasibility analysis has been completed, it is submitted to the approval committee, along with a revised system request. The committee then decides whether to approve the project, decline the project, or table it until additional information is available. The project selection process uses portfolio management to take into account all the projects in the organization. The approval committee weighs many factors and makes trade-offs before a project is selected.

Traditional Project Management Tools

Even though object-oriented systems development projects are significantly different from traditional systems development projects, a set of useful traditional project management tools can be used to manage object-oriented systems development projects. These tools include work breakdown structures, Gantt charts, and network diagrams. Work breakdown structures can be structured either by phase or by product. Gantt charts are drawn using horizontal bars to represent the duration of each task, and as people work on tasks, the appropriate bars are filled in proportionately to how much of the task is finished. Network diagrams are the best way to communicate task dependencies because they lay out the tasks as a flowchart in the order in which they need to be completed. The longest path from the project inception to completion is referred to as the critical path.

Estimating Project Effort

Use-case points are an effort-estimation technique that is based on the unique characteristics of a use-case–driven systems development method. Use-case points are founded on the two primary constructs associated with use-case analysis: actors and use cases. Use-case points have a set of factors used to modify their raw value: technical complexity factors and environmental factors. Technical complexity factors address the complexity of the project under consideration, whereas the environmental factors deal with the level of experience of the development staff. Based on the number of use-case points, the estimated effort required can be computed.

Creating and Managing the Workplan

Once a project manager has a general idea of the effort to develop the project, he or she creates a workplan, which is a dynamic schedule that records and keeps track of all the tasks that need to be accomplished over the course of the project. To create an iterative workplan, the project manager first begins with an evolutionary work breakdown structure that allows the project manager to provide more realistic estimates for each iteration, or build, of a system. Iterative workplans are decoupled from the architecture of the system, thus allowing projects to be comparable. By supporting comparability among projects, evolutionary WBSs enable organizational learning to take place.

Scope creep has always been a problem with systems development projects. Essentially, the farther along the development process, the better the understanding of the underlying problem and the technology being used becomes. Estimating what an IS development project will cost, how long it will take, and what the final system will actually do tends to follow a hurricane model. Iterative workplans support changing requirements by simply supporting development in an incremental and iterative manner. One approach that has been used quite successfully to address scope creep is timeboxing. Timeboxing sets a fixed deadline for a project and delivers the system by that deadline no matter what, even if functionality must be reduced. Finally, managing risks through the development process is essential. A risk assessment is used to help mitigate risk because it identifies potential risks and evaluates the likelihood of risk and its potential impact on the project.

Staffing the Project

Staffing involves determining how many people should be assigned to the project, assigning project roles to team members, developing a reporting structure for the team, and matching people's skills with the needs of the project. Staffing also includes motivating the team to meet the project's objectives and minimizing conflict among team members. Both motivation and cohesiveness have been found to greatly influence performance of team members in project situations. Team members are motivated most by such nonmonetary rewards as recognition, achievement, and the work itself. Clearly defining the roles on a project and holding team members accountable for their tasks can minimize conflict. Some managers create a project charter that lists the project's norms and ground rules.

Environment and Infrastructure Management

The environment workflow supports the development team throughout the development process by ensuring that the team has access to the appropriate set of CASE tools and standards that will be used during the development process. Case tools are a type of software that automates all or part of the development process. Standards are formal rules or guidelines that project teams must follow during the project to allow projects to be comparable. The infrastructure management workflow deals with setting up the project documentation and identifying possible reusable components, frameworks, libraries, and patterns.

KEY TERMS

Actor, 74
Adjusted use-case points (UCP), 77
Application program
   interface (API), 74
Approval committee, 50
Average actors, 74
Average use case, 74
Break-even point, 61
Business need, 51
Business requirement, 52
Business value, 52
Call option, 63
Cash flow method, 59
Champion, 65
Compatibility, 56
Complex actors, 74
Complex use case, 74
Computer-aided software
   engineering (CASE), 96
CASE repository, 97
Cost–benefit analysis, 56
Critical path method, 72

Critical task, 72
Development costs, 57
Documentation, 98
Economic feasibility, 56
Effort, 78
Emerging Technology, 51
Environmental factor (EF), 77
Environmental factor value
   (EFactor), 77
Estimation, 73
Evolutionary WBS, 80
Familiarity with the functional area, 55
Familiarity with the technology, 56
Feasibility analysis, 50
Feasibility study, 55
First mover, 51
Functional lead, 90
Functionality, 52
Gantt chart, 71
Group cohesiveness, 94
Hurricane model, 86
Intangible benefits, 57

Intangible costs, 57
Intangible value, 52
Integrated CASE, 96
Iterative workplan, 80
Interpersonal skills, 91
Lower CASE, 96
Methodology, 79
Milestone, 70
Motivation, 93
Net present value (NPV), 60
Network Diagram, 71
Node, 72
Operational costs, 57
Option pricing models (OPMs), 63
Organizational feasibility, 64
Organizational management, 65
Person-hours multiplier (PHM), 78
Program evaluation and review
   technique (PERT), 71
Portfolio management, 66
Project, 49
Project binder, 98

QUESTIONS

1. Give three examples of business needs for a system.
2. What is the purpose of an approval committee? Who is usually on this committee?
3. Why should the system request be created by a business person as opposed to an IS professional?
4. What is the difference between intangible value and tangible value? Give three examples of each.
5. What are the purposes of the system request and the feasibility analysis? How are they used in the project selection process?
6. Describe two special issues that may be important to list on a system request.
7. Describe the three techniques for feasibility analysis.
8. Describe a risky project in terms of technical feasibility. Describe a project that would *not* be considered risky.
9. What are the steps for assessing economic feasibility? Describe each step.
10. List two intangible benefits. Describe how these benefits can be quantified.
11. List two tangible benefits and two operational costs for a system. How would you determine the values that should be assigned to each item?
12. Explain the net present value and return on investment for a cost–benefit analysis. Why would these calculations be used?
13. What is the break-even point for the project? How is it calculated?

14. What is stakeholder analysis? Discuss three stakeholders that would be relevant for most projects.
15. Why do many projects end up having unreasonable deadlines? How should a project manager react to unreasonable demands?
16. What are the trade-offs that project managers must manage?
17. Compare and contrast the Gantt chart with the network diagram.
18. Some companies hire consulting firms to develop the initial project plans and manage the project, but use their own analysts and programmers to develop the system. Why do you think some companies do this?
19. What is a use-case point? For what is it used?
20. What process do we use to estimate systems development based on use cases?
21. Name two ways to identify the tasks that need to be accomplished over the course of a project.
22. What are the problems associated with conventional WBSs?
23. What is an evolutionary WBS? How does it address the problems associated with a conventional WBS?
24. What is an iterative workplan?
25. What is scope creep, and how can it be managed?
26. What is timeboxing, and why is it used?
27. Describe the hurricane model.
28. Create a list of potential risks that could affect the outcome of a project.

29. Describe the differences between a technical lead and a functional lead. How are they similar?
30. Describe three technical skills and three interpersonal skills that are very important to have on any project.
31. What are the best ways to motivate a team? What are the worst ways?
32. List three techniques to reduce conflict.
33. What is the difference between upper CASE and lower CASE?
34. Describe three types of standards and provide examples of each.
35. What belongs in the project binder? How is the project binder organized?

EXERCISE

A. Locate a news article in an IT trade magazine (e.g., *Computerworld*) about an organization that is implementing a new computer system. Describe the tangible and intangible value that the organization is likely to realize from the new system.

B. Car dealers have realized how profitable it can be to sell automobiles using the web. Pretend you work for a local car dealership that is part of a large chain such as CarMax. Create a system request you might use to develop a web-based sales system. Remember to list special issues that are relevant to the project.

C. Suppose that you are interested in buying a new computer. Create a cost–benefit analysis that illustrates the return on investment that you would receive from making this purchase. Computer-related websites (e.g., Apple, Dell, HP) should have real tangible costs that you can include in your analysis. Project your numbers out to include a three-year period and provide the net present value of the final total.

D. The Amazon.com website originally sold books; then the management of the company decided to extend their web-based system to include other products. How would you have assessed the feasibility of this venture when the idea first came up? How risky would you have considered the project that implemented this idea? Why?

E. Interview someone who works in a large organization and ask him or her to describe the approval process that exists for approving new development projects. What do they think about the process? What are the problems? What are the benefits?

F. Reread Your Turn 2-1 (Identify Tangible and Intangible Value). Create a list of the stakeholders that should be considered in a stakeholder analysis of this project.

G. Visit a project management website, such as the Project Management Institute (www.pmi.org). Most have links to project management software products, white papers, and research. Examine some of the links for project management to better understand a variety of Internet sites that contain information related to this chapter.

H. Select a specific project management topic such as CASE, project management software, or timeboxing and search for information on that topic using the web. Any search engine (e.g., Bing, Google) can provide a starting point for your efforts.

I. Pretend that the career services office at your university wants to develop a system that collects student résumés and makes them available to students and recruiters over the web. Students should be able to input their résumé information into a standard résumé template. The information then is presented in a résumé format, and it also is placed in a database that can be queried using an online search form. You have been put in charge of the project. Develop a plan for estimating the project. How long do you think it would take for you and three other students to complete the project? Provide support for the schedule that you propose.

J. Refer to the situation in exercise I. You have been told that recruiting season begins a month from today and that the new system must be used. How would you approach this situation? Describe what you can do as the project manager to make sure that your team does not burn out from unreasonable deadlines and commitments.

K. Consider the system described in exercise I. Create a workplan listing the tasks that will need to be completed to meet the project's objectives. Create a Gantt chart and a network diagram in a project management tool (e.g., Microsoft Project) or using a spreadsheet package to graphically show the high-level tasks of the project.

L. Suppose that you are in charge of the project that is described in exercise I and the project will be staffed by members of your class. Do your classmates have all the right skills to implement such a project? If not, how will you go about making sure that the proper skills are available to get the job done?

M. Complete a use-case point worksheet to estimate the effort to build the system described in exercises I, J, K, and L. You will need to make assumptions regarding the actors, the use cases, and the technical complexity and environmental factors.

N. Consider the application that is used at your school to register for classes. Complete a use-case point worksheet to estimate the effort to build such an application. You will need to make some assumptions about the application's interfaces and the various factors that affect its complexity.

O. Read Your Turn 2-6. Create a risk assessment that lists the potential risks associated with performing the project, along with ways to address the risks.

P. Pretend that your instructor has asked you and two friends to create a web page to describe the course to potential students and provide current class information (e.g., syllabus, assignments, readings) to current students. You have been assigned the role of leader, so you will need to coordinate your activities and those of your classmates until the project is completed. Describe how you would apply the project manage-ment techniques that you have learned in this chapter in this situation. Include descriptions of how you would create a workplan, staff the project, and coordi-nate all activities—yours and those of your classmates.

Q. Select two project management software packages and research them using the web or trade magazines. Describe the features of the two packages. If you were a project manager, which one would you use to help support your job? Why?

R. In 1997, Oxford Health Plans had a computer problem that caused the company to overestimate revenue and underestimate medical costs. Problems were caused by the migration of its claims processing system from the Pick operating system to a UNIX-based system that uses Oracle database software and hardware from Pyramid Technology. As a result, Oxford's stock price plummeted, and fixing the system became the number one priority for the company. Suppose that you have been placed in charge of managing the repair of the claims processing system. Obviously, the project team will not be in good spirits. How will you motivate team members to meet the project's objectives?

MINICASES

1. Jerry Larsen is a project manager at BrightFuture Insur-ance, an international insurance company with sub-sidiaries on four continents—America, Asia, Africa, and Europe—and over 25 countries, including US, Canada, Germany, Japan, India, and China. Recently, the company made the strategic decision to convert his web-based customer subscription system to the new mobile technology platforms. It has been determined that the mobile implementation should support Google's Android, Apple's iOS, and Microsoft's Windows Phone 7 devices. Jerry has been charged with this project and needs to come up with a project proposal that would be eligible for the strategic plan of BrightFuture Insurance. The situation is complicated by the fact that there is little experience in mobile development within the company; only one developer has expertise in mobile and she is a new employee fresh out of school. Furthermore, out of the team that developed the web-based version of the customer subscription system, only two of the analysts and one programmer are still with BrightFuture Insur-ance. How should Jerry proceed with the preparation of this project proposal? Who should he involve in this process? Why? What are the main risk elements and potential pitfalls of this project?

2. The Seisakusho Company operates a fleet of 20 trucks and crews that provide a variety of repair services to residential customers in Tokyo. Currently, it takes on average about six hours before a service team responds to a request. Each truck and crew averages 10 service calls per week, and the average revenue earned per service call is ¥20,000. Each truck is in service 50 weeks per year. Due to difficulty in scheduling and routing, there is considerable slack time for each truck and crew during a typical week.

To schedule the trucks and crews more efficiently and improve their productivity, the Seisakusho manage-ment is evaluating the installation of a routing and scheduling software package. The benefits would include reduced response time to service requests and result in more productive service teams. However, the manage-ment has difficulty quantifying these benefits.

One approach is to make an estimate of how much service response time will decrease with the new system, which then can be used to project the increase in the number of service calls made each week. For example, if the system permits the average service response time to fall to four hours, then each truck will be able to make 15 service calls per week on average—an increase of five calls per week. With each truck making five additional calls per week and the average revenue per call at ¥20,000, the revenue increase per truck per week is ¥100,000 (5 × ¥20,000). With 20 trucks in service 50 weeks per year, the average annual revenue increase will be ¥100,000,000 (¥100,000 × 20 × 50).

The Seisakusho management is unsure if the new system will enable response time to fall to four hours on average. Therefore, it has compiled the following data on possible outcomes of the new system:

New Response Time	# Calls/Truck/Week	Likelihood
3 hours	18	30%
4 hours	15	50%
5 hours	12	20%

Given these figures, prepare a spreadsheet model that computes the expected value of the annual revenues to be produced by this new system.

3. Jessica is a young American student from Wisconsin visiting as part of a semester abroad program at the Vrije Universiteit Amsterdam (Free University of Amsterdam). She has elected to take Prof. Vermeulen's project development course within the Computing Department, a much appreciated capstone class that, among other things, strives to reproduce for its students an environment that is as close as possible to what graduates are to expect in the real world. The students in this class work in teams on projects provided by the industry partners of the department who play the role of clients. The students are expected to interact directly with the representatives of their clients, organize their own team, manage and develop their project (from requirement identification to documentation and deployment), and deliver an application by the end of the semester-long activity. Aside from providing some very general guidelines on professional conduct (such as dressing appropriately and showing up on time for meetings), Prof. Vermeulen is priding himself on limiting his interventions in the relationship the students have with their clients. While Jessica enjoyed, and was appreciative of, this whole experience she also had her doubts, which she expressed to Prof. Vermeulen.

"Yeah, this is all good," Jessica said, "But it is still not what real life is going to be."

"Very well," said Prof. Vermeulen, "Why don't you write a brief essay highlighting your concerns on this issue?"

What are the main ideas around which Jessica should develop her essay?

4. HCL Technologies is a leading software outsourcing firm in India. It has been asked to bid on a small project by one of its US clients. The project will require HCL to write a custom production scheduling component to the large Enterprise Resource Planning (ERP) system that the client uses. Based on preliminary discussions with the client, HCL has developed this list of system elements:

Inputs: 1, low complexity; 2, medium complexity; 3, high complexity

Outputs: 4, medium complexity

Queries: 1, low complexity; 4, medium complexity; 4, high complexity

Files: 2, medium complexity

Program Interfaces: 3, medium complexity

Assume that an adjusted program complexity of 1.3 is appropriate for this project. Calculate the total adjusted function points for this project.

PART ONE

ANALYSIS MODELING

Analysis modeling answers the questions of *who* will use the system, *what* the system will do, and *where* and *when* it will be used. During analysis, detailed requirements are identified and a system proposal is created. The team then produces the functional model (use-case diagram, activity diagrams, and use-case descriptions), structural model (CRC cards and class diagram, and object diagrams), and behavioral models (sequence diagrams, communication diagrams, behavioral state machines, and a CRUDE matrix).

CHAPTER 3 ▦
REQUIREMENTS
DETERMINATION

CHAPTER 4 ▦
BUSINESS PROCESS
AND FUNCTIONAL
MODELING

CHAPTER 5 ▦
STRUCTURAL
MODELING

CHAPTER 6 ▦
BEHAVIORAL
MODELING

Requirements Definition

System Proposal

Use-Case Diagrams

Use-Case Descriptions

Activity Diagrams

CRC Cards

Object Diagrams

Class Diagrams

Sequence Diagrams

CRUDE Matrix

Communication Diagrams

Behavioral State Machines

CHAPTER 3

REQUIREMENTS DETERMINATION

One of the first activities of an analyst is to determine the business requirements for a new system. This chapter begins by presenting the requirements definition, a document that lists the new system's capabilities. It then describes how to analyze requirements using business process automation, business process improvement, and business process reengineering techniques and how to gather requirements using interviews, JAD sessions, questionnaires, document analysis, and observation. The chapter also describes a set of alternative requirements-documentation techniques and describes the system proposal document that pulls everything together.

OBJECTIVES

- Understand how to create a requirements definition
- Become familiar with requirements-analysis techniques
- Understand when to use each requirements-analysis technique
- Understand how to gather requirements using interviews, JAD sessions, questionnaires, document analysis, and observation
- Understand the use of concept maps, story cards, and task lists as requirements-documentation techniques
- Understand when to use each requirements-gathering technique
- Be able to begin creating a system proposal

CHAPTER OUTLINE

Introduction
Requirements Determination
  Defining a Requirement
  Requirements Definition
  Determining Requirements
  Creating a Requirements Definition
  Real-World Problems with
    Requirements Determination
Requirements Analysis Strategies
  Business Process Automation
  Business Process Improvement
  Business Process Reengineering
  Selecting the Appropriate
    Strategies

Requirements-Gathering Techniques
  Interviews
  Joint Application Development (JAD)
  Questionnaires
  Document Analysis
  Observation
  Selecting Appropriate Techniques
Alternative Requirements-Documentation
  Techniques
  Concept Maps
  Story Cards and Task Lists
The System Proposal
Applying the Concepts at CD Selections
Summary

INTRODUCTION

The systems development process aids an organization in moving from the current system (often called the *as-is system*) to the new system (often called the *to-be system*). The output of planning, discussed in Chapter 2, is the system request, which provides general ideas for the to-be system, defines the project's scope, and provides the initial workplan. Analysis takes the general ideas in the system request and refines them into a detailed requirements definition (this chapter), functional models (Chapter 4), structural models (Chapter 5), and behavioral models (Chapter 6) that together form the *system proposal*. The system proposal also includes revised project management deliverables, such as the feasibility analysis and the workplan (Chapter 2).

The system proposal is presented to the approval committee, who decides if the project is to continue. This usually happens at a system *walkthrough*, a meeting at which the concept for the new system is presented to the users, managers, and key decision makers. The goal of the walkthrough is to explain the system in moderate detail so that the users, managers, and key decision makers clearly understand it, can identify needed improvements, and can make a decision about whether the project should continue. If approved, the system proposal moves into design, and its elements (requirements definition and functional, structural, and behavioral models) are used as inputs to the steps in design. This further refines them and defines in much more detail how the system will be built.

The line between analysis and design is very blurry. This is because the deliverables created during analysis are really the first step in the design of the new system. Many of the major design decisions for the new system are found in the analysis deliverables. In fact, a better name for analysis is really analysis and initial design, but because this is a rather long name and because most organizations simply call it analysis, we do too. Nonetheless, it is important to remember that the deliverables from analysis are really the first step in the design of the new system.

In many ways, because it is here that the major elements of the system first emerge, the requirements-determination step is the single most critical step of the entire system development process. During requirements determination, the system is easy to change because little work has been done yet. As the system moves through the system development process, it becomes harder and harder to return to requirements determination and to make major changes because of all of the rework that is involved. Several studies have shown that more than half of all system failures are due to problems with the requirements.[1] This is why the iterative approaches of many object-oriented methodologies are so effective—small batches of requirements can be identified and implemented in incremental stages, allowing the overall system to evolve over time. In this chapter, we focus on the requirements workflow of the Unified Process. We begin by explaining what a requirement is and the overall process of requirements gathering and requirements analysis. We then present a set of techniques that can be used to analyze and gather requirements.

REQUIREMENTS DETERMINATION

The purpose of *requirements determination* is to turn the very high-level explanation of the business requirements stated in the system request into a more precise list of requirements that can be used as inputs to the rest of analysis (creating functional, structural, and behavioral models). This expansion of the requirements ultimately leads to the design of the system. However, the most difficult aspect of determining the actual requirements is analogous to the story of the blind men and the elephant (see Figure 3-1). In this story, depending on which part of the elephant each blind man touches, each "sees" the elephant

---

[1] For example, see *The Scope of Software Development Project Failures* (Dennis, MA: The Standish Group, 1995).

It was six men of Indostan
To learning much inclined,
Who went to see the Elephant
(Though all of them were blind),
That each by observation
Might satisfy his mind

The First approached the Elephant,
And happening to fall
Against his broad and sturdy side,
At once began to bawl:
God bless me! but the Elephant
Is very like a wall!

The Second, feeling of the tusk,
Cried, Ho! what have we here
So very round and smooth and sharp?
To me tis mighty clear
This wonder of an Elephant
Is very like a spear!

The Third approached the animal,
And happening to take
The squirming trunk within his hands,
Thus boldly up and spake:
I see, quoth he, the Elephant
Is very like a snake!

The Fourth reached out an eager hand,
And felt about the knee.
What most this wondrous beast is like
Is mighty plain, quoth he;
'Tis clear enough the Elephant
Is very like a tree!

The Fifth, who chanced to touch the ear,
Said: Even the blindest man
Can tell what this resembles most;
Deny the fact who can
This marvel of an Elephant
Is very like a fan!?

The Sixth no sooner had begun
About the beast to grope,
Than, seizing on the swinging tail
That fell within his scope,
I see, quoth he, the Elephant
Is very like a rope!

And so these men of Indostan
Disputed loud and long,
Each in his own opinion
Exceeding stiff and strong,
Though each was partly in the right,
And all were in the wrong!

Moral:
So oft in theologic wars,
The disputants, I ween,
Rail on in utter ignorance
Of what each other mean,
And prate about an Elephant
Not one of them has seen!

– John Godfrey Saxe

FIGURE 3-1

The Blind Men and the
Elephant

differently. In many ways, the analyst is like one of the blind men. Depending on which part of the proverbial elephant the analyst touches, the analyst sees the requirements differently. Also, like the blind men, the analyst may only be able to perceive the individual part in a biased manner. Therefore, the analyst must be on guard to prevent the poor elephant (requirements) from being misrepresented.

Defining a Requirement

A *requirement* is simply a statement of what the system must do or what characteristic it must have. During analysis, requirements are written from the perspective of the businessperson, and they focus on the "what" of the system. Because they focus on the needs of the business user, they are usually called *business requirements* (and sometimes *user requirements*). Later in design, business requirements evolve to become more technical, and they describe how the system will be implemented. Requirements in design are written from the developer's perspective, and they are usually called *system requirements.*

Before we continue, we want to stress that there is no black-and-white line dividing a business requirement and a system requirement—and some companies use the terms interchangeably. The important thing to remember is that a requirement is a statement of what the system must do, and requirements will change over time as the project moves from inception to elaboration to construction. Requirements evolve from detailed statements of the business capabilities that a system should have to detailed statements of the technical way the capabilities will be implemented in the new system.

Requirements can be either functional or nonfunctional in nature. A *functional requirement* relates directly to a process a system has to perform or information it needs to contain. For example, requirements that state that a system must have the ability to search for available inventory or to report actual and budgeted expenses are functional requirements. Functional requirements flow directly into the creation of functional, structural, and behavioral models that represent the functionality of the evolving system.

Nonfunctional requirements refer to behavioral properties that the system must have, such as performance and usability. The ability to access the system using a Web browser is considered a nonfunctional requirement. Nonfunctional requirements can influence the rest of analysis (functional, structural, and behavioral models) but often do so only indirectly; nonfunctional requirements are used primarily in design when decisions are made about the user interface, the hardware and software, and the system's underlying physical architecture.

Figure 3-2 lists different kinds of nonfunctional requirements and examples of each kind. Notice that the nonfunctional requirements describe a variety of characteristics regarding the system: operational, performance, security, and cultural and political. For example, the project team needs to know if a system must be highly secure, requires sub-second response time, or has to reach a multicultural customer base.

These characteristics do not describe business processes or information, but they are very important in understanding what the final system should be like. Nonfunctional requirements primarily affect decisions that will be made during the design of a system. We will return to this topic later in the book when we discuss design. The goal in this chapter is to identify any major issues.

Four topics that have influenced information system requirements are the Sarbanes-Oxley Act, COBIT (Control OBjectives for Information and related Technology) compliance, ISO 9000 compliance, and Capability Maturity Model compliance. Depending on the system being considered, these four topics could affect the definition of a system's functional requirements, nonfunctional requirements, or both. The Sarbanes-Oxley Act, for example, mandates additional functional and nonfunctional requirements. These include

Nonfunctional Requirement	Description	Examples
Operational	The physical and technical environments in which the system will operate	■ The system should be able to fit in a pocket or purse. ■ The system should be able to integrate with the existing inventory system. ■ The system should be able to work on any web browser.
Performance	The speed, capacity, and reliability of the system	■ Any interaction between the user and the system should not exceed 2 seconds. ■ The system should receive updated inventory information every 15 minutes. ■ The system should be available for use 24 hours per day, 365 days per year.
Security	Who has authorized access to the system under what circumstances	■ Only direct managers can see personnel records of staff. ■ Customers can see their order history only during business hours.
Cultural and political	Cultural, political factors, and legal requirements that affect the system	■ The system should be able to distinguish between United States and European currency. ■ Company policy says that we buy computers only from Dell. ■ Country managers are permitted to authorize customer user interfaces within their units. ■ The system shall comply with insurance industry standards.

Source: The Atlantic Systems Guild, http://www.systemsguild.com/GuildSite/Robs/Template.html

FIGURE 3-2   Nonfunctional Requirements

additional security concerns (nonfunctional) and specific information requirements that management must now provide (functional). When developing financial information systems, information system developers should be sure to include Sarbanes-Oxley expertise in the development team. In another example, a client could insist on COBIT compliance, ISO 9000 compliance, or that a specific Capability Maturity Model level had been reached for the firm to be considered as a possible vendor to supply the system under consideration. Obviously, these types of requirements add to the nonfunctional requirements. Further discussion of these topics is beyond the scope of this book.[2]

Another recent topic that influences requirements for some systems is the whole area of globalization. The idea of having a global information supply chain brings to bear a large number of additional nonfunctional requirements. For example, if the necessary operational environments do not exist for a mobile solution to be developed, it is important to adapt the solution to the local environment. Or, it may not be reasonable to expect to deploy a high-technology-based solution in an area that does not have the necessary power and communications infrastructure. In some cases, we may need to consider some parts of the global information supply chain to be supported with manual—rather than automated—information systems.

[2]A concise discussion of the Sarbanes-Oxley Act is presented in G. P. Lander, *What is Sarbanes-Oxley?* (New York: McGraw-Hill, 2004). A good reference for Sarbanes-Oxley Act–based security requirements is D. C. Brewer, *Security Controls for Sarbanes-Oxley Section 404 IT Compliance: Authorization, Authentication, and Access* (Indianapolis, IN: Wiley, 2006). For detailed information on COBIT, see www.isaca.org; for ISO 9000, see www.iso.org; and for details on the Capability Maturity Model, see www.sei.cmu.edu/cmmi/.

YOUR TURN

3-1 Identifying Requirements

One of the most common mistakes by new analysts is to confuse functional and nonfunctional requirements. Pretend that you received the following list of requirements for a sales system.

Requirements for Proposed System:
The system should

1. be accessible to the web users;
2. include the company standard logo and color scheme;
3. restrict access to profitability information;
4. include actual and budgeted cost information;
5. provide management reports;
6. include sales information that is updated at least daily;
7. have two-second maximum response time for predefined queries and ten-minute maximum response time for ad hoc queries;
8. include information from all company subsidiaries;
9. print subsidiary reports in the primary language of the subsidiary;
10. provide monthly rankings of salesperson performance.

Questions

1. Which requirements are functional business requirements? Provide two additional examples.
2. Which requirements are nonfunctional business requirements? What kind of nonfunctional requirements are they? Provide two additional examples.

Manual systems have an entirely different set of requirements that create different performance expectations and additional security concerns. Furthermore, cultural and political concerns are potentially paramount. A simple example that affects the design of user interfaces is the proper use of color on forms (on a screen or paper). Different cultures interpret different colors differently. In other words, in a global, multicultural business environment, addressing cultural concerns goes well beyond simply having a multilingual user interface. We must be able to adapt the global solution to the local realities. Friedman

CONCEPTS IN ACTION

3-A What Can Happen If You Ignore Nonfunctional Requirements

I once worked on a consulting project in which my manager created a requirements definition without listing nonfunctional requirements. The project was then estimated based on the requirements definition and sold to the client for $5,000. In my manager's mind, the system that we would build for the client would be a very simple stand-alone system running on current technology. It shouldn't take more than a week to analyze, design, and build.

Unfortunately, the clients had other ideas. They wanted the system to be used by many people in three different departments, and they wanted the ability for any number of people to work on the system concurrently. The technology they had in place was antiquated; nonetheless, they wanted the system to run effectively on the existing equipment. Because we didn't set the project scope properly by including our assumptions about nonfunctional requirements in the requirements definition, we basically had to do whatever they wanted.

The capabilities they wanted took weeks to design and program. The project ended up taking four months, and the final project cost was $250,000. Our company had to pick up the tab for everything except the agreed-upon $5,000. This was by far the most frustrating project situation I ever experienced.

—*Barbara Wixom*

Nonfunctional Requirements

1. Operational Requirements
1.1. The system will operate in Windows environment.
1.2. The system should be able to connect to printers wirelessly.
1.3. The system should automatically back up at the end of each day.

2. Performance Requirements
2.1. The system will store a new appointment in 2 seconds or less.
2.2. The system will retrieve the daily appointment schedule in 2 seconds or less.

3. Security Requirements
3.1. Only doctors can set their availability.
3.2. Only a manager can produce a schedule.

4. Cultural and Political Requirements
4.1. No special cultural and political requirements are anticipated.

Functional Requirements

1. Manage Appointments
1.1. Patient makes new appointment.
1.2. Patient changes appointment.
1.3. Patient cancels appointment.

2. Produce Schedule
2.1. Office Manager checks daily schedule.
2.2. Office Manager prints daily schedule.

3. Record Doctor Availability
3.1. Doctor updates schedule

FIGURE 3-3
Sample Requirements
Definition

refers to these concerns as glocalization.[3] Otherwise, we will simply create another example of a failed information system development project.

Requirements Definition

The requirements definition report—usually just called the *requirements definition*—is a straightforward text report that simply lists the functional and nonfunctional requirements in an outline format. Figure 3-3 shows a sample requirements definition for an appointment system for a typical doctor's office. Notice it contains both functional and nonfunctional requirements. The functional requirements include managing appointments, producing schedules, and recording the availability of the individual doctors. The nonfunctional

---

[3] T. L. Friedman, *The World is Flat: A Brief History of the Twenty-First Century, Updated and Expanded Edition.* (New York: Farrar, Straus, and Giroux, 2006). For a criticism of Friedman's view, see R. Aronica and M. Ramdoo, *The World is FLAT? A Critical Analysis of Thomas L. Friedman's New York Times Bestseller* (Tampa, FL: Meghan-Kiffer Press, 2006).

requirements includes items such as the expected amount of time that it takes to store a new appointment, the need to support wireless printing, and which types of employees have access to the different parts of the system.

The requirements are numbered in a legal or outline format so that each requirement is clearly identified. The requirements are first grouped into functional and nonfunctional requirements; within each of those headings, they are further grouped by the type of nonfunctional requirement or by function.

Sometimes, business requirements are prioritized on the requirements definition. They can be ranked as having high, medium, or low importance in the new system, or they can be labeled with the version of the system that will address the requirement (e.g., release 1, release 2, release 3). This practice is particularly important when using object-oriented methodologies because they deliver requirements in batches by developing incremental versions of the system.

The most obvious purpose of the requirements definition is to provide the information needed by the other deliverables in analysis, which include functional, structural, and behavioral models, and to support activities in design. The most important purpose of the requirements definition, however, is to define the scope of the system. The document describes to the analysts exactly what the system needs to end up doing. When discrepancies arise, the document serves as the place to go for clarification.

Determining Requirements

Determining requirements for the requirements definition is both a business task and an information technology task. In the early days of computing, there was a presumption that the systems analysts, as experts with computer systems, were in the best position to define how a computer system should operate. Many systems failed because they did not adequately address the true business needs of the users. Gradually, the presumption changed so that the users, as the business experts, were seen as being in the best position to define how a computer system should operate. However, many systems failed to deliver performance benefits because users simply automated an existing inefficient system, and they failed to incorporate new opportunities offered by technology.

A good analogy is building a house or an apartment. We have all lived in a house or apartment, and most of us have some understanding of what we would like to see in one. However, if we were asked to design one from scratch, it would be a challenge because we lack appropriate design skills and technical engineering skills. Likewise, an architect acting alone would probably miss some of our unique requirements.

Therefore, the most effective approach is to have both business people and analysts working together to determine business requirements. Sometimes, however, users don't know exactly what they want, and analysts need to help them discover their needs. Three kinds of strategies have become popular to help analysts do this: *business process automation (BPA)*, *business process improvement (BPI)*, and *business process reengineering (BPR)*. Analysts can use these tools when they need to guide the users in explaining what is wanted from a system.

The three kinds of strategies work similarly. They help users critically examine the current state of systems and processes (the as-is system), identify exactly what needs to change, and develop a concept for a new system (the to-be system). A different amount of change is associated with each technique; BPA creates a small amount of change, BPI creates a moderate amount of change, and BPR creates significant change that affects much of the organization.

Although BPA, BPI, and BPR enable the analyst to help users create a vision for the new system, they are not sufficient for extracting information about the detailed business requirements that are needed to build it. Therefore, analysts use a portfolio of requirements-gathering techniques to acquire information from users. The analyst has many techniques from which to choose: interviews, questionnaires, observation, joint application development (JAD), and document analysis. The information gathered using these techniques is critically analyzed and used to craft the requirements definition report. A later section of this chapter describes each of the requirements-gathering techniques in greater depth.

Creating a Requirements Definition

Creating a requirements definition is an iterative and ongoing process whereby the analyst collects information with requirements-gathering techniques (e.g., interviews, document analysis), critically analyzes the information to identify appropriate business requirements for the system, and adds the requirements to the requirements definition report. The requirements definition is kept up to date so that the project team and business users can refer to it and get a clear understanding of the new system.

To create a requirements definition, the project team first determines the kinds of functional and nonfunctional requirements that they will collect about the system (of course, these may change over time). These become the main sections of the document. Next, the analysts use a variety of requirements-gathering techniques (e.g., interviews, observation) to collect information, and they list the business requirements that were identified from that information. Finally, the analysts work with the entire project team and the business users to verify, change, and complete the list and to help prioritize the importance of the requirements that were identified.

This process continues throughout analysis, and the requirements definition evolves over time as new requirements are identified and as the project moves into later phases of the Unified Process. Beware: The evolution of the requirements definition must be carefully managed. The project team cannot keep adding to the requirements definition, or the system will keep growing and growing and never get finished. Instead, the project team carefully identifies requirements and evaluates which ones fit within the scope of the system. When a requirement reflects a real business need but is not within the scope of the current system or current release, it is either added on a list of future requirements or given a low priority. The management of requirements (and system scope) is one of the hardest parts of managing a project.

Real-World Problems with Requirements Determination

Avison and Fitzgerald provide us with a set of problems that can arise with regard to determining the set of requirements to be dealt with. [4] First, the analyst might not have access to the correct set of users to uncover the complete set of requirements. This can lead to requirements being missed, misrepresented, and/or overspecified. This is analogous to the blind men and the elephant metaphor described earlier. Second, the specification of the requirements may be inadequate. This can be especially true with the lightweight techniques associated with agile methodologies. Third, some requirements are simply unknowable at the beginning of a development process. However, as the system is developed, the users and analysts will get a better understanding of both the domain issues and the applicable technology. This can cause new functional and nonfunctional requirements to be identified and current requirements to evolve or be canceled. Iterative and incremental-based development methodologies, such as the Unified Process and agile, can help in this

---

[4] See D. Avison and G. Fitzgerald, *Information Systems Development: Methodologies, Techniques, & Tools*, 4th Ed. (London: McGraw-Hill, 2006).

case. Fourth, verifying and validating of requirements can be very difficult. We take up this topic in the chapters that deal with the creation of functional (Chapter 4), structural (Chapter 5), and behavioral (Chapter 6) models.

REQUIREMENTS ANALYSIS STRATEGIES

Before the project team can determine what requirements are appropriate for a given system, they need to have a clear vision of the kind of system that will be created and the level of change that it will bring to the organization. The basic process of *analysis* is divided into three steps: understanding the as-is system, identifying improvements, and developing requirements for the to-be system.

Sometimes the first step (i.e., understanding the as-is system) is skipped or is performed in a cursory manner. This happens when no current system exists, if the existing system and processes are irrelevant to the future system, or if the project team is using a RAD or agile development methodology in which the as-is system is not emphasized. Users of traditional design methods such as waterfall and parallel development (see Chapter 1) typically spend significant time understanding the as-is system and identifying improvements before moving to capture requirements for the to-be system. However, newer RAD, agile, and object-oriented methodologies, such as phased development, prototyping, throwaway prototyping, extreme programming, and Scrum (see Chapter 1) focus almost exclusively on improvements and the to-be system requirements, and they spend little time investigating the current as-is system.

Three requirements analysis strategies—business process automation, business process improvement, and business process reengineering—help the analyst lead users through the analysis steps so that the vision of the system can be developed. Requirements analysis strategies and requirements-gathering techniques go hand in hand. Analysts need to use requirements-gathering techniques to collect information; requirements analysis strategies drive the kind of information that is gathered and how it is ultimately analyzed. Although we now focus on the analysis strategies and then discuss requirements gathering at the end of the chapter, they happen concurrently and are complementary activities.

The choice of analysis technique to be used is based on the amount of change the system is meant to create in the organization. BPA is based on small change that improves process efficiency, BPI creates process improvements that lead to better effectiveness, and BPR revamps the way things work so that the organization is transformed on some level.

To move the users from here to there, an analyst needs strong *critical thinking skills.* Critical thinking is the ability to recognize strengths and weaknesses and recast an idea in an improved form, and critical thinking skills are needed to really understand issues and develop new business processes. These skills are also needed to thoroughly examine the results of requirements gathering, to identify business requirements, and to translate those requirements into a concept for the new system.

Business Process Automation (BPA)

BPA leaves the basic way the organization operates unchanged and uses computer technology to do some of the work. BPA can make the organization more efficient but has the least impact on the business. Planners in BPA projects spend a significant time understanding the current as-is system before moving on to improvements and to-be system requirements. Problem analysis and root cause analysis are two popular BPA techniques.

Problem Analysis   The most straightforward (and probably the most commonly used) requirements analysis technique is *problem analysis*. Problem analysis means asking the users and managers to identify problems with the as-is system and to describe how to solve them in the to-be system. Most users have a very good idea of the changes they would like to see, and most are quite vocal about suggesting them. Most changes tend to solve problems rather than capitalize on opportunities, but the latter is possible as well. Improvements from problem analysis tend to be small and incremental (e.g., provide more space in which to type the customer's address; provide a new report that currently does not exist).

This type of improvement often is very effective at improving a system's efficiency or ease of use. However, it often provides only minor improvements in business value—the new system is better than the old, but it may be hard to identify significant monetary benefits from the new system.

Root Cause Analysis   The ideas produced by problem analysis tend to be solutions to problems. All solutions make assumptions about the nature of the problem, assumptions that might or might not be valid. In our experience, users (and most people in general) tend to quickly jump to solutions without fully considering the nature of the problem. Sometimes the solutions are appropriate, but many times they address a *symptom* of the problem, not the true problem or *root cause* itself.[5]

For example, suppose a firm notices that its users report that inventory stock-outs are common. The cost of inventory stock-outs can be quite significant. For example, in this case, because they happen frequently, customers could find another source for the items that they are purchasing from the firm. It is in the firm's interest to determine the underlying cause and not simply provide a knee-jerk reaction such as arbitrarily increasing the amount of inventory kept on hand. In the business world, the challenge lies in identifying the root cause—few real-world problems are simple. The users typically propose a set of causes for the problem under consideration. The solutions that users propose can address either symptoms or root causes, but without a careful analysis, it is difficult to tell which one is addressed. The analyst must keep in mind the parable of the blind men and the elephant, where the blind men, in this case, are the users.

Root cause analysis, therefore, focuses on problems, not solutions. The analyst starts by having the users generate a list of problems with the current system and then prioritize the problems in order of importance. Starting with the most important, the users and/or the analysts then generate all the possible root causes for the problems. Each possible root cause is investigated (starting with the most likely or easiest to check) until the true root causes are identified. If any possible root causes are identified for several problems, those should be investigated first, because there is a good chance they are the real root causes influencing the symptom problems. In our example, there are several possible root causes:

- The firm's supplier might not be delivering orders to the firm in a timely manner.
- There could be a problem with the firm's inventory controls.
- The reorder level and quantities could be set wrong.

Sometimes, using a hierarchical chart to represent the causal relationships helps with the analysis. As Figure 3-4 shows, there are many possible root causes that underlie the higher-level causes identified. The key point in root cause analysis is always to challenge the obvious.

---

[5] Two good books that discuss the difficulty in finding the root causes to problems are: E. M. Goldratt and J. Cox, *The Goal* (Croton-on-Hudson, NY: North River Press, 1986); and E. M. Goldratt, *The Haystack Syndrome* (Croton-on-Hudson, NY: North River Press, 1990).

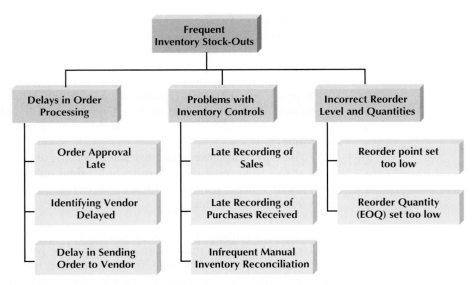

FIGURE 3-4   Root Cause Analysis for Inventory Stock-Outs

CONCEPTS   **3-B Success from Failure**

IN ACTION

*I*n the gold rush days of the late 1990s, getting on the Internet was a hot topic. Many companies (many of which no longer exist) created computers for the home Internet market, many with built-in dial-up connectivity and contracts for that connectivity. The AtHome company made such an Internet appliance. Taken out of the box, connected to a phone line, and provided with initial start-up, it "phoned home" (made the connection to an Internet service provider).

But, the days of the Internet appliance were short lived. Consumers wanted more than just Internet access—they wanted to be able to have and share files, photos, and materials with others. The basic AtHome Internet Appliance did not have any storage and was useful only for connecting to the Internet (by phone modem) and browsing the Internet.

The stock price dropped and sales dropped; even with prices that were comparable to giving away the device, consumers were no longer interested as 2001 came to an end.

When faced with such a situation, what does a company do? The company faced a real challenge—go out of business or reorganize. In this case, AtHome, which had expertise in hardware and telecommunications, restructured into a security company. With September 11, 2001, bringing calls for better security, AtHome scrambled to create a hardware device that would sit between the Internet connection and a business network. To keep their stock listed on the New York Stock Exchange, AtHome did a 15-to-1 reverse stock split and changed their name to indicate a new focus. Having been burned by consumers' whims, they set their sights on capturing major corporate business. It took two long years before their device started to be noticed—and just paying the employees almost ate up the available funds before sales of the new device started to kick in. The reorganized company is now recognized as a leader in the intrusion-prevention field.

A company that falls from consumer favor cannot always restructure itself to become successful in an alternative area. In this case, there was success from failure.

Questions

1. When should a company that has lost in the consumer marketplace re-create itself for the corporate market?
2. How might a systems analyst for the AtHome company learn to change with the times and adapt to the new environment?

Business Process Improvement (BPI)

BPI makes moderate changes to the way the organization operates in order to take advantage of new opportunities offered by technology or to copy what competitors are doing. BPI can improve efficiency (i.e., doing things right) and improve effectiveness (i.e., doing the right things). Planners of BPI projects also spend time understanding the as-is system, but much less time than with BPA projects; their primary focus is on improving business processes, so time is spent on the as-is only to help with the improvement analyses and the to-be system requirements. *Duration analysis, activity-based costing,* and *informal benchmarking* are three popular BPI activities.

Duration Analysis   Duration analysis requires a detailed examination of the amount of time it takes to perform each process in the current as-is system. The analysts begin by determining the total amount of time it takes, on average, to perform a set of business processes for a typical input. They then time each of the individual steps (or subprocesses) in the business process. The time to complete the basic steps are then totaled and compared to the total for the overall process. A significant difference between the two—and in our experience the total time often can be 10 or even 100 times longer than the sum of the parts—indicates that this part of the process is badly in need of a major overhaul.

For example, suppose that the analysts are working on a home mortgage system and discover that on average, it takes thirty days for the bank to approve a mortgage. They then look at each of the basic steps in the process (e.g., data entry, credit check, title search, appraisal) and find that the total amount of time actually spent on each mortgage is about eight hours. This is a strong indication that the overall process is badly broken, because it takes thirty days to perform one day's work.

These problems probably occur because the process is badly fragmented. Many different people must perform different activities before the process finishes. In the mortgage example, the application probably sits on many people's desks for long periods of time before it is processed.

Processes in which many different people work on small parts of the inputs are prime candidates for *process integration* or *parallelization*. Process integration means changing the fundamental process so that fewer people work on the input, which often requires changing the processes and retraining staff to perform a wider range of duties. Process parallelization means changing the process so that all the individual steps are performed at the same time. For example, in the mortgage application case, there is probably no reason that the credit check cannot be performed at the same time as the appraisal and title check.

CONCEPTS   **3-C Duration Analysis**

IN ACTION

A group of executives from a Fortune 500 company used duration analysis to discuss their procurement process. Using a huge wall of Velcro and a handful of placards, a facilitator mapped out the company's process for procuring a $50 software upgrade. Having quantified the time it took to complete each step, she then assigned costs based on the salaries of the employees involved. The fifteen-minute exercise left the group stunned. Their procurement process had gotten so convoluted that it took eighteen days, countless hours of paperwork, and nearly $22,000 in employee time to get the product ordered, received, and up and running on the requester's desktop.

Source: "For Good Measure" Debby Young, *CIO Magazine* (March 1, 1999).

Activity-Based Costing   Activity-based costing is a similar analysis; it examines the cost of each major process or step in a business process rather than the time taken.[6] The analysts identify the costs associated with each of the basic functional steps or processes, identify the most costly processes, and focus their improvement efforts on them.

Assigning costs is conceptually simple. Analysts simply examine the direct cost of labor and materials for each input. Materials costs are easily assigned in a manufacturing process, whereas labor costs are usually calculated based on the amount of time spent on the input and the hourly cost of the staff. However, as you may recall from a managerial accounting course, there are indirect costs such as rent, depreciation, and so on, that also can be included in activity costs.

Informal Benchmarking   Benchmarking refers to studying how other organizations perform a business process in order to learn how your organization can do something better. Benchmarking helps the organization by introducing ideas that employees may never have considered but that have the potential to add value.

Informal benchmarking is fairly common for customer-facing business processes (i.e., processes that interact with the customer). With informal benchmarking, the managers and analysts think about other organizations or visit them as customers to watch how the business process is performed. In many cases, the business studied may be a known leader in the industry or simply a related firm. For example, suppose the team is developing a website for a car dealer. The project sponsor, key managers, and key team members would likely visit the websites of competitors as well as those of others in the car industry (e.g., manufacturers, accessories suppliers) and those in other industries that have won awards for their websites.

Business Process Reengineering

BPR means changing the fundamental way the organization operates, obliterating the current way of doing business and making major changes to take advantage of new ideas and new technology. Planners of BPR projects spend little time understanding the as-is, because their goal is to focus on new ideas and new ways of doing business. Outcome analysis, technology analysis, and activity elimination are three popular BPR activities.

Outcome Analysis   *Outcome analysis* focuses on understanding the fundamental outcomes that provide value to customers. Although these outcomes sound as though they should be obvious, they often are not. For example, consider an insurance company. One of its customers has just had a car accident. What is the fundamental outcome from the *customer's* perspective? Traditionally, insurance companies have answered this question by assuming the customer wants to receive the insurance payment quickly. To the customer, however, the payment is only a *means* to the real outcome: a repaired car. The insurance company might benefit by extending its view of the business process past its traditional boundaries to include not paying for repairs but performing the repairs or contracting with an authorized body shop to do them.

With this approach, system analysts encourage the managers and project sponsor to pretend they are customers and to think carefully about what the organization's products and services enable the customers to do—and what they *could* enable the customer to do.

Technology Analysis   Many major changes in business since the turn of the century have been enabled by new technologies. *Technology analysis* starts by having the analysts and

---

[6] Many books have been written on activity-based costing. Useful ones include K. B. Burk and D. W. Webster, *Activity-Based Costing* (Fairfax, VA: American Management Systems, 1994); and D. T. Hicks, *Activity-Based Costing: Making It Work for Small and Mid-sized Companies* (New York: Wiley, 1998). The two books by Eli Goldratt mentioned previously (*The Goal* and *The Haystack Syndrome*) also offer unique insights into costing.

managers develop a list of important and interesting technologies. Then the group systematically identifies how every technology could be applied to the business process and identifies how the business would benefit.

For example, one useful technology is the Internet. Saturn, the car manufacturer, took this idea and developed an extranet application for its suppliers. Rather than ordering parts for its cars, Saturn made its production schedule available electronically to its suppliers, who shipped the parts Saturn needed so that they arrived at the plant just in time. This saved Saturn significant costs because it eliminated the need for people to monitor the production schedule and issue purchase orders.

Activity Elimination   *Activity elimination* is exactly what it sounds like. The analysts and managers work together to identify how the organization could eliminate each activity in the business process, how the function could operate without it, and what effects are likely to occur. Initially, managers are reluctant to conclude that processes can be eliminated, but this is a force-fit exercise in that they must eliminate each activity. In some cases, the results are silly; nonetheless, participants must address every activity in the business process.

For example, in the home mortgage approval process discussed earlier, the managers and analysts would start by eliminating the first activity, entering the data into the mortgage company's computer. This leads to two obvious possibilities: eliminate the use of a computer system or make someone else do the data entry (e.g., the customer over the web). They would then eliminate the next activity, the credit check. Silly, right? After all, making sure the applicant has good credit is critical in issuing a loan. Not really. The real answer depends upon how many times the credit check identifies bad applications. If all or almost all applicants have good credit and are seldom turned down by a credit check, then the cost of the credit check might not be worth the cost of the few bad loans it prevents. Eliminating it might actually result in lower costs, even with the cost of bad loans.

Selecting Appropriate Strategies

Each technique discussed in this chapter has its own strengths and weaknesses (see Figure 3-5). No one technique is inherently better than the others, and in practice, most projects use a combination of techniques.

Potential Business Value   *Potential business value* varies with the analysis strategy. Although BPA has the potential to improve the business, most of the benefits from BPA are tactical and small. Because BPA does not seek to change the business processes, it can only improve their efficiency. BPI usually offers moderate potential benefits, depending upon the scope of the project, because it seeks to change the business in some way. It can increase both efficiency and effectiveness. BPR creates large *potential* benefits because it seeks to radically improve the nature of the business.

Project Cost   *Project cost* is always important. In general, BPA has the lowest cost because it has the narrowest focus and seeks to make the fewest changes. BPI can be moderately expensive, depending upon the scope of the project. BPR is usually expensive, because of the amount of time required of senior managers and the amount of redesign to business processes.

Breadth of Analysis   *Breadth of analysis* refers to the scope of analysis, or whether analysis includes business processes within a single business function, processes that cross the organization, or processes that interact with those in customer or supplier organizations. BPR takes a broad perspective, often spanning several major business processes, even across multiple organizations. BPI has a much narrower scope that usually includes one or several business functions. BPA typically examines a single process.

YOUR TURN

3-2 IBM Credit

*I*BM Credit was a wholly owned subsidiary of IBM responsible for financing mainframe computers sold by IBM. Although some customers bought mainframes outright or obtained financing from other sources, financing computers provided significant additional profit.

When an IBM sales representative made a sale, he or she would immediately call IBM Credit to obtain a financing quote. The call was received by a credit officer, who would record the information on a request form. The form would then be sent to the credit department to check the customer's credit status. This information would be recorded on the form, which was then sent to the business practices department, who would write a contract (sometimes reflecting changes requested by the customer). The form and the contract would then go to the pricing department, which used the credit information to establish an interest rate and recorded it on the form. The form and contract were then sent to the clerical group, where an administrator would prepare a cover letter quoting the interest rate and send the letter and contract via Federal Express to the customer.

The problem at IBM Credit was a major one. Getting a financing quote took anywhere from four to eight days (six days on average), giving the customer time to rethink the order or find financing elsewhere. While the quote was being prepared, sales representatives would often call to find out where the quote was in the process so they could tell the customer when to expect it. However, no one at IBM Credit could answer the question because the paper forms could be in any department, and it was impossible to locate one without physically walking through the departments and going through the piles of forms on everyone's desk.

IBM Credit examined the process and changed it so that each credit request was logged into a computer system and each department could record an application's status as they completed it and sent it to the next department. In this way, sales representatives could call the credit office and quickly learn the status of each application. IBM used some sophisticated management science queuing theory analysis to balance workloads and staff across the different departments so none would be overloaded. They also introduced performance standards for each department (e.g., the pricing decision had to be completed within one day after that department received an application).

However, process times got worse, even though each department was achieving almost 100% compliance on its performance goals. After some investigation, managers found that when people got busy, they conveniently found errors that forced them to return credit requests to the previous department for correction, thereby removing it from their time measurements.

Questions

1. What techniques can you use to identify improvements?
2. Choose one technique and apply it to this situation. What improvements did you identify?

Source: M. Hammer and J. Champy, *Reengineering the Corporation* (1993). New York, NY: Harper Business.

Risk   One final issue is *risk* of failure, which is the likelihood of failure due to poor design, unmet needs, or too much change for the organization to handle. BPA and BPI have low to moderate risk because the to-be system is fairly well defined and well understood, and its potential impact on the business can be assessed before it is implemented. BPR projects, on the other hand, are less predictable. BPR is extremely risky and is not something to be

	Business Process Automation	Business Process Improvement	Business Process Reengineering
Potential business value	Low–moderate	Moderate	High
Project cost	Low	Low–moderate	High
Breadth of analysis	Narrow	Narrow–moderate	Very broad
Risk	Low–moderate	Low–moderate	Very high

FIGURE 3-5

Characteristics of Analysis Strategies

undertaken unless the organization and its senior leadership are committed to making significant changes. Mike Hammer, the father of BPR, estimates that 70% of BPR projects fail.

YOUR

TURN

3-3 Analysis Strategy

Suppose you are the analyst charged with developing a new website for a local car dealer who wants to be very innovative and try new things. What analysis strategies would you recommend? Why?

CONCEPTS

IN ACTION

3-D Implementing a Satellite Data Network

A major retail store recently spent $24 million on a large private satellite communication system. The system provides state-of-the-art voice, data, and video transmission between stores and regional headquarters. When an item is sold, the scanner software updates the inventory system in real time. As a result, store transactions are passed on to regional and national headquarters instantly, which keeps inventory records up to date. One of their major competitors has an older system, where transactions are uploaded at the end of a business day. The first company feels such instant communication and feedback allows them to react more quickly to changes in the market and gives them a competitive advantage. For example, if an early winter snowstorm causes stores across the upper Midwest to start selling high-end (and high-profit) snowblowers, the nearest warehouse can quite quickly prepare next-day shipments to maintain a good inventory balance, whereas the competitor might not move quite as quickly and thus will lose out on such quick inventory turnover.

Questions

1. Do you think a $24 million investment in a private satellite communication system could be justified by a cost–benefit analysis? Could this be done with a standard communication line (with encryption)?
2. How might the competitor in this example attempt to close the information gap?

REQUIREMENTS-GATHERING TECHNIQUES

An analyst is very much like a detective (and business users are sometimes like elusive suspects). He or she knows that there is a problem to be solved and therefore must look for clues that uncover the solution. Unfortunately, the clues are not always obvious (and are often missed), so the analyst needs to notice details, talk with witnesses, and follow leads just as Sherlock Holmes would have done. The best analysts thoroughly gather requirements using a variety of techniques and make sure that the current business processes and the needs for the new system are well understood before moving into design. Analysts don't want to discover later that they have key requirements wrong—such surprises late in the development process can cause all kinds of problems.

The requirements-gathering process is used for building political support for the project and establishing trust and rapport between the project team building the system and the users who ultimately will choose to use or not use the system. Involving someone in the process implies that the project teams view that person as an important resource and value his or her opinions. All the key stakeholders (the people who can affect the system or who will be affected by the system) must be included in the requirements-gathering process.

The stakeholders might include managers, employees, staff members, and even some customers and suppliers. If a key person is not involved, that individual might feel slighted, which can cause problems during implementation (e.g., How could they have developed the system without my input?).

The second challenge of requirements gathering is choosing the way(s) information is collected. There are many techniques for gathering requirements that vary from asking people questions to watching them work. In this section, we focus on the five most commonly used techniques: interviews, JAD sessions (a special type of group meeting), questionnaires, document analysis, and observation. Each technique has its own strengths and weaknesses, many of which are complementary, so most projects use a combination of techniques.[7]

Interviews

An interview is the most commonly used requirements-gathering technique. After all, it is natural— if you need to know something, you usually ask someone. In general, interviews are conducted one-on-one (one interviewer and one interviewee), but sometimes, owing to time constraints, several people are interviewed at the same time. There are five basic steps to the interview process: selecting interviewees, designing interview questions, preparing for the interview, conducting the interview, and postinterview follow-up.[8]

1. Select Interviewees

The first step in interviewing is to create an *interview schedule* listing all the people who will be interviewed, when, and for what purpose (see Figure 3-6). The schedule can be an informal list that is used to help set up meeting times or a formal list that is incorporated into the workplan. The people who appear on the interview schedule are selected based on the analyst's information needs. The project sponsor, key business users, and other members of the project team can help the analyst determine who in the organization can best provide important information about requirements. These people are listed on the interview schedule in the order in which they should be interviewed.

People at different levels of the organization have different perspectives on the system, so it is important to include both managers, who manage the processes, and staff, who actually perform the processes, to gain both high-level and low-level perspectives on an issue. Also, the kinds of interview subjects needed can change over time. For example, at the start of the project, the analyst has a limited understanding of the as-is business process. It is common to begin by interviewing one or two senior managers to get a strategic view and then to move to midlevel managers, who can provide broad, overarching information about the business process and the expected role of the system being developed. Once the analyst has a good understanding of the big picture, lower-level managers and staff members can fill in the exact details of how the process works. Like most other things about systems analysis, this is an iterative process—starting with senior managers, moving to midlevel managers, then staff members, back to midlevel managers, and so on, depending upon what information is needed along the way.

It is quite common for the list of interviewees to grow, often by 50 to 75 percent. As people are interviewed, more information that is needed and additional people who can provide the information will probably be identified.

---

[7] Some excellent books that address the importance of gathering requirements and various techniques include Alan M. Davis, *Software Requirements: Objects, Functions, & States, Revision* (Englewood Cliffs, NJ: Prentice Hall, 1993); Gerald Kotonya and Ian Sommerville, *Requirements Engineering* (Chichester, England: Wiley, 1998); and Dean Leffingwell and Don Widrig, *Managing Software Requirements: A Unified Approach* (Reading, MA: Addison-Wesley, 2000).

[8] A good book on interviewing is that by Brian James, *The Systems Analysis Interview* (Manchester, England: NCC Blackwell, 1989).

Name	Position	Purpose of Interview	Meeting
Andria McClellan	Director, Accounting	Strategic vision for new accounting system	Mon., March 1 8:00–10:00 AM
Jennifer Draper	Manager, Accounts Receivable	Current problems with accounts receivable process; future goals	Mon., March 1 2:00–3:15 PM
Mark Goodin	Manager, Accounts Payable	Current problems with accounts payable process; future goals	Mon., March 1 4:00–5:15 PM
Anne Asher	Supervisor, Data Entry	Accounts receivable and payable processes	Wed., March 3 10:00–11:00 AM
Fernando Merce	Data Entry Clerk	Accounts receivable and payable processes	Wed., March 3 1:00–3:00 PM

FIGURE 3-6
Sample Interview Schedule

---

2. Design Interview Questions

There are three types of interview questions: closed-ended questions, open-ended questions, and probing questions. *Closed-ended questions* are those that require a specific answer. They are similar to multiple-choice or arithmetic questions on an exam (see Figure 3-7). Closed-ended questions are used when an analyst is looking for specific, precise information (e.g., how many credit card requests are received per day). In general, precise questions are best. For example, rather than asking, Do you handle a lot of requests? it is better to ask, How many requests do you process per day? Closed-ended questions enable analysts to control the interview and obtain the information they need. However, these types of questions don't uncover *why* the answer is the way it is, nor do they uncover information that the interviewer does not think to ask for ahead of time.

Open-ended questions are those that leave room for elaboration on the part of the interviewee. They are similar in many ways to essay questions that you might find on an exam (see Figure 3-7 for examples). Open-ended questions are designed to gather rich information and give the interviewee more control over the information that is revealed during the interview. Sometimes the information that the interviewee chooses to discuss uncovers information that is just as important as the answer (e.g., if the interviewee talks only about other departments when asked for problems, it may suggest that he or she is reluctant to admit his or her own problems).

The third type of question is the *probing question*. Probing questions follow up on what has just been discussed in order to learn more, and they often are used when the interviewer is unclear about an interviewee's answer. They encourage the interviewee to expand on or to confirm information from a previous response, and they signal that the interviewer is listening and is interested in the topic under discussion. Many beginning analysts are reluctant to use probing questions because they are afraid that the interviewee might be offended at being challenged or because they believe it shows that they didn't understand what the interviewee said. When done politely, probing questions can be a powerful tool in requirements gathering.

In general, an interviewer should not ask questions about information that is readily available from other sources. For example, rather than asking what information is used to perform to a task, it is simpler to show the interviewee a form or report (see the section on document analysis) and ask what information on it is used. This helps focus the interviewee on the task and saves time, because the interviewee does not need to describe the information detail—he or she just needs to point it out on the form or report.

No type of question is better than another, and a combination of questions is usually used during an interview. At the initial stage of an IS development project, the as-is process can be unclear, so the interview process begins with *unstructured interviews,* interviews that seek broad and roughly defined information. In this case, the interviewer has a general sense of the information needed but has few closed-ended questions to ask. These are the most challenging interviews to conduct because they require the interviewer to ask open-ended questions and probe for important information on the fly.

As the project progresses, the analyst comes to understand the business process much better and needs very specific information about how business processes are performed (e.g., exactly how a customer credit card is approved). At this time, the analyst conducts *structured interviews,* in which specific sets of questions are developed before the interviews. There usually are more closed-ended questions in a structured interview then in the unstructured approach.

No matter what kind of interview is being conducted, interview questions must be organized into a logical sequence so that the interview flows well. For example, when trying to gather information about the current business process, it can be useful to move in logical order through the process or from the most important issues to the least important.

There are two fundamental approaches to organizing the interview questions: top down or bottom up (see Figure 3-8). With the *top-down interview*, the interviewer starts with broad, general issues and gradually works toward more specific ones. With the *bottom-up interview*, the interviewer starts with very specific questions and moves to broad questions. In practice, analysts mix the two approaches, starting with broad, general issues, moving to specific questions, and then returning to general issues.

The top-down approach is an appropriate strategy for most interviews (it is certainly the most common approach). The top-down approach enables the interviewee to become accustomed to the topic before he or she needs to provide specifics. It also enables the interviewer to understand the issues before moving to the details because the interviewer might not have sufficient information at the start of the interview to ask very specific questions. Perhaps most importantly, the top-down approach enables the interviewee to raise a set of big-picture issues before becoming enmeshed in details, so the interviewer is less likely to miss important issues.

One case in which the bottom-up strategy may be preferred is when the analyst already has gathered a lot of information about issues and just needs to fill in some holes with details. Bottom-up interviewing may be appropriate if lower-level staff members feel threatened or are unable to answer high-level questions. For example, How can we improve customer service? might be too broad a question for a customer service clerk, whereas a specific question is readily answerable (e.g., How can we speed up customer returns?). In any event, all interviews should begin with noncontroversial questions and then gradually move into more contentious issues after the interviewer has developed some rapport with the interviewee.

Types of Questions	Examples
Closed-ended questions	• How many telephone orders are received per day? • How do customers place orders? • What information is missing from the monthly sales report?
Open-ended questions	• What do you think about the current system? • What are some of the problems you face on a daily basis? • What are some of the improvements you would like to see in a new system?
Probing questions	• Why? • Can you give me an example? • Can you explain that in a bit more detail?

FIGURE 3-7

Three Types of Questions

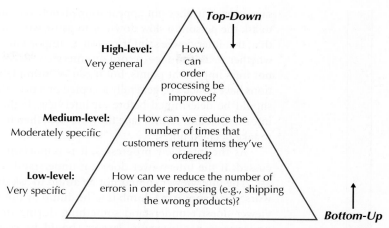

FIGURE 3-8   Top-Down and Bottom-Up Questioning Strategies

3. Prepare for the Interview

It is important to prepare for the interview in the same way that you would prepare to give a presentation. The interviewer should have a general interview plan listing the questions to be asked in the appropriate order, should anticipate possible answers and provide follow-up with them, and should identify segues between related topics. The interviewer should confirm the areas in which the interviewee has knowledge so as not to ask questions that the interviewee cannot answer. Review the topic areas, the questions, and the interview plan, and clearly decide which have the greatest priority in case time runs short.

In general, structured interviews with closed-ended questions take more time to prepare than unstructured interviews. Some beginning analysts prefer unstructured interviews, thinking that they can wing it. This is very dangerous and often counterproductive, because any information not gathered in the first interview will require follow-up efforts, and most users do not like to be interviewed repeatedly about the same issues.

The interviewer should be sure to prepare the interviewee as well. When the interview is scheduled, the interviewee should be told the reason for the interview and the areas that will be discussed far enough in advance so that he or she has time to think about the issues and organize his or her thoughts. This is particularly important when the interviewer is an outsider to the organization and for lower-level employees, who often are not asked for their opinions and who may be uncertain about why they are being interviewed.

4. Conduct the Interview

In starting the interview, the first goal is to build rapport with the interviewee, so that he or she trusts the interviewer and is willing to tell the whole truth, not just give the answers that he or she thinks are wanted. The interviewer should appear to be professional and an unbiased, independent seeker of information. The interview should start with an explanation of why the interviewer is there and why he or she has chosen to interview the person; then the interviewer should move into the planned interview questions.

It is critical to carefully record all the information that the interviewee provides. In our experience, the best approach is to take careful notes—write down *everything* the interviewee says, even if it does not appear immediately relevant. The interviewer shouldn't be afraid to ask the person to slow down or to pause while writing, because this is a clear indication that the interviewee's information is important. One potentially controversial issue is whether or not to tape-record an interview. Recording ensures that the interviewer does not miss important points, but it can be intimidating for the interviewee. Most organizations have policies or generally accepted practices about the recording of interviews, so they should be determined before an interview. If the interviewer is worried about missing information and cannot tape the interview, then he or she can bring along a second person to take detailed notes.

As the interview progresses, it is important to understand the issues that are discussed. If the interviewer does not understand something, he or she should be sure to ask. The interviewer should not be afraid to ask dumb questions, because the only thing worse than appearing dumb is to be dumb by not understanding something. If the interviewer doesn't understand something during the interview, he or she certainly won't understand it afterwards. Jargon should be recognized and defined; any jargon not understood should be clarified. One good strategy to increase understanding during an interview is to periodically summarize the key points that the interviewee is communicating. This avoids misunderstandings and also demonstrates that the interviewer is listening.

Finally, facts should be separated from opinion. The interviewee may say, for example, We process too many credit card requests. This is an opinion, and it is useful to follow this up with a probing question requesting support for the statement (e.g., Oh, how many do you process in a day?). It is helpful to check the facts because any differences between the facts and the interviewee's opinions can point out key areas for improvement. Suppose the interviewee complains about a high or increasing number of errors, but the logs show that errors have been decreasing. This suggests that errors are viewed as a very important problem that should be addressed by the new system, even if they are declining.

As the interview draws to a close, the interviewee should have time to ask questions or provide information that he or she thinks is important but was not part of the interview plan. In most cases, the interviewee has no additional concerns or information, but in some cases this leads to unanticipated, but important, information. Likewise, it can be useful to ask the interviewee if there are other people who should be interviewed. The interview should end on time (if necessary, some topics can be omitted or another interview can be scheduled).

As a last step in the interview, the interviewer should briefly explain what will happen. The interviewer shouldn't prematurely promise certain features in the new system or a specific delivery date, but he or she should reassure the interviewee that his or her time was well spent and very helpful to the project.

5. Post-Interview Follow-up

After the interview is over, the analyst needs to prepare an *interview report* that describes the information from the interview (Figure 3-9). The report contains *interview notes*, information that was collected over the course of the interview and is summarized in a useful format. In general, the interview report should be written within forty-eight hours of the interview, because the longer the interviewer waits, the more likely he or she is to forget information.

Often, the interview report is sent to the interviewee with a request to read it and inform the analyst of clarifications or updates. The interviewee needs to be convinced that the interviewer genuinely wants his or her corrections to the report. Usually there are few changes, but the need for any significant changes suggests that a second interview will be required. Never distribute someone's information without prior approval.

Interview Notes Approved by: Linda Estey

Person Interviewed: Linda Estey,
    Director, Human Resources

Interviewer: Barbara Wixom

Purpose of Interview:
- Understand reports produced for Human Resources by the current system
- Determine information requirements for future system

Summary of Interview:
- Sample reports of all current HR reports are attached to this report. The information that is not used and missing information are noted on the reports.
- Two biggest problems with the current system are:
  1. The data are too old (the HR Department needs information within two days of month end; currently information is provided to them after a three-week delay)
  2. The data are of poor quality (often reports must be reconciled with departmental HR database)
- The most common data errors found in the current system include incorrect job level information and missing salary information.

Open Items:
- Get current employee roster report from Mary Skudrna (extension 4355).
- Verify calculations used to determine vacation time with Mary Skudrna.
- Schedule interview with Jim Wack (extension 2337) regarding the reasons for data quality problems.

Detailed Notes: See attached transcript.

FIGURE 3-9   Interview Report

CONCEPTS

IN ACTION

3-E Selecting the Wrong People

*I*n 1990, I led a consulting team for a major development project for the U.S. Army. The goal was to replace eight existing systems used on virtually every Army base across the United States. The as-is process and data models for these systems had been built, and our job was to identify improvement opportunities and develop to-be process models for each of the eight systems.

    For the first system, we selected a group of midlevel managers (captains and majors) recommended by their commanders as being the experts in the system under construction. These individuals were the first- and second-line managers of the business function. The individuals were expert at managing the process but did not know the exact details of how the process worked. The resulting to-be process model was very general and nonspecific.

—Alan Dennis

Question

Suppose you were in charge of the project. What interview schedule for the remaining seven projects would you use?

YOUR TURN

3-4 Interview Practice

*I*nterviewing is not as simple as it first appears. Select two people from class to go to the front of the room to demonstrate an interview. (This also can be done in groups.) Have one person be the interviewer and the other be the interviewee. The interviewer should conduct a five-minute interview regarding the school's course registration system. Gather information about the existing system and how the system can be improved. If there is time, repeat with another pair.

Questions

1. What was the body language of the interview pair like?
2. What kind of interview was conducted?
3. What kinds of questions were asked?
4. What was done well? How could the interview be improved?

PRACTICAL TIP

3-1 Developing Interpersonal Skills

*I*nterpersonal skills are skills that enable you to develop rapport with others, and they are very important for interviewing. They help you to communicate with others effectively. Some people develop good interpersonal skills at an early age; they simply seem to know how to communicate and interact with others. Other people are less lucky and need to work hard to develop their skills.

Interpersonal skills, like most skills, can be learned. Here are some tips:

- **Don't worry, be happy.** Happy people radiate confidence and project their feelings on others. Try interviewing someone while smiling and then interviewing someone else while frowning and see what happens.
- **Pay attention.** Pay attention to what the other person is saying (which is harder than you might think). See how many times you catch yourself with your mind on something other than the conversation at hand.
- **Summarize key points.** At the end of each major theme or idea that someone explains, repeat the key points back to the speaker (e.g., "Let me make sure I

understand. The key issues are. . . ."). This demonstrates that you consider the information important, and it also forces you to pay attention (you can't repeat what you didn't hear).

- **Be succinct.** When you speak, be succinct. The goal in interviewing (and in much of life) is to learn, not to impress. The more you speak, the less time you give to others.
- **Be honest.** Answer all questions truthfully, and if you don't know the answer, say so.
- **Watch body language (yours and theirs).** The way a person sits or stands conveys much information. In general, a person who is interested in what you are saying sits or leans forward, makes eye contact, and often touches his or her face. A person leaning away from you or with an arm over the back of a chair is uninterested. Crossed arms indicate defensiveness or uncertainty, and steepling (sitting with hands raised in front of the body with fingertips touching) indicates a feeling of superiority.

Joint Application Development (JAD)

JAD is an information-gathering technique that allows the project team, users, and management to work together to identify requirements for the system. IBM developed the JAD technique in the late 1970s, and it is often the most useful method for collecting information from users.[9]

[9] More information on JAD can be found in J. Wood and D. Silver, *Joint Application Development* (New York: Wiley, 1989); and Alan Cline, "Joint Application Development for Requirements Collection and Management," http://www.carolla.com/wp-jad.htm.

Capers Jones claims that JAD can reduce scope creep by 50%, and it prevents the system's requirements from being too specific or too vague, both of which cause trouble during later stages of the development process.[10]

JAD is a structured process in which ten to twenty users meet together under the direction of a *facilitator* skilled in JAD techniques. The facilitator is a person who sets the meeting agenda and guides the discussion but does not join in the discussion as a participant. He or she does not provide ideas or opinions on the topics under discussion so as to remain neutral during the session. The facilitator must be an expert in both group-process techniques and systems-analysis and design techniques. One or two *scribes* assist the facilitator by recording notes, making copies, and so on. Often the scribes use computers and CASE tools to record information as the JAD session proceedings.

The JAD group meets for several hours, several days, or several weeks until all the issues have been discussed and the needed information is collected. Most JAD sessions take place in a specially prepared meeting room, away from the participants' offices so that they are not interrupted. The meeting room is usually arranged in a U-shape so that all participants can easily see each other (see Figure 3-10). At the front of the room (the open part of the U), are a whiteboard, flip chart, and/or overhead projector for use by the facilitator leading the discussion.

One problem with JAD is that it suffers from the traditional problems associated with groups: Sometimes people are reluctant to challenge the opinions of others (particularly their boss), a few people often dominate the discussion, and not everyone participates. In a fifteen-member group, for example, if everyone participates equally, then each person can talk for only four minutes each hour and must listen for the remaining fifty-six minutes— not a very efficient way to collect information.

A new form of JAD called *electronic JAD*, or *e-JAD*, attempts to overcome these problems by using groupware. In an e-JAD meeting room, each participant uses special software on a networked computer to send anonymous ideas and opinions to everyone else. In this way, all participants can contribute at the same time without fear of reprisal from people with differing opinions. Initial research suggests that e-JAD can reduce the time required to run JAD sessions by 50 to 80 percent.[11] A good JAD approach follows a set of five steps.

1. Select Participants

First, selecting JAD participants is done in the same basic way as selecting interview participants. Participants are selected based on the information they can contribute in order to provide a broad mix of organizational levels and to build political support for the new system. The need for all JAD participants to be away from their office at the same time can be a major problem. The office might need to be closed or operate with a skeleton staff until the JAD sessions are complete.

Ideally, the participants who are released from regular duties to attend the JAD sessions should be the very best people in that business unit. However, without strong management support, JAD sessions can fail because those selected to attend the JAD session are people who are less likely to be missed (i.e., the least competent people).

The facilitator should be someone who is an expert in JAD or e-JAD techniques and, ideally, someone who has experience with the business under discussion. In many cases, the JAD facilitator is a consultant external to the organization because the organization might not have a recurring need for JAD or e-JAD expertise. Developing and maintaining this expertise in-house can be expensive.

[10] See Kevin Strehlo, "Catching up with the Jones and 'Requirement' Creep," *Infoworld* (July 29, 1996); and Kevin Strehlo, "The Makings of a Happy Customer: Specifying Project X," *Infoworld* (November 11, 1996).

[11] For more information on e-JAD, see A. R. Dennis, G. S. Hayes, and R. M. Daniels, "Business Process Modeling with Groupware," *Journal of Management Information Systems* 15, no. 4 (1999): 115–142.

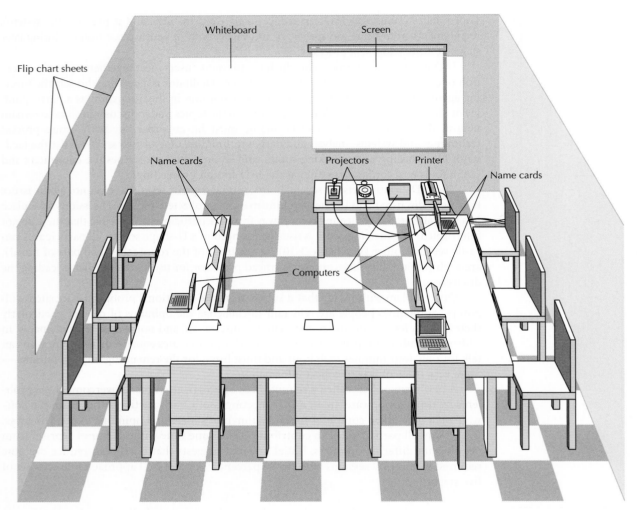

FIGURE 3-10   JAD Meeting Room

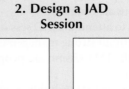

2. Design a JAD Session

Second, JAD sessions can run from as little as half a day to several weeks, depending upon the size and scope of the project. In our experience, most JAD sessions tend to last five to ten days, spread over a three-week period. Most e-JAD sessions tend to last one to four days in a one-week period. JAD and e-JAD sessions usually go beyond collecting information and move into analysis. For example, the users and the analysts collectively can create analysis deliverables, such as the functional models or the requirements definition.

As with interviewing, success depends upon a careful plan. JAD sessions usually are designed and structured using the same principles as interviews. Most JAD sessions are designed to collect specific information from users, and this requires developing a set of questions before the meeting. One difference between JAD and interviewing is that all JAD sessions are structured—they *must* be carefully planned. In general, closed-ended questions are seldom used because they do not spark the open and frank discussion that is typical of JAD. In our experience, it is better to proceed top down in JAD sessions when gathering information. Typically thirty minutes is allocated to each separate agenda item, and frequent breaks are scheduled throughout the day because participants tire easily.

3. Preparing for a JAD Session

Third, as with interviewing, it is important to prepare the analysts and participants for a JAD session. Because the sessions can go beyond the depth of a typical interview and are usually conducted off-site, participants can be more concerned about how to prepare. It is important that the participants understand what is expected of them. If the goal of the JAD session, for example, is to develop an understanding of the current system, then participants can bring procedure manuals and documents with them. If the goal is to identify improvements for a system, then before they come to the JAD session they can think about how they would improve the system.

4. Conducting a JAD Session

Fourth, most JAD sessions try to follow a formal agenda, and most have formal *ground rules* that define appropriate behavior. Common ground rules include following the schedule, respecting others' opinions, accepting disagreement, and ensuring that only one person talks at a time.

The role of a JAD facilitator can be challenging. Many participants come to a JAD session with strong feelings about the system to be discussed. Channeling these feelings so that the session moves forward in a positive direction and getting participants to recognize and accept—but not necessarily agree on—opinions and situations different from their own requires significant expertise in systems analysis and design, JAD, and interpersonal skills. Few systems analysts attempt to facilitate JAD sessions without being trained in JAD techniques, and most apprentice with a skilled JAD facilitator before they attempt to lead their first session.

The JAD facilitator performs three key functions. First, he or she ensures that the group sticks to the agenda. The only reason to digress from the agenda is when it becomes clear to the facilitator, project leader, and project sponsor that the JAD session has produced some new information that is unexpected and requires the JAD session (and perhaps the project) to move in a new direction. When participants attempt to divert the discussion away from the agenda, the facilitator must be firm but polite in leading discussion back to the agenda and getting the group back on track.

Second, the facilitator must help the group understand the technical terms and jargon that surround the system-development process and help the participants understand the specific analysis techniques used. Participants are experts in their area, or their part of the business, but they are not experts in systems analysis. The facilitator must, therefore, minimize the learning required and teach participants how to effectively provide the right information.

Third, the facilitator records the group's input on a public display area, which can be a whiteboard, flip chart, or computer display. He or she structures the information that the group provides and helps the group recognize key issues and important solutions. Under no circumstance should the facilitator insert his or her opinions into the discussion. The facilitator must remain neutral at all times and simply help the group through the process. The moment the facilitator offers an opinion on an issue, the group will see him or her not as a neutral party but rather as someone who could be attempting to sway the group into some predetermined solution.

However, this does not mean that the facilitator should not try to help the group resolve issues. For example, if two items appear to be the same to the facilitator, the facilitator should not say, "I think these may be similar." Instead, the facilitator should ask, "Are these similar?" If the group decides they are, the facilitator can combine them and move on. However, if the group decides they are not similar (despite what the facilitator believes), the facilitator should accept the decision and move on. The group is *always* right, and the facilitator has no opinion.

5. Post-JAD Follow-up

Fifth, as with interviews, a JAD *post-session report* is prepared and circulated among session attendees. The post-session report is essentially the same as the interview report in Figure 3-9. Because the JAD sessions are longer and provide more information, it usually takes a week or two after the JAD session before the report is complete.

PRACTICAL TIP

Managing Problems in JAD Sessions

I have run more than a hundred JAD sessions and have learned several standard "facilitator tricks." Here are some common problems and some ways to deal with them.

- **Domination.** The facilitator should ensure that no one person dominates the group discussion. The only way to deal with someone who dominates is head on. During a break, approach the person, thank him or her for his or her insightful comments, and ask the person to help you make sure that others also participate.

- **Noncontributors.** Drawing out people who have participated very little is challenging because you want to bring them into the conversation so that they will contribute again. The best approach is to ask a direct factual question that you are certain they can answer. And it helps to ask the question in a long way to give them time to think. For example, "Pat, I know you've worked shipping orders a long time. You've probably been in the shipping department longer than anyone else. Could you help us understand exactly what happens when an order is received in shipping?"

- **Side discussions.** Sometimes participants engage in side conversations and fail to pay attention to the group. The easiest solution is simply to walk close to the people and continue to facilitate right in front of them. Few people will continue a side conversion when you are two feet from them and the entire group's attention is on you and them.

- **Agenda merry-go-round.** The merry-go-round occurs when a group member keeps returning to the same issue every few minutes and won't let go. One solution is to let the person have five minutes to ramble on about the issue while you carefully write down every point on a flip chart or computer file. This flip chart or file is then posted conspicuously on the wall. When the person brings up the issue again, you interrupt them, walk to the paper and ask them what to add. If they mention some-

thing already on the list, you quickly interrupt, point out that it is there, and ask what other information to add. Don't let them repeat the same point, but write any new information.

- **Violent agreement.** Some of the worst disagreements occur when participants really agree on the issues, but don't realize that they agree because they are using different terms. An example is arguing whether a glass is half empty or half full; they agree on the facts but can't agree on the words. In this case, the facilitator has to translate the terms into different words and find common ground so the parties recognize that they really agree.

- **Unresolved conflict.** In some cases, participants don't agree and can't understand how to determine what alternatives are better. You can help by structuring the issue. Ask for criteria by which the group will identify a good alternative (e.g., "Suppose this idea really did improve customer service. How would I recognize the improved customer service?"). Then once you have a list of criteria, ask the group to assess the alternatives using them.

- **True conflict.** Sometimes, despite every attempt, participants just can't agree on an issue. The solution is to postpone the discussion and move on. Document the issue as an open issue and list it prominently on a flip chart. Have the group return to the issue hours later. Often the issue will have resolved itself by then and you haven't wasted time on it. If the issue cannot be resolved later, move it to the list of issues to be decided by the project sponsor or some other more senior member of management.

- **Humor.** Humor is one of the most powerful tools a facilitator has and thus must be used judiciously. The best JAD humor is always in context; never tell jokes but take the opportunity to find the humor in the situation.

—Alan Dennis

Questionnaires

A questionnaire is a set of written questions used to obtain information from individuals. Questionnaires are often used when there is a large number of people from whom information and opinions are needed. In our experience, questionnaires are a common technique with systems intended for use outside the organization (e.g., by customers or vendors) or for

YOUR
TURN

3-5 JAD Practice

*O*rganize yourselves into groups of four to seven people, and pick one person in each group to be the JAD facilitator. Using a blackboard, whiteboard or flip chart, gather information about how the group performs some process (e.g., working on a class assignment, making a sandwich, paying bills, getting to class).

Questions

1. How did the JAD session go?
2. Based on your experience, what are pros and cons of using JAD in a real organization?

systems with business users spread across many geographic locations. Most people automatically think of paper when they think of questionnaires, but today more questionnaires are being distributed in electronic form, either via e-mail or on the web. Electronic distribution can save a significant amount of money as compared to distributing paper questionnaires. A good process to use when using questionnaires follows four steps.

1. Select Participants

First, as with interviews and JAD sessions, the first step is to identify the individuals to whom the questionnaire will be sent. However, it is not usual to select every person who could provide useful information. The standard approach is to select a *sample*, or subset, of people who are representative of an entire group. Sampling guidelines are discussed in most statistics books, and most business schools include courses that cover the topic, so we do not discuss it here. The important point in selecting a sample, however, is to realize that not everyone who receives a questionnaire will actually complete it. On average, only 30 to 50 percent of paper and e-mail questionnaires are returned. Response rates for web-based questionnaires tend to be significantly lower (often only 5 to 30 percent).

2. Designing a Questionnaire

Second, because the information on a questionnaire cannot be immediately clarified for a confused respondent, developing good questions is critical for questionnaires. Questions on questionnaires must be very clearly written and leave little room for misunderstanding, so closed-ended questions tend to be most commonly used. Questions must clearly enable the analyst to separate facts from opinions. Opinion questions often ask respondents the extent to which they agree or disagree (e.g., Are network problems common?), whereas factual questions seek more precise values (e.g., How often does a network problem occur: once an hour, once a day, once a week?). See Figure 3-11 for guidelines on questionnaire design.

Perhaps the most obvious issue—but one that is sometimes overlooked—is to have a clear understanding of how the information collected from the questionnaire will be analyzed and used. This issue must be addressed before the questionnaire is distributed, because it is too late afterward.

Questions should be relatively consistent in style, so that the respondent does not have to read instructions for each question before answering it. It is generally good practice to group related questions together to make them simpler to answer. Some experts suggest that questionnaires should start with questions important to respondents, so that the questionnaire immediately grabs their interest and induces them to answer it. Perhaps the most important step is to have several colleagues review the questionnaire and then pretest it with a few people drawn from the groups to whom it will be sent. It is surprising how often seemingly simple questions can be misunderstood.

- Begin with nonthreatening and interesting questions.
- Group items into logically coherent sections.
- Do not put important items at the very end of the questionnaire.
- Do not crowd a page with too many items.
- Avoid abbreviations.
- Avoid biased or suggestive items or terms.
- Number questions to avoid confusion.
- Pretest the questionnaire to identify confusing questions.
- Provide anonymity to respondents.

FIGURE 3-11
Good Questionnaire
Design

3. Administering the Questionnaire

Third, the key issue in administering the questionnaire is getting participants to complete the questionnaire and send it back. Dozens of marketing research books have been written about ways to improve response rates. Commonly used techniques include clearly explaining why the questionnaire is being conducted and why the respondent has been selected; stating a date by which the questionnaire is to be returned; offering an inducement to complete the questionnaire (e.g., a free pen); and offering to supply a summary of the questionnaire responses. Systems analysts have additional techniques to improve response rates inside the organization, such as personally handing out the questionnaire and personally contacting those who have not returned them after a week or two, as well as requesting the respondents' supervisors to administer the questionnaires in a group meeting.

4. Questionnaire Follow-up

Fourth, it is helpful to process the returned questionnaires and develop a questionnaire report soon after the questionnaire deadline. This ensures that the analysis process proceeds in a timely fashion and that respondents who requested copies of the results receive them promptly.

YOUR TURN

3-6 Questionnaire Practice

*O*rganize yourselves into small groups. Have each person develop a short questionnaire to collect information about how often group members perform some process (e.g., working on a class assignment, making a sandwich, paying bills, getting to class), how long it takes them, how they feel about the process, and opportunities for improving the process.

Once everyone has completed his or her questionnaire, ask each member to pass it to the right and then complete his or her neighbor's questionnaire. Pass the questionnaire back to the creator when it is completed.

Questions

1. How did the questionnaire you completed differ from the one you created?
2. What are the strengths of each questionnaire?
3. How would you analyze the survey results if you had received fifty responses?
4. What would you change about the questionnaire that you developed?

Document Analysis

Project teams often use document analysis to understand the as-is system. Under ideal circumstances, the project team that developed the existing system will have produced documentation that was then updated by all subsequent projects. In this case, the project team can start by reviewing the documentation and examining the system itself.

Unfortunately, most systems are not well documented because project teams fail to document their projects along the way, and when the projects are over, there is no time to go back and document. Therefore, there might not be much technical documentation about the current systems available, or it might not contain updated information about recent system changes. However, many helpful documents do exist in an organization: paper reports, memorandums, policy manuals, user-training manuals, organization charts, forms, and, of course, the user interface with the existing system.

But these documents tell only part of the story. They represent the *formal system* that the organization uses. Quite often, the real, or *informal, system* differs from the formal one, and these differences, particularly large ones, give strong indications of what needs to be changed. For example, forms or reports that are never used should probably be eliminated. Likewise, boxes or questions on forms that are never filled in (or are used for other purposes) should be rethought. See Figure 3-12 for an example of how a document can be interpreted.

The most powerful indication that the system needs to be changed is when users create their own forms or add additional information to existing ones. Such changes clearly demonstrate the need for improvements to existing systems. Thus, it is useful to review both blank and completed forms to identify these deviations. Likewise, when users access multiple reports to satisfy their information needs, it is a clear sign that new information or new information formats are needed.

CONCEPTS

IN ACTION

3-F Publix Credit-Card Forms

At my neighborhood Publix grocery store, the cashiers always hand-write the total amount of the charge on every credit-card charge form, even though it is printed on the form. Why? Because the "back office" staff people who reconcile the cash in the cash drawers with the amount sold at the end of each shift find it hard to read the small print on the credit-card forms. Writing in large print makes it easier for them to add the values up. However, cashiers sometimes make mistakes and write the wrong amount on the forms, which causes problems.

Questions

1. What does the credit-card charge form indicate about the existing system?
2. How can you make improvements with a new system?

—*Barbara Wixom*

Observation

Observation, the act of watching processes being performed, is a powerful tool for gathering information about the as-is system because it enables the analyst to see the reality of a situation, rather than listening to others describe it in interviews or JAD sessions. Several research studies have shown that many managers really do not remember how they work and how they allocate their time. (Quick, how many hours did you spend last week on each of your courses?) Observation is a good way to check the validity of information gathered from indirect sources such as interviews and questionnaires.

In many ways, the analyst becomes an anthropologist as he or she walks through the organization and observes the business system as it functions. The goal is to keep a low profile, to not interrupt those working, and to not influence those being observed. Nonetheless, it is important to understand that what analysts observe may not be the normal day-to-day routine because people tend to be extremely careful in their behavior when they are being watched. Even though normal practice may be to break formal organizational rules, the observer is unlikely to see this. (Remember how you drove the last time a police car followed you?) Thus, what you see might *not* be what you get.

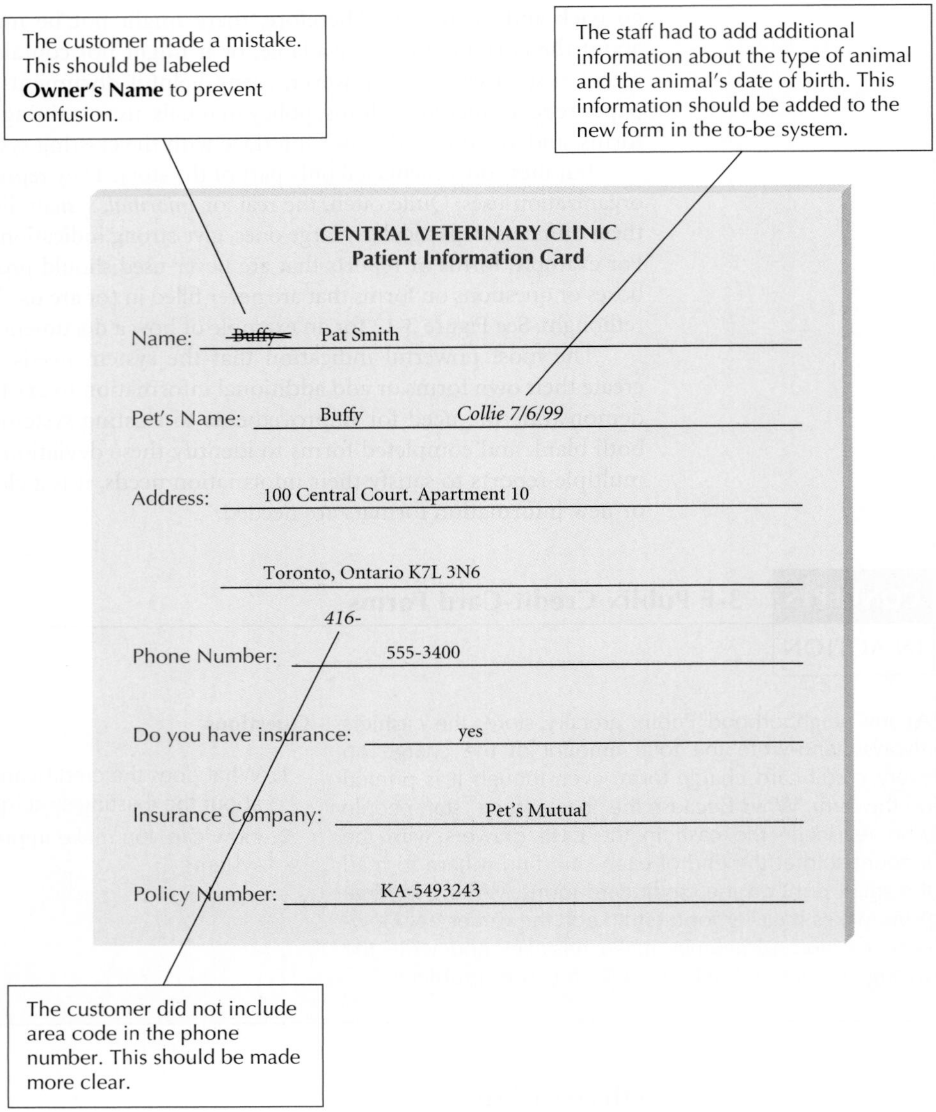

The customer made a mistake. This should be labeled **Owner's Name** to prevent confusion.

The staff had to add additional information about the type of animal and the animal's date of birth. This information should be added to the new form in the to-be system.

CENTRAL VETERINARY CLINIC
Patient Information Card

Name: ~~Buffy~~ Pat Smith

Pet's Name: Buffy *Collie 7/6/99*

Address: 100 Central Court. Apartment 10

Toronto, Ontario K7L 3N6

416-

Phone Number: 555-3400

Do you have insurance: yes

Insurance Company: Pet's Mutual

Policy Number: KA-5493243

The customer did not include area code in the phone number. This should be made more clear.

FIGURE 3-12
Performing a
Document Analysis

YOUR TURN

3-7 Observation Practice

*V*isit the library at your college or university and observe how the book checkout process occurs. First watch several students checking books out, and then check one out yourself. Prepare a brief summary report of your observations.

When you return to class, share your observations with others.

Questions

1. Why might the reports present different information?
2. How would the information be different had you used the interview or JAD technique?

	Interviews	Joint Application Design	Questionnaires	Document Analysis	Observation
Type of information	As-is, improvements, to-be	As-is, improvements, to-be	As-is, improvements	As-is	As-is
Depth of information	High	High	Medium	Low	Low
Breadth of information	Low	Medium	High	High	Low
Integration of information	Low	High	Low	Low	Low
User involvement	Medium	High	Low	Low	Low
Cost	Medium	Low–Medium	Low	Low	Low to Medium

FIGURE 3-13 Table of Requirements-Gathering Techniques

Observation is often used to supplement interview information. The location of a person's office and its furnishings give clues to the person's power and influence in the organization and can be used to support or refute information given in an interview. For example, an analyst might become skeptical of someone who claims to use the existing computer system extensively if the computer is never turned on while the analyst visits. In most cases, observation supports the information that users provide in interviews. When it does not, it is an important signal that extra care must be taken in analyzing the business system.

Selecting the Appropriate Techniques

Each of the requirements-gathering techniques discussed earlier has strengths and weaknesses. No one technique is always better than the others, and in practice, most projects use a combination of techniques. Thus, it is important to understand the strengths and weaknesses of each technique and when to use each (see Figure 3-13). One issue not discussed is that of the analysts' experience. In general, document analysis and observation require the least amount of training, whereas JAD sessions are the most challenging.

Type of Information   The first characteristic is type of information. Some techniques are more suited for use at different stages of the analysis process, whether understanding the as-is system, identifying improvements, or developing the to-be system. Interviews and JAD are commonly used in all three stages. In contrast, document analysis and observation usually are most helpful for understanding the as-is, although occasionally, they provide information about current problems that need to be improved. Questionnaires are often used to gather information about the as-is system as well as general information about improvements.

Depth of Information   The depth of information refers to how rich and detailed the information is that the technique usually produces and the extent to which the technique is useful for obtaining not only facts and opinions, but also an understanding of *why* those facts and opinions exist. Interviews and JAD sessions are very useful for providing a good depth of rich and detailed information and helping the analyst to understand the reasons behind them. At the other extreme, document analysis and observation are useful for obtaining facts, but little beyond that. Questionnaires can provide a medium depth of information, soliciting both facts and opinions with little understanding of why they exist.

Breadth of Information   Breadth of information refers to the range of information and information sources that can be easily collected using the chosen technique. Questionnaires and document analysis are both easily capable of soliciting a wide range of information from a large number of information sources. In contrast, interviews and observation require the analyst to visit each information source individually and therefore take more time. JAD sessions are in the middle because many information sources are brought together at the same time.

Integration of Information   One of the most challenging aspects of requirements gathering is integrating the information from different sources. Simply put, different people can provide conflicting information. Combining this information and attempting to resolve differences in opinions or facts is usually very time-consuming because it means contacting each information source in turn, explaining the discrepancy, and attempting to refine the information. In many cases, the individual wrongly perceives that the analyst is challenging his or her information, when, in fact, it is another user in the organization who is doing so. This can make the user defensive and make it hard to resolve the differences.

All techniques suffer integration problems to some degree, but JAD sessions are designed to improve integration because all information is integrated when it is collected, not afterward. If two users provide conflicting information, the conflict becomes immediately obvious, as does the source of the conflict. The immediate integration of information is the single most important benefit of JAD that distinguishes it from other techniques, and this is why most organizations use JAD for important projects.

User Involvement   User involvement refers to the amount of time and energy the intended users of the new system must devote to the analysis process. It is generally agreed that as users become more involved in the analysis process, the chance of success increases. However, user involvement can have a significant cost, and not all users are willing to contribute valuable time and energy. Questionnaires, document analysis, and observation place the least burden on users, whereas JAD sessions require the greatest effort.

Cost   Cost is always an important consideration. In general, questionnaires, document analysis, and observation are low-cost techniques (although observation can be quite time consuming). The low cost does not imply that they are more or less effective than the other techniques. Interviews and JAD sessions generally have moderate costs. In general, JAD sessions are much more expensive initially, because they require many users to be absent from their offices for significant periods of time, and they often involve highly paid consultants. However, JAD sessions significantly reduce the time spent in information integration and thus can cost less in the long term.

Combining Techniques   In practice, requirements gathering combines a series of different techniques. Most analysts start by using interviews with senior manager(s) to gain an understanding of the project and the big-picture issues. From these interviews, it becomes clear whether large or small changes are anticipated. These interviews are often followed with analysis of documents and policies to gain some understanding of the as-is system. Usually interviews come next to gather the rest of the information needed for the as-is picture.

In our experience, identifying improvements is most commonly done using JAD sessions because the JAD session enables the users and key stakeholders to work together through an analysis technique and come to a shared understanding of the possibilities for the to-be system. Occasionally, these JAD sessions are followed by questionnaires sent to a much wider set of users or potential users to see whether the opinions of those who participated in the JAD sessions are widely shared.

Developing the concept for the to-be system is often done through interviews with senior managers, followed by JAD sessions with users of all levels to make sure the key needs of the new system are well understood.

CONCEPTS

IN ACTION

3-G Campus Technology Updates

Colleges and universities need to stay current with technologies. Many campuses have adopted laptop programs, where students are expected to purchase or lease a particular model of laptop that will be preloaded with appropriate software and used for the students' collegiate careers. Likewise, the campuses need to update their infrastructure—such as increasing bandwidth (to handle more video, such as YouTube)—and to provide wireless communication.

The University of Northern Wisconsin is a campus that is trying to remain current with technology. Campus budgets are almost always tight. UNW offers programs from its Superior, Wisconsin, main campus as well as programs on two satellite campuses in Ashland and Rhinelander. Users on the two satellite campuses frequently do not get the same level of service as students on the main campus. Internet access is generally slower and not all the software is the same. For example, students at the main campus have access to Bloomberg systems for analysis of financial trading data. The campus opted to build an Internet portal for all students to get to the same software and systems, set up by student ID and student profiles and permissions.

Questions

1. What technologies would be needed to make your campus a premier technology-oriented school?
2. How might a college campus be like a business with multiple locations and software needs?

ALTERNATIVE REQUIREMENTS DOCUMENTATION TECHNIQUES

Some other very useful requirements-gathering and documentation techniques include throwaway prototyping, use cases, role-playing CRC cards with use case–based scenarios, concept mapping, recording user stories on story cards, and task lists. Throwaway prototyping was described in Chapter 1. In essence, throwaway prototypes are created to better understand some aspect of the new system. In many cases, they are used to test out some technical aspect of a nonfunctional requirement, such as connecting a client workstation to a server. If you have never done this before, it will be a lot easier to develop a very small example system to test out the necessary design of the connection from the client workstation to the server, instead of trying to do it the first time with the full-blown system. Throwaway prototyping is very useful when designing the physical architecture of the system (see Chapter 11). Throwaway prototyping can also be very useful in designing user interfaces (see Chapter 10).

Use cases, as described in Chapter 1, are the fundamental approach that the Unified Process and Unified Modeling Language (UML) use to document and gather functional requirements. We describe them in Chapter 4. Role-playing CRC cards with use case–based scenarios are very useful when creating functional (see Chapter 4), structural (see Chapter 5), and behavioral (see Chapter 6) models. We describe this approach in Chapter 5. The remainder of this section describes the use of concept mapping, story cards, and task lists.

Concept Maps

Concept maps represent meaningful relationships between concepts. They are useful for focusing individuals on the small number of key ideas on which they should concentrate. A concept map is essentially a node-and-arc representation, where the nodes represent the individual requirements and the arcs represent the relationships among the requirements. Each arc is labeled with a relationship name. Concept maps also have been recommended as a possible technique to support modeling requirements for object-oriented systems development and knowledge-management systems.[12] *Concept mapping* is an educational psychology technique that has been used in schools, corporations, and health-care agencies to facilitate learning, understanding, and knowledge creation.[13] The advantage of the concept-mapping approach to representing requirements over the typical textual approach (see Figure 3-3) is that a concept map is not limited to a hierarchical representation. Concept maps allow the relationships among the functional and nonfunctional requirements to be explicitly represented. Figure 3-14 shows a concept map that portrays the information contained in the requirements definition shown in Figure 3-3. By using a concept map to represent the requirements instead of the textual approach, the relationship between the functional and nonfunctional requirements can be made explicit. For example, the two security requirements, Only Doctors Set Availability and Only Managers Can Produce Schedule are explicitly linked to the Record Doctor Availability and Produce Schedule functional requirements, respectively. This is very difficult to represent in a text-only version of the requirements definition. Also, by having the user and analyst focus on the graphical layout of the map, additional requirements can be discovered. One obvious issue with this approach is that if the number of requirements become many and the relationships between them become complex, then the number of nodes and arcs will become so intertwined that the advantage of being able to explicitly see the relationships will be lost. However, by combining both text and concept-map representations, it is possible to leverage the strength of both textual and graphical representations to more completely represent the requirements.

Story Cards and Task Lists

The use of *story cards* and *task lists* is associated with the agile development approaches. From an agile perspective, documentation is only a necessary evil and should be minimized. Both story cards and task lists are considered to be lightweight approaches to documenting and gathering requirements.[14] A story card is typically an index card with a single requirement (functional or nonfunctional) written on it. For example, with regard to the doctor's office appointment example, a story card could simply have "Make Appointment" written on it, while another could have "Back up Schedule Daily" written on it (see Figure 3-15). Once the requirement is written down, it is discussed to determine the amount of effort it will take to implement it. During the discussion, a task list is created for

---

[12] See B. Henderson-Sellers, A. Simons, and H. Younessi, *The OPEN Toolbox of Techniques* (Harlow, England: Addison-Wesley, 1998).

[13] For more information on concept mapping, see J. D. Novak and D. B. Gowin, *Learning How to Learn* (Cambridge, UK: Cambridge University Press, 1984); and J. D. Novak, *Learning, Creating, and Using Knowledge: Concept Maps*™ *as Facilitative Tools in Schools and Corporations* (Mahwah, NJ: Lawrence Erlbaum Associates, Publishers, 1998). Also, a free concept mapping tool is available from the Institute of Human and Machine Cognition at cmap.ihmc.us.

[14] For more information on story cards and task lists, see M. Lippert, S. Roock, H. Wolf, *eXtreme Programming in Action: Practical Experiences from Real World Projects* (Chichester, England: Wiley & Sons, Ltd., 2002); and C. Larman, *Agile & Iterative Development: A Manager's Guide* (Boston, MA: Addison-Wesley, 2004).

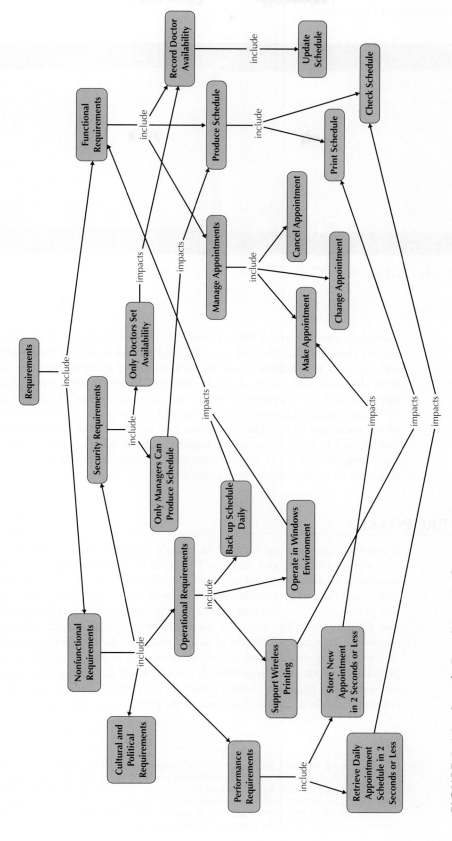

FIGURE 3-14 Sample Requirements Concept Map

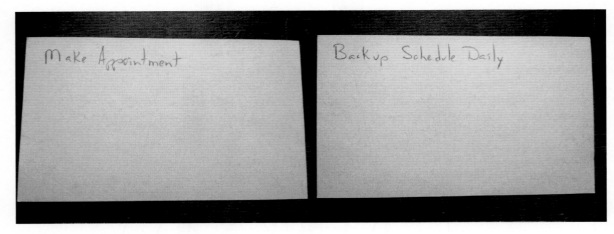

FIGURE 3-15   Sample Story Cards

the requirement (story). If the requirement is deemed to be too large—for example, there are too many tasks on the task list—the requirement is split up into multiple story cards and the tasks are allocated across the new stories. In many shops, once a set of tasks has been identified with a story, the story and its tasks are taped on a wall together so that all members of the development team can see the requirements. The story can be prioritized by importance by placing a rating on the card. The story can also be evaluated for the level of risk associated with it. The importance level and amount of risk associated with the story can be used to help choose which requirements to implement first. Advantages of using story cards and task lists to document requirements is that they are very low tech, high touch, easily updatable, and very portable.

THE SYSTEM PROPOSAL

A system proposal brings together in a single comprehensive document the material created during planning and analysis. The system proposal typically includes an executive summary, the system request, the workplan, the feasibility analysis, the requirements definition, and the evolving models that describe the new system. The evolving models include functional models (see Chapter 4), structural models (see Chapter 5), and behavioral models (see Chapter 6).[15] The executive summary provides all critical information in a very concise form. It can be thought of as a summary of the complete proposal. Its purpose is to allow a busy executive to quickly read through it and determine which parts of the proposal he or she needs to go through more thoroughly. The executive summary is typically no more than a single page long. Figure 3-16 provides a template for a system proposal and references to where the other sections of the proposal are described.

[15] Depending on the client, much more detailed specifications may be required; for example Department of Defense, NASA, IEEE/ANSI, and the Naval Research Laboratory all have very specific formats that must be followed. For more information on these more detailed specifications, see A. M Davis, *Software Requirements, Revision* (Upper Saddle River, NJ: Prentice Hall, 1993); G. Kotonya and I. Sommerville, *Requirements Engineering* (Chichester, England: Wiley, 1998); and R. H. Thayer and M. Dorfman (eds.), *Software Requirements Engineering*, 2nd ed. (Los Alamitos, CA: IEEE Computer Society Press, 1997).

---

1. **Table of Contents**

2. **Executive Summary**

   A summary of all the essential information in the proposal so a busy executive can read it quickly and decide what parts of the proposal to read in more depth.

3. **System Request**

   The revised system request form (see Chapter 2).

4. **Workplan**

   The original workplan, revised after having completed analysis (see Chapter 2).

5. **Feasibility Analysis**

   A revised feasibility analysis, using the information from analysis (see Chapter 2).

6. **Requirements Definition**

   A list of the functional and nonfunctional business requirements for the system (this chapter).

7. **Functional Model**

   An activity diagram, a set of use-case descriptions, and a use-case diagram that illustrate the basic processes or external functionality that the system needs to support (see Chapter 4).

8. **Structural Models**

   A set of CRC cards, class diagram, and object diagrams that describe the structural aspects of the to-be system (see Chapter 5). This may also include structural models of the current as-is system that will be replaced.

9. **Behavioral Models**

   A set of sequence diagrams, communication diagrams, behavioral-state machines, and a CRUDE matrix that describe the internal behavior of the to-be system (see Chapter 6). This may include behavioral models of the as-is system that will be replaced.

10. **Appendices**

    These contain additional material relevant to the proposal, often used to support the recommended system. This might include results of a questionnaire survey or interviews, industry reports and statistics, and so on.

FIGURE 3-16
System Proposal Template

APPLYING THE CONCEPTS AT CD SELECTIONS

In this chapter, we introduced how the requirements are determined in object-oriented systems development projects. Specifically, we described what a requirement is, how to create a requirements definition, and a set of problems that can arise when determining requirements. Next, we reviewed three different requirements analysis strategies, along with a set of techniques that can be used in conjunction with the strategies. After that, we reviewed a set of generic requirements-gathering techniques and a couple of alternative techniques that can be used with an object-oriented system development project. Finally, we showed how the results of the requirements determination processes, along with an updated system request, feasibility analysis, and workplan, are organized into and documented by a system proposal. In this installment of the CD Selections case, we see how Alec and Margaret work through all of these topics with regards to the web-based solution that they hope to create.

SUMMARY

Requirements Determination

Requirements determination is the part of analysis whereby the project team turns the very high-level explanation of the business requirements stated in the system request into a more precise list of requirements. A requirement is simply a statement of what the system must do or what characteristic it needs to have. Business requirements describe the "what" of the systems, and system requirements describe how the system will be implemented. A functional requirement relates directly to a process the system has to perform or information it needs to contain. Nonfunctional requirements refer to behavioral properties that the system must have, such as performance and usability. All the functional and nonfunctional business requirements that fit within the scope of the system are written in the requirements definition, which is used to create other analysis deliverables and leads to the initial design for the new system.

Requirements Analysis Strategies

The basic process of analysis is divided into three steps: understanding the as-is system, identifying improvements, and developing requirements for the to-be system. Three requirements analysis strategies—BPA, BPI, and BPR—help the analyst lead users through the analysis steps so that the vision of the system can be developed. BPA means leaving the basic way the organization operates unchanged and using computer technology to do some of the work. Problem analysis and root cause analysis are two popular BPA techniques. BPI means making moderate changes to the way the organization operates to take advantage of new opportunities offered by technology or to copy what competitors are doing. Duration analysis, activity-based costing, and information benchmarking are three popular BPI activities. BPR means changing the fundamental way the organization operates. Outcome analysis, technology analysis, and activity elimination are three popular BPR activities.

Requirements-Gathering Techniques

Five techniques can be used to gather the business requirements for the proposed system: interviews, joint application development, questionnaires, document analysis, and observation. Interviews involve meeting one or more people and asking them questions. There are five basic steps in the interview process: selecting interviewees, designing interview questions, preparing for the interview, conducting the interview, and performing postinterview follow-up. JAD allows the project team, users, and management to work together to identify requirements for the system. Electronic JAD attempts to overcome common problems associated with groups by using groupware. A questionnaire is a set of written questions for obtaining information from individuals. Questionnaires are often used when information and opinions are needed from a large number of people. Document analysis entails reviewing the documentation and examining the system itself. It can provide insights into the formal and informal system. Observation, the act of watching processes being performed, is a powerful tool for gathering information about the as-is system because it enables the analyst to see the reality of a situation firsthand.

Alternative Requirements Documentation Techniques

In addition to the five traditional approaches to gathering and documenting requirements, a set of alternative approaches may be useful. Concept maps are not limited to supporting hierarchical relationships; they support networked or web-based relationships. Concept maps, therefore, can provide a more complete picture of the relationships among the functional and nonfunctional requirements. Story cards and task lists from the agile methodologies provide a low-tech, high-touch, easily updatable, and very portable approach that users find simple and intuitive to use to document both functional and nonfunctional requirements.

The System Proposal

The system proposal documents the results of the planning and analysis activities in a single comprehensive document. The actual format of the system proposal depends somewhat on the client. For example, the federal government has very specific requirements that a system proposal must meet, whereas a small, locally owned bike shop would be willing to use a much simpler format.

KEY TERMS

Activity elimination, 123
Activity-based costing, 121
Analysis, 118
As-is system, 110
Benchmarking, 121
Bottom-up interview, 128
Breadth of analysis, 123
Business process automation
    (BPA), 116
Business process improvement
    (BPI), 116
Business process reengineering
    (BPR), 116
Business requirements, 112
Closed-ended question, 127
Concept mapping, 144
Concept maps, 144
Critical thinking skills, 118
Document analysis, 138
Duration analysis, 121
Electronic JAD (e-JAD), 133
Facilitator, 133

Formal system, 139
Functional requirements, 112
Ground rules, 135
Informal benchmarking, 121
Informal system, 139
Interpersonal skills, 132
Interview, 126
Interview notes, 130
Interview report, 130
Interview schedule, 126
JAD (joint application
    development), 132
Nonfunctional requirements, 112
Observation, 139
Open-ended question, 127
Outcome analysis,122
Parallelization, 121
Post-session report, 135
Potential business value, 123
Probing question, 127
Problem analysis, 119

Project cost, 123
Questionnaire, 136
Requirement, 112
Requirements definition, 115
Requirements determination, 110
Risk, 124
Root cause, 119
Root cause analysis, 119
Sample, 137
Scribe, 133
Story cards, 144
Structured interview, 128
System proposal, 110
System requirements, 112
Task lists, 144
Technology analysis, 122
To-be system, 110
Top-down interview, 128
Unstructured interview, 128
Walkthrough, 110

QUESTIONS

1. What are the key deliverables that are created during analysis? What is the final deliverable from analysis, and what does it contain?

2. What is the difference between an as-is system and a to-be system?

3. What is the purpose of the requirements definition?

4. What are the three basic steps of the analysis process? Which step is sometimes skipped or done in a cursory fashion? Why?

5. Compare and contrast the business goals of BPA, BPI, and BPR.

6. Compare and contrast problem analysis and root cause analysis. Under what conditions would you use problem analysis? Under what conditions would you use root cause analysis?

7. Compare and contrast duration analysis and activity-based costing.

8. Assuming time and money were not important concerns, would BPR projects benefit from additional time spent understanding the as-is system? Why or why not?

9. What are the important factors in selecting an appropriate analysis strategy?

10. Describe the five major steps in conducting interviews.
11. Explain the differences among a closed-ended question, an open-ended question, and a probing question. When would you use each?
12. Explain the differences between unstructured interviews and structured interviews. When would you use each approach?
13. Explain the difference between a top-down and bottom-up interview approach. When would you use each approach?
14. How are participants selected for interviews and JAD sessions?
15. How can you differentiate between facts and opinions? Why can both be useful?
16. Describe the five major steps in conducting JAD sessions.
17. How does a JAD facilitator differ from a scribe?
18. What are the three primary things that a facilitator does in conducting the JAD session?
19. What is e-JAD and why might a company be interested in using it?
20. How does designing questions for questionnaires differ from designing questions for interviews or JAD sessions?
21. What are typical response rates for questionnaires and how can you improve them?
22. What is document analysis?
23. How does the formal system differ from the informal system? How does document analysis help you understand both?
24. What are the key aspects of using observation in the information-gathering process?
25. Explain factors that can be used to select information-gathering techniques.
26. What is the primary advantage that concept maps have over traditional textual requirements documents techniques?
27. What are some of the advantages of using story cards and task lists as a requirements-gathering and documentation technique?
28. What information is typically included in a system proposal?
29. What is the purpose of the executive summary of the system proposal?

EXERCISE

A. Review the Amazon.com website. Develop the requirements definition for the site. Create a list of functional business requirements that the system meets. What different kinds of nonfunctional business requirements does the system meet? Provide examples for each kind.
B. Suppose you are going to build a new system that automates or improves the interview process for the career services department of your school. Develop a requirements definition for the new system. Include both functional and nonfunctional system requirements. Pretend you will release the system in three different versions. Prioritize the requirements accordingly.
C. Describe in very general terms the as-is business process for registering for classes at your university. What BPA technique would you use to identify improvements? With whom would you use the BPA technique? What requirements-gathering technique would help you apply the BPA technique? List some examples of improvements that you would expect to find.
D. Describe in very general terms the as-is business process for registering for classes at your university. What BPI technique would you use to identify improvements?

With whom would you use the BPI technique? What requirements-gathering technique would help you apply the BPI technique? List some examples of improvements that you would expect to find.
E. Describe in very general terms the as-is business process for registering for classes at your university. What BPR technique would you use to identify improvements? With whom would you use the BPR technique? What requirements-gathering technique would help you apply the BPR technique? List some examples of improvements that you would expect to find.
F. Suppose your university is having a dramatic increase in enrollment and is having difficulty finding enough seats in courses for students. Perform a technology analysis to identify new ways to help students complete their studies and graduate.
G. Suppose you are the analyst charged with developing a new system for the university bookstore so students can order books online and have them delivered to their dorms or off-campus housing. What requirements-gathering techniques will you use? Describe in detail how you would apply the techniques.

H. Suppose you are the analyst charged with developing a new system to help senior managers make better strategic decisions. What requirements-gathering techniques will you use? Describe in detail how you would apply the techniques.

I. Find a partner and interview each other about what tasks each did in the last job you held (full-time, part-time, past, or current). If you haven't worked before, then assume your job is being a student. Before you do this, develop a brief interview plan. After your partner interviews you, identify the type of interview, interview approach, and types of questions used.

J. Find a group of students and run a 60-minute JAD session on improving alumni relations at your university. Develop a brief JAD plan, select two techniques that will help identify improvements, and then develop an agenda. Conduct the session using the agenda, and write your post-session report.

K. Find a questionnaire on the web that has been created to capture customer information. Describe the purpose of the survey, the way questions are worded, and how the questions have been organized. How can it be improved? How will the responses be analyzed?

L. Develop a questionnaire that will help gather information regarding processes at a popular restaurant or the college cafeteria (e.g., ordering, customer service). Give the questionnaire to ten to fifteen students, analyze the responses, and write a brief report that describes the results.

M. Contact the career services department at your university and find all the pertinent documents designed to help students find permanent and/or part-time jobs. Analyze the documents and write a brief report.

MINICASES

1. The Victoria Public Service Council is a small organization with a membership of 1,000 public sector workers in the Australian state of Victoria. Most members work in small towns across the state. The Council was formed in 1922 with the goal of providing education and training to small town government employees (e.g., city managers, police chiefs, firefighters). The Council provides online materials and also organizes a conference each year, bringing together members from all over the state. Members' dues are billed annually. The Council's bookkeeping is handled by the elected treasurer, Jack Archibald. Jack runs unopposed each year at the election, because no one wants to take over the tedious and time-consuming job of tracking memberships for the small $4,000 annual stipend. Several Council members feel that the Council needs to improve its systems before Jack decides to step down from his position.

   The system that is currently used to track the billing and receipt of funds was developed many years ago and is not user-friendly. The company that wrote the system is no longer in business. Questions from members concerning their statements cannot be easily answered. Usually, Jack just jots down the enquiry and calls back the member with the answer. Sometimes he has to calculate manually because the system was not programmed to handle certain types of queries. For example, the membership report does not alphabetize members by city; only the names of cities are listed in alphabetical order.

   What requirements analysis strategy or strategies would you recommend for this situation? Explain your answer.

2. David graduated with a Computing and Information Systems (CIS) major two years ago and has worked as a freelancer ever since. He worked on several, rather small projects for various clients, mostly from his home, while communicating with his clients via the Internet. David's friend, Brian, has just finished his Business and Management major, and decided to start his own DVD rental business. He asked for David's assistance in creating a computer application to automate the activities in the new store. David found the challenge quite interesting, and since he was quite excited about the perspective of helping his good friend, he decided to do so. Enthusiastically, David and Brian went on with their project. They figured that David's knowledge from his OO S&A class and his experience as a freelance software developer, along with Brian's skills acquired as a business major are just sufficient to get the job done. They spent many hours together figuring out all the aspects of the DVD rental business, discussing new ideas, and debating various perspectives of Brian's view about how his business should function. After several months of effort, David managed to build the new system and Brian finally opened his DVD rental store. However, their initially satisfaction of a job well done was short lived as only weeks after the grand opening, problems started to emerge. It turns out both David and Brian

had failed to anticipate some of the problems that came up soon enough in practice. Some were simple ones and relatively easy to fix, like tracking the faulty customers or accounting for multiple copies of the same DVD title. David was able to take care of them without much effort. Other problems turned out to be more difficult, such as DVD title searches that took too long. These got both David and Brian frustrated and disappointed about the outcome of their work. What was the mistake David made with this project? How should he have proceeded in the first place?

3. The Morning Star Tribune is a popular local newspaper with a population of loyal readers consisting mostly of subscribers who renew their subscription on a monthly basis. The newspaper uses a desktop application to register the subscriptions. People can either call the editor's office or simply walk in to renew their subscription. When the Morning Star Tribune's management decided to upgrade their desktop application to a web-based one, they contacted the very same software company that developed the existing application. The new web-based application was expected to improve Morning Star Tribune's business practices by allowing loyal readers to renew their subscriptions online without the hassle of a phone call, or the need to actually pay a visit to the editor's office. There were also other complaints related to the desktop application, such as the inexplicable but steady slowdown of the most frequent operations the application performed. Mr. Pendleton is an experienced system analyst looking toward his retirement within the next four years or so. He was one of the developers who worked on the old application and was now designated to lead the team charged with the upgrade. As expected, Mr. Pendleton proceeded with a comprehensive evaluation of the as-is system, and with the definition of the requirements for the system to be. His analysis also revealed the cause of the "inexplicable" slowdown of the existing application. It turns out that whenever a person renewed his or her subscription, a completely new record of that person's data (first-last name, birth date, address, and other) was redundantly created by the application. Given the situation, Mr. Pendleton concluded that a unique identification feature for each subscriber needed to be added, especially in the web-based version where end users might add an uncontrollable number of spurious records. He first recommended that the person's social security

number be used; when the Morning Star Tribune's management rejected the idea, he proposed a unique 5-digit subscriber ID, which the subscriber would have to keep secret. Needless to say, this idea wasn't accepted either. Mr. Pendleton argued that this was a necessary improvement in the newspaper business practices that would not only prevent the new application from slowing down over time, but would also facilitate new features like giving a one-month bonus subscription to people who subscribed for at least twelve consecutive months. To Mr. Pendleton's surprise, none of his arguments were convincing, and as the relationship with the Morning Star Tribune's management deteriorated, the contract with the software company got cancelled and Mr. Pendleton was fired. What did Mr. Pendleton do wrong? Which of the changes suggested above can be considered as a business practice improvement, and which ones cannot? Explain why. Would you have fired Mr. Pendleton?

4. Anne has been given the task of conducting a survey of sales clerks who will be using a new order-entry system being developed for a household products catalog company. The goal of the survey is to identify the clerks' opinions on the strengths and weaknesses of the current system. There are about 50 clerks who work in three different cities, so a survey seemed like an ideal way of gathering the needed information from the clerks.

Anne developed the questionnaire carefully and pretested it on several sales supervisors who were available at corporate headquarters. After revising it based on their suggestions, she sent a paper version of the questionnaire to each clerk, asking that it be returned within one week. After one week, she had only three completed questionnaires returned. After another week, Anne received just two more completed questionnaires. Feeling somewhat desperate, Anne then sent out an e-mail version of the questionnaire, again to all the clerks, asking them to respond to the questionnaire by e-mail as soon as possible. She received two e-mail questionnaires and three messages from clerks who had completed the paper version expressing annoyance at being bothered with the same questionnaire a second time. At this point, Anne has just a 14% response rate, which she is sure will not please her team leader. What suggestions do you have that could have improved Anne's response rate to the questionnaire?

CHAPTER 4

BUSINESS PROCESS AND FUNCTIONAL MODELING

Functional models describe business processes and the interaction of an information system with its environment. In object-oriented systems development, two types of models are used to describe the functionality of an information system: use cases and activity diagrams. Use cases are used to describe the basic functions of the information system. Activity diagrams support the logical modeling of business processes and workflows. Both can be used to describe the current as-is system and the to-be system being developed. This chapter describes business process and functional modeling as a means to document and understand requirements and to understand the functional or external behavior of the system.

OBJECTIVES

- Understand the process used to identify business processes and use cases.
- Understand the process used to create use-case diagrams.
- Understand the process used to model business processes with activity diagrams.
- Understand the rules and style guidelines for activity diagrams.
- Understand the process used to create use-case descriptions.
- Understand the rules and style guidelines for use-case descriptions.
- Be able to create functional models of business processes using use-case diagrams, activity diagrams, and use-case descriptions.

CHAPTER OUTLINE

INTRODUCTION

The previous chapter discussed the more popular requirements-gathering techniques, such as interviewing, JAD, and observation. Using these techniques, the analyst determined the requirements and created a requirements definition. The requirements definition defined what the system is to do. In this chapter, we discuss how the information that is gathered using these techniques is organized and presented in the form of use-case and activity diagrams and use-case descriptions. Because Unified Modeling Language (UML) has been accepted as the standard notation by the Object Management Group (OMG), almost all object-oriented development projects today use these models to document and organize the requirements that are obtained during the analysis workflow.[1]

A *use case* is a formal way of representing the way a business system interacts with its environment. It illustrates the activities performed by the users of the system. Use-case modeling is often thought of as an external or functional view of a business process in that it shows how the users view the process rather than the internal mechanisms by which the process and supporting systems operate. Use cases can document the current system (i.e., as-is system) or the new system being developed (i.e., to-be system).

An *activity diagram* can be used for any type of process-modeling activity.[2] In this chapter, we describe their use in the context of business process modeling. *Process models* depict how a business system operates. They illustrate the processes or activities that are performed and how objects (data) move among them. A process model can be used to document a current system (i.e., as-is system) or a new system being developed (i.e., to-be system), whether computerized or not. Many different process-modeling techniques are in use today:[3]

Activity diagrams and use cases are *logical models*—models that describe the business domain's activities without suggesting how they are conducted. Logical models are sometimes referred to as *problem domain models*. Reading a use-case or activity diagram, in principle, should not indicate if an activity is computerized or manual, if a piece of information is collected by paper form or via the web, or if information is placed in a filing cabinet or a large database. These physical details are defined during design when the logical models are refined into *physical models*. These models provide information that is needed to ultimately build the system. By focusing on logical activities first, analysts can focus on how the business should run without being distracted with implementation details.

As a first step, the project team gathers requirements from the users (see Chapter 3). Next, using the gathered requirements, the project team identifies the business processes and their environment using use cases and *use-case diagrams*. Use cases are the discrete

---

[1] Other, similar techniques that are commonly used in non-UML projects are task modeling and scenario-based design. For task modeling, see Ian Graham, *Migrating to Object Technology* (Reading, MA: Addison-Wesley, 1995); and Ian Graham, Brian Henderson-Sellers, and Houman Younessi, *The OPEN Process Specification*, (Reading, MA: Addison-Wesley, 1997). For scenario-based design—see John M. Carroll, *Scenario-Based Design: Envisioning Work and Technology in System Development* (New York: Wiley, 1995).

[2] We actually used an activity diagram to describe a simple process in Chapter 1 (see Figure 1-1).

[3] Another commonly used process-modeling technique is IDEF0. IDEF0 is used extensively throughout the U.S. federal government. For more information about IDEF0, see FIPS 183: Integration Definition for Function Modeling (IDEF0), Federal Information Processing Standards Publications (Washington, DC: U.S. Department of Commerce, 1993). From an object-oriented perspective, a good book that uses the UML to address business process modeling is Hans-Erik Eriksson and Magnus Penker, *Business Modeling with UML* (New York: Wiley, 2000). Finally, a new process-modeling technique is BPMN (Business Process Modeling Notation). A good book that compares the notation and use of BPMN to UML's activity diagram is Martin Schedlbauer, *The Art of Business Process Modeling: The Business Analysts Guide to Process Modeling with UML & BPMN* (Sudbury, MA: The Cathris Group, 2010).

activities that the users perform, such as selling CDs, ordering CDs, and accepting returned CDs from customers. Next, users work closely with the team to model the business processes in the form of activity diagrams. Next, the team documents the business processes described in the use-case and activity diagrams by creating a *use-case description* for each use case. Finally, the team verifies and validates the current understanding of the business processes by ensuring that all three models (use-case diagram, activity diagram(s), and use-case descriptions) agree with one another. Once the current understanding of the business processes are fully documented in the functional models, the team is ready to move on to structural modeling (see Chapter 5).

In this chapter, we first describe business process identification using use cases and use-case diagrams. Second, we describe business process modeling with activity diagrams. Third, we describe use-case descriptions, their elements, and a set of guidelines for creating them. Fourth, we describe the process of verification and validation of the business process and functional models.

BUSINESS PROCESS IDENTIFICATION WITH USE CASES AND USE-CASE DIAGRAMS

In the previous chapter, we learned about different strategies and techniques that are useful in identifying the different business processes of a system so that a requirements definition could be created. In this section, we learn how to begin modeling business processes with use cases and the use-case diagram. An analyst can employ use cases and the use-case diagram to better understand the functionality of the system at a very high level. Typically, because a use-case diagram provides a simple, straightforward way of communicating to the users exactly what the system will do, a use-case diagram is drawn when gathering and defining requirements for the system. In this manner, the use-case diagram can encourage the users to provide additional high-level requirements. A use-case diagram illustrates in a very simple way the main functions of the system and the different kinds of users that will interact with it. Figure 4-1 describes the basic syntax rules for a use-case diagram. Figure 4-2 presents a use-case diagram for the doctor's office appointment system introduced in the previous chapter. We can see from the diagram that patients, doctors, and management personnel will use the appointment system to manage appointments, record availability, and produce schedules, respectively. In this section, we describe how to identify the major use cases (business processes) for the new system. However, before we do this, we introduce the elements of the use-case diagram.

Elements of Use-Case Diagrams

The elements of a use-case diagram include actors, use cases, subject boundaries, and a set of relationships among actors, actors and use cases, and use cases. These relationships consist of association, include, extend, and generalization relationships. Each of these elements is described next.

Actors   The stick figures on the diagram represent actors (see Figure 4-1). An *actor* is not a specific user but, instead, is a role that a user can play while interacting with the system. An actor can also represent another system in which the current system interacts. In this case, the actor optionally can be represented by a rectangle containing <<actor>> and the name of the system. Basically, actors represent the principal elements in the environment in which the system operates. Actors can provide input to the system, receive output from

An actor: ■ Is a person or system that derives benefit from and is external to the subject. ■ Is depicted as either a stick figure (default) or, if a nonhuman actor is involved, as a rectangle with <<actor>> in it (alternative). ■ Is labeled with its role. ■ Can be associated with other actors using a specialization/superclass association, denoted by an arrow with a hollow arrowhead. ■ Is placed outside the subject boundary.	**Actor/Role** **<<actor>>** **Actor/Role**
A use case: ■ Represents a major piece of system functionality. ■ Can extend another use case. ■ Can include another use case. ■ Is placed inside the system boundary. ■ Is labeled with a descriptive verb–noun phrase.	**Use Case**
A subject boundary: ■ Includes the name of the subject inside or on top. ■ Represents the scope of the subject, e.g., a system or an individual business process.	**Subject**
An association relationship: ■ Links an actor with the use case(s) with which it interacts.	* *
An include relationship: ■ Represents the inclusion of the functionality of one use case within another. ■ Has an arrow drawn from the base use case to the used use case.	<<include>> ←--------------
An extend relationship: ■ Represents the extension of the use case to include optional behavior. ■ Has an arrow drawn from the extension use case to the base use case.	<<extend>> --------------→
A generalization relationship: ■ Represents a specialized use case to a more generalized one. ■ Has an arrow drawn from the specialized use case to the base use case.	⬆

FIGURE 4-1   Syntax for Use-Case Diagram

the system, or both. The diagram in Figure 4-2 shows that three actors will interact with the appointment system (a patient, a doctor, and management).

Sometimes an actor plays a specialized role of a more general type of actor. For example, there may be times when a new patient interacts with the system in a way that is somewhat different from a general patient. In this case, a *specialized actor* (i.e., new patient) can be placed on the model, shown using a line with a hollow triangle at the end of the more-general actor (i.e., patient). The specialized actor inherits the behavior of the more general

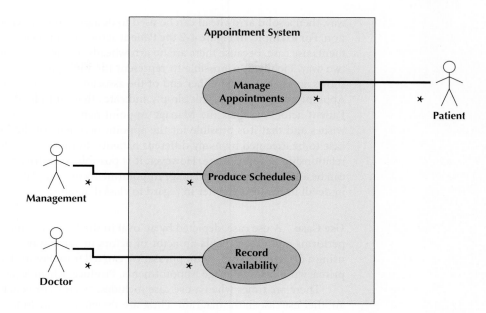

FIGURE 4-2
Use-Case Diagram
for the Appointment
System

actor and extends it in some way (see Figure 4-3). Can you think of some ways a new patient might behave differently from an existing patient?

Association   Use cases are connected to actors through association relationships; these relationships show with which use cases the actors interact (see Figure 4-1). A line drawn from an actor to a use case depicts an association. The association typically represents two-way communication between the use case and the actor. If the communication is only one

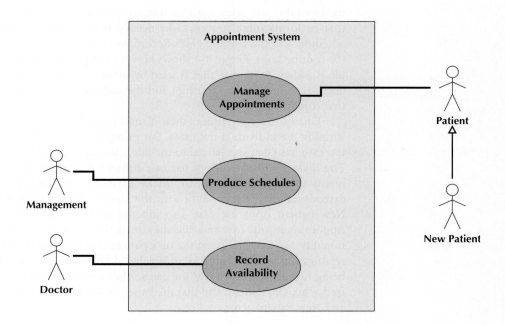

FIGURE 4-3
Use-Case Diagram
with a Specialized
Actor

way, then a solid arrowhead can be used to designate the direction of the flow of information. For example, in Figure 4-2 the Patient actor communicates with the Manage Appointments use case. Because there are no arrowheads on the association, the communication is two-way. Finally, it is possible to represent the multiplicity of the association. Figure 4-2 shows an asterisk (*) at either end of the association between the Patient and the Manage Appointments use case. This simply indicates that an individual patient (instance of the Patient actor) executes the Manage Appointments use case as many times as he or she wishes and that it is possible for the appointment part of the Manage Appointments use case to be executed by many different patients. In most cases, this type of many-to-many relationship is appropriate. However, it is possible to restrict the number of patients that can be associated with the Manage Appointments use case. We discuss the multiplicity issue in detail in the next chapter in regard to class diagrams.

Use Case A use case, depicted by an oval in the UML, is a major process that the system performs and that benefits an actor or actors in some way (see Figure 4-1); it is labeled using a descriptive verb–noun phrase. We can tell from Figure 4-2 that the system has three primary use cases: Manage Appointments, Produce Schedule, and Record Availability.

There are times when a use case includes, extends, or generalizes the functionality of another use case in the diagram. These are shown using include, extend, and generalization relationships. To increase the ease of understanding a use-case diagram, higher-level use cases are normally drawn above the lower-level ones. It may be easier to understand these relationships with the help of examples. Let's assume that every time a patient makes an appointment, the patient is asked to verify payment arrangements. However, it is occasionally necessary to actually make new payment arrangements. Therefore, we may want to have a use case called Make Payment Arrangements that *extends* the Manage Appointments use case to include this additional functionality. In Figure 4-4, an arrow labeled with *extend* was drawn between the Make Payment Arrangements use case and the Manage Appointment use case to denote this special use-case relationship. The Make Payment Arrangements use case was drawn lower than the Manage Appointments use case.

Similarly, there are times when a single use case contains common functions that are used by other use cases. For example, suppose there is a use case called Manage Schedule that performs some routine tasks needed to maintain the doctor's office appointment schedule, and the two use cases Record Availability and Produce Schedule both perform the routine tasks. Figure 4-4 shows how we can design the system so that Manage Schedule is a shared use case that is used by others. An arrow labeled with *include* is used to denote the include relationship, and the included use case is drawn below the use cases that contain it.

Finally, there are times when it makes sense to use a generalization relationship to simplify the individual use cases. For example in Figure 4-4, the Manage Appointments use case has been specialized to include a use case for an Old Patient and a New Patient. The Make Old Patient Appt use case inherits the functionality of the Manage Appointments use case (including the Make Payment Arrangements use-case extension) and extends its own functionality with the Update Patient Information use case. The Make New Patient Appt use case also inherits all the functionality of the generic Manage Appointments use case and calls the Create New Patient use case, which includes the functionality necessary to insert the new patient into the patient database. The generalization relationship is represented as an unlabeled hollow arrow with the more general use case being higher than the lower use cases. Also, notice that we have added a second specialized actor, Old Patient, and that the Patient actor is now simply a generalization of the Old and New Patient actors.

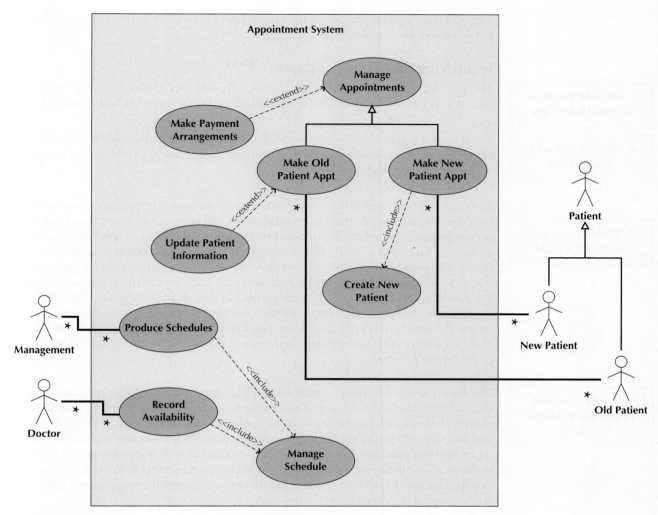

FIGURE 4-4 Extend and Include Relationships

Subject Boundary The use cases are enclosed within a *subject boundary,* which is a box that defines the scope of the system and clearly delineates what parts of the diagram are external or internal to it (see Figure 4-1). One of the more difficult decisions to make is where to draw the subject boundary. A subject boundary can be used to separate a software system from its environment, a subsystem from other subsystems within the software system, or an individual process in a software system. They also can be used to separate an information system, including both software and internal actors, from its environment. Care should be taken to decide what the scope of the information system is to be.

The name of the subject can appear either inside or on top of the box. The subject boundary is drawn based on the scope of the system. In the appointment system, we assumed that the Management and Doctor actors are outside of the scope of the system, that is, they use the system. We could have included a receptionist as an actor. However, in this case, we assumed that the receptionist is an internal actor who is part of the Manage

Appointments use case with which the Patient actor interacts. Therefore, the receptionist is not drawn on the diagram.[4]

Identifying the Major Use Cases

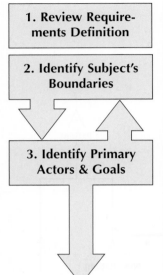

The first step is to review the requirements definition (see Figure 3-3). This helps the analyst to get a complete overview of the underlying business process being modeled.

The second step is to identify the subject's boundaries. This helps the analyst to identify the scope of the system. However, as we work through the development process, the boundary of the system most likely will change.

The third step is to identify the primary actors and their goals. The primary actors involved with the system comes from a list of stakeholders and users. Recall that a stakeholder is a person, group, or organization that can affect (or will be affected by) a new system, whereas an actor is a role that a stakeholder or user plays, not a specific user (e.g., doctor, not Dr. Jones). The goals represent the functionality that the system must provide the actor for the system to be a success. Identifying the tasks that each actor must perform can facilitate this. For example, does the actor need to create, read, update, delete, or execute (CRUDE)[5] any information currently in the system, are there any external changes of which an actor must inform the system, or is there any information that the system should give the actor? Steps 2 and 3 are intertwined. As actors are identified and their goals are uncovered, the boundary of the system will change.

The fourth step is to simply identify the business processes and major use cases. Rather than jumping into one use case and describing it completely at this point, we only want to identify the use cases. Identifying only the major use cases at this time prevents the users and analysts from forgetting key business processes and helps the users explain the overall set of business processes for which they are responsible. It is important at this point to understand and define acronyms and jargon so that the project team and others from outside the user group can clearly understand the use cases. Again, the requirements definition is a very useful beginning point for this step.

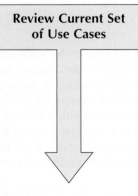

The fifth step is to carefully review the current set of use cases. It may be necessary to split some of them into multiple use cases or merge some of them into a single use case. Also, based on the current set, a new use case may be identified. You should remember that identifying use cases is an iterative process, with users often changing their minds about what a use case is and what it includes. It is very easy to get trapped in the details at this point, so you need to remember that the goal at this step is to only identify the *major* use cases. For example, in the doctor's office example in Figure 4-2, we defined one use case as Manage Appointments. This use case included the cases for both new patients and existing patients, as well as for when a patient changes or cancels an appointment. We could have defined each of these activities (makes an appointment, changes an appointment, or cancels an appointment) as separate use cases, but this would have created a huge set of small use cases.

---

[4] In other non-UML approaches to object-oriented systems development, it is possible to represent external actors along with internal actors. In this example, the receptionist would be considered an internal actor (see Graham, *Migrating to Object Technology*, and Graham, Henderson-Sellers, and Younessi, *The OPEN Process Specification*).

[5] We describe the use of CRUDE analysis and matrices in Chapter 6.

The trick is to select the right size so that you end up with three to nine use cases in each system. If the project team discovers many more than eight use cases, this suggests that the use cases are too small or that the system boundary is too big. If more than nine use cases exist, the use cases should be grouped together into *packages* (i.e., logical groups of use cases) to make the diagrams easier to read and keep the models at a reasonable level of complexity. It is simple at that point to sort the use cases and group together these small use cases into larger use cases that include several small ones or to change the system boundaries.[6]

Creating a Use-Case Diagram

Basically, drawing the use-case diagram is very straightforward once use cases have been detailed. The actual use-case diagram encourages the use of information hiding. The only parts drawn on the use-case diagram are the system boundary, the use cases themselves, the actors, and the various associations between these components. The major strength of the use-case diagram is that it provides the user with an overview of the business processes. However, remember that any time a use case changes, it could affect the use-case diagram. There are four major steps in drawing a use-case diagram.

1. Place & Draw Use Cases

First, we place and draw the use cases on the diagram. These are taken directly from the major use cases previously identified. Special use-case associations (include, extend, or generalization) are also added to the model at this point. Be careful in laying out the diagram. There is no formal order to the use cases, so they can be placed in whatever fashion is needed to make the diagram easy to read and to minimize the number of lines that cross. It often is necessary to redraw the diagram several times with use cases in different places to make the diagram easy to read. Also, for understandability purposes, there should be no more than three to nine use cases on the model so the diagram is as simple as possible. These include use cases that have been factored out and now are associated with another use case through the include, extend, or generalization relationships.

2. Place & Draw Actors

Second, the actors are placed and drawn on the diagram. Like use-case placement, to minimize the number of lines that cross on the diagram, the actors should be placed near the use cases with which they are associated.

3. Draw Subject Boundary

Third, the subject boundary is drawn. This forms the border of the subject, separating use cases (i.e., the subject's functionality) from actors (i.e., the roles of the external users).

4. Add Associations

The fourth and last step is to add associations by drawing lines to connect the actors to the use cases with which they interact. No order is implied by the diagram, and the items added along the way do not have to be placed in a particular order; therefore, it might help to rearrange the symbols a bit to minimize the number of lines that cross, making the diagram less confusing.

YOUR TURN   4-1 Use–Case Diagram

*L*ook at the use-case diagram in Figure 4-4. Consider if a use case were added to maintain patient insurance information.   Make assumptions about the details of this use case and add it to the existing use-case diagram.

---

[6] For those familiar with structured analysis and design, packages serve a similar purpose as the leveling and balancing processes used in data flow diagramming. Packages are described in Chapter 7.

YOUR TURN

4-2 Campus Housing

Identify a set of major use cases for the following high-level business processes in a housing system run by the campus housing service. The campus housing service helps students find apartments. Apartment owners fill in information forms about the rental units they have available (e.g., location, number of bedrooms, monthly rent), which are then entered into a database. Students can search through this database via the web to find apartments that meet their needs (e.g., a two-bedroom apartment for $400 or less per month within a half mile of campus). They then contact the apartment owners directly to see the apartment and possibly rent it. Apartment owners call the service to delete their listing when they have rented their apartment(s).

Based on those use cases, create a use-case diagram.

Example   The functional requirements for an automated university library circulation system include the need to support searching, borrowing, and book-maintenance activities. The system should support searching by title, author, keywords, and ISBN. Searching the library's collection database should be available on terminals in the library and available to potential borrowers via the web. If the book of interest is currently checked out, a valid borrower should be allowed to request the book to be returned. Once the book has been checked back in, the borrower requesting the book should be notified of the book's availability.

The borrowing activities are built around checking books out and returning books by borrowers. There are three types of borrowers: students, faculty or staff, and guests. Regardless of the type of borrower, the borrower must have a valid ID card. If the borrower is a student, having the system check with the registrar's student database validates the ID card. If the borrower is a faculty or staff member, having the system check with the personnel office's employee database validates the ID card. If the borrower is a guest, the ID card is checked against the library's own borrower database. If the ID card is valid, the system must also check to determine whether the borrower has any overdue books or unpaid fines. If the ID card is invalid, or if the borrower has overdue books or unpaid fines, the system must reject the borrower's request to check out a book; otherwise, the borrower's request should be honored. If a book is checked out, the system must update the library's collection database to reflect the book's new status.

The book-maintenance activities deal with adding and removing books from the library's book collection. This requires a library manager to both logically and physically add and remove the book. Books being purchased by the library or books being returned in a damaged state typically cause these activities. If a book is determined to be damaged when it is returned and it needs to be removed from the collection, the last borrower will be assessed a fine. However, if the book can be repaired, depending on the cost of the repair, the borrower might not be assessed a fine. Every Monday, the library sends reminder emails to borrowers who have overdue books. If a book is overdue more than two weeks, the borrower is assessed a fine. Depending on how long the book remains overdue, the borrower can be assessed additional fines every Monday.

To begin with, we need to identify the major use cases and create a use-case diagram that represents the high-level business processes in the business situation just described. Based on the steps to identify the major use cases, we need to review the requirements definition and identify the boundaries (scope) of the problem. Based on the description of the problem, it is obvious that the system to be created is limited to managing the library's book collection. The next thing we need to do is to identify the primary actors and business processes that need to be supported by the system. Based on the functional requirements described, the primary

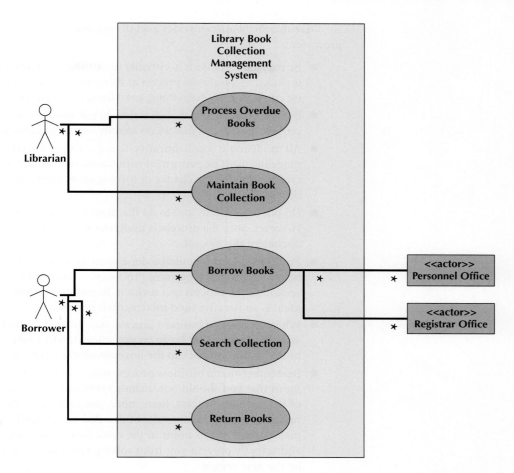

FIGURE 4-5

Library Book Collection
Management System
Use-Case Diagram

actors are borrowers and librarians, whereas the primary business processes are borrowing books, returning books, searching the book collection, maintaining the book collection, and processing overdue books. Now that we have identified all of the actors and major use cases, we can draw the use-case diagram that represents an overview of the library's book collection management system (see Figure 4-5). Notice the nonhuman actors (Personnel Office and Registrar Office) that were added.

BUSINESS PROCESS MODELING WITH ACTIVITY DIAGRAMS

Business process models describe the different activities that, when combined, support a business process. Business processes typically cut across functional departments (e.g., the creation of a new product involves many different activities that combine the efforts of many employees in many departments). From an object-oriented perspective, they cut across multiple objects. Many of the earlier object-oriented systems development approaches tended to ignore business process modeling. However, today we realize that modeling business processes themselves is a very constructive activity that can be used to make sense of the gathered requirements (see Chapter 3). The one potential problem of building business process models, from an object-oriented systems development perspective, is that they tend to reinforce a functional decomposition mindset. However, as long as they are used properly, business process models are very powerful tools for communicating the analyst's current understanding of the requirements to the user.

Martin Schedlbauer provides a set of best practices to follow when modeling business processes.[7]

- Be realistic, because it is virtually impossible to identify everything that is included in a business process at this point in the evolution of the system. Even if we could identify everything, everything is not equally important.

- Be agile because even though we might not identify every single feature of a business process, the features that we do identify should be identified in a rigorous manner.

- All modeling is a collaborative/social activity. Therefore, business process modeling must be performed with teams, not by individuals. When an individual creates a model, the chance of mixing up or omitting important tasks is greatly increased.

- Do not use a CASE tool to do the modeling but use whiteboards instead. However, once the process is understood, it is a good idea to use a CASE tool to document the process.

- Process modeling should be done in an iterative manner. In other words, as you better understand a business process, you will need to return to the documented version of the process and revise it. Remember, object-oriented system development is an iterative (and incremental) process.

- When modeling a business process, stay focused on that specific process. If tasks associated with other business processes are identified, simply record them on a to-do list and get back to the business process that you are currently modeling.

- Remember that a business process model is an abstraction of reality. By that, we mean that you should not include every minor task in the current description of the business process. Remember, you cannot afford to lose sight of the proverbial forest for the sake of detailed understanding of a single tree. Too many details at this point in the evolution of the system can cause confusion and actually prevent you from solving the underlying problem being addressed by the new system.

In this section of the chapter, we introduce the use of UML's activity diagrams as a means to document business process models. Activity diagrams are used to model the behavior in a business process independent of objects. In many ways, activity diagrams can be viewed as sophisticated data flow diagrams that are used in conjunction with structured analysis; however, unlike data flow diagrams, activity diagrams include notation that addresses the modeling of parallel, concurrent activities and complex decision processes.[8] Activity diagrams can be used to model everything from a high-level business workflow that involves many different use cases, to the details of an individual use case, all the way down to the specific details of an individual method. In a nutshell, activity diagrams can be used to model any type of process.[9] In this chapter, we restrict our coverage of activity diagrams to documenting and modeling high-level business processes.

---

[7] Martin Schedlbauer, *The Art of Business Process Modeling: The Business Analysts Guide to Process Modeling with UML & BPMN* (Sudbury, MA: The Cathris Group, 2010).

[8] For a good introduction to data flow diagrams and structured approaches to systems analysis and design, see Alan Dennis, Barbara Haley Wixom, and Roberta M. Roth, *Systems Analysis & Design*, 4th ed. (New York: Wiley, 2009).

[9] Technically speaking, activity diagrams combine process-modeling ideas from many different techniques including event models, statecharts, and Petri nets. However, UML 2.0's activity diagram has more in common with Petri nets than the other process-modeling techniques. For a good description of using Petri nets to model business workflows, see Wil van der Aalst and Kees van Hee, *Workflow Management: Models, Methods, and Systems* (Cambridge, MA: MIT Press, 2002).

Elements of an Activity Diagram

Activity diagrams portray the primary activities and the relationships among the activities in a process. Figure 4-6 shows the syntax of an activity diagram. Figure 4-7 presents a simple activity diagram that represents the Manage Appointments use case of the appointment system for the doctor's office example.[10]

Actions and Activities   *Actions* and *activities* are performed for some specific business reason. Actions and activities can represent manual or computerized behavior. They are depicted in an activity diagram as a rounded rectangle (see Figure 4-6). They should have a name that begins with a verb and ends with a noun (e.g., Get Patient Information or Make Payment Arrangements). Names should be short, yet contain enough information so that the reader can easily understand exactly what they do. The only difference between an action and an activity is that an activity can be decomposed further into a set of activities and/or actions, whereas an action represents a simple nondecomposable piece of the overall behavior being modeled. Typically, only activities are used for business process or workflow modeling. In most cases, each activity is associated with a use case. The activity diagram in Figure 4-7 shows a set of separate but related activities for the Manage Appointments use case (see Figures 4-2, 4-3, and 4-4): Get Patient Information, Update Patient Information, Create New Patient, Make Payment Arrangements, Make New Appointment, Change Appointment, and Cancel Appointment. Notice that the Make Payment Arrangements and Make New Appointment activities appear twice in the diagram; once for an "old" patient and once for a "new" patient.

Object Nodes   Activities and actions typically modify or transform objects. *Object nodes* model these objects in an activity diagram. Object nodes are portrayed in an activity diagram as rectangles (see Figure 4-6). The name of the class of the object is written inside the rectangle. Essentially, object nodes represent the flow of information from one activity to another activity. The simple appointment system portrayed in Figure 4-7 shows object nodes flowing from Get Patient Information activity.

Control Flows and Object Flows   There are two different types of flows in activity diagrams: control and object (see Figure 4-6). *Control flows* model the paths of execution through a business process. A control flow is portrayed as a solid line with an arrowhead on it showing the direction of flow. Control flows can be attached only to actions or activities. Figure 4-7 portrays a set of control flows through the doctor's office's appointment system. *Object flows* model the flow of objects through a business process. Because activities and actions modify or transform objects, object flows are necessary to show the actual objects that flow into and out of the actions or activities.[11] An object flow is depicted as a dashed line with an arrowhead on it showing the direction of flow. An individual object flow must be attached to an action or activity on one end and an object node on the other end. Figure 4-9 portrays a set of control and object flows through the appointment system of a doctor's office.

---

[10] Owing to the actual complexity of the syntax of activity diagrams, we follow a minimalist philosophy in our coverage [see John M. Carrol, *The Nurnberg Funnel: Designing Minimalist Instruction for Practical Computer Skill* (Cambridge, MA: MIT Press, 1990)]. However, the material contained in this section is based on the *Unified Modeling Language: Superstructure Version 2.5, ptc/2010-11-14* (www.uml.org). Additional useful references include Michael Jesse Chonoles and James A. Schardt, *UML 2 for Dummies* (Indianapolis, IN: Wiley, 2003); Hans-Erik Eriksson, Magnus Penker, Brian Lyons, and David Fado, *UML 2 Toolkit* (Indianapolis: Wiley, 2004); and Kendall Scott, *Fast Track UML 2.0* (Berkeley, CA: Apress, 2004). For a complete description of all diagrams, see www.uml.org.

[11] These are identical to data flows in data flow diagrams.

An action: ■ Is a simple, nondecomposable piece of behavior. ■ Is labeled by its name.	**Action**
An activity: ■ Is used to represent a set of actions. ■ Is labeled by its name.	**Activity**
An object node: ■ Is used to represent an object that is connected to a set of object flows. ■ Is labeled by its class name.	**Class Name**
A control flow: ■ Shows the sequence of execution.	⟶
An object flow: ■ Shows the flow of an object from one activity (or action) to another activity (or action).	⤑
An initial node: ■ Portrays the beginning of a set of actions or activities.	●
A final-activity node: ■ Is used to stop all control flows and object flows in an activity (or action).	⊗
A final-flow node: ■ Is used to stop a specific control flow or object flow.	◉
A decision node: ■ Is used to represent a test condition to ensure that the control flow or object flow only goes down one path. ■ Is labeled with the decision criteria to continue down the specific path.	[Decision Criteria] [Decision Criteria]
A merge node: ■ Is used to bring back together different decision paths that were created using a decision node.	
A fork node: ■ Is used to split behavior into a set of parallel or concurrent flows of activities (or actions).	
A join node: ■ Is used to bring back together a set of parallel or concurrent flows of activities (or actions).	
A swimlane: ■ Is used to break up an activity diagram into rows and columns to assign the individual activities (or actions) to the individuals or objects that are responsible for executing the activity (or action). ■ Is labeled with the name of the individual or object responsible.	**Swimlane**

FIGURE 4-6　Syntax for an Activity Diagram

Control Nodes　There are seven different types of *control nodes* in an activity diagram: initial, final-activity, final-flow, decision, merge, fork, and join (see Figure 4-6). An *initial node* portrays the beginning of a set of actions or activities.[12] An initial node is shown as a small, filled-in circle. A *final-activity node* is used to stop the process being modeled. Any time a final-activity

[12] For those familiar with IBM flowcharts, this is similar to the start node.

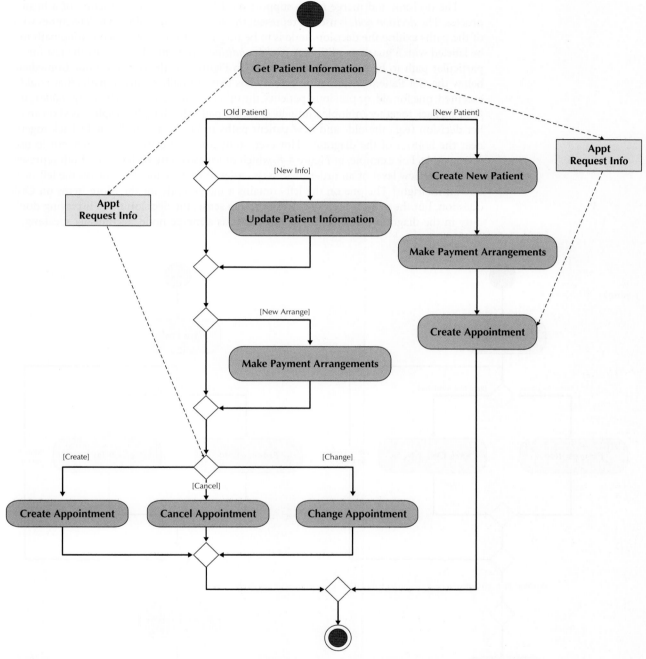

FIGURE 4-7   Activity Diagram for the Manage Appointments Use Case

node is reached, all actions and activities are ended immediately, regardless of whether they are completed. A final-activity node is represented as a circle surrounding a small, filled-in circle, making it resemble a bull's-eye. A *final-flow* node is similar to a final-activity node, except that it stops a specific path of execution through the business process but allows the other concurrent or parallel paths to continue. A final-flow node is shown as a small circle with an X in it.

The decision and merge nodes support modeling the decision structure of a business process. The *decision node* is used to represent the actual test condition that determines which of the paths exiting the decision node is to be traversed. In this case, each exiting path must be labeled with a guard condition. A *guard condition* represents the value of the test for that particular path to be executed. For example, in Figure 4-7, the decision node immediately below the Get Patient Information activity has two mutually exclusive paths that could be executed: one for old, or previous, patients, the other for new patients. The *merge node* is used to bring back together multiple mutually exclusive paths that have been split based on an earlier decision (e.g., the old- and new-patient paths in Figure 4-7 are brought back together near the bottom of the diagram). However, sometimes, for clarity, it is better not to use a merge node. For example, in Figure 4-8, which of the two activity diagrams, both representing an overview level of an order process, is easier to understand, the one on the left or the one on the right? The one on the left contains a merge node for the More Items on Order question, but the one on the right does not. In a sense, the decision node is playing double duty in the diagram on the right: It also serves as a merge node. Technically speaking, we

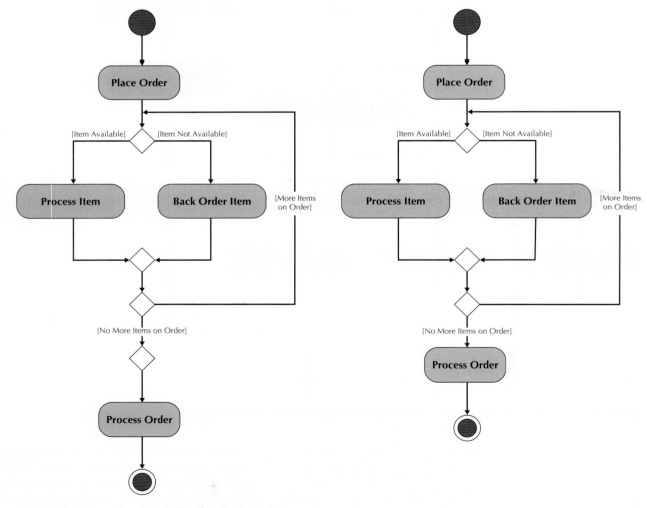

FIGURE 4-8   Two Very Similar Activity Diagrams

should not omit the merge node; however, sometimes being technically correct according to the UML's diagramming rules actually causes the diagram to become confusing. From a business process modeling perspective, a good deal of common sense can go a long way.

The fork and join nodes allow parallel and concurrent processes to be modeled (see Figure 4-6). The *fork node* is used to split the behavior of the business process into multiple parallel or concurrent flows. Unlike the decision node, the paths are not mutually exclusive (i.e., both paths are executed concurrently). For example, in Figure 4-9, the fork node

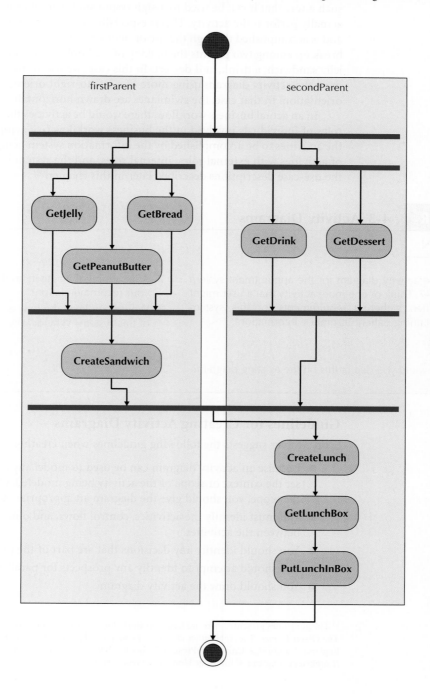

FIGURE 4-9

Activity Diagram for Making a School Box Lunch

is used to show that two concurrent, parallel processes are to be executed. In this case, each process is executed by two separate processors (parents). The purpose of the join node is similar to that of the merge node. The *join node* simply brings back together the separate parallel or concurrent flows in the business process into a single flow.

Swimlanes   Activity diagrams can model a business process independent of any object implementation. However, there are times when it helps to break up an activity diagram in such a way that it can be used to assign responsibility to objects or individuals who would actually perform the activity. This is especially useful when modeling a business workflow and is accomplished through the use of *swimlanes*. In Figure 4-9, the swimlanes are used to break up among two parents the making of a school lunch comprising a peanut butter and jelly sandwich, a drink, and dessert. In this case, we use vertical swimlanes. We could also draw the activity diagram using more of a left-to-right orientation instead of a top-down orientation. In that case, the swimlanes are drawn horizontally.

In an actual business workflow, there would be activities that should be associated with roles of individuals involved in the business workflow (e.g., employees or customers) and the activities to be accomplished by the information system being created. This association of activities with external roles, internal roles, and the system is very useful when creating the use-case descriptions described later in this chapter.

YOUR
TURN

4-3 Activity Diagrams

*L*ook at the activity diagram for the appointment system in Figure 4-7. Think of one more activity that a user might ask for when gathering requirements for this system (e.g., maintaining patient insurance information).

Questions

1. How would you depict this on the existing diagram?

2. After adding the activity to the diagram, what is your recommendation?

3. Would you keep the new activity within the scope of this system? Why or why not?

Guidelines for Creating Activity Diagrams

Scott Ambler suggests the following guidelines when creating activity diagrams:[13]

- Because an activity diagram can be used to model any kind of process, you should set the context or scope of the activity being modeled. Once you have determined the scope, you should give the diagram an appropriate title.
- You must identify the activities, control flows, and object flows that occur between the activities.
- You should identify any decisions that are part of the process being modeled.
- You should attempt to identify any prospects for parallelism in the process.
- You should draw the activity diagram.

---

[13] The guidelines presented here are based on work done by Scott Ambler. For more details, see Scott W. Ambler, *The Object Primer: The Application Developer's Guide to Object Orientation and the UML,* 2nd ed. (Cambridge, England: Cambridge University Press/SIGS Books, 2001); and Scott W. Ambler, *The Elements of UML Style* (Cambridge, England: Cambridge University Press, 2003).

When drawing an activity diagram, the diagram should be limited to a single initial node that starts the process being modeled. This node should be placed at the top or top left of the diagram, depending on the complexity of the diagram. For most business processes, there should only be a single final-activity node. This node should be placed at the bottom or bottom right of the diagram (see Figures 4-7, 4-8, and 4-9). Because most high-level business processes are sequential, not parallel, the use of a final-flow node should be limited.

When modeling high-level business processes or workflows, only the more important decisions should be included in the activity diagrams. In those cases, the guard conditions associated with the outflows of the decision nodes should be mutually exclusive. The outflows and guard conditions should form a complete set (i.e., all potential values of the decision are associated with one of the flows).

As in decision modeling, forks and joins should be included only to represent the more important parallel activities in the process. For example, an alternative version of Figure 4-9 might not include the forks and joins associated with the Get Jelly, Get Bread, Get Peanut Butter, Get Drink, and Get Dessert activities. This would greatly simplify the diagram.[14]

When laying out the activity diagram, line crossings should be minimized to enhance the readability of the diagram. The activities on the diagram should also be laid out in a left-to-right and/or top-to-bottom order based on the order in which the activities are executed. For example, in Figure 4-9, the Create Sandwich activity takes place before the Create Lunch activity.

Swimlanes should be used only to simplify the understanding of an activity diagram. Furthermore, the swimlanes should enhance the readability of a diagram. For example, when using a horizontal orientation for swimlanes, the top swimlane should represent the most important object or individual involved with the process. The order of the remaining swimlanes should be based on minimizing the number of flows crossing the different swimlanes. Also, when there are object flows among the activities associated with the different individuals (swimlanes) executing the activities of the process, it is useful to show the actual object flowing from one individual to another individual by including an object node between the two individuals (i.e., between the two swimlanes). This, of course, affects how the swimlanes should be placed on the diagram.

Finally, any activity that does not have any outflows or any inflows should be challenged. Activities with no outflows are referred to as *black-hole activities*. If the activity is truly an end point in the diagram, the activity should have a control flow from it to a final-activity or final-flow node. An activity that does not have any inflow is known as a *miracle activity*. In this case, the activity is missing an inflow either from the initial node of the diagram or from another activity.

Creating Activity Diagrams

1. Choose a Business Process

There are five steps in creating an activity diagram to document and model a business process. First, you must choose a business process that was previously identified to model. To do this, you should review the requirements definition (see Figure 3-3) and the use-case diagram (see Figures 4-2, 4-3, and 4-4) created to represent the requirements. You should also review all of the documentation created during the requirements-gathering process (see Chapter 3), for example, reports created that documented interviews or observations, any output from any JAD sessions, any analysis of any questionnaires used, and any story cards or task lists created. In most cases, the use cases on the use-case diagram will be the

---

[14] In fact, the only reason we depicted the diagram in Figure 4-9 with the multiple fork and join nodes was to demonstrate that it could be done.

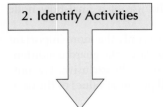

best place to start. For example, in the appointment system, we had identified three primary use cases: Manage Appointments, Produce Schedule, and Record Doctor Availability. We also identified a whole set of minor use cases (these will be useful in identifying the elements of the activity diagram).

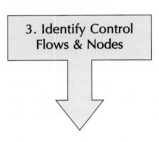

Second, identify the set of activities necessary to support the business process. For example, in Figure 3-3, three processes are identified as being part of the Manage Appointments business process. Also, by reviewing the use-case diagram (see Figure 4-4), we see that five minor use cases are associated with the Manage Appointments major use case. Based on this information, we can identify a set of activities. In this case, the activities are Update Patient Information, Make Payment Arrangements, Create New Patient, Create Appointment, Cancel Appointment, and Change Appointment.

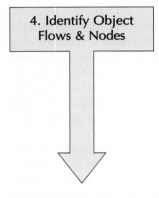

Third, identify the control flows and nodes necessary to document the logic of the business process. For example, in Figure 4-4, the Make Payment Arrangements and Update Patient Information use cases are extensions to the Manage Appointments and Make Old Patient Appt use cases. We know that these use cases are executed only in certain circumstances. From this we can infer that the activity diagram must include some decision and merge nodes. Based on the requirements definition (see Figure 3-3), we can infer another set of decision and merge nodes based on the Create Appointment, Cancel Appointment, and Change Appointment activities identified in the previous step.

Fourth, identify the object flows and nodes necessary to support the logic of the business process. Typically, object nodes and flows are not shown on many activity diagrams used to model a business process. The primary exception is if information captured by the system in one activity is used in an activity that is performed later, but *not* immediately after the activity that captured the information. In the appointment example, it is obvious that we need to be able to determine whether the patient is an old or new patient and the type of action that the patient would like to have performed (create, cancel, or change an appointment). It is obvious that a new patient cannot cancel or change an appointment because the patient is, by definition, a new patient. Obviously, we need to capture this type of information at the beginning of the business process and use it when required. For example, in the appointment problem, we need to have a Get Patient Information activity that captures the appropriate information and makes it available at the appropriate time in the process.

Fifth, lay out and draw the activity diagram to document the business process. For esthetic and understandability reasons, just as when drawing a use-case diagram, you should attempt to minimize potential line crossings. Based on the previous steps and carefully laying out the diagram, the activity diagram in Figure 4-7 was created to document the Manage Appointments business process.

Example   The first step is to choose a business process to model. In this case, we want to create an activity diagram for the Borrow Books use case (see Figure 4-5). The functional requirements for this use case were:

> The borrowing activities are built around checking books out and returning books by borrowers. There are three types of borrowers: students, faculty or staff, and guests. Regardless of the type of borrower, the borrower must have a valid ID card. If the borrower is a student, having the system check with the registrar's student database validates the ID card. If the borrower is a faculty or staff member, having the system check with the personnel office's employee database validates the ID card.

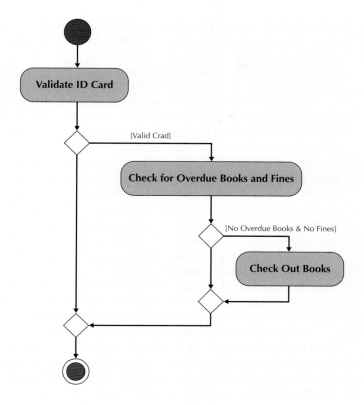

FIGURE 4-10

Activity Diagram for the Borrow Books Use Case

If the borrower is a guest, the ID card is checked against the library's own borrower database. If the ID card is valid, the system must also check to determine whether the borrower has any overdue books or unpaid fines. If the ID card is invalid, or if the borrower has overdue books or unpaid fines, the system must reject the borrower's request to check out a book; otherwise, the borrower's request should be honored.

The second step to model a business process is to identify the activities that make up the process. Based on the requirements for the Borrow Books use case, we can identify three major activities: Validate ID Card, Check for Overdue Books and Fines, and Check Out Books. The third step is to identify the control flows and control nodes necessary to model the decision logic of the business process. In this case, there obviously will have to be an initial node, a final-flow node, and a set of decision and merge nodes for each decision to be made. The fourth step is to identify the object flows and object nodes necessary to complete the description of the business process. In this case, there really is no need to include object nodes and flows. Finally, we can lay out the diagram (see Figure 4-10).

BUSINESS PROCESS DOCUMENTATION WITH USE CASES AND USE-CASE DESCRIPTIONS

Use-case diagrams provided a bird's-eye view of the basic functionality of the business processes contained in the evolving system. Activity diagrams, in a sense, open up the black box of each business process by providing a more-detailed graphical view of the underlying activities that support each business process. Use-case descriptions provide a means to more fully document the different aspects of each individual use case.[15] The use-case descriptions are based on the identified requirements, use-case diagram, and the activity diagram descriptions of the business processes. Use-case descriptions contain all the information needed to document the functionality of the business processes.[16]

Use cases are the primary drivers for all the UML diagramming techniques. A use case communicates at a high level what the system needs to do, and all the UML diagramming techniques build on this by presenting the use-case functionality in a different way for a different purpose. Use cases are the building blocks by which the system is designed and built.

Use cases capture the typical interaction of the system with the system's users (end users and other systems). These interactions represent the external, or functional, view of the system from the perspective of the user. Each use case describes one and only one function in which users interact with the system.[17] Although a use case may contain several paths that a user can take while interacting with the system (e.g., when searching for a book in a Web bookstore, the user might search by subject, by author, or by title), each possible execution path through the use case is referred to as a *scenario*. Another way to look at a scenario is as if a scenario is an instantiation of a specific use case. Scenarios are used extensively in behavioral modeling (see Chapter 6). Finally, by identifying all scenarios and trying to execute them through role-playing CRC cards (see Chapter 5), you will be testing the clarity and completeness of your evolving understanding of the system being developed.[18]

When creating use-case descriptions, the project team must work closely with the users to fully document the functional requirements. Organizing the functional requirements and documenting them in a use-case description is a relatively simple process, but it takes considerable practice to ensure that the descriptions are complete enough to use in structural (Chapter 5) and behavioral (Chapter 6) modeling. The best place to begin is to review the use-case and activity diagrams. The key thing to remember is that each use case is associated with one and only one *role* that users have in the system. For example, a receptionist in a doctor's office may play multiple roles—he or she can make appointments, answer the telephone, file medical records, welcome patients, and so on. It is possible that

---

[15] For a more detailed description of use-case modeling, see Alistair Cockburn, *Writing Effective Use Cases* (Reading, MA: Addison-Wesley, 2001).

[16] Nonfunctional requirements, such as reliability requirements and performance requirements, are often documented outside of the use case through more traditional requirements documents. See Gerald Kotonya and Ian Sommerville, *Requirements Engineering* (Chichester, England: Wiley, 1998); Benjamin L. Kovitz, *Practical Software Requirements: A Manual of Content & Style* (Greenwich, CT: Manning, 1999); Dean Leffingwell and Don Widrig, *Managing Software Requirements: A Unified Approach* (Reading, MA: Addison-Wesley, 2000); and Richard H. Thayer, M. Dorfman, and Sidney C. Bailin (eds.), *Software Requirements Engineering*, 2nd ed. (Los Alamitos, CA: IEEE Computer Society, 1997).

[17] This is one key difference between traditional structured analysis and design approaches and object-oriented approaches. If you have experience using traditional structured approaches (or have taken a course on them), then this is an important change for you. If you have no experience with structured approaches, then you can skip this footnote.

The traditional structured approach is to start with one overall view of the system and to model processes via functional decomposition—the gradual decomposition of the overall view of the system into the smaller and smaller parts that make up the whole system. On the surface, this is similar to business-process modeling using activity diagrams. However, functional decomposition is *not* used with object-oriented approaches. Instead, each of the use cases documents one individual piece of the system; there is no overall use case that documents the entire system in the same way that a level 0 data flow diagram attempts to document the entire system. By removing this overall view, object-oriented approaches make it easier to decouple the system's objects so they can be designed, developed, and reused independently of the other parts of the system. Although this lack of an overall view might prove unsettling initially, it is very liberating over the long term.

[18] For presentation purposes, we defer discussion of role-playing to Chapter 5.

multiple users will play the same role. Therefore, use cases should be associated with the roles played by the users and not with the users themselves.

Types of Use Cases

There are many different types of use cases. We suggest two separate dimensions on which to classify a use case based on the purpose of the use case and the amount of information that the use case contains: overview versus detail and essential versus real.

An *overview use case* is used to enable the analyst and user to agree on a high-level overview of the requirements. Typically, overview use cases are created very early in the process of understanding the system requirements, and they document only basic information about the use case, such as its name, ID number, primary actor, type, a brief description, and the relationships among the actors, actors and use cases, and use cases. These can easily be created immediately after the creation of the use-case diagram.

Once the user and the analyst agree upon a high-level overview of the requirements, the overview use cases can be converted to detail use cases. A *detail use case* typically documents, as far as possible, all the information needed for the use case. These can be based on the activities and control flows contained in the activity diagrams.

An *essential use case* is one that describes only the minimum essential issues necessary to understand the required functionality. A *real use case* goes farther and describes a specific set of steps. For example, an essential use case in a doctor office might say that the receptionist should attempt to match the patient's desired appointment times with the available times, whereas a real use case might say that the receptionist should look up the available dates on the calendar using MS Exchange to determine if the requested appointment times were available. The primary difference is that essential use cases are implementation independent, whereas real use cases are detailed descriptions of how to use the system once it is implemented. Thus real use cases tend to be used only in the design, implementation, and testing.

Elements of a Use-Case Description

A use-case description contains all the information needed to build the structural (Chapter 5) and behavioral (Chapter 6) diagrams that follow, but it expresses the information in a less-formal way that is usually simpler for users to understand. Figure 4-11 shows a sample use-case description.[19] A use-case description has three basic parts: overview information, relationships, and the flow of events.

Overview Information   The overview information identifies the use case and provides basic background information about the use case. The *use-case name* should be a verb–noun phrase (e.g., Make Old Patient Appt). The *use-case ID number* provides a unique way to find every use case and also enables the team to trace design decisions back to a specific requirement. The *use-case type* is either overview or detail and essential or real. The *primary actor* is usually the trigger of the use case—the person or thing that starts the execution of the use case. The primary purpose of the use case is to meet the goal of the primary actor. The *brief description* is typically a single sentence that describes the essence of the use case.

---

[19] Currently there is no standard set of elements for a use case. The elements described in this section are based on recommendations contained in Alistair Cockburn, *Writing Effective Use Cases* (Reading, MA: Addison-Wesley, 2001); Craig Larman, *Applying UML and Patterns: An Introduction to Object-Oriented Analysis and Design and the Unified Process,* 2nd ed. (Upper Saddle River, NJ: Prentice Hall, 2002); and Brian Henderson-Sellers and Bhuvan Unhelkar, *OPEN Modeling with UML* (Reading, MA: Addison-Wesley, 2000). Also see Graham, *Migrating to Object Technology.*

Use Case Name: Make Old Patient Appt		ID: 2	Importance Level: High
Primary Actor: Old Patient		Use Case Type: Detail, Essential	

Stakeholders and Interests:
Old Patient – wants to make, change, or cancel an appointment
Doctor – wants to ensure patient's needs are met in a timely manner

Brief Description: This use case describes how we make an appointment as well as changing or canceling
an appointment for a previously seen patient.

Trigger: Patient calls and asks for a new appointment or asks to cancel or change an existing appointment

Type: External

Relationships:
    Association: Old Patient
    Include:
    Extend: Update Patient Information
    Generalization: Manage Appointments

Normal Flow of Events:
1. The Patient contacts the office regarding an appointment.
2. The Patient provides the Receptionist with his or her name and address.
3. If the Patient's information has changed
    Execute the Update Patient Information use case.
4. If the Patient's payment arrangements has changed
    Execute the Make Payments Arrangements use case.
5. The Receptionist asks Patient if he or she would like to make a new appointment, cancel an existing appointment, or change
an existing appointment.
    If the patient wants to make a new appointment,
        the S-1: new appointment subflow is performed.
    If the patient wants to cancel an existing appointment,
        the S-2: cancel appointment subflow is performed.
    If the patient wants to change an existing appointment,
        the S-3: change appointment subflow is performed.
6. The Receptionist provides the results of the transaction to the Patient.

SubFlows:
    S-1: New Appointment
        1. The Receptionist asks the Patient for possible appointment times.
        2. The Receptionist matches the Patient's desired appointment times with available dates and
        times and schedules the new appointment.
    S-2: Cancel Appointment
        1. The Receptionist asks the Patient for the old appointment time.
        2. The Receptionist finds the current appointment in the appointment file and cancels it.
    S-3: Change Appointment
        1. The Receptionist performs the S-2: cancel appointment subflow.
        2. The Receptionist performs the S-1: new appointment subflow.

Alternate/Exceptional Flows:
    S-1, 2a1: The Receptionist proposes some alternative appointment times based on what is available in the
        appointment schedule.
    S-1, 2a2: The Patient chooses one of the proposed times or decides not to make an appointment.

TEMPLATE
can be found at
www.wiley.com
/go/global/
tegarden

FIGURE 4-11   Sample Use-Case Description

The *importance level* can be used to prioritize the use cases. The importance level
enables the users to explicitly prioritize which business functions are most important and
need to be part of the first version of the system and which are less important and can wait
until later versions if necessary. The importance level can use a fuzzy scale, such as high,
medium, and low (e.g., in Figure 4-11, we have assigned an importance level of high to the

Make Old Patient Appt use case). It can also be done more formally using a weighted average of a set of criteria. For example, Larman[20] suggests rating each use case over the following criteria using a scale from zero to five:

- The use case represents an important business process.
- The use case supports revenue generation or cost reduction.
- Technology needed to support the use case is new or risky and therefore requires considerable research.
- Functionality described in the use case is complex, risky, and/or time critical. Depending on a use case's complexity, it may be useful to consider splitting its implementation over several different versions.
- The use case could increase understanding of the evolving design relative to the effort expended.

A use case may have multiple *stakeholders* that have an interest in the use case. Each use case lists each of the stakeholders with each one's interest in the use case (e.g., Old Patient and Doctor). The stakeholders' list always includes the primary actor (e.g., Old Patient).

Each use case typically has a *trigger* —the event that causes the use case to begin (e.g., Old Patient calls and asks for a new appointment or asks to cancel or change an existing appointment). A trigger can be an *external trigger,* such as a customer placing an order or the fire alarm ringing, or it can be a *temporal trigger,* such as a book being overdue at the library or the need to pay the rent.

Relationships   Use-case relationships explain how the use case is related to other use cases and users. There are four basic types of *relationships:* association, extend, include, and generalization. An *association relationship* documents the communication that takes place between the use case and the actors that use the use case. An actor is the UML representation for the role that a user plays in the use case. For example, in Figure 4-11, the Make Old Patient Appt use case is associated with the actor Old Patient (see Figure 4-4). In this case, a patient makes an appointment. All actors involved in the use case are documented with the association relationship.

An *include relationship* represents the mandatory inclusion of another use case. The include relationship enables *functional decomposition*—the breaking up of a complex use case into several simpler ones. For example, in Figure 4-4, the Manage Schedule use case was considered to be complex and complete enough to be factored out as a separate use case that could be executed by the Produce Schedules and Record Availability use cases. The include relationship also enables parts of use cases to be reused by creating them as separate use cases.

An *extend relationship* represents the extension of the functionality of the use case to incorporate optional behavior. In Figure 4-11, the Make Old Patient Appt use case conditionally uses the Update Patient Information use case. This use case is executed only if the patient's information has changed.

The *generalization relationship* allows use cases to support *inheritance.* For example, the use case in Figure 4-4, the Manage Appointments use case, was specialized so that a new patient would be associated with the Make New Patient Appt and an old patient could be associated with a Make Old Patient Appt. The common, or generalized, behavior that both the Make New Patient Appointment and Make Old Patient Appointment use

---

[20] Larman, *Applying UML and Patterns: An Introduction to Object-Oriented Analysis and Design.*

cases contain would be placed in the generalized Manage Appointments use case. In other words, the Make New Patient Appointment and Make Old Patient Appointment use cases would inherit the common functionality from the Manage Appointments use case. The specialized behavior would be placed in the appropriate specialized use case. For example, the extend relationship to the Update Patient Information use case would be placed with the specialized Make Old Patient Appointment use case.

Flow of Events   Finally, the individual steps within the business process are described. Three different categories of steps, or *flows of events,* can be documented: normal flow of events, subflows, and alternative, or exceptional, flows:

- The *normal flow of events* includes only steps that normally are executed in a use case. The steps are listed in the order in which they are performed. In Figure 4-11, the patient and the receptionist have a conversation regarding the patient's name, address, and action to be performed.

- In some cases, the normal flow of events should be decomposed into a set of *subflows* to keep the normal flow of events as simple as possible. In Figure 4-11, we have identified three subflows: Create Appointment, Cancel Appointment, and Change Appointment. Each of the steps of the subflows is listed. These subflows are based on the control flow logic in the activity diagram representation of the business process (see Figure 4-6). Alternatively, we could replace a subflow with a separate use case that could be incorporated via the include relationships (see the earlier discussion). However, this should be done only if the newly created use case makes sense by itself. For example, in Figure 4-11, does it make sense to factor out a Create Appointment, Cancel Appointment, and/or Change Appointment use case? If it does, then the specific subflow(s) should be replaced with a call to the related use case, and the use case should be added to the include relationship list.

- *Alternative or exceptional flows* are ones that do happen but are not considered to be the norm. These must be documented. For example, in Figure 4-11, we have identified two alternative or exceptional flows. The first one simply addresses the situation that occurs when the set of requested appointment times are not available. The second one is simply a second step to the alternative flow. Like the subflows, the primary purpose of separating out alternate or exceptional flows is to keep the normal flow of events as simple as possible. Again, as with the subflows, replace the alternate or exceptional flows with separate use cases that could be integrated via the extend relationship (see the earlier discussion).

When should events be factored out from the normal flow of events into subflows? Or when should subflows and/or alternative or exceptional flows be factored out into separate use cases? Or when should things simply be left alone? The primary criteria should be based on the level of complexity that the use case entails. The more difficult it is to understand the use case, the more likely events should be factored out into subflows, or subflows and/or alternative or exceptional flows should be factored out into separate use cases that are called by the current use case. This, of course, creates more use cases. Therefore, the use-case diagram will become more cluttered. Practically speaking, we must decide which makes more sense. This varies greatly, depending on the problem and the client. Remember, we are trying to represent, in a manner as complete and concise as possible, our understanding of the business processes that we are investigating so that the client can validate the requirements that we are modeling. Therefore, there really is no single right answer. It really depends on the analyst, the client, and the problem.

Optional Characteristics   Other characteristics of use cases can be documented by use-case descriptions. These include the level of complexity of the use case, the estimated amount of time it takes to execute the use case, the system with which the use case is associated, specific data flows between the primary actor and the use case, any specific attribute, constraint, or operation associated with the use case, any preconditions that must be satisfied for the use case to execute, or any guarantees that can be made based on the execution of the use case. As we noted at the beginning of this section, there is no standard set of characteristics of a use case that must be captured. In this book, we suggest that the information contained in Figure 4-11 is the minimal amount to be captured.

Guidelines for Creating Use-Case Descriptions

The essence of a use case is the flow of events. Writing the flow of events in a manner that is useful for later stages of development generally comes with experience. Figure 4-12 provides a set of guidelines that have proved to be useful.[21]

First, write each individual step in the form subject–verb–direct object and, optionally, preposition–indirect object. This form has become known as *SVDPI* sentences. This form of sentence has proved to be useful in identifying classes and operations (see Chapter 5). For example, in Figure 4-11, the first step in the normal flow of events, the Patient contacts the office regarding an appointment, suggests the possibility of three classes of objects: Patient, Office, and Appointment. This approach simplifies the process of identifying the classes in the structural model (see Chapter 5). SVDPI sentences cannot be used for all steps, but they should be used whenever possible.

Second, make clear who or what is the initiator of the action and who or what is the receiver of the action in each step. Normally, the initiator should be the subject of the sentence and the receiver should be the direct object of the sentence. For example, in Figure 4-11, the second step, Patient provides the Receptionist with his or her name and address, clearly portrays the Patient as the initiator and the Receptionist as the receiver.

Third, write the step from the perspective of an independent observer. To accomplish this, each step might have to be written first from the perspective of both the initiator and the receiver. Based on the two points of view, the bird's eye view version can then be written. For example, in Figure 4-11, the Patient provides the Receptionist with his or her name and address; neither the patient's nor the receptionist's perspective is represented.

Fourth, write each step at the same level of abstraction. Each step should make about the same amount of progress toward completing the use case as each of the other steps. On high-level use cases, the amount of progress could be very substantial, whereas in a low-level use case, each step could represent only incremental progress. For example, in Figure 4-11, each step represents about the same amount of effort to complete.

---

1. Write each set in the form of subject–verb–direct object (and sometimes preposition–indirect object).
2. Make sure it is clear who the initiator of the step is.
3. Write the steps from the perspective of the independent observer.
4. Write each step at about the same level of abstraction.
5. Ensure the use case has a sensible set of steps.
6. Apply the KISS principle liberally.
7. Write repeating instructions after the set of steps to be repeated.

FIGURE 4-12
Guidelines for Writing Effective Use-Case Descriptions

---

[21] These guidelines are based on Cockburn, *Writing Effective Use Cases,* and Graham, *Migrating to Object Technology.*

Fifth, ensure that the use case contains a sensible set of actions. Each use case should represent a transaction. Therefore, each use case should comprise four parts:

1. The primary actor initiates the execution of the use case by sending a request (and possibly data) to the system.
2. The system ensures that the request (and data) is valid.
3. The system processes the request (and data) and possibly changes its own internal state.
4. The system sends the primary actor the result of the processing.

For example, in Figure 4-11, the patient requests an appointment (steps 1 and 2), the receptionist determines whether any of the patient's information has changed or not (step 3), the receptionist determines whether the patient's payments arrangements has changed or not (step 4), the receptionist sets up the appointment transaction (step 5), and the receptionist provides the results of the transaction to the patient (step 6).

The sixth guideline is the KISS principle. If the use case becomes too complex and/or too long, the use case should be decomposed into a set of use cases. Furthermore, if the normal flow of events of the use case becomes too complex, subflows should be used. For example, in Figure 4-11, the fifth step in the normal flow of events was sufficiently complex to decompose it into three separate subflows. However, care must be taken to avoid the possibility of decomposing too much. Most decomposition should be done with classes (see Chapter 5).

The seventh guideline deals with repeating steps. Normally, in a programming language such as Visual Basic or C, we put loop definition and controls at the beginning of the loop. However, because the steps are written in simple English, it is normally better to simply write Repeat steps A through E until some condition is met after step E. It makes the use case more readable to people unfamiliar with programming.

YOUR TURN   **4-4 Use-Case Descriptions**

Look at the activity diagram for the appointment system in Figure 4-7 and the use-case description that was created in Figure 4-11. Create your own use-case description for the Make New Patient Appt or the activity that you created in Your Turn 4-3. Use Figure 4-11 to guide your efforts.

Creating Use-Case Descriptions

Use cases provide a bird's-eye view of the business processes contained in the evolving system. The use-case diagram depicts the communication path between the actors and the system. Use cases and their use-case description documentation tend to be used to model both the contexts of the system and the detailed requirements for the system. Even though the primary purpose of use cases is to document the functional requirements of the system, they also are used as a basis for testing the evolving system. In this section, we provide a set of steps that can be used to guide the actual creation of a use-case description for each use case in the use-case diagram based on the requirements definition and the use-case and activity diagrams.[22] These steps are performed in order, but of course the analyst often cycles among them in an iterative fashion as he or she moves from one use case to another use case.

---

[22] The approach in this section is based on the work of Cockburn, *Writing Effective Use Cases*; Graham, *Migrating to Object Technology*; George Marakas and Joyce Elam,"Semantic Structuring in Analyst Acquisition and Representation of Facts in Requirements Analysis," *Information Systems Research* 9, no. 1 (1998): 37–63; and Alan Dennis, Glenda Hayes, and Robert Daniels, "Business Process Modeling with Group Support Systems," *Journal of Management Information Systems* 15, no. 4 (1999): 115–142.

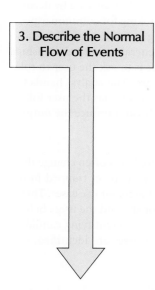

1. Choose a Use Case

The first step is to choose one of the use cases to document with a use-case description. Using the importance level of the use case can help do this. For example, in Figure 4-11, the Make Old Patient Appt use case has an importance level of high. As such, it should be one of the earlier use cases to be expanded. The criteria suggested by Larman[23] can also be used to set the prioritization of the use cases, as noted earlier. An alternative approach suggests that each use case should be voted on by each member of the development team. In this approach, each team member is given a set of "dots" that they can use to vote on the use cases. They can use all of their dots to vote for a single use case, or they can spread them over a set of use cases. The use cases then can be ranked in order of the number of dots received. Use case descriptions are created for the individual use cases based on the rank order.[24]

2. Create Overview Description

The second step is to create an overview description of the use case, that is, name the primary actor, set the type for the use case, list all of the identified stakeholders and their interests in the use case, identify the level of importance of the use case, give a brief description of the use case, give the trigger information for the use case, and list the relationships in which the use case participates.

3. Describe the Normal Flow of Events

The third step is to fill in the steps of the normal flow of events required to describe each use case. The steps focus on what the business process does to complete the use case, as opposed to what actions the users or other external entities do. In general, the steps should be listed in the order in which they are performed, from first to last. Remember to write the steps in an SVDPI form whenever possible. In writing the use case, remember the seven guidelines described earlier. The goal at this point is to describe how the chosen use case operates. One of the best ways to begin to understand how an actor works through a use case is to visualize performing the steps in the use case—that is, role play. The techniques of visualizing how to interact with the system and of thinking about how other systems work (informal benchmarking) are important techniques that help analysts and users understand how systems work and how to write a use case. Both techniques (visualization and informal benchmarking) are common in practice. It is important to remember that at this point in the development of a use case, we are interested only in the typical successful execution of the use case. If we try to think of all of the possible combinations of activities that could go on, we will never get anything written down. At this point, we are looking only for the three to seven *major* steps. Focus only on performing the typical process that the use case represents.

4. Check the Normal Flow of Events

The fourth step is to ensure that the steps listed in the normal flow of events are not too complex or too long. Each step should be about the same size as the others. For example, if we were writing steps for preparing a meal, steps such as "take fork out of drawer" and "put fork on table" are much smaller than "prepare cake using mix". If we end up with more than seven steps, or steps that vary greatly in size, we should go back and review each step carefully, and possibly rewrite the steps.

One good approach to produce the steps for a use case is to have the users visualize themselves actually performing the use case and to have them write down the steps as if they were writing a recipe for a cookbook. In most cases, the users will be able to quickly define what they do in the as-is model. Defining the steps for to-be use cases might take a bit more coaching. In our experience, the descriptions of the steps change greatly as users

[23] Larman, *Applying UML and Patterns: An Introduction to Object-Oriented Analysis and Design.*
[24] C. Larman, *Agile & Iterative Development: A Manager's Guide* (Boston, MA: Addison-Wesley, 2004).

work through a use case. Our advice is to use a blackboard or whiteboard (or paper with pencil) that can be easily erased to develop the list of steps, and then write the list on the use-case form. It should be written on the use-case form only after the set of steps is fairly well defined.

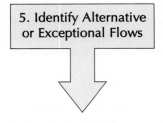

5. Identify Alternative or Exceptional Flows

The fifth step focuses on identifying and writing the alternative or exceptional flows. Alternative or exceptional flows are flows of success that represent optional or exceptional behavior. They tend to occur infrequently or as a result of a normal flow failing. They should be labeled so that there is no doubt as to which normal flow of events it is related. For example, in Figure 4-11, alternative/exceptional flow S-1, 2a1 executes when step 2 of subflow S-1 fails (i.e., the requested appointment time was not available). Like the normal flows and subflows, alternative or exceptional flows should be written in the SVDPI form whenever possible.

6. Review the Use-Case Description

The sixth step is to carefully review the use-case description and confirm that the use case is correct as written, which means reviewing the use case with the users to make sure each step is correct.[25] The review should look for opportunities to simplify a use case by decomposing it into a set of smaller use cases, merging it with others, looking for common aspects in both the semantics and syntax of the use cases, and identifying new use cases. This is also the time to look into adding the include, extend, and/or generalization relationships between use cases. The most powerful way to confirm a use case is to ask the user to role-play, or execute the process using the written steps in the use case. The analyst hands the user pieces of paper labeled with the major inputs to the use case and has the user follow the written steps like a recipe to make sure that those steps really can produce the outputs defined for the use case using its inputs.

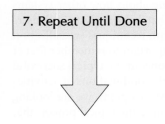

7. Repeat Until Done

The seventh and final step is to *iterate* the entire set of steps again. Users often change their minds about what is a use case and what it includes. It is very easy to get trapped in the details at this point, so remember that the goal is to just address the major use cases. Therefore, the analyst should continue to iterate these steps until he or she and the users believe that a sufficient number of use cases has been documented to begin identifying candidate classes for the structural model (see Chapter 5). As candidate classes are identified, it is likely that additional use cases will be uncovered.

Example   The first step to document business processes with use-case descriptions is to choose a use case. Because we previously chose the Borrow Books use case in the Library Collection Management System example, we will stay with it. Next, we need to create the overview description. In this case, we have to go back and look at the use case diagram (see Figure 4-5) that describes the external behavior of the Library Collection Management System and the activity diagram (see Figure 4-10) that describes the functionality of the Borrow Books use case. It also is a good idea to refer back, once again, to the functional requirements that drove the creation of the Borrow Books use case. Here they are:

> The borrowing activities are built around checking books out and returning books by borrowers. There are three types of borrowers: students, faculty or staff, and guests. Regardless of the type of borrower, the borrower must have a valid ID card.

[25] This process is related to role-playing, which is discussed in Chapter 5.

Use Case Name: Borrow Books		ID: 2	Importance Level: <u>High</u>
Primary Actor: Borrower		Use Case Type: Detail, Essential	
Stakeholders and Interests: Borrower - wants to check out books Librarian - wants to ensure borrower only gets books deserved			
Brief Description: This use case describes how books are checked out of the library.			
Trigger: Borrower brings books to checkout desk. Type: External			

Relationships:
    Association: Borrower, Personnel Office, Registrar's Office
    Include:
 	
    Extend:
    Generalization:			

FIGURE 4-13   Overview Description for the Borrow Books Use Case

If the borrower is a student, having the system check with the registrar's student database validates the ID card. If the borrower is a faculty or staff member, having the system check with the personnel office's employee database validates the ID card. If the borrower is a guest, the ID card is checked against the library's own borrower database. If the ID card is valid, the system must also check to determine whether the borrower has any overdue books or unpaid fines. If the ID card is invalid, or if the borrower has overdue books or unpaid fines, the system must reject the borrower's request to check out a book; otherwise, the borrower's request should be honored.

Based on these three critical pieces of information and using the use-case description template (see Figure 4-11), we can create the overview description of the Borrow Books use case (see Figure 4-13).

By carefully reviewing the functional requirements (above) and the activity diagram (Figure 4-10), we can easily identify the Normal Flow of Events for the Borrow Books use case. Furthermore, it is possible to decide whether any of the events contained in the Normal Flow of Events list should be decomposed using Subflows or other use cases that would need to be included. In the latter case, we would have to modify the Relationships section of the overview description and modify the use-case diagram to reflect this addition. Also, based on the logic structure of the activity diagram, it is possible to identify the alternative/ exceptional flows to the normal flow of events for the Borrow Books use case. Based on the overall simplicity of the Borrow Books use case, we decided not to decompose the process using either subflows or included use cases. However, due to the logic structure laid out in the activity diagram, there were two alternate/exceptional flows identified. Figure 4-14 depicts the Normal Flow of Events, Subflows, and Alternative/Exceptional Flows sections of the Borrow Books use-case description.

Normal Flow of Events:
1. The Borrower brings books to the Librarian at the checkout desk.
2. The Borrower provides Librarian their ID card.
3. The Librarian checks the validity of the ID Card.
    If the Borrower is a Student Borrower, Validate ID Card against Registrar's Database.
    If the Borrower is a Faculty/Staff Borrower, Validate ID Card against Personnel Database.
    If the Borrower is a Guest Borrower, Validate ID Card against Library's Guest Database.
4. The Librarian checks whether the Borrower has any overdue books and/or fines.
5. The Borrower checks out the books.

SubFlows:

Alternate/Exceptional Flows:
4a The ID Card is invalid, the book request is rejected.
5a The Borrower either has overdue books, fines, or both, the book request is rejected.

FIGURE 4-14   Flow Descriptions for the Borrow Books Use Case

VERIFYING AND VALIDATING THE BUSINESS PROCESSES AND FUNCTIONAL MODELS[26]

Before we move on to structural (Chapter 5) and behavioral (Chapter 6) modeling, we need to verify and validate the current set of functional models to ensure that they faithfully represent the business processes under consideration. This includes testing the fidelity of each model; for example, we must be sure that the activity diagram(s), use-case descriptions, and use-case diagrams all describe the same functional requirements. Before we describe the specific tests to consider, we describe walkthroughs, a manual approach that supports verifying and validating the evolving models.[27]

Verification and Validation through Walkthroughs

A *walkthrough* is essentially a peer review of a product. In the case of the functional models, a walkthrough is a review of the different models and diagrams created during functional modeling. This review typically is completed by a team whose members come from the development team and the client. The purpose of a walkthrough is to thoroughly *test* the fidelity of the functional models to the functional requirements and to ensure that the models are consistent. That is, a walkthrough uncovers *errors* or *faults* in the evolving specification. However, a walkthrough does not correct errors—it simply identifies them. Error correction is to be accomplished by the team after the walkthrough is completed.

Walkthroughs are very interactive. As the presenter walks through the representation, members of the walkthrough team should ask questions regarding the representation. For

---

[26] The material in this section has been adapted from E. Yourdon, *Modern Structured Analysis* (Englewood Cliffs, NJ: Prentice Hall, 1989). Verifying and validating are types of testing.

[27] Even though many modern CASE tools can automate much of the verifying and validating of the analysis models, we feel that it is paramount that systems analysts understand the principles of verification and validation. Furthermore, some tools, such as Visio, that support UML diagramming are only diagramming tools. Regardless, the analyst is expected to perform all diagramming correctly.

example, if the presenter is walking through an activity diagram, another member of the team could ask why certain activities or objects were not included. The actual process of simply presenting the representation to a new set of eyes can uncover obvious misunderstandings and omissions. In many cases, the representation creator can get lost in the proverbial trees and not see the forest.[28] In fact, many times the act of walking through the representation causes a presenter to see the error himself or herself. For psychological reasons, hearing the representation helps the analyst to see the representation more completely.[29] Therefore, the representation creators should regularly do a walkthrough of the models themselves by reading the representations out loud to themselves, regardless of how they think it might make them look.

There are specified roles that different members of the walkthrough team can play. The first is the *presenter* role. This should be played by the person who is primarily responsible for the specific representation being reviewed. This individual presents the representation to the walkthrough team. The second role is *recorder,* or *scribe.* The recorder should be a member of the analysis team. This individual carefully takes the minutes of the meeting by recording all significant events that occur during the walkthrough. In particular, all errors that are uncovered must be documented so that the analysis team can address them. Another important role is to have someone who raises issues regarding maintenance of the representation. Yourdon refers to this individual as a *maintenance oracle.*[30] Owing to the emphasis on reusability in object-oriented development, this role becomes particularly crucial. Finally, someone must be responsible for calling, setting up, and running the walkthrough meetings.

For a walkthrough to be successful, the members of the walkthrough team must be fully prepared. All materials to be reviewed must be distributed with sufficient time for the team members to review them before the actual meeting. All team members should be expected to mark up the representations so that during the walkthrough meeting, all relevant issues can be discussed. Otherwise, the walkthrough will be inefficient and ineffective. During the actual meeting, as the presenter is walking through the representation, the team members should point out any potential errors or misunderstandings. In many cases, the errors and misunderstandings are caused by invalid assumptions that would not be uncovered without the walkthrough.

One potential danger of walkthroughs is when management decides the results of uncovering errors in the representation are a reflection of an analyst's capability. This must be avoided at all costs. Otherwise, the underlying purpose of the walkthrough—to improve the fidelity of the representation—will be thwarted. Depending on the organization, it may be necessary to omit management from the walkthrough process. If not, the walkthrough process could break down into a slugfest to make some team members look good by destroying the presenter. To say the least, this is obviously counterproductive.

Functional Model Verification and Validation

In this book, we have suggested three different representations for the functional model: activity diagrams, use-case descriptions, and use-case diagrams In this section,

---

[28] In fact, all joking aside, in many cases the developer is down at the knothole level and can't even see the tree, let alone the forest.

[29] This has to do with using different senses. Because our haptic senses are the most sensitive, touching the representation would be best. However, it is not clear how one can touch a use case or a class.

[30] See Appendix D of Yourdon, *Modern Structured Analysis.*

we describe a set of rules to ensure that these three representations are consistent among themselves.

First, when comparing an activity diagram to a use-case description, there should be at least one event recorded in the normal flow of events, subflows, or alternative/exceptional flows of the use-case description for each activity or action that is included on an activity diagram, and each event should be associated with an activity or action. For example, in Figure 4-4, there is an activity labeled Get Patient Information that is associated with the first two events contained in the normal flow of events of the use-case description shown in Figure 4-11.

Second, all objects portrayed as an object node in an activity diagram must be mentioned in an event in the normal flow of events, subflows, or alternative/exceptional flows of the use-case description. For example, the activity diagram in Figure 4-4 portrays an Appt object, and the use-case description refers to a new appointment and changing or canceling an existing appointment.

Third, sequential order of the events in a use-case description should occur in the same sequential order of the activities contained in an activity diagram. For example, in Figures 4-4 and 4-11, the events associated with the Get Patient Information activity (events 1 and 2) should occur before the events associated with the Make Payment Arrangements activity (event 4).

Fourth, when comparing a use-case description to a use-case diagram, there must be one and only one use-case description for each use case, and vice versa. For example, Figure 4-11 portrays the use-case description of the Make Old Patient Appt use case. However, the use-case diagram shown in Figures 4-4, the activity diagram shown in Figure 4-7, and the use-case description given in Figure 4-11 are inconsistent with each other. In this case, the use-case diagram implies that the Make Payment Arrangements use case is optional regardless of whether the patient is a new or old patient. However, when we review the activity diagram, we see that it is an optional activity for old patients, but a required activity for a new patient. Therefore, only one of the diagrams is correct. In this instance, the use-case diagram needs to be corrected. The new corrected use-case diagram is shown in Figure 4-15.

Fifth, all actors listed in a use-case description must be portrayed on the use-case diagram. Each actor must have an association link that connects it to the use case and must be listed with the association relationships in the use-case description. For example, the Old Patient actor is listed in the use-case description of the Make Old Patient Appt use case (see Figure 4-11), it is listed with the association relationships in the Make Old Patient Appt use-case description, and it is connected to the use case in the use-case diagram (see Figure 4-15).

Sixth, in some organizations, we should also include the stakeholders listed in the use-case description as actors in the use-case diagram. For example, there could have been an association between the Doctor actor and the Make Old Patient Appt use case (see Figures 4-11 and 4-15). However, in this case, it was decided not to include this association because the Doctor never participates in the Make Old Patient Appt use case.[31]

Seventh, all other relationships listed in a use-case description (include, extend, and generalization) must be portrayed on a use-case diagram. For example, in Figure 4-11,

---

[31] Another possibility could have been to include a Receptionist actor. However, we had previously decided that the Receptionist was, in fact, part of the Appointment System and not simply a user of the system. If UML supported the idea of internal actors, or actor-to-actor associations, this implicit association could easily be made explicit by having the Patient actor communicate with the Receptionist actor directly, regardless of whether the Receptionist actor was part of the system or not. See footnote 4.

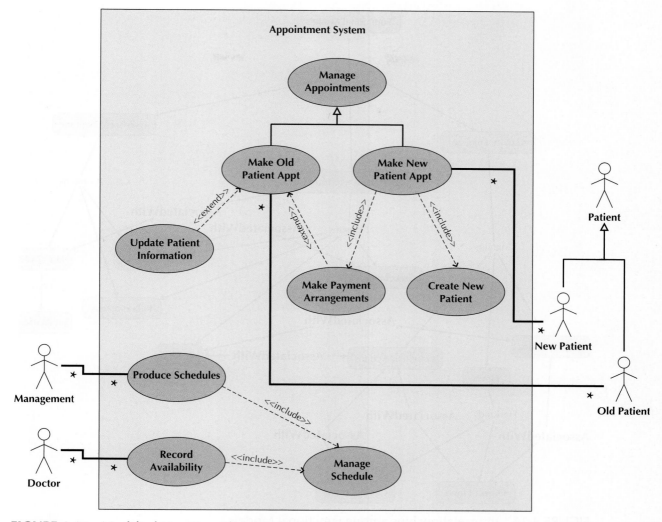

FIGURE 4-15   Modified Use-Case Diagram for the Appointment System

there is an extend relationship listed with the Update Patient Information use case, and in Figure 4-15, we see that it appears on the diagram between the two use cases.

Finally, there are many diagram-specific requirements that must be enforced. For example, in an activity diagram a decision node can be connected to activity or action nodes only with a control flow, and for every decision node there should be a matching merge node. Every type of node and flow has different restrictions. However, the complete restrictions for all the UML diagrams are beyond the scope of this text.[32] The concept map in Figure 4-16 portrays the associations among the functional models.

---

[32] A good reference for these types of restrictions is S.W. Ambler, *The Elements of UML 2.0 Style* (Cambridge, UK: Cambridge University Press, 2005).

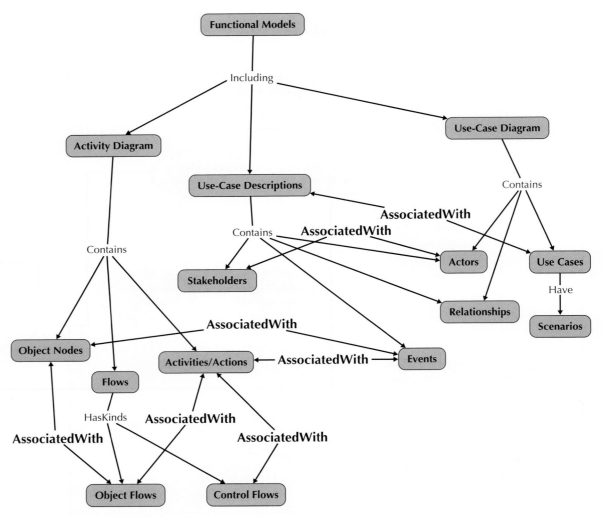

FIGURE 4-16   Interrelationships among Functional Models

APPLYING THE CONCEPTS AT CD SELECTIONS

In this chapter, we introduced how business processes are identified, modeled, and documented using the functional models of the UML. Specifically, we described how the functional requirements of business processes are identified by use cases and use-case diagrams. We described how activity diagrams model business processes and we described how use-case descriptions are used to more fully document the business processes. Finally, we described how to verify and validate the evolving representations of the business processes contained in the functional models. In this installment of the CD Selections case, we see how Alec and Margaret work through all of these topics with regard to the web-based solution that they hope to create.

SUMMARY

Business Process Identification with Use Cases and Use-Case Diagrams

Use-case diagrams are simply a graphical overview of the set of use cases contained in the system. They illustrate the main functionality of a system and the actors that interact with the system. The diagram includes actors, which are people or things that derive value from the system, and use cases that represent the functionality of the system. The actors and use cases are separated by a system boundary and connected by lines representing associations. At times, actors are specialized versions of more general actors. Similarly, use cases can extend or include other use cases. Creating use-case diagrams is a four-step process whereby the analyst draws the use cases, places the actors near the associated use cases, adds the system boundary, and finally, draws the associations to connect use cases and actors.

Business Process Modeling with Activity Diagrams

Even from an object-oriented systems development point of view, business process modeling has been shown to be valuable. The one major hazard of business process modeling is that it focuses the systems development process in a functional decomposition direction. However, if used carefully, it can enhance the development of object-oriented systems. UML supports process modeling using an activity diagram. Activity diagrams comprise activities or actions, objects, control flows, object flows, and a set of seven different control nodes (initial, final-activity, final-flow, decision, merge, fork, and join). Swimlanes can be used to enhance the readability of the diagrams. An activity diagram is very useful for helping the analyst to identify the relevant use cases for the information system being developed.

Business Process Documentation with Use Cases and Use-Case Descriptions

Use cases are the primary method of documenting requirements for object-oriented systems. They represent a functional view of the system being developed. There are overview and detailed use cases. Overview use cases comprise the use-case name, ID number, primary actor, type, a brief description, and the relationships in which the use case participates (association, extend, generalization, and include). Detailed use cases extend the overview use case with the identification and description of the stakeholders and their interest, the trigger and its type, the normal flow of events, the subflows, and any alternative or exception flows to the normal flow of events. There are seven guidelines and seven steps for writing effective use-case descriptions.

Verifying and Validating the Business Processes and Functional Models

Before actually adding system environment details to the analysis models, the various representations need to be verified and validated. One very useful approach to test the fidelity of the representations is to perform a walkthrough in which developers walk through the representations by presenting the different models to members of the development team and representatives of the client. The walkthrough must validate each model to be sure that the different representations within the model all agree with one another; for example, the activity diagrams, use-case descriptions, and use-case diagrams must be consistent with one another. Care must be taken during the walkthroughs to ensure that the presenter is not simply degraded and destroyed.

KEY TERMS

Action, 165
Activity, 165
Activity diagram, 154
Actor, 155
Alternative flows, 178
Association relationship, 177
Black-hole activities, 171
Brief description, 175
Control flow, 165
Control node, 166
Decision node, 168
Detail use case, 175
Errors, 184
Essential use case, 175
Exceptional flows, 178
Extend relationship, 177
External trigger, 177
Faults, 184
Final-activity node, 166
Final-flow node, 167
Flow of events, 178
Fork node, 169
Functional decomposition, 177

Generalization relationship, 177
Guard condition, 168
Importance level, 176
Include relationship, 177
Inheritance, 177
Initial node, 166
Iterate, 182
Join node, 170
Logical model, 154
Maintenance oracle, 185
Merge node, 168
Miracle activity, 171
Normal flow of events, 178
Object flow, 165
Object node, 165
Overview use cases, 175
Packages, 161
Physical model, 154
Presenter, 185
Primary actor, 175
Process models, 154
Real use case, 175
Recorder, 185

Relationships, 177
Role, 174
Scenario, 174
Scribe, 185
Specialized actor, 156
Stakeholders, 177
Subflows, 178
Subject boundary, 159
SVDPI, 179
Swimlanes, 170
Temporal trigger, 177
Test, 184
Trigger, 177
Use case, 154
Use-case description, 155
Use-case diagram, 154
Use-case ID number, 175
Use-case name, 175
Use-case type, 175
Validation, 184
Verification, 184
Walkthrough, 184

QUESTIONS

1. Why is business process modeling important?
2. How do you create use cases?
3. Why do we strive to have about three to nine major use cases in a business process?
4. How do you create use-case diagrams?
5. How is use-case diagramming related to functional modeling?
6. Explain the following terms. Use layperson's language, as though you were describing them to a user: actor, use case, system boundary, relationship.
7. Every association must be connected to at least one _____ and one _____. Why?

8. What are some heuristics for creating a use-case diagram?
9. Why is iteration important in creating use cases?
10. What is the purpose of an activity diagram?
11. What is the difference between an activity and an action?
12. What is the purpose of a fork node?
13. What are the different types of control nodes?
14. What is the difference between a control flow and an object flow?
15. What is an object node?
16. How does a detail use case differ from an overview use case?

17. How does an essential use case differ from a real use case?
18. What are the major elements of an overview use case?
19. What are the major elements of a detail use case?
20. What is the viewpoint of a use case, and why is it important?
21. What are some guidelines for designing a set of use cases? Give two examples of the extend associations on a use-case diagram. Give two examples for the include associations.
22. Which of the following could be an actor found on a use-case diagram? Why?

    Ms. Mary Smith
    Supplier
    Customer
    Internet customer
    Mr. John Seals
    Data entry clerk
    Database administrator

23. What is CRUD? Why is it useful?
24. What is a walkthrough? How does it relate to verification and validation?
25. What are the different roles played during a walkthrough? What are their purposes?
26. How are the different functional models related and how does this affect the verification and validation of the models?

EXERCISES

A. Investigate the website for Rational Software (www.ibm.com/software/rational/) and its repository of information about UML. Write a paragraph news brief on the current state of UML (e.g., the current version and when it will be released, future improvements).

B. Investigate the Object Management Group. Write a brief memo describing what it is, its purpose, and its influence on UML and the object approach to systems development. (*Hint:* A good resource is www.omg.org.)

C. Draw a use-case diagram and a set of activity diagrams for the process of buying glasses from the viewpoint of the patient. The first step is to see an eye doctor who will give you a prescription. Once you have a prescription, you go to an optical dispensary, where you select your frames and place the order for your glasses. Once the glasses have been made, you return to the store for a fitting and pay for the glasses.

D. Create a set of detailed use-case descriptions for the process of buying glasses in exercise C.

E. Draw a use-case diagram and a set of activity diagrams for the following doctor's office system. Whenever new patients are seen for the first time, they complete a patient information form that asks their name, address, phone number, and brief medical history, which are stored in the patient information file. When a patient calls to schedule a new appointment or change an existing appointment, the receptionist checks the appointment file for an available time. Once a good time is found for the patient, the appointment is scheduled. If the patient is a new patient, an incomplete entry is made in the patient's file; the full information will be collected when the patient arrives for the appointment. Because appointments are often made far in advance, the receptionist usually mails a reminder postcard to each patient two weeks before the appointment.

F. Create a set of detail use-case descriptions for the dentist's office system in exercise E.

G. Draw a use-case diagram and a set of activity diagrams for an online university registration system. The system should enable the staff of each academic department to examine the courses offered by their department, add and remove courses, and change the information about them (e.g., the maximum number of students permitted). It should permit students to examine currently available courses, add and drop courses to and from their schedules, and examine the courses for which they are enrolled. Department staff should be able to print a variety of reports about the courses and the students enrolled in them. The system should ensure that no student takes too many courses and that students who have any unpaid fees are not permitted to register (assume that fees data are maintained by the university's financial office, which the registration system accesses but does not change).

H. Create a set of detailed use-case descriptions for the online university registration system in exercise G.

I. Draw a use-case diagram and a set of activity diagrams for the following system. A Real Estate Inc. (AREI) sells houses. People who want to sell their houses sign a contract with AREI and provide information on their house. This information is kept in a database by AREI, and a subset of this information is sent to the citywide multiple-listing service used by all real estate agents. AREI works with two types of potential buyers. Some buyers have an interest in one specific house. In this case, AREI prints information from its database, which the real estate agent uses to help show the house to the buyer (a process beyond the scope of the system to be modeled). Other buyers seek AREI's advice in finding a house that meets their needs. In this case, the buyer completes a buyer information form that is entered into a buyer database, and AREI real estate agents use its information to search AREI's database and the multiple-listing service for houses that meet their needs. The results of these searches are printed and used to help the real estate agent show houses to the buyer.

J. Create a set of detailed use-case descriptions for the real estate system in exercise I.

K. Perform a verification and validation walkthrough of the functional models of the real estate system described in exercises I and J.

L. Draw a use-case diagram and a set of activity diagrams for the following system. A Video Store (AVS) runs a series of fairly standard video stores. Before a video can be put on the shelf, it must be cataloged and entered into the video database. Every customer must have a valid AVS customer card in order to rent a video. Customers rent videos for three days at a time. Every time a customer rents a video, the system must ensure that he or she does not have any overdue videos. If so, the overdue videos must be returned and an overdue fee paid before the customer can rent more videos. Likewise, if the customer has returned overdue videos but has not paid the overdue fee, the fee must be paid before new videos can be rented. Every morning, the store manager prints a report that lists overdue videos. If a video is two or more days overdue, the manager calls the customer to remind him or her to return the video. If a video is returned in damaged condition, the manager removes it from the video database and may sometimes charge the customer.

M. Create a set of detailed use-case descriptions for the video system in exercise L.

N. Perform a verification and validation walkthrough of the functional models of the video store system described in exercises L and M.

O. Draw a use-case diagram and a set of activity diagrams for a gym membership system. When members join the gym, they pay a fee for a certain length of time. Most memberships are for one year, but memberships as short as two months are available. Throughout the year, the gym offers a variety of dis-

counts on their regular membership prices (e.g., two memberships for the price of one for Valentine's day). It is common for members to pay different amounts for the same length of membership. The gym wants to mail out reminder letters to members asking them to renew their memberships one month before their memberships expire. Some members have become angry when asked to renew at a much higher rate than their original membership contract, so the club wants to track the prices paid so that the manager can override the regular prices with special prices when members are asked to renew. The system must track these new prices so that renewals can be processed accurately. One of the problems in the industry is the high turnover rate of members. Although some members remain active for many years, about half of the members do not renew their memberships. This is a major problem, because the gym spends a lot in advertising to attract each new member. The manager wants the system to track each time a member comes into the gym. The system will then identify the heavy users and generate a report so the manager can ask them to renew their memberships early, perhaps offering them a reduced rate for early renewal. Likewise, the system should identify members who have not visited the gym in more than a month, so the manager can call them and attempt to reinterest them in the gym.

P. Create a set of detailed use-case descriptions for the system in exercise O.

Q. Perform a verification and validation walkthrough of the functional models of the gym membership system described in exercises O and P.

R. Draw a use-case diagram and a set of activity diagrams for the following system. Picnics R Us (PRU) is a small catering firm with five employees. During a typical summer weekend, PRU caters fifteen picnics with twenty to fifty people each. The business has grown rapidly over the past year and the owner wants to install a new computer system for managing the ordering and buying process. PRU has a set of ten standard menus. When potential customers call, the receptionist describes the menus to them. If the customer decides to book a picnic, the receptionist records the customer information (e.g., name, address, phone number) and the information about the picnic (e.g., place, date, time, which one of the standard menus, total price) on a contract. The customer is then faxed a copy of the contract, and must sign and return it along with a deposit (often a credit card or by debit card) before the picnic is officially booked. The remaining money is collected when the picnic is delivered. Sometimes, the customer wants something special (e.g., birthday cake). In this case, the receptionist takes the information and gives it to the owner, who determines the cost; the receptionist then calls the customer back with the price information. Sometimes, the customer accepts the price; at other times, the customer requests some changes that have to go back to the owner for a new cost estimate. Each week, the owner looks through the picnics scheduled for that weekend and orders the supplies (e.g., plates) and food (e.g., bread, chicken) needed to make them. The owner would like to use the system for marketing as well. It should be able to track how customers learned about PRU and identify repeat customers, so that PRU can mail special offers to them. The owner also wants to track the picnics for which PRU sent a contract, but the customer never signed the contract and actually booked a picnic.

S. Create a set of detailed use-case descriptions for the system in exercise R.

T. Perform a verification and validation walkthrough of the functional models of the catering system described in exercises R and S.

U. Draw a use-case diagram and a set of activity diagrams for the following system: Of-the-Month Club (OTMC) is an innovative, young firm that sells memberships to

people who have an interest in certain products. People pay membership fees for one year and each month receive a product by mail. For example, OTMC has a coffee-of-the-month club that sends members one pound of special coffee each month. OTMC currently has six memberships (coffee, wine, beer, cigars, flowers, and computer games), each of which costs a different amount. Customers usually belong to just one, but some belong to two or more. When people join OTMC, the telephone operator records the name, mailing address, phone number, e-mail address, credit card information, start date, and membership service(s) (e.g., coffee). Some customers request a double or triple membership (e.g., two pounds of coffee, three cases of beer). The computer game membership operates a bit differently from the others. In this case, the member must also select the type of game (action, arcade, fantasy/science fiction, educational, etc.) and age level. OTMC is planning to greatly expand the number of memberships it offers (e.g., video games, movies, toys, cheese, fruit, and vegetables), so the system needs to accommodate this future expansion. OTMC is also planning to offer three-month and six-month memberships.

V. Create a set of detailed use-case descriptions for the system in exercise U.

W. Perform a verification and validation walkthrough of the functional models of the Of-the-Month Club system described in exercises U and V.

MINICASES

1. Kunsan Hanvit Inc. is a small custom-manufacturing firm located in Kunsan, South Korea. When Chung-Hee Ko, the owner, first brought computers into the business office, the firm was very small and simple. He was able to use an inexpensive PC-based accounting system to handle the basic information-processing needs of the firm. As time went on, the firm grew and the work has become much more complex. The firm's business contracts are as complex as the custom products it manufactures. The simple accounting software is no longer able to keep track of many of the company's sophisticated contracts with its customers.

Mr. Ko has a staff of four in the business office who are familiar with the intricacies of the company's record-keeping requirements. He recently discussed with them his plans to hire an information systems (IS) consultancy to evaluate Kunsan Hanvit's IS needs and propose an upgrading of its

computer system. The staff are excited about the prospect of a new system, because the current system has caused much aggravation. However, they are wary of the consultants who will be conducting the project.

Assume that you are a systems analyst on the consulting team assigned to the Kunsan Hanvit Inc. project. At your first meeting with the staff, you want to be sure that they understand the work that your team will be performing and how they will participate in that work.

a. Explain, in clear, nontechnical terms, the goals of analyzing the project.

b. Explain, in clear, nontechnical terms, how use cases and a use-case diagram will be used by the project team. Explain what these models are, what they represent in the system, and how they will be used by the team.

2. Professional and Scientific Staff Management (PSSM) is a unique type of temporary staffing agency. Many organizations today hire highly skilled technical employees on a short-term, temporary basis to assist with special projects or to provide a needed technical skill. PSSM negotiates contracts with its client companies in which it agrees to provide temporary staff in specific job categories for a specified cost. For example, PSSM has a contract with an oil and gas exploration company in which it agrees to supply geologists with at least a master's degree for $5,000 per week. PSSM has contracts with a wide range of companies and can place almost any type of professional or scientific staff members, from computer programmers to geologists to astrophysicists.

When a PSSM client company determines that it will need a temporary professional or scientific employee, it issues a staffing request against the contract it had previously negotiated with PSSM. When PSSM's contract manager receives a staffing request, the contract number referenced on the staffing request is entered into the contract database. Using information from the database, the contract manager reviews the terms and conditions of the contract and determines whether the staffing request is valid. The staffing request is valid if the contract has not expired, the type of professional or scientific employee requested is listed on the original contract, and the requested fee falls within the negotiated fee range. If the staffing request is not valid, the contract manager sends the staffing request back to the client with a letter stating why the staffing request cannot be filled, and a copy of the letter is filed. If the staffing request is valid, the contract manager enters the staffing request into the staffing request database as an outstanding staffing request. The staffing request is then sent to the PSSM placement department.

In the placement department, the type of staff member, experience, and qualifications requested on the staffing request are checked against the database of available professional and scientific staff. If a qualified individual is found, he or she is marked "reserved" in the staff database. If a qualified individual cannot be found in the database or is not immediately available, the placement department creates a memo that explains the inability to meet the staffing request, and attaches it to the staffing request. All staffing requests are then sent to the arrangements department.

In the arrangements department, the prospective temporary employee is contacted and asked to agree to the placement. After the placement details have been worked out and agreed to, the staff member is marked "placed" in the staff database. A copy of the staffing request and a bill for the placement fee is sent to the client. Finally, the staffing request, the "unable-to-fill" memo (if any), and a copy of the placement fee bill is sent to the contract manager. If the staffing request was filled, the contract manager closes the open staffing request in the staffing request database. If the staffing request could not be filled, the client is notified. The staffing request, placement fee bill, and unable-to-fill memo are then filed in the contract office.

a. Create a use-case diagram for the system described here.

b. Create an activity diagram for the business process described here.

c. Develop a use-case description for each major use case.

d. Verify and validate the functional models.

CHAPTER 5

STRUCTURAL MODELING

A structural, or conceptual, model describes the structure of the objects that supports the business processes in an organization. During analysis, the structural model presents the logical organization of the objects without indicating how they are stored, created, or manipulated so that analysts can focus on the business, without being distracted by technical details. Later during design, the structural model is updated to reflect exactly how the objects will be stored in databases and files. This chapter describes *class–responsibility–collaboration (CRC)* cards, *class diagrams,* and *object diagrams,* which are used to create the structural model.

OBJECTIVES

- Understand the rules and style guidelines for creating CRC cards, class diagrams, and object diagrams.
- Understand the processes used to create CRC cards, class diagrams, and object diagrams.
- Be able to create CRC cards, class diagrams, and object diagrams.
- Understand the relationship among the structural models.
- Understand the relationship between the structural and functional models.

CHAPTER OUTLINE

INTRODUCTION

During analysis, analysts create functional models to represent how the business system will behave. At the same time, analysts need to understand the information that is used and created by the business system (e.g., customer information, order information). In this chapter, we discuss how the objects underlying the behavior modeled in the use cases are organized and presented.

A *structural model* is a formal way of representing the objects that are used and created by a business system. It illustrates people, places, or things about which information is captured and how they are related to one another. The structural model is drawn using an iterative process in which the model becomes more detailed and less conceptual over time. In analysis, analysts draw a *conceptual model*, which shows the logical organization of the objects without indicating how the objects are stored, created, or manipulated. Because this model is free from any implementation or technical details, the analysts can focus more easily on matching the model to the real business requirements of the system.

In design, analysts evolve the conceptual structural model into a design model that reflects how the objects will be organized in databases and software. At this point, the model is checked for redundancy and the analysts investigate ways to make the objects easy to retrieve. The specifics of the design model are discussed in detail in the design chapters.

In this chapter, we focus on creating a conceptual structural model of the objects using CRC cards and class diagrams. Using these techniques, it is possible to show all the objects of a business system. We first describe structural models and their elements. Next, we describe a set of useful approaches that have been used to identify potential objects. Then we describe CRC cards, class diagrams, and object diagrams. Next, we describe how to create structural models using CRC cards and class diagrams, and how the structural model relates to the functional models that we learned about in Chapter 4. Finally, we describe the process to verify and validate the objects in the structural model.

STRUCTURAL MODELS

Every time a systems analyst encounters a new problem to solve, the analyst must learn the underlying problem domain. The goal of the analyst is to discover the key data contained in the problem domain and to build a structural model of the objects. Object-oriented modeling allows the analyst to reduce the semantic gap between the underlying problem domain and the evolving structural model. However, the real world and the world of software are very different. The real world tends to be messy, whereas the world of software must be neat and logical. Thus, an exact mapping between the structural model and the problem domain may not be possible. In fact, it might not even be desirable.

One of the primary purposes of the structural model is to create a vocabulary that can be used by the analyst and the users. Structural models represent the things, ideas, or concepts—that is, the objects—contained in the domain of the problem. They also allow the representation of the relationships among the things, ideas, or concepts. By creating a structural model of the problem domain, the analyst creates the vocabulary necessary for the analyst and users to communicate effectively.

One important thing to remember is that at this stage of development, the structural model does not represent software components or classes in an object-oriented programming language, even though the structural model does contain analysis classes, attributes, operations, and relationships among the analysis classes. The refinement of these initial classes into programming-level objects comes later. Nonetheless, the structural model at this point should represent the responsibilities of each class and the collaborations among the classes. Typically, structural models are depicted using CRC cards, class diagrams, and, in some cases, object diagrams. However, before describing CRC cards, class diagrams, and object diagrams, we describe the basic elements of structural models: classes, attributes, operations, and relationships.

Classes, Attributes, and Operations

A *class* is a general template that we use to create specific instances, or *objects,* in the problem domain. All objects of a given class are identical in structure and behavior but contain different data in their attributes. There are two different general kinds of classes of interest during analysis: concrete and abstract. Normally, when an analyst describes the application domain classes, he or she is referring to concrete classes; that is, *concrete classes* are used to create objects. *Abstract classes* do not actually exist in the real world; they are simply useful abstractions. For example, from an employee class and a customer class, we may identify a generalization of the two classes and name the abstract class *person*. We might not actually instantiate the person class in the system itself, instead creating and using only employees and customers.[1]

A second classification of classes is the type of real-world thing that a class represents. There are domain classes, user-interface classes, data structure classes, file structure classes, operating environment classes, document classes, and various types of multimedia classes. At this point in the development of our evolving system, we are interested only in domain classes. Later in design and implementation, the other types of classes become more relevant.

An *attribute* of an analysis class represents a piece of information that is relevant to the description of the class within the application domain of the problem being investigated. An attribute contains information the analyst or user feels the system should store. For example, a possible relevant attribute of an employee class is employee name, whereas one that might not be as relevant is hair color. Both describe something about an employee, but hair color is probably not all that useful for most business applications. Only attributes that are important to the task should be included in the class. Finally, only attributes that are primitive or atomic types (i.e., integers, strings, doubles, date, time, Boolean, etc.) should be added. Most complex or compound attributes are really placeholders for relationships between classes. Therefore, they should be modeled as relationships, not as attributes (see the next section).

The behavior of an analysis class is defined in an *operation* or service. In later phases, the operations are converted to *methods*. However, because methods are more related to implementation, at this point in the development we use the term *operation* to describe the actions to which the *instances* of the class are capable of responding. Like attributes, only problem domain–specific operations that are relevant to the problem being investigated should be considered. For example, it is normally required that classes provide means of creating instances, deleting instances, accessing individual attribute values, setting individual attribute values, accessing individual relationship values, and removing individual relationship values. However, at this point in the development of the evolving system, the analyst should avoid cluttering up the definition of the class with these basic types of operations and focus only on relevant problem domain–specific operations.

Relationships

There are many different types of relationships that can be defined, but all can be classified into three basic categories of data-abstraction mechanisms: generalization relationships, aggregation relationships, and association relationships. These data-abstraction mechanisms

---

[1] Because abstract classes are essentially not necessary and are not instantiated, arguments have been made that it would be better not to include any of them in the description of the evolving system at this stage of development (see J. Evermann and Y. Wand, "Towards Ontologically Based Semantics for UML Constructs," in H. S. Junii, S. Jajodia, and A. Solvberg (eds.) *ER 2001, Lecture Notes in Computer Science 2224* (Berlin: Springer-Verlag, 2001): 354–367. However, because abstract classes traditionally have been included at this stage of development, we also include them.

allow the analyst to focus on the important dimensions while ignoring nonessential dimensions. Like attributes, the analyst must be careful to include only relevant relationships.

Generalization Relationships   The generalization abstraction enables the analyst to create classes that inherit attributes and operations of other classes. The analyst creates a *superclass* that contains the basic attributes and operations that will be used in several *subclasses*. The subclasses inherit the attributes and operations of their superclass and can also contain attributes and operations that are unique just to them. For example, a customer class and an employee class can be generalized into a person class by extracting the attributes and operations they have in common and placing them into the new superclass, *person*. In this way, the analyst can reduce the redundancy in the class definitions so that the common elements are defined once and then reused in the subclasses. Generalization is represented with the *a-kind-of* relationship, so that we say that an employee is a-kind-of person.

The analyst also can use the flip side of generalization, specialization, to uncover additional classes by allowing new subclasses to be created from an existing class. For example, an employee class can be specialized into a secretary class and an engineer class. Furthermore, generalization relationships between classes can be combined to form generalization hierarchies. Based on the previous examples, a secretary class and an engineer class can be subclasses of an employee class, which, in turn, could be a subclass of a person class. This would be read as a secretary and an engineer are a-kind-of employee, and a customer and an employee are a-kind-of person.

The generalization data abstraction is a very powerful mechanism that encourages the analyst to focus on the properties that make each class unique by allowing the similarities to be factored into superclasses. However, to ensure that the semantics of the subclasses are maintained, the analyst should apply the principle of *substitutability*. By this, we mean that the subclass should be capable of substituting for the superclass anywhere that uses the superclass (e.g., anywhere we use the employee superclass, we could also logically use its secretary subclass). By focusing on the a-kind-of interpretation of the generalization relationship, the principle of substitutability is applied.

Aggregation Relationships   There have been many different types of aggregation or composition relationships proposed in data modeling, knowledge representation, and linguistics, for example, a-part-of (logically or physically), a-member-of (as in set membership), contained-in, related-to, and associated-with. However, generally speaking, all aggregation relationships relate *parts* to *wholes* or parts to *assemblies*. For our purposes, we use the *a-part-of* or *has-parts* semantic relationship to represent the aggregation abstraction. For example, a door is a-part-of a car, an employee is a-part-of a department, or a department is a-part-of an organization. Like the generalization relationship, aggregation relationships can be combined into aggregation hierarchies. For example, a piston is a-part-of an engine, and an engine is a-part-of a car.

Aggregation relationships are bidirectional. The flip side of aggregation is *decomposition*. The analyst can use decomposition to uncover parts of a class that should be modeled separately. For example, if a door and an engine are a-part-of a car, then a car has-parts door and engine. The analyst can bounce around between the various parts to uncover new parts. For example, the analyst can ask, What other parts are there to a car? or To which other assemblies can a door belong?

Association Relationships   There are other types of relationships that do not fit neatly into a generalization (a-kind-of) or aggregation (a-part-of) framework. Technically speaking, these relationships are usually a weaker form of the aggregation relationship. For example, a patient schedules an appointment. It could be argued that a patient is a-part-of an appointment. However, there is a clear semantic difference between this type of relationship and one that models the relationship between doors and cars or even workers and unions. Thus, they are simply considered to be *associations* between instances of classes.

OBJECT IDENTIFICATION

Many different approaches have been suggested to aid the analyst in identifying a set of candidate objects for the structural model. The four most common approaches are textual analysis, brainstorming, common object lists, and patterns. Most analysts use a combination of the techniques to make sure that no important objects and object attributes, operations, and relationships have been overlooked.

Textual Analysis

Textual analysis is an analysis of the text in the use-case descriptions. The analyst starts by reviewing the use-case descriptions and the use-case diagrams. The text in the descriptions is examined to identify potential objects, attributes, operations, and relationships. The nouns in the use case suggest possible classes, and the verbs suggest possible operations. Figure 5-1 presents a summary of guidelines we have found useful. The textual analysis of use-case descriptions has been criticized as being too simple, but because its primary purpose is to create an initial rough-cut structural model, its simplicity is a major advantage. For example, if we applied these rules to the Make Old Patient Appt use case described in Chapter 4 and replicated in Figure 5-2, we can easily identify potential objects for an old patient, doctor, appointment, patient, office, receptionist, name, address, patient information, payment, date, and time. We also can easily identify potential operations that can be associated with the identified objects. For example, patient contacts office, makes a new appointment, cancels an existing appointment, changes an existing appointment, matches requested appointment times and dates with requested times and dates, and finds current appointment.

- A common or improper noun implies a class of objects.
- A proper noun or direct reference implies an instance of a class.
- A collective noun implies a class of objects made up of groups of instances of another class.
- An adjective implies an attribute of an object.
- A doing verb implies an operation.
- A being verb implies a classification relationship between an object and its class.
- A having verb implies an aggregation or association relationship.
- A transitive verb implies an operation.
- An intransitive verb implies an exception.
- A predicate or descriptive verb phrase implies an operation.
- An adverb implies an attribute of a relationship or an operation.

Source: These guidelines are based on Russell J. Abbott, "Program Design by Informal English Descriptions," *Communications of the ACM* 26, no. 11 (1983): 882–894; Peter P-S Chen, "English Sentence Structure and Entity-Relationship Diagrams," *Information Sciences: An International Journal* 29, no. 2–3 (1983): 127–149; and Ian Graham, *Migrating to Object Technology* (Reading, MA: Addison Wesley Longman, 1995).

FIGURE 5-1
Textual Analysis
Guidelines

Use Case Name:	Make Old Patient Appt		ID:	2	Importance Level:	High
Primary Actor:	Old Patient		Use Case Type:	Detail, Essential		

Stakeholders and Interests:
Old Patient – wants to make, change, or cancel an appointment
Doctor – wants to ensure patient's needs are met in a timely manner

Brief Description: This use case describes how we make an appointment as well as changing or canceling an appointment for a previously seen patient.

Trigger: Patient calls and asks for a new appointment or asks to cancel or change an existing appointment.

Type: External

Relationships:
> Association: Old Patient
> Include:
> Extend: Update Patient Information
> Generalization: Manage Appointments

Normal Flow of Events:
1. The Patient contacts the office regarding an appointment.
2. The Patient provides the Receptionist with his or her name and address.
3. If the Patient's information has changed
> Execute the Update Patient Information use case.
4. If the Patient's payment arrangements has changed
> Execute the Make Payments Arrangements use case.
5. The Receptionist asks Patient if he or she would like to make a new appointment, cancel an existing appointment, or change an existing appointment.
> If the patient wants to make a new appointment,
> > the S-1: new appointment subflow is performed.
> If the patient wants to cancel an existing appointment,
> > the S-2: cancel appointment subflow is performed.
> If the patient wants to change an existing appointment,
> > the S-3: change appointment subflow is performed.
6. The Receptionist provides the results of the transaction to the Patient.

SubFlows:
> S-1: New Appointment
> > 1. The Receptionist asks the Patient for possible appointment times.
> > 2. The Receptionist matches the Patient's desired appointment times with available dates and times and schedules the new appointment.
> S-2: Cancel Appointment
> > 1. The Receptionist asks the Patient for the old appointment time.
> > 2. The Receptionist finds the current appointment in the appointment file and cancels it.
> S-3: Change Appointment
> > 1. The Receptionist performs the S-2: cancel appointment subflow.
> > 2. The Receptionist performs the S-1: new appointment subflow.

Alternate/Exceptional Flows:
> S-1, 2a1: The Receptionist proposes some alternative appointment times based on what is available in the appointment schedule.
> S-1, 2a2: The Patient chooses one of the proposed times or decides not to make an appointment.

TEMPLATE
can be found at
www.wiley.com
/go/global/
tegarden

FIGURE 5-2   Use-Case Description (Figure 4-11)

Brainstorming

Brainstorming is a discovery technique that has been used successfully in identifying candidate classes. Essentially, in this context, brainstorming is a process of a set of individuals sitting around a table and suggesting potential classes that could be useful for the problem under consideration. Typically, a brainstorming session is kicked off by a facilitator who asks the set of individuals to address a specific question or statement that frames the session. For example, using the appointment problem described previously, the facilitator could ask the development team and users to think about their experiences of making appointments and to identify candidate classes based on their past experiences. Notice that this approach does not use the functional models developed earlier. It simply asks the participants to identify the objects with which they have interacted. For example, a potential set of objects that come to mind are doctors, nurses, receptionists, appointments, illnesses, treatments, prescriptions, insurance cards, and medical records. Once a sufficient number of candidate objects have been identified, the participants should discuss and select which of the candidate objects should be considered further. Once these have been identified, further brainstorming can take place to identify potential attributes, operations, and relationships for each of the identified objects.

Bellin and Simone[2] have suggested a set of useful principles to guide a brainstorming session. First, all suggestions should be taken seriously because they could be good suggestions. At this point in the development of the system, it is much better to have to delete something later than to accidentally leave something critical out. Second, all participants should think fast and furiously first. After everything is out on the proverbial table, then the participants can be encouraged to ponder the candidate classes they have identified. Third, based on the second principle, the facilitator must manage the process. Otherwise, the process will be chaotic. Furthermore, the facilitator should ensure all participants are involved and that a few participants do not dominate the process. You want to get a view of the problem that is as complete as possible. One way that this can be executed is to use a round-robin approach to suggesting candidate classes where each participant takes a turn in suggesting a class. Another approach is to use an electronic brainstorming tool that supports anonymity.[3] Fourth, the facilitator can use humor to break the ice so that all participants can feel comfortable in making suggestions.

Common Object Lists

As its name implies, a *common object list* is simply a list of objects common to the business domain of the system. Several categories of objects have been found to help the analyst in creating the list, such as physical or tangible things, incidents, roles, and interactions.[4] Analysts should first look for physical, or *tangible, things* in the business domain. These could include books, desks, chairs, and office equipment. Normally, these types of objects are the easiest to identify. *Incidents* are events that occur in the business domain, such as meetings, flights, performances, or accidents. Reviewing the use cases can readily identify the *roles* that the people play in the problem, such as doctor, nurse, patient, or receptionist. Typically, an *interaction* is a transaction that takes place in the business domain, such as a sales transaction. Other types of objects that can be identified include places, containers, organizations,

[2] D. Bellin and S. S. Simone, *The CRC Card Book* (Reading, MA: Addison-Wesley, 1997).

[3] A.R. Dennis, J.S. Valacich, T. Connolly, and B.E. Wynne, "Process Structuring in Electronic Brainstorming," *Information Systems Research* 7, no. 2 (June 1996): 268-277.

[4] For example, see C. Larman, *Applying UML and Patterns: An Introduction to Object-Oriented Analysis and Design* (Englewood Cliffs, NJ: Prentice Hall, 1998); and S. Shlaer and S. J. Mellor, *Object-Oriented Systems Analysis: Modeling the World in Data* (Englewood Cliffs, NJ: Yourdon Press, 1988).

business records, catalogs, and policies. In rare cases, processes themselves may need information stored about them. In these cases, processes may need an object, in addition to a use case, to represent them. Finally, there are libraries of reusable objects that have been created for different business domains. For example, with regard to the appointment problem, the Common Open Source Medical Objects[5] could be useful to investigate for potential objects that should be included.

Patterns

The idea of using patterns is a relatively new area in object-oriented systems development.[6] There have been many definitions of exactly what a pattern is. From our perspective, a *pattern* is simply a useful group of collaborating classes that provide a solution to a commonly occurring problem. Because patterns provide a solution to commonly occurring problems, they are reusable.

An architect, Christopher Alexander, has inspired much of the work associated with using patterns in object-oriented systems development. According to Alexander and his colleagues,[7] it is possible to make very sophisticated buildings by stringing together commonly found patterns, rather than creating entirely new concepts and designs. In a very similar manner, it is possible to put together commonly found object-oriented patterns to form elegant, object-oriented information systems. For example, many business transactions involve the same types of objects and interactions. Virtually all transactions would require a transaction class, a transaction line item class, an item class, a location class, and a participant class. By simply reusing these existing patterns of classes, we can more quickly and more completely define the system than if we start with a blank piece of paper.

Many different types of patterns have been proposed, ranging from high-level business-oriented patterns to more low-level design patterns. For example, Figure 5-3 depicts a set of useful analysis patterns[8]. Figure 5-4 portrays a class diagram that we created by merging the patterns contained in Figure 5-3 into a single reusable pattern. In this case, we merged the Account–Entry–Transaction pattern (located at the bottom left of Figure 5-3) with the Place–Transaction–Participant–Transaction Line Item–Item pattern (located at the top left of Figure 5-3) on the common Transaction class. Next, we merged the Party–Person–Organization (located at the top right of Figure 5-3) by merging the Participant and Party classes. Finally, we extended the Item class by merging the Item class with the Product class of the Product–Good–Service pattern (located at the bottom right of Figure 5-3).

---

[5] See sourceforge.net/projects/cosmos/.

[6] Many books have been devoted to this topic. For example, see P. Coad, D. North, and M. Mayfield, *Object Models: Strategies, Patterns, & Applications*, 2nd Ed. (Englewood Cliffs, NJ: Prentice Hall, 1997); H-E. Eriksson and M. Penker, *Business Modeling with UML: Business Patterns at Work* (New York: Wiley, 2000); M. Fowler, *Analysis Patterns: Reusable Object Models* (Reading, MA: Addison-Wesley, 1997); E. Gamma, R. Helm, R. Johnson, and J. Vlissides, *Design Patterns: Elements of Reusable Object-Oriented Software* (Reading, MA: Addison-Wesley, 1995); David C. Hay, *Data Model Patterns: Conventions of Thought* (New York: Dorset House, 1996), and L. Silverston, *The Data Model Resource Book: A Library of Universal Data Models for All Enterprises, Volume 1, Revised Ed.* (New York, NY; Wiley, 2001).

[7] C. Alexander, S. Ishikawa, M. Silverstein, M. Jacobson, I. Fiksdahl-King, and S. Angel, *A Pattern Language* (New York: Oxford University Press, 1977).

[8] The patterns are portrayed using UML Class Diagrams. We describe the syntax of the diagrams later in this chapter. The specific patterns shown have been adapted from patterns described in P. Coad, D. North, and M. Mayfield, *Object Models: Strategies, Patterns, & Applications*, 2nd Ed.; M. Fowler, *Analysis Patterns: Reusable Object Models*; and L. Silverston, *The Data Model Resource Book: A Library of Universal Data Models for All Enterprises, Volume 1, Revised Edition.*

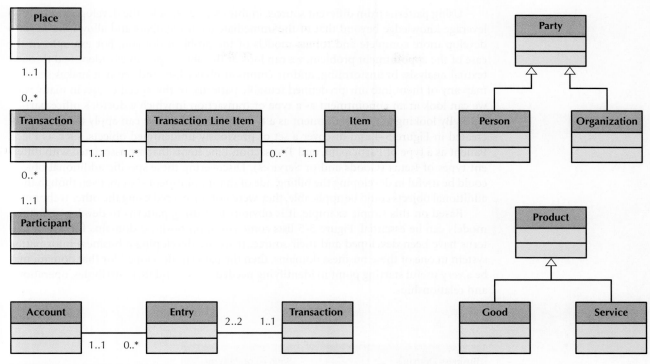

FIGURE 5-3   Sample Patterns

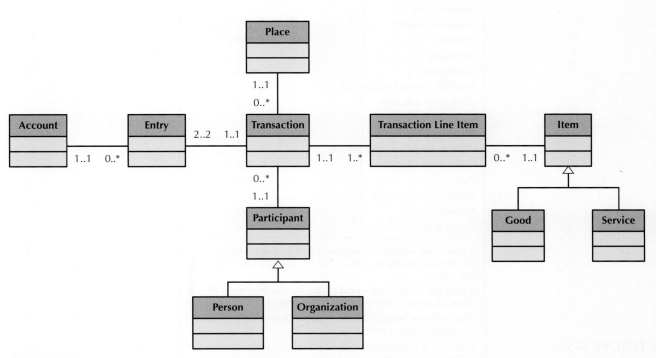

FIGURE 5-4   Sample Integration of Sample Patterns

Using patterns from different sources in this manner enables the development team to leverage knowledge beyond that of the immediate team members and allows the team to develop more complete and robust models of the problem domain. For example, in the case of the appointment problem, we can look at the objects previously identified through textual analysis, brainstorming, and/or common object lists and see if it makes sense to map any of them into any predefined reusable patterns. In this specific case, in many ways we can look at an appointment as a type of transaction in which a doctor's office participates. By looking at an appointment as a type of transaction, we can apply the pattern we created in Figure 5-4 and discover a set of previously unidentified objects, such as Place, Patient as a type of Participant, and Transaction Line Items that are associated with different types of Items (Goods and/or Services). Discovering these specific additional objects could be useful in developing the billing side of the appointment system. Even though these additional objects could be applicable, they were not uncovered using the other techniques.

Based on this simple example, it is obvious that using patterns to develop structural models can be essential. Figure 5-5 lists some common business domains for which patterns have been developed and their source. If we are developing a business information system in one of these business domains, then the patterns developed for that domain may be a very useful starting point in identifying needed classes and their attributes, operations, and relationships.

Business Domains	Sources of Patterns
Accounting	3, 4
Actor-Role	2
Assembly-Part	1
Container-Content	1
Contract	2, 4
Document	2, 4
Employment	2, 4
Financial Derivative Contracts	3
Geographic Location	2, 4
Group-Member	1
Interaction	1
Material Requirements Planning	4
Organization and Party	2, 3
Plan	1, 3
Process Manufacturing	4
Trading	3
Transactions	1, 4

1. Peter Coad, David North, and Mark Mayfield, *Object Models: Strategies, Patterns, and Applications,* 2nd ed. (Englewood Cliffs, NJ: Prentice Hall, 1997).

2. Hans-Erik Eriksson and Magnus Penker, *Business Modeling with UML: Business Patterns at Work* (New York: Wiley, 2000).

3. Martin Fowler, *Analysis Patterns: Reusable Object Models* (Reading, MA: Addison-Wesley, 1997).

4. David C. Hay, *Data Model Patterns: Conventions of Thought* (New York, NY, Dorset House, 1996).

FIGURE 5-5
Useful Patterns

CRC CARDS

CRC (Class–Responsibility–Collaboration) cards are used to document the responsibilities and collaborations of a class. In some object-oriented systems-development methodologies, CRC cards are seen to be an alternative competitor to the Unified Process employment of use cases and class diagrams. However, we see them as a useful, low-tech approach that can compliment a typical high-tech Unified Process approach that uses CASE tools. We use an extended form of the CRC card to capture all relevant information associated with a class.[9] We describe the elements of our CRC cards later, after we explain responsibilities and collaborations.

Responsibilities and Collaborations

Responsibilities of a class can be broken into two separate types: knowing and doing. *Knowing responsibilities* are those things that an instance of a class must be capable of knowing. An instance of a class typically knows the values of its attributes and its relationships. *Doing responsibilities* are those things that an instance of a class must be capable of doing. In this case, an instance of a class can execute its operations or it can request a second instance, which it knows about, to execute one of its operations on behalf of the first instance.

The structural model describes the objects necessary to support the business processes modeled by the use cases. Most use cases involve a set of several classes, not just one class. These classes form *collaborations*. Collaborations allow the analyst to think in terms of clients, servers, and contracts.[10] A *client* object is an instance of a class that sends a request to an instance of another class for an operation to be executed. A *server* object is the instance that receives the request from the client object. A *contract* formalizes the interactions between the client and server objects. For example, a patient makes an appointment with a doctor. This sets up an obligation for both the patient and doctor to appear at the appointed time. Otherwise, consequences, such as billing the patient for the appointment regardless of whether he or she appears, can be dealt out. Also, the contract should spell out what the benefits of the contract will be, such as a treatment being prescribed for whatever ails the patient and a payment to the doctor for the services provided. Chapter 8 provides a more detailed explanation of contracts and examples of their use.

An analyst can use the idea of class responsibilities and client–server–contract collaborations to help identify the classes, along with the attributes, operations, and relationships, involved with a use case. One of the easiest ways to use CRC cards in developing a structural model is through anthropomorphism—pretending that the classes have human characteristics. Members of the development team can either ask questions of themselves or be asked questions by other members of the team. Typically the questions asked are of the form:

> Who or what are you?
> What do you know?
> What can you do?

---

[9] Our CRC cards are based on the work of D. Bellin and S. S. Simone, *The CRC Card Book* (Reading, MA: Addison-Wesley, 1997); I. Graham, *Migrating to Object Technology* (Wokingham, England: Addison-Wesley, 1995); and B. Henderson-Sellers and B. Unhelkar, *OPEN modeling with UML* (Harlow, England: Addison-Wesley, 2000).

[10] For more information, see K. Beck and W. Cunningham, "A Laboratory for Teaching Object-Oriented Thinking," *Proceedings of OOPSLA, SIGPLAN Notices*, 24, no. 10 (1989): 1–6; B. Henderson-Sellers and B. Unhelkar, *OPEN Modeling with UML* (Harlow, England: Addison-Wesley, 2000); C. Larman, *Applying UML and Patterns: An Introduction to Object-Oriented Analysis and Design* (Englewood Cliffs, NJ: Prentice Hall, 1998); B. Meyer, *Object-Oriented Software Construction* (Englewood Cliffs, NJ: Prentice Hall, 1994); and R. Wirfs-Brock, B. Wilkerson, and L. Wiener, *Designing Object-Oriented Software* (Englewood Cliffs, NJ, Prentice Hall, 1990).

The answers to the questions are then used to add detail to the evolving CRC cards. For example, in the appointment problem, a member of the team can pretend that he or she is an appointment. In this case, the appointment would answer that he or she knows about the doctor and patient who participate in the appointment and they would know the date and time of the appointment. Furthermore, an appointment would have to know how to create itself, delete itself, and to possibly change different aspects of itself. In some cases, this approach will uncover additional objects that have to be added to the evolving structural model.

Elements of a CRC Card

The set of CRC cards contains all the information necessary to build a logical structural model of the problem under investigation. Figure 5-6 shows a sample CRC card. Each CRC card captures and describes the essential elements of a class. The front of the card contains

Front:		
Class Name: Old Patient	**ID:** 3	**Type:** Concrete, Domain
Description: An individual that needs to receive or has received medical attention		**Associated Use Cases:** 2

Responsibilities	Collaborators
Make appointment	Appointment
Calculate last visit	
Change status	
Provide medical history	Medical history

Back:

Attributes:
Amount (double)
Insurance carrier (text)

Relationships:

Generalization (a-kind-of): Person

Aggregation (has-parts): Medical History

Other Associations: Appointment

FIGURE 5-6
Sample CRC Card

the class's name, ID, type, description, associated use cases, responsibilities, and collaborators. The name of a class should be a noun (but not a proper noun, such as the name of a specific person or thing). Just like the use cases, in later stages of development, it is important to be able to trace back design decisions to specific requirements. In conjunction with the list of associated use cases, the ID number for each class can be used to accomplish this. The description is simply a brief statement that can be used as a textual definition for the class. The responsibilities of the class tend to be the operations that the class must contain (i.e., the doing responsibilities).

The back of a CRC card contains the attributes and relationships of the class. The attributes of the class represent the knowing responsibilities that each instance of the class has to meet. Typically, the data type of each attribute is listed with the name of the attribute (e.g., the amount attribute is double and the insurance carrier is text). Three types of relationships typically are captured at this point: generalization, aggregation, and other associations. In Figure 5-6, we see that a Patient is a-kind-of Person and that a Patient is associated with Appointments.

CRC cards are used to document the essential properties of a class. However, once the cards are filled out, the analyst can use the cards and anthropomorphisms in role-playing (described in the next section) to uncover missing properties by executing the different scenarios associated with the use cases (see Chapter 4). Role-playing also can be used as a basis to test the clarity and completeness of the evolving representation of the system.

Role-Playing CRC Cards with Use Cases[11, 12]

In addition to the object identification approaches described earlier (textual analysis, brainstorming, common object lists, and patterns), CRC cards can be used in a *role-playing* exercise that has been shown to be useful in discovering additional objects, attributes, relationships, and operations. In general, members of the team perform roles associated with the actors and objects previously identified with the different use cases. Technically speaking, the members of the team perform the different steps associated with a specific scenario of a use case. Remember, a scenario is a single, unique execution path through a use case. A useful place to look for the different scenarios of a use case is the activity diagrams (e.g., see Figures 4-7, 4-8, 4-9, and 4-10). A different scenario exists for each time a decision node causes a split in the execution path of the use case. Also, scenarios can be identified from the alternative/exceptional flows in a use-case description. Even though the activity diagrams and use-case descriptions should contain the same information and given the incremental and iterative nature of object-oriented systems development, at this point in the evolution of the system, we suggest that you review both representations to ensure that you do not miss any relevant scenarios.

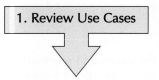

The first step is to review the use-case descriptions (see Figure 5-2). This allows the team to pick a specific use case to role-play. Even though it is tempting to try to complete as many use cases as possible in a short time, the team should not choose the easiest use cases first. Instead, at this point in the development of the system, the team should choose the use case that is the most important, the most complex, or the least understood.

---

[11] This step is related to the verification and validation of the analysis models (functional, structural, and behavioral). Because this deals with verification and validation that takes place between the models, in this case, functional and structural, we will return to this topic in Chapter 7.

[12] Our role-playing approach is based on the work of D. Bellin and S. S. Simone, *The CRC Card Book* (Reading, MA: Addison-Wesley, 1997).

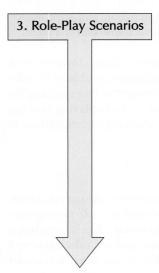

The second step is to identify the relevant roles that are to be played. Each role is associated with either an actor or an object. To choose the relevant objects, the team reviews each of the CRC cards and picks the ones that are associated with the chosen use case. For example, in Figure 5-6, we see that the CRC card that represents the Old Patient class is associated with Use Case number 2. So if we were going to role-play the Make Old Patient Appt use case (see Figure 5-2), we would need to include the Old Patient CRC card. By reviewing the use-case description, we can easily identify the Old Patient and Doctor actors (see Primary Actor and Stakeholders section of the use-case description in Figure 5-2). By reading the event section of the use-case description, we identify the internal actor role of Receptionist. After identifying all of the relevant roles, we assign each one to a different member of the team.

The third step is to role-play scenarios of the use case by having the team members perform each one. To do this, each team member must pretend that they are an instance of the role assigned to them. For example, if a team member was assigned the role of the Receptionist, then he or she would have to be able to perform the different steps in the scenario associated with the Receptionist. In the case of the change appointment scenario, this would include steps 2, 5, 6, S-3, S-1, and S-2. However, when this scenario is performed (role-played), it would be discovered that steps 1, 3, and 4 were incomplete. For example, in Step 1, what actually occurs? Does the Patient make a phone call? If so, who answers the phone? In other words, a lot of information contained in the use-case description is only identified in an implicit, not explicit, manner. When the information is not identified explicitly, there is a lot of room for interpretation, which requires the team members to make assumptions. It is much better to remove the need to make an assumption by making each step explicit. In this case, step 1 of the Normal Flow of Events should be modified. Once the step has been fixed, the scenario is tried again. This process is repeated until the scenario can be executed to a successful conclusion. Once the scenario has successfully concluded, the next scenario is performed. This is repeated until all of the scenarios of the use case can be performed successfully. [13]

The fourth step is to simply repeat steps 1 through 3 for the remaining use cases.

CLASS DIAGRAMS

A *class diagram* is a *static model* that shows the classes and the relationships among classes that remain constant in the system over time. The class diagram depicts classes, which include both behaviors and states, with the relationships between the classes. The following sections present the elements of the class diagram, different approaches that can be used to simplify a class diagram, and an alternative structure diagram: the object diagram.

Elements of a Class Diagram

Figure 5-7 shows a class diagram that was created to reflect the classes and relationships associated with the appointment system. This diagram is based on the classes uncovered

---

[13] In some cases, some scenarios are only executed in very rare circumstances. So, from a practical perspective, each scenario could be prioritized individually and only "important" scenarios would have to be implemented for the first release of the system. Only those scenarios would have to be tested at this point in the evolution of the system.

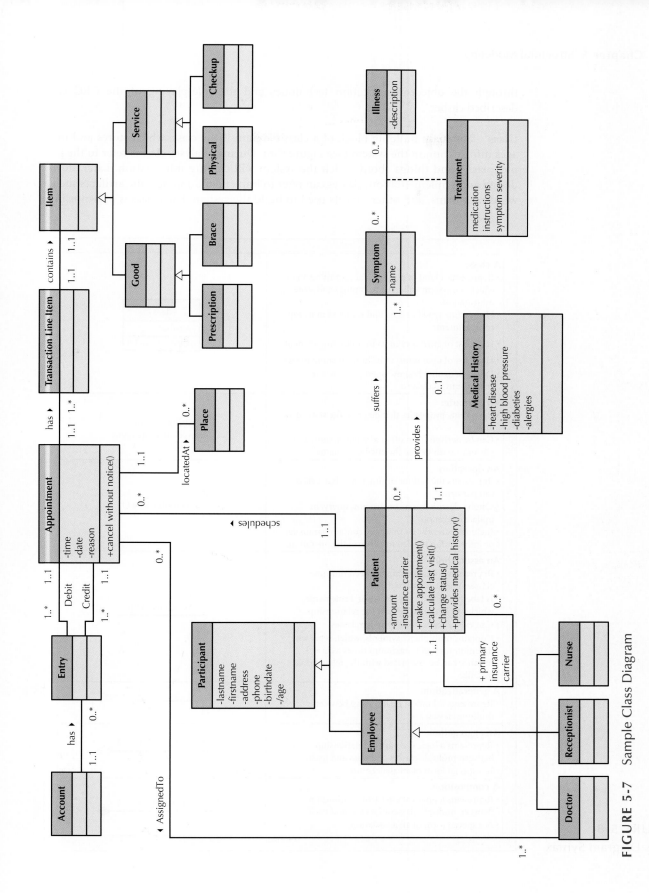

FIGURE 5-7 Sample Class Diagram

through the object-identification techniques and the role-playing of the CRC cards described earlier.

Class   The main building block of a class diagram is the class, which stores and manages information in the system (see Figure 5-8). During analysis, classes refer to the people, places, and things about which the system will capture information. Later, during design and implementation, classes can refer to implementation-specific artifacts such as windows, forms, and other objects used to build the system. Each class is drawn using a

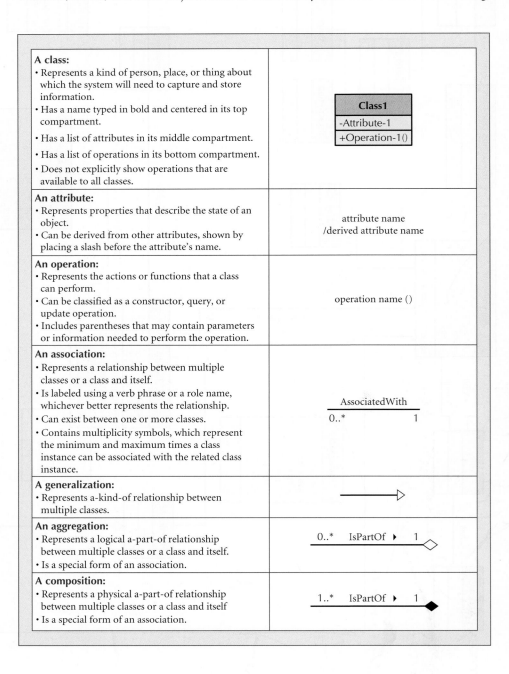

A class:
- Represents a kind of person, place, or thing about which the system will need to capture and store information.
- Has a name typed in bold and centered in its top compartment.
- Has a list of attributes in its middle compartment.
- Has a list of operations in its bottom compartment.
- Does not explicitly show operations that are available to all classes.

An attribute:
- Represents properties that describe the state of an object.
- Can be derived from other attributes, shown by placing a slash before the attribute's name.

An operation:
- Represents the actions or functions that a class can perform.
- Can be classified as a constructor, query, or update operation.
- Includes parentheses that may contain parameters or information needed to perform the operation.

An association:
- Represents a relationship between multiple classes or a class and itself.
- Is labeled using a verb phrase or a role name, whichever better represents the relationship.
- Can exist between one or more classes.
- Contains multiplicity symbols, which represent the minimum and maximum times a class instance can be associated with the related class instance.

A generalization:
- Represents a-kind-of relationship between multiple classes.

An aggregation:
- Represents a logical a-part-of relationship between multiple classes or a class and itself.
- Is a special form of an association.

A composition:
- Represents a physical a-part-of relationship between multiple classes or a class and itself
- Is a special form of an association.

FIGURE 5-8
Class Diagram Syntax

three-part rectangle, with the class's name at top, attributes in the middle, and operations at the bottom. We can see that the classes identified earlier, such as Participant, Doctor, Patient, Receptionist, Medical History, Appointment, and Symptom, are included in Figure 5-7. The attributes of a class and their values define the *state* of each object created from the class, and the behavior is represented by the operations.

Attributes are properties of the class about which we want to capture information (see Figure 5-8). Notice that the Participant class in Figure 5-7 contains the attributes lastname, firstname, address, phone, and birthdate. At times, you might want to store *derived attributes*, which are attributes that can be calculated or derived; these special attributes are denoted by placing a slash (/) before the attribute's name. Notice how the person class contains a derived attribute called /age, which can be derived by subtracting the patient's birth date from the current date. It is also possible to show the *visibility* of the attribute on the diagram. Visibility relates to the level of information hiding to be enforced for the attribute. The visibility of an attribute can be public (+), protected (#), or private (−). A *public* attribute is one that is not hidden from any other object. As such, other objects can modify its value. A *protected* attribute is one that is hidden from all other classes except its immediate subclasses. A *private* attribute is one that is hidden from all other classes. The default visibility for an attribute is normally private.

Operations are actions or functions that a class can perform (see Figure 5-8). The functions that are available to all classes (e.g., create a new instance, return a value for a particular attribute, set a value for a particular attribute, delete an instance) are not explicitly shown within the class rectangle. Instead, only operations unique to the class are included, such as the cancel without notice operation in the Appointment class and the calculate last visit operation in the Patient class in Figure 5-7. Notice that both the operations are followed by parentheses, which contain the parameter(s) needed by the operation. If an operation has no parameters, the parentheses are still shown but are empty. As with attributes, the visibility of an operation can be designated public, protected, or private. The default visibility for an operation is normally public.

There are four kinds of operations that a class can contain: constructor, query, update, and destructor. A *constructor operation* creates a new instance of a class. For example, the Patient class may have a method called insert (), which creates a new patient instance as patients are entered into the system. As we just mentioned, if an operation implements one of the basic functions (e.g., create a new instance), it is normally not explicitly shown on the class diagram, so typically, we do not see constructor methods explicitly on the class diagram.

A *query operation* makes information about the state of an object available to other objects, but it does not alter the object in any way. For instance, the calculate last visit () operation that determines when a patient last visited the doctor's office will result in the object's being accessed by the system, but it will not make any change to its information. If a query method merely asks for information from attributes in the class (e.g., a patient's name, address, phone), then it is not shown on the diagram because we assume that all objects have operations that produce the values of their attributes.

An *update operation* changes the value of some or all the object's attributes, which may result in a change in the object's state. Consider changing the status of a patient from new to current with a method called change status() or associating a patient with a particular appointment with make appointment (appointment).

A *destructor operation* simply deletes or removes the object from the system. For example, if an employee object no longer represents an actual employee associated with the firm, the employee could need to be removed from the employee database, and a destructor operation would be used to implement this behavior. However, deleting an object is one of the basic functions and therefore would not be included on the class diagram.

Relationships   A primary purpose of a class diagram is to show the relationships, or associations, that classes have with one another. These are depicted on the diagram by drawing lines between classes (see Figure 5-8). When multiple classes share a relationship (or a class shares a relationship with itself), a line is drawn and labeled with either the name of the relationship or the roles that the classes play in the relationship. For example, in Figure 5-7, the two classes, Patient and Appointment, are associated with one another whenever a patient schedules an appointment. Thus, a line labeled schedules connects Patient and Appointment, representing exactly how the two classes are related to each other. Also, notice that there is a small solid triangle beside the name of the relationship. The triangle allows a direction to be associated with the name of the relationship. In Figure 5-7, the schedules relationship includes a triangle, indicating that the relationship is to be read as "patient schedules appointment." Inclusion of the triangle simply increases the readability of the diagram. In Figure 5-9, three additional examples of associations are portrayed: An Invoice is AssociatedWith a Purchase Order (and vice versa), a Pilot Flies an Aircraft, and a Spare Tire IsLocatedIn a Trunk.

Sometimes a class is related to itself, as in the case of a patient being the primary insurance carrier for other patients (e.g., spouse, children). In Figure 5-7, notice that a line was drawn between the Patient class and itself and called *primary insurance carrier* to depict the role that the class plays in the relationship. Notice that a plus (+) sign is placed before the label to communicate that it is a role, as opposed to the name of the relationship. When labeling an association, we use either a relationship name or a role name (not both), whichever communicates a more thorough understanding of the model.

Relationships also have *multiplicity*, which documents how an instance of an object can be associated with other instances. Numbers are placed on the association path to denote the minimum and maximum instances that can be related through the association in the format minimum number.. maximum number (see Figure 5-10). The numbers specify the relationship from the class at the far end of the relationship line to the end with the number. For example, in Figure 5-7, there is a 0..* on the appointment end of the patient schedules appointment relationship. This means that a patient can be associated

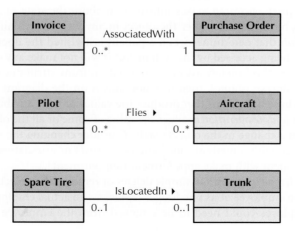

FIGURE 5-9
Sample Association

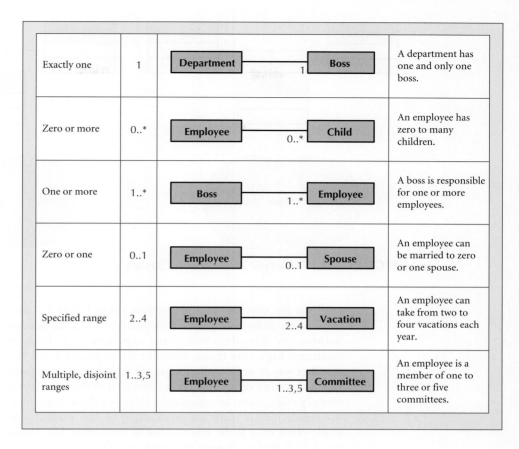

FIGURE 5-10
Multiplicity

with zero through many different appointments. At the patient end of this same relationship there is a 1..1, meaning that an appointment must be associated with one and only one (1) patient. In Figure 5-9, we see that an instance of the Invoice class must be AssociatedWith one instance of the Purchase Order class and that an instance of the Purchase Order class may be AssociatedWith zero or more instances of the Invoice class, that an instance of the Pilot class Flies zero or more instances of the Aircraft class, and that an instance of the Aircraft class may be flown by zero or more instances of the Pilot class. Finally, we see that an instance the Spare Tire class IsLocatedIn zero or one instance of the Trunk class, whereas an instance of the Trunk class can contain zero or one instance of the Spare Tire class.

There are times when a relationship itself has associated properties, especially when its classes share a many-to-many relationship. In these cases, a class called an *association class* is formed, which has its own attributes and operations.[14] It is shown as a rectangle attached by a dashed line to the association path, and the rectangle's name matches the label of the association. Think about the case of capturing information about illnesses and symptoms. An illness (e.g., the flu) can be associated with many symptoms (e.g., sore throat, fever), and a symptom (e.g., sore throat) can be associated with many illnesses

---

[14] For those familiar with data modeling, associative classes serve a purpose similar to the one the associative entity serves in ER diagramming.

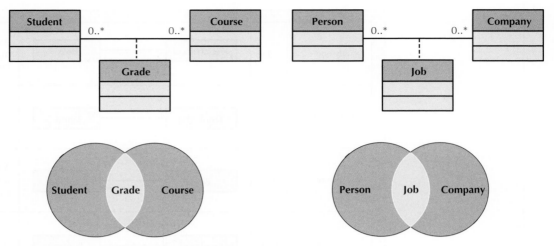

FIGURE 5-11   Sample Association Classes

(e.g., the flu, strep throat, the common cold). Figure 5-7 shows how an association class can capture information about remedies that change depending on the various combinations. For example, a sore throat caused by strep throat requires antibiotics, whereas treatment for a sore throat from the flu or a cold could be throat lozenges or hot tea. Another way to decide when to use an association class is when attributes that belong to the intersection of the two classes involved in the association must be captured. We can visually think about an association class as a Venn diagram. For example, in Figure 5-11, the Grade idea is really an intersection of the Student and Course classes, because a grade exists only at the intersection of these two ideas. Another example shown in Figure 5-11 is that a job may be viewed as the intersection between a Person and a Company. Most often, classes are related through a normal association; however, there are two special cases of an association that you will see appear quite often: generalization and aggregation.

Generalization and Aggregation Associations   A *generalization association* shows that one class (subclass) inherits from another class (superclass), meaning that the properties and operations of the superclass are also valid for objects of the subclass. The generalization path is shown with a solid line from the subclass to the superclass and a hollow arrow pointing at the superclass (see Figure 5-8). For example, Figure 5-7 communicates that doctors, nurses, and receptionists are all kinds of employees and those employees and patients are kinds of participants. Remember that the generalization relationship occurs when you need to use words like "is a kind of" to describe the relationship. Some additional examples of generalization are given in Figure 5-12. For example, Cardinal is a-kind-of Bird, which is a-kind-of Animal; a General Practitioner is a-kind-of Physician, which is a-kind-of Person; and a Truck is a-kind-of Land Vehicle, which is a-kind-of Vehicle.

An *aggregation association* is used when classes actually comprise other classes. For example, think about a doctor's office that has decided to create health-care teams that include doctors, nurses, and administrative personnel. As patients enter the office, they are assigned to a health-care team, which cares for their needs during their visits. We could include this new knowledge in Figure 5-7 by adding two new classes (Administrative Personnel and Health Team) and aggregation relationships from the Doctor, the Nurse, and

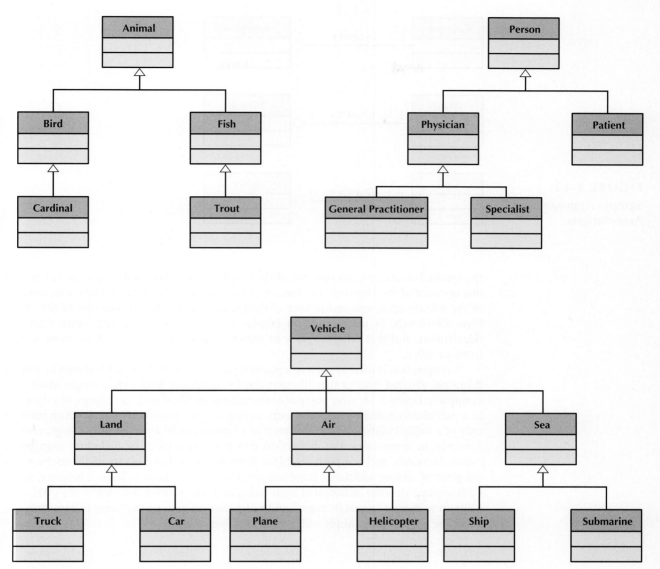

FIGURE 5-12   Sample Generalizations

the new Administrative Personnel classes to the new Health Team class. A diamond is placed nearest the class representing the aggregation (health-care team), and lines are drawn from the diamond to connect the classes that serve as its parts (doctors, nurses, and administrative personnel). Typically, you can identify these kinds of associations when you need to use words like "is a part of" or "is made up of" to describe the relationship. However, from a UML perspective, there are two types of aggregation associations: aggregation and composition (see Figure 5-8).

   Aggregation is used to portray logical a-part-of relationships and is depicted on a UML class diagram by a hollow or white diamond. For example, in Figure 5-13, three logical aggregations are shown. Logical implies that it is possible for a part to be associated with multiple wholes or that it is relatively simple for the part to be removed from

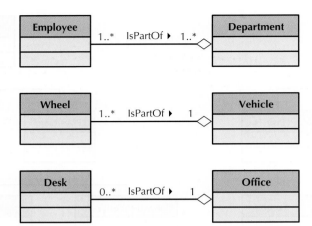

FIGURE 5-13

Sample Aggregation
Associations

the whole. For example, an instance of the Employee class IsPartOf an instance of at least one instance of the Department class, an instance of the Wheel class IsPartOf an instance of the Vehicle class, and an instance of the Desk class IsPartOf an instance of the Office class. Obviously, in many cases an employee can be associated with more than one department, and it is relatively easy to remove a wheel from a vehicle or move a desk from an office.

Composition is used to portray a physical part of relationships and is shown by a black diamond. *Physical* implies that the part can be associated with only a single whole. For example in Figure 5-14, three physical compositions are illustrated: an instance of a door can be a part of only a single instance of a car, an instance of a room can be a part of an instance only of a single building; and an instance of a button can be a part of only a single mouse. However, in many cases, the distinction that you can achieve by including aggregation (white diamonds) and composition (black diamonds) in a class diagram might not be worth the price of adding additional graphical notation for the client to learn. Therefore, many UML experts view the inclusion of aggregation and composition notation to the UML class diagram as simply "syntactic sugar" and not necessary because the same information can always be portrayed by simply using the association syntax.

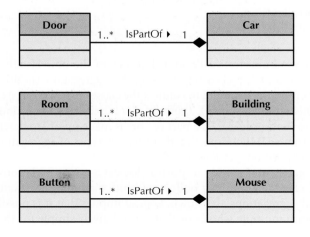

FIGURE 5-14

Sample Composition
Associations

Simplifying Class Diagrams

When a class diagram is fully populated with all the classes and relationships for a real-world system, the class diagram can become very difficult to interpret (i.e., can be very complex). When this occurs, it is sometimes necessary to simplify the diagram. One way to simplify the class diagram is to show only concrete classes.[15] However, depending on the number of associations that are connected to abstract classes—and thus inherited down to the concrete classes—this particular suggestion could make the diagram more difficult to comprehend.

A second way to simplify the class diagram is through the use of a *view* mechanism. Views were developed originally with relational database management systems and are simply a subset of the information contained in the database. In this case, the view would be a useful subset of the class diagram, such as a use-case view that shows only the classes and relationships relevant to a particular use case. A second view could be to show only a particular type of relationship: aggregation, association, or generalization. A third type of view is to restrict the information shown with each class, for example, show only the name of the class, the name and attributes, or the name and operations. These view mechanisms can be combined to further simplify the diagram.

A third approach to simplifying a class diagram is through the use of *packages* (i.e., logical groups of classes). To make the diagrams easier to read and keep the models at a reasonable level of complexity, the classes can be grouped together into packages. Packages are general constructs that can be applied to any of the elements in UML models. In Chapter 4, we introduced the package idea to simplify use-case diagrams. In the case of class diagrams, it is simple to sort the classes into groups based on the relationships that they share.[16]

Object Diagrams

Although class diagrams are necessary to document the structure of the classes, there are times when a second type of *static structure diagram*, called an object diagram, can be useful. An *object diagram* is essentially an instantiation of all or part of a class diagram. *Instantiation* means to create an instance of the class with a set of appropriate attribute values.

Object diagrams can be very useful when trying to uncover details of a class. Generally speaking, it is easier to think in terms of concrete objects (instances) rather than abstractions of objects (classes). For example in Figure 5-15, a portion of the class diagram in Figure 5-7 has been copied and instantiated. The top part of the figure simply is a copy of a small view of the overall class diagram. The lower portion is the object diagram that instantiates that subset of classes. By reviewing the actual instances involved, John Doe, Appt1, Symptom1, and Dr. Smith, we may discover additional relevant attributes, relationships, and/or operations or possibly misplaced attributes, relationships, and/or operations. For example, an appointment has a reason attribute. Upon closer examination, the reason attribute might have been better modeled as an association with the Symptom class. Currently, the Symptom class is associated with the Patient class. After reviewing the object diagram, this seems to be in error. Therefore, we should modify the class diagram to reflect this new understanding of the problem.

---

[15] See footnote 1.

[16] For those familiar with structured analysis and design, packages serve a purpose similar to the leveling and balancing processes used in data flow diagramming. Packages and package diagrams are described in more detail in Chapter 7.

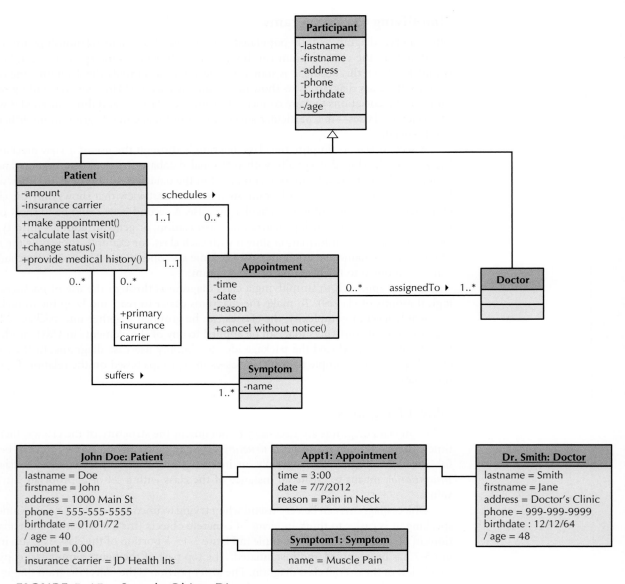

FIGURE 5-15   Sample Object Diagram

CREATING STRUCTURAL MODELS USING CRC CARDS AND CLASS DIAGRAMS

Creating a structural model is an incremental and iterative process whereby the analyst makes a rough cut of the model and then refines it over time. Structural models can become quite complex—in fact, there are systems that have class diagrams containing hundreds of classes. It is important to remember that CRC cards and class diagrams can be used to describe both the as-is and to-be structural models of the evolving system, but they are most often used for the to-be model. There are many different ways to identify a set of candidate objects and to create CRC cards and class diagrams. Today most object identification begins with the use cases identified for the problem (See Chapter 4). In this section, we describe a use-case–driven process that can be used to create the structural model of a problem domain.

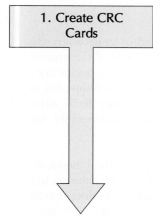

We could begin creating the structural model with a class diagram instead of CRC cards. However, owing to the low-tech nature and the ease of role-playing use-case scenarios with CRC cards, we prefer to create the CRC cards first and then transfer the information from the CRC cards into a class diagram later. As a result, the first step of our recommended process is to create CRC cards. Performing textual analysis on the use-case descriptions does this. If you recall, the normal flow of events, subflows, and alternative/exceptional flows of the use-case description were written in a special form called Subject–Verb–Direct-Object–Preposition–Indirect object (*SVDPI*). By writing the use-case events in this form, it is easier to use the guidelines for textual analysis in Figure 5-1 to identify the objects. Reviewing the primary actors, stakeholders and interests, and brief descriptions of each use case allows additional candidate objects to be identified. It is useful to go back and review the original requirements to look for information that was not included in the text of the use cases. Record all the uncovered information for each candidate object on a CRC card.

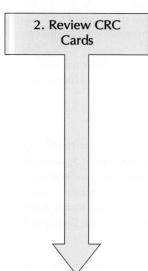

The second step is to review the CRC cards to determine if additional candidate objects, attributes, operations, and relationships are missing. In conjunction with this review, using the brainstorming and common object list approaches described earlier can aid the team in identifying missing classes, attributes, operations, and relationships. For example, the team could start a brainstorming session with a set of questions such as:

- What are the tangible things associated with the problem?
- What are the roles played by the people in the problem domain?
- What incidents and interactions take place in the problem domain?

As you can readily see, by beginning with the use-case descriptions, many of these questions already have partial answers. For example, the primary actors and stakeholders are the roles that are played by the people in the problem domain. However, it is possible to uncover additional roles not thought of previously. This obviously would cause the use-case descriptions, and possibly the use-case diagram, to be modified and possibly expanded. As in the previous step, be sure to record all the uncovered information onto the CRC cards. This includes any modifications uncovered for any previously identified candidate objects and any information regarding any new candidate objects identified.

3. Role-Play the CRC Cards

The third step is to role-play each use-case scenario using the CRC cards. Each CRC card should be assigned to an individual, who will perform the operations for the class on the CRC card. As the performers play out their roles, the system tends to break down. When this occurs, additional objects, attributes, operations, or relationships will be identified. Again, as in the previous steps, any time any new information is discovered, new CRC cards are created or modifications to existing CRC cards are made.

4. Create Class Diagram

The fourth step is to create the class diagram based on the CRC cards. Information contained on the CRC cards is simply transferred to the class diagrams. The responsibilities are transferred as operations, the attributes are drawn as attributes, and the relationships are drawn as generalization, aggregation, or association relationships. However, the class diagram also requires that the visibility of the attributes and operations be known. As a general rule, attributes are private and operations are public. Therefore, unless the analyst has a good reason to change the default visibility of these properties, then the defaults should be accepted. Finally, the analyst should examine the model for additional opportunities to use aggregation or generalization relationships. These types of relationships can simplify the individual class descriptions. As in the previous steps, all changes must be recorded on the CRC cards.

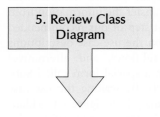

The fifth step is to review the structural model for missing and/or unnecessary classes, attributes, operations, and relationships. Up until this step, the focus of the process has been on adding information to the evolving model. At this point, the focus begins to switch from simply adding information to also challenging the reasons for including the information contained in the model. One very useful approach here is to play devil's advocate, where a team member, just for the sake of being a pain in the neck, challenges the reasoning for including all aspects of the model.

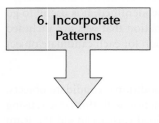

The sixth step is to incorporate useful patterns into the evolving structural model. A useful pattern is one that would allow the analyst to more fully describe the underlying domain of the problem being investigated. Looking at the collection of patterns available (Figure 5-5) and comparing the classes contained in the patterns with those in the evolving class diagram does this. After identifying the useful patterns, the analyst incorporates the identified patterns into the class diagram and modifies the affected CRC cards. This includes adding and removing classes, attributes, operations, and/or relationships.

The seventh and final step is to validate the structural model, including both the CRC cards and the class diagram. We discuss this content in the next section of the chapter and in Chapter 7.

Example

The first step is to create the CRC cards that represent the classes in the structural model. In the previous chapter, we used the Library Book Collection Management System example to describe the process of creating the functional models (use-case and activity diagrams and use-case descriptions). In this chapter, we follow the same familiar example. Because we are following a use-case–driven approach to object-oriented systems development, we first review the events described in the use-case descriptions (see Figure 5-16).

Normal Flow of Events:
  1. The Borrower brings books to the Librarian at the checkout desk.
  2. The Borrower provides Librarian their ID card.
  3. The Librarian checks the validity of the ID Card.
         If the Borrower is a Student Borrower, Validate ID Card against Registrar's Database.
         If the Borrower is a Faculty/Staff Borrower, Validate ID Card against Personnel Database.
         If the Borrower is a Guest Borrower, Validate ID Card against Library's Guest Database.
  4. The Librarian checks whether the Borrower has any overdue books and/or fines.
  5. The Borrower checks out the books.

SubFlows:

Alternate/Exceptional Flows:
     4a. The ID Card is invalid, the book request is rejected.
     5a. The Borrower either has overdue books fines, or both, the book request is rejected.

FIGURE 5-16   Flow Descriptions for the Borrow Books Use Case (Figure 4-14)

Use Case Name: Borrow Books		ID: 2	Importance Level: High
Primary Actor: Borrower		Use Case Type: Detail, Essential	

Stakeholders and Interests:
Borrower—wants to check out books
Librarian—wants to ensure borrower only gets books deserved

Brief Description: This use case describes how books are checked out of the library.

Trigger: Borrower brings books to checkout desk.

Type: External

Relationships:
    Association: Borrower, Personnel Office, Registrar's Office
    Include:
    Extend:
    Generalization :

FIGURE 5-17   Overview Description for the Borrow Books Use Case (Figure 4-13)

Next, we perform textual analysis on the events by applying the textual analysis rules described in Figure 5-1. In this case, we can quickly identify the need to include classes for Borrower, Books, Librarian, Checkout Desk, ID Card, Student Borrower, Faculty/Staff Borrower, Guest Borrower, Registrar's Database, Personnel Database, Library's Guest Database, Overdue Books, Fines, Book Request. We also can easily identify provide operations to "check the validity" of a book request, to "check out" the books, and to "reject" a book request. Furthermore, the events suggest a "brings" relationship between Borrower and Books and a "provides" relationship between Borrower and Librarian. This step also suggests that we should review the overview section of the use-case description (see Figure 5-17). In this case, the only additional information gleaned from the use-case description is the possible inclusion of classes for Personnel Office and Registrar's Office. This same process would also be completed for the remaining use cases contained in the functional model: Process Overdue Books, Maintain Book Collection, Search Collection, and Return Books (see Figure 4-5). Because we did discuss these use cases in the previous chapter, we will review the problem description as a basis for beginning the next step (see Figure 5-18).

The second step is to review the CRC cards to determine if there is any information missing. In the case of the library system, because we only used the Borrow Books use-case description, some information is obviously missing. By reviewing Figure 5-18, we see that we need to include the ability to search the book collection by title, author, keywords, and ISBN. This obviously implies a Book Collection class with four different search operations: Search By Title, Search By Author, Search By Keywords, and Search By ISBN. Interestingly, the description also implies either a set of subclasses or states for the Book class: Checked Out, Overdue, Requested, Available, and Damaged. We will return to the issue of states versus subclasses in the next chapter. The description implies many additional operations including Returning Books, Requesting Books,

The functional requirements for an automated university library circulation system include the need to support searching, borrowing, and book-maintenance activities. The system should support searching by title, author, keywords, and ISBN. Searching the library's collection database should be available on terminals in the library and available to potential borrowers via the World Wide Web. If the book of interest is currently checked out, a valid borrower should be allowed to request the book to be returned. Once the book has been checked back in, the borrower requesting the book should be notified of the book's availability.

The borrowing activities are built around checking books out and returning books by borrowers. There are three types of borrowers: students, faculty and staff, and guests. Regardless of the type of borrower, the borrower must have a valid ID card. If the borrower is a student, having the system check with the registrar's student database validates the ID card. If the borrower is a faculty or staff member, having the system check with the personnel office's employee database validates the ID card. If the borrower is a guest, the ID card is checked against the library's own borrower database. If the ID card is valid, the system must also check to determine whether the borrower has any overdue books or unpaid fines. If the ID card is invalid, or if the borrower has overdue books or unpaid fines, the system must reject the borrower's request to check out a book; otherwise, the borrower's request should be honored. If a book is checked out, the system must update the library's collection database to reflect the book's new status.

The book-maintenance activities deal with adding and removing books from the library's book collection. This requires a library manager to both logically and physically add and remove the book. Books being purchased by the library or books being returned in a damaged state typically cause these activities. If a book is determined to be damaged when it is returned and it needs to be removed from the collection, the last borrower will be assessed a fine. However, if the book can be repaired, depending on the cost of the repair, the borrower might not be assessed a fine. Finally, every Monday, the library sends reminder emails to borrowers who have overdue books. If a book is overdue more than two weeks, the borrower is assessed a fine. Depending on how long the book remains overdue, the borrower can be assessed additional fines every Monday.

FIGURE 5-18
Overview Description of the Library Book Collection Management System

Adding Books, Removing Books, Repairing Books, Fining Borrowers, and Emailing Reminders.

Next, we should use our own library experience to brainstorm potential additional classes, attributes, operations, and relationships that could be useful to include in the Library Book Collection Management System. In our library, there is also the need to Retrieve Books From Storage, Move Books To Storage, Request Books from the Interlibrary Loan System, Return Books to the Interlibrary Loan System, and deal with E-Books. You also could include classes for Journals, DVDs, and other media. As you can see, with very little information, many classes, attributes, operations, and relationships can be identified.

The third step, role-playing the CRC cards, requires us to apply the three role-playing steps described earlier:

- Review Use Cases
- Identify Relevant Actors and Objects
- Role Play Scenarios

For our purposes, we will use the Borrow Books use case to demonstrate. The relevant actors include Student Borrowers, Faculty/Staff Borrowers, Guest Borrowers, Librarians, Personnel Office, and Registrar's Office. These can be easily gleaned from the overview section of the use-case description (see Figure 5-17) and the use-case diagram (see Figure 4-5). The relevant objects seem to include Books, Borrower, and ID Card.

Finally, to role-play the scenarios, we need to assign the roles to the different members of the team and try to perform each of the paths through the events of the use case (see Figure 5-16). Based on the Events of the use case and the use case's activity diagram (see Figure 5-19.), we can quickly identify nine scenarios, three for each type of Borrower (Student, Faculty/Staff, and Guest): Valid ID and No Overdue Books & No Fines, Valid ID only, and no Valid ID. When role-playing these scenarios, one question should come up. What happens to the books that are requested when the request is rejected? Based on the current functional and structural models, the books are left sitting on the checkout desk. That doesn't quite seem right. In reality, the books are reshelved. In fact, the notion of reshelving books is also relevant to when books are checked back in or after books have been repaired. Furthermore, the idea of adding books to the collection should also include the operation of shelving the books. As you should readily see, building structural models will also help uncover behavior that was omitted when building the functional models. Remember, object-oriented systems development is not only use-case–driven but also is incremental and iterative.

The fourth step is to put everything together and to draw the class diagram. Figure 5-20 represents the first cut at drawing the class diagram for the Library Book Collection Management System. The classes identified in the previous steps have been hooked up with other classes via association, aggregation, and generalization relationships. For simplicity

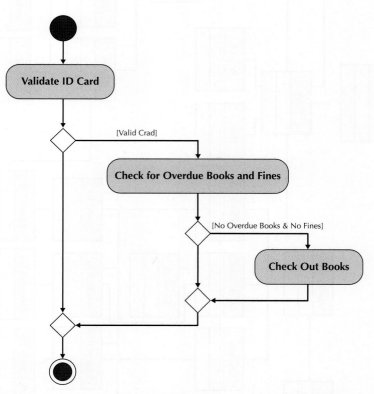

FIGURE 5-19   Activity Diagram for the Borrow Books
Use Case (Figure 4-10)

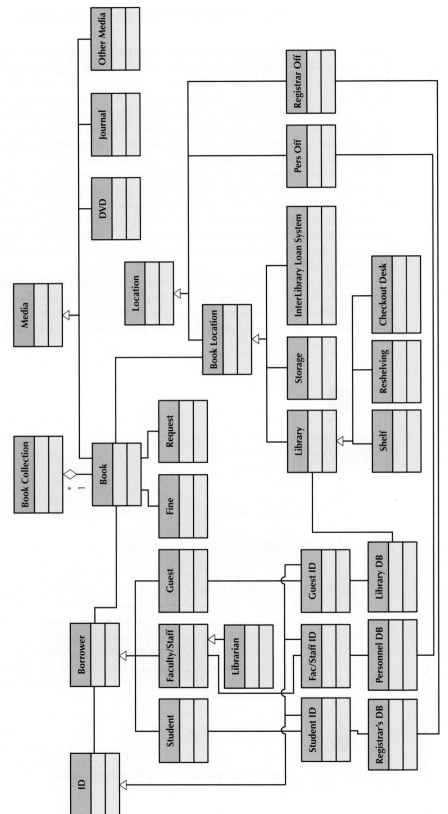

FIGURE 5-20   First-Cut Class Diagram for the Library Book Collection System

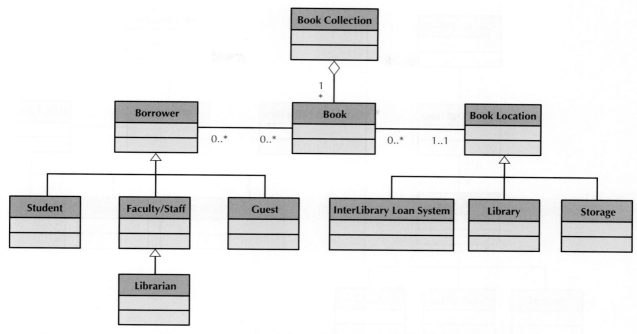

FIGURE 5-21   Second-Cut Class Diagram for the Library Book Collection System

purposes, we only show the classes and their relationships; not their attributes, operations, or even the multiplicities on the association relationships.

The fifth step is to take a step back and carefully review what has been created. Not only should you see if there are any missing classes, attributes, operations, and/or relationships, but you should also challenge every aspect of the current model. Specifically, are there classes, attributes, operations, and/or relationships that should be removed from the model? In this case, there seem to be many classes on the diagram that should have been modeled as attributes. For example, the whole idea of IDs should have been attributes. Furthermore, because this is supposed to be a book collection management system, the inclusion of other media seems to be inappropriate. Finally, the Personnel Office and Registrar's Office were actually only actors in the system, not objects. Based on all of these deletions, a new version of the class diagram was drawn (see Figure 5-21). This diagram is much simpler and easier to understand.

The sixth step, incorporating useful patterns, enables us to take advantage of knowledge that was developed elsewhere. In this case, the pattern used in the library problem includes too many ideas that are not relevant to the current problem. However, by looking back to Figure 5-3, we see that one of the original patterns (the one in the top left of the figure) is relevant. We incorporate that pattern into the class diagram by replacing Place by Checkout Desk, Participant by Borrower, Transaction by Check Out Trans, and Item by Book (Figure 5-22). Technically speaking, each of these replacements is simply a version that is customized to the problem at hand. We also then add the Transaction Line Item class that we had missed in the original structural model.

The seventh step is to review the current state of the structural model. Needless to say, the CRC card version and the class diagram version are no longer in agreement with each other. We return to this step in the next section of the chapter.

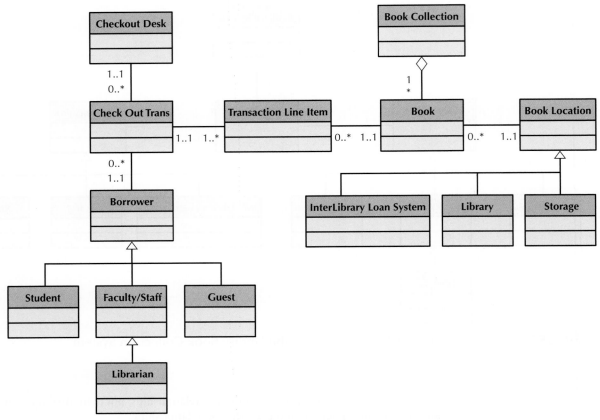

FIGURE 5-22 Class Diagram with Incorporated Pattern for the Library Book Collection System

---

YOUR TURN 5-1 Appointment System

*U*sing Figure 5-6 as a template, complete the CRC cards for the remaining identified classes in Figure 5-7.

YOUR TURN 5-2 Campus Housing

*I*n "Your Turn 4-4," you created a set of use cases from the campus housing service that helps students find apartments. Using the same use cases, create a structural model (CRC cards and class diagram). See if you can identify at least one potential derived attribute, aggregation relationship, generalization relationship, and association relationship for the model.

CONCEPTS

IN ACTION

5-A Health and Insurance Medical Provider—Implementing an EIM System

A large direct health and insurance medical provider needed an Enterprise Information Management (EIM) system to enable enterprise-wide information management and support the effective use of data for critical cross-functional decision making. In addition, the client needed to resolve issues related to data redundancy, inconsistency, and unnecessary expenditure. The client faced information challenges such as these: The company data resided in multiple sources and was developed for department-specific use, with limited enterprise access. In addition, departments created varied data definitions and data were being managed by multiple departments within the company.

Source: http://www.deloitte.com/dtt/case_study/0,1005,sid%
253D26562%02526cid%253D132760,00.html

Questions

1. Should the company assess their current information management?
2. How would a structural model aid the firm in understanding their current information management? What solution would you propose?

VERIFYING AND VALIDATING THE STRUCTURAL MODEL[17]

Before we move on to creating behavioral models (see Chapter 6) of the problem domain, we need to verify and validate the structural model. In the previous chapter, we introduced the notion of walkthroughs as a way to verify and validate business processes and functional models. In this chapter, we combine walkthroughs with the powerful idea of role-playing as a way to more completely verify and validate the structural model that will underlie the business processes and functional models. Because we have already introduced the idea of role-playing the CRC cards, in this section we focus on performing walkthroughs.

In this case, the verification and validation of the structural model is accomplished during a formal review meeting using a walkthrough approach in which an analyst presents the model to a team of developers and users. The analyst walks through the model, explaining each part of the model and all the reasoning that went into the decision to include each of the classes in the structural model. This explanation includes justifications for the attributes, operations, and relationships associated with the classes. Each class should be linked back to at least one use case; otherwise, the purpose of including the class in the structural model will not be understood. It is often best if the review team also includes people outside the development team that produced the model because these individuals can bring a fresh perspective to the model and uncover missing objects.

Previously, we suggested three representations that could be used for structural modeling: CRC cards, class diagrams, and object diagrams. Because an object diagram is simply an instantiation of some part of a class diagram, we limit our discussion to CRC cards and class diagrams. As in the previous chapter regarding verification and validation of business process and functional models, we provide a set of rules that will test

---

[17] The material in this section has been adapted from E. Yourdon, *Modern Structured Analysis* (Englewood Cliffs, NJ: Prentice Hall, 1989).

the consistency within the structural models. For example purposes, we use the appointment problem described in Chapter 4 and in this chapter. An example CRC card for the Old Patient class is shown in Figure 5-6 and the associated class diagram is portrayed in Figure 5-7.

First, every CRC card should be associated with a class on the class diagram, and vice versa. For example, the Old Patient class represented by the CRC card does not seem to be included on the class diagram. However, there is a Patient class on the class diagram (see Figures 5-6 and 5-7). The Old Patient CRC card most likely should be changed to simply Patient.

Second, the responsibilities listed on the front of the CRC card must be included as operations in a class on a class diagram, and vice versa. For example, the make appointment responsibility on the new Patient CRC card also appears as the make appointment() operation in the Patient class on the class diagram. Every responsibility and operation must be checked.

Third, collaborators on the front of the CRC card imply some type of relationship on the back of the CRC card and some type of association that is connected to the associated class on the class diagram. For example, the appointment collaborator on the front of the CRC card also appears as an other association on the back of the CRC card and as an association on the class diagram that connects the Patient class with the Appointment class.

Fourth, attributes listed on the back of the CRC card must be included as attributes in a class on a class diagram, and vice versa. For example, the amount attribute on the new Patient CRC card is included in the attribute list of the Patient class on the class diagram.

Fifth, the object type of the attributes listed on the back of the CRC card and with the attributes in the attribute list of the class on a class diagram implies an association from the class to the class of the object type. For example, technically speaking, the amount attribute implies an association with the double type. However, simple types such as int and double are never shown on a class diagram. Furthermore, depending on the problem domain, object types such as Person, Address, or Date might not be explicitly shown either. However, if we know that messages are being sent to instances of those object types, we probably should include these implied associations as relationships.

Sixth, the relationships included on the back of the CRC card must be portrayed using the appropriate notation on the class diagram. For example in Figure 5-6, instances of the Patient class are *a-kind-of* Person, it has instances of the Medical History class as part of it, and it has an association with instances of the Appointment class. Thus, the association from the Patient class to the Person class should indicate that the Person class is a generalization of its subclasses, including the Patient class; the association from the Patient class to the Medical History class should be in the form of an aggregation association (a white diamond); and the association between instances of the Patient class and instances of the Appointment class should be a simple association. However, when we review the class diagram in Figure 5-7, this is not what we find. If you recall, we included in the class diagram the transaction pattern portrayed in Figure 5-4. When we did this, many changes were made to the classes contained in the class diagram. All of these changes should have been cascaded back through all of the CRC cards. In this case, the CRC card for the Patient class should show that a Patient is a-kind-of Participant (not Person) and that the relationship from Patient to Medical History should be a simple association (see Figure 5-23).

Seventh, an association class, such as the Treatment class in Figure 5-7, should be created only if there is indeed some unique characteristic (attribute, operation, or relationship)

Front:

Class Name: Patient	ID: 3	Type: Concrete, Domain
Description: An individual that needs to recieve or has received medical attention		Associated Use Cases: 2

Responsibilities	Collaborators
Make appointment	Appointment
Calculate last visit	
Change status	
Provide medical history	Medical history

Back:

Attributes:

Amount (double)

Insurance carrier (text)

Relationships:

Generalization (a-kind-of): Participant

Aggregation (has-parts):

Other Associations: Appointment, Medical History

FIGURE 5-23
Patient CRC Card

about the intersection of the connecting classes. If no unique characteristic exists, then the association class should be removed and only an association between the two connecting classes should be displayed.

Finally, as in the functional models, specific representation rules must be enforced. For example, a class cannot be a subclass of itself. The Patient CRC card cannot list Patient with the generalization relationships on the back of the CRC card, nor can a generalization relationship be drawn from the Patient class to itself. Again, all the detailed restrictions for each representation are beyond the scope of this book.[18] Figure 5-24 portrays the associations among the structural models.

[18] A good reference for these types of restrictions is S.W. Ambler, *The Elements of UML 2.0 Style* (Cambridge, UK: Cambridge University Press, 2005).

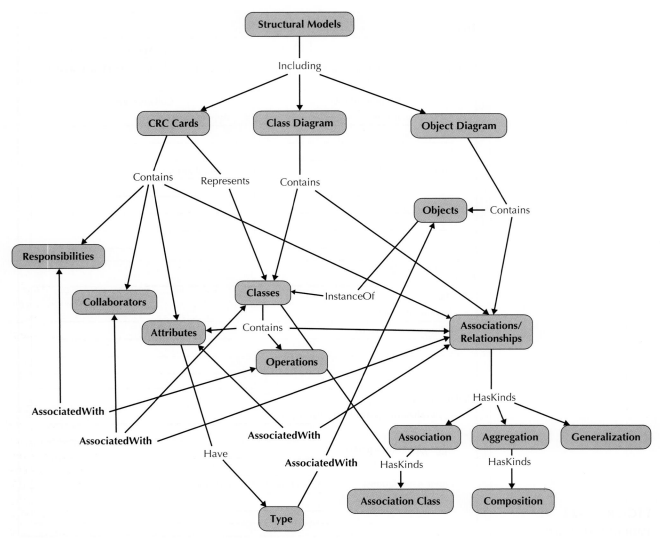

FIGURE 5-24   Interrelationships among Structural Models

APPLYING THE CONCEPTS AT CD SELECTIONS

In the previous chapter's installment of the CD Selections case, we saw how Alec, Margaret, and the team worked through building functional models of the business processes contained in their evolving web-based solution. In this chapter, we introduced how structural models using CRC cards and class and object diagrams could be created, verified, and validated. In this installment of the CD Selections case, we see how Alec and Margaret work through creating, verifying, and validating the structural models of the web-based solution that they hope to create.

SUMMARY

Structural Models

Structural models describe the underlying structure of an object-oriented system. Whereas functional models provide an external view of the evolving system (i.e., what the system does), structural models provide an internal static view of the evolving system (i.e., how the objects are organized in the system). At this point in the development of the system, the structural model represents only a logical model of the underlying problem domain. One of the primary purposes of the structural model is to create a vocabulary that allows the users and developers to communicate effectively about the problem under investigation. Structural models comprise classes, attributes, operations, and relationships. Three basic types of relationships that normally are depicted on structural models: aggregation, generalization, and association. Structural models typically are represented by CRC cards, class diagrams, and, in some cases, object diagrams.

Object Identification

There are many different ways to identify the objects that make up a problem domain. Four of the more popular approaches include textual analysis of use-case descriptions, brainstorming, using common object lists, and using patterns. By combining these approaches, it is relatively straightforward to uncover the objects for a given problem domain.

CRC Cards

CRC cards model the classes, their responsibilities, and their collaborations. There are two different types of responsibilities: knowing and doing. Knowing responsibilities are associated mostly with attributes of instances of the class, whereas doing responsibilities are associated mostly with operations of instances of the class. Collaborations support the concepts of clients, servers, and contracts between objects. CRC cards capture all the essential elements of the instances of a class. The front of the card contains the class's name, ID, type, description, list of associated use cases, responsibilities, and collaborators, and the back of the card contains the attributes and relationships. CRC cards are used to role-play the use-case scenarios. Through role-playing, additional classes, attributes, operations, and relationships can be discovered.

Class Diagrams

A class diagram is a graphical description of the information contained on the CRC cards. It shows the classes and the relationships among the classes. The class diagram portrays additional information not included on the CRC cards, such as the visibility of the attributes and operations and the multiplicity of the relationships. When relationship itself contains information, an associated class is created. There are special arcs for each of the relationships (aggregation, generalization, and association) contained in the diagram.

In real-world systems that can have hundreds of classes, the class diagram can become overly complicated. To simplify the diagram, a view mechanism can be used. A view restricts the amount of information portrayed on the diagram. Some useful views include hiding all information about the class except for its name and relationships, showing only the classes that are associated with a particular use case, and limiting the relationships included to only one specific type (aggregation, generalization, and association).

When attempting to uncover additional information about the details of a class, it can be useful to portray specific instances of a class instead of the classes themselves. In those cases, an object diagram is used to depict a set of objects that represent an instantiation of all or part of a class diagram.

Creating Structural Models Using CRC Cards and Class Diagrams

Creating a structural model of a problem domain is an incremental and iterative process. The process described includes seven steps. The recommended process begins with creating CRC cards through the textual analysis of the use-case descriptions. Next, the CRC cards are reviewed. During the review, team members brainstorm and review common object lists for additional classes, attributes, operations, and relationships. At this point, the team role-plays the CRC cards using the use-case scenarios. After this, a class diagram is drawn and reviewed. Finally, useful patterns are incorporated into the class diagram and the diagram is verified and validated.

Verifying and Validating the Structural Model

Verifying and validating the structural model is very similar to verifying and validating the functional model. In this case, the CRC cards, class diagrams, and, in some cases, object diagrams are carefully checked for consistency. This is done, typically, through role-play, walkthroughs, or both.

KEY TERMS

A-kind-of, 198
A-part-of, 198
Abstract class, 197
Aggregation association, 214
Assemblies, 198
Association, 199
Association class, 213
Attribute, 197
Brainstorming, 201
Class, 197
Class diagram, 195
Client, 205
Collaboration, 205
Common object list, 201
Conceptual model, 196
Concrete class, 197
Constructor operation, 211
Contract, 205
Class–Responsibility–
  Collaboration (CRC), 195
CRC cards, 205

Decomposition, 198
Derived attribute, 211
Doing responsibility, 205
Destructor operation, 211
Generalization association, 214
Has-parts, 198
Incidents, 201
Instance, 197
Interactions, 201
Knowing responsibility, 205
Method, 197
Multiplicity, 212
Object, 197
Object diagram, 195
Operation, 197
Package, 217
Parts, 198
Pattern, 202
Private, 211
Protected, 211
Public, 211

Query operation, 211
Responsibility, 205
Role-playing, 207
Roles, 201
Server, 205
State, 211
Static model, 208
Static structure diagram, 217
Structural model, 196
Subclass, 198
Substitutability, 198
Superclass, 198
SVDPI, 219
Tangible things, 201
Textual analysis, 199
Update operation, 211
View, 217
Visibility, 211
Wholes, 198

QUESTIONS

1. Describe to a businessperson the multiplicity of a relationship between two classes.
2. Why are assumptions important to a structural model?
3. What is an association class?
4. Contrast the following sets of terms: object, class, method, attribute, superclass, subclass, concrete class, abstract class.
5. Give three examples of derived attributes that may exist on a class diagram. How would they be denoted on the class diagram?
6. What are the different types of visibility? How would they be denoted on a class diagram?
7. Draw the relationships that are described by the following business rules. Include the multiplicities for each relationship.

A patient must be assigned to only one doctor and a doctor can have one or many patients.

An employee has one phone extension, and a unique phone extension is assigned to an employee.

A movie theater shows at least one movie, and a movie can be shown at up to four other movie theaters around town.

A movie either has one star, two costars, or more than ten people starring together. A star must be in at least one movie.

8. How do you designate the reading direction of a relationship on a class diagram?
9. For what is an association class used in a class diagram? Give an example of an association class that may be found in a class diagram that captures students and the courses that they have taken.
10. Give two examples of aggregation, generalization, and association relationships. How is each type of association depicted on a class diagram?
11. Identify the following operations as constructor, query, or update. Which operations would not need to be shown in the class rectangle?

   Calculate employee raise (raise percent)
   Calculate sick days ()
   Increment number of employee vacation days ()
   Locate employee name ()
   Place request for vacation (vacation day)
   Find employee address ()
   Insert employee ()
   Change employee address ()
   Insert spouse ()

12. How are the different structural models related and how does this affect verification and validation of the model?

EXERCISES

A. Create a CRC card for each of the following classes:

   Movie (title, producer, length, director, genre)
   Ticket (price, adult or child, showtime, movie)
   Patron (name, adult or child, age)

B. Create a class diagram based on the CRC cards you created for exercise A.

C. Create a CRC card for each of the following classes. Consider that the entities represent a system for a patient billing system. Include only the attributes that would be appropriate for this context.

   Patient (age, name, hobbies, blood type, occupation, insurance carrier, address, phone)
   Insurance carrier (name, number of patients on plan, address, contact name, phone)
   Doctor (specialty, provider identification number, golf handicap, age, phone, name)

D. Create a class diagram based on the CRC cards you created for exercise C.

E. Create a class diagram showing the following relationships:

1. A patient must be assigned to only one doctor and a doctor can have many patients.
2. An employee has one phone extension, and a unique phone extension is assigned to an employee.
3. A movie theater shows many different movies, and the same movie can be shown at different movie theaters around town.

F. Draw a class diagram for each of the following situations:

1. Whenever new patients are seen for the first time, they complete a patient information form that asks their name, address, phone number and insurance carrier, which are stored in the patient information file. Patients can be signed up with only one carrier, but they must be signed up to be seen by the doctor. Each time a patient visits the doctor, an insurance

claim is sent to the carrier for payment. The claim must contain information about the visit, such as the date, purpose, and cost. It would be possible for a patient to submit two claims on the same day.

2. The state of Georgia is interested in designing a system that will track its researchers. Information of interest includes researcher name, title, position, researcher's university name, university location, university enrollment, and researcher's research interests. Researchers are associated with one institution, and each researcher has several research interests.

3. A department store has a wedding registry. This registry keeps information about the customer (usually the bride), the products that the store carries, and the products for which each customer registers. Customers typically register for a large number of products and many customers register for the same products.

4. Jim Smith's dealership sells Fords, Hondas, and Toyotas. The dealership keeps information about each car manufacturer with whom they deal so that they can get in touch with them easily. The dealership also keeps information about the models of cars that they carry from each manufacturer. They keep information such as list price, the price the dealership paid to obtain the model, and the model name and series (e.g., Honda Civic LX). They also keep information about all sales that they have made (for instance, they record the buyer's name, the car they bought, and the amount they paid for the car). To contact the buyers in the future, contact information is also kept (e.g., address, phone number).

G. Create object diagrams based on the class diagrams you drew for exercise F.

H. Examine the class diagrams that you created for exercise F. How would the models change (if at all) based on these new assumptions?

1. Two patients have the same first and last names.
2. Researchers can be associated with more than one institution.
3. The store would like to keep track of purchase items.
4. Many buyers have purchased multiple cars from Jim over time because he is such a good dealer.

I. Visit a website that allows customers to order a product over the web (e.g., Amazon.com). Create a structural model (CRC cards and class diagram) that the site must need to support its business process. Include classes to show what they need information about. Be sure to include the attributes and operations to represent the type of information they use and create. Finally, draw relationships, making assumptions about how the classes are related.

J. Using the seven-step process described in this chapter, create a structural model (CRC cards and class diagram) for exercise C in Chapter 4.

K. Perform a verification and validation walkthrough for the structural model created for exercise J.

L. Using the seven-step process described in this chapter, create a structural model for exercise E in Chapter 4.

M. Perform a verification and validation walkthrough for the structural model created for exercise L.

N. Using the seven-step process described in this chapter, create a structural model for exercise G in Chapter 4.

O. Perform a verification and validation walkthrough for the structural model created for exercise N.

P. Using the seven-step process described in this chapter, create a structural model for exercise I in Chapter 4.

Q. Perform a verification and validation walkthrough for the structural model created for exercise P.

R. Using the seven-step process described in this chapter, create a structural model for exercise L in Chapter 4.

S. Perform a verification and validation walkthrough for the structural model created for exercise R.

T. Using the seven-step process described in this chapter, create a structural model for exercise O in Chapter 4.

U. Perform a verification and validation walkthrough for the structural model created for exercise T.

V. Using the seven-step process described in this chapter, create a structural model for exercise R in Chapter 4.

W. Perform a verification and validation walkthrough for the structural model created for exercise V.

X. Using the seven-step process described in this chapter, create a structural model for exercise U in Chapter 4.

Y. Perform a verification and validation walkthrough for the structural model created for exercise X.

MINICASES

1. You are the project manager of an IS consulting firm that has been working with Hock Hai Tan, the general manager of Singapore Importers Pte Ltd. Singapore Importers acts as an intermediary between Western firms and low-cost suppliers in China. It has extensive knowledge of Chinese industry and can offer low-cost manufacturing services to small- and medium-size Western firms that lack knowledge of Chinese companies, but want to do business in China. Singapore Importers operates 10 offices across China to identify potential suppliers and monitor the performance of the companies they deal with.

   Your team has been hard at work on a project that will eventually link all the Singapore Importers' offices into one unified, networked system. The team has developed a use-case diagram of the current system. This model has been carefully checked. Last week, the team invited several system users to role-play the various use cases, which were refined to the users' satisfaction. Right now, you feel confident that the as-is system has been adequately represented in the use-case diagram.

   Mr. Tan sat in on the role-playing of the use cases and was very pleased by the thorough job your team had done in developing the model. He made it clear to you that he was anxious to see your team begin work on the use cases for the to-be system. He was skeptical that your team had to spend time modeling the current system in the first place, but grudgingly admitted that your team seemed to understand the business after having gone through that phase.

   The methodology you are following, however, specifies that the team should now turn its attention to developing the structural models for the as-is system. When you mentioned this to Mr. Tan, he seemed confused and a little irritated. "You are going to spend even more time looking at the current system? I thought you were done with that! Why is this necessary? I want to see some progress on the way things will work in the future!"

   What is your response to Mr. Tan? Why do we perform structural modeling? Is there any benefit to developing a structural model of the current system at all? How do the use cases and use-case diagram help us develop the structural model?

2. Professor Takato is teaching databases at the City University of Osaka. He is also the president of the Foundation for Research of Digital Information (FRDI), a non-profit organization that aims to promote all forms of non-conventional research in computing. One of the activities FRDI is involved in is to organize computing conferences with widespread international participation. In order to support the organization of peer-reviewed conferences, Prof. Takato decided to build his own web-based application, a Conference Management System (CMS) through which, among other things, authors could register their papers for the conference and reviewers could access and provide feedback on the papers assigned to them for review. What are the two use cases Prof. Takato's system intends to support? Name and describe in one statement each of the two use cases. What classes of the structural model corresponding to the two use cases can you identity? Considering the following facts: (1) an author could be a reviewer and the other way around, (2) an author may submit more than one paper and one paper may have several authors, and (3) a reviewer evaluates several papers and a paper is evaluated by several reviewers. Professor Takato created a class diagram with a total of six classes as follows: (1) Author, (2) Reviewer, (3) Paper, (4) Participant as a generalization of Author and reviewer, (5) Author_Paper as the association class to represent the many-to-many relationship between Author and Paper, and, last but not least, (6) Paper_Reviewer as the association class to represent the many-to-many relationship between Paper and Reviewer. What do you think of Prof. Takato's approach? Are all the six classes listed above necessary? Discuss and explain, if and under what circumstances can the class diagram be simplified. What would be a minimal number of classes for this diagram? Under what assumptions would that be adequate?

CHAPTER 6

BEHAVIORAL MODELING

Behavioral models describe the internal dynamic aspects of an information system that supports the business processes in an organization. During analysis, behavioral models describe what the internal logic of the processes is without specifying how the processes are to be implemented. Later, in the design and implementation phases, the detailed design of the operations contained in the object is fully specified. In this chapter, we describe three Unified Modeling Language (UML) diagrams that are used in behavioral modeling (sequence diagrams, communication diagrams, and behavioral state machines) and CRUDE (create, read, update, delete, execute) matrices.

OBJECTIVES

- Understand the rules and style guidelines for sequence and communication diagrams and behavioral state machines
- Understand the processes used to create sequence and communication diagrams, behavioral state machines, and CRUDE matrices
- Be able to create sequence and communication diagrams, behavioral state machines, and CRUDE matrices
- Understand the relationship between the behavioral models and the structural and functional models

CHAPTER OUTLINE

INTRODUCTION

The previous two chapters discussed functional models and structural models. Systems analysts use functional models to describe the external behavioral view of an information system, and they use structural models to depict the static view of an information system.

In this chapter, we discuss how analysts use *behavioral models* to represent the internal behavior or dynamic view of an information system.

There are two types of behavioral models. First, there are behavioral models used to represent the underlying details of a business process portrayed by a use-case model. In UML, interaction diagrams (sequence and communication) are used for this type of behavioral model. Practically speaking, interaction diagrams allow the analyst to model the distribution of the behavior of the system over the actors and objects in the system. In this way, we can easily see how actors and objects collaborate to provide the functionality defined in a use case. Second, a behavioral model is used to represent the changes that occur in the underlying data. UML uses behavioral state machines for this.

During analysis, analysts use behavioral models to capture a basic understanding of the dynamic aspects of the underlying business process. Traditionally, behavioral models have been used primarily during design, where analysts refine the behavioral models to include implementation details (see Chapter 8). For now, our focus is on *what* the dynamic view of the evolving system is and not on *how* the dynamic aspect of the system will be implemented.

In this chapter, we concentrate on creating behavioral models of the underlying business process. Using the interaction diagrams (sequence and communication diagrams) and behavioral state machines, it is possible to give a complete view of the dynamic aspects of the evolving business information system. We first describe behavioral models and their components. We then describe each of the diagrams, how they are created, and how they are related to the functional and structural models described in Chapters 4 and 5.

BEHAVIORAL MODELS

When an analyst is attempting to understand the underlying application domain of a problem, he or she must consider both structural and behavioral aspects of the problem. Unlike other approaches to the development of information systems, object-oriented approaches attempt to view the underlying application domain in a holistic manner. By viewing the problem domain as a set of *use cases* that are supported by a set of collaborating objects, object-oriented approaches allow an analyst to minimize the semantic gap between the real-world set of objects and the evolving object-oriented model of the problem domain. However, as we pointed out in the previous chapter, the real world tends to be messy, which makes perfect modeling of the application domain practically impossible in software. This is because software must be neat and logical to work.

One of the primary purposes of behavioral models is to show how the underlying objects in a problem domain will work together to form a *collaboration* to support each of the *use cases*. Whereas structural models represent the objects and the relationships between them, behavioral models depict the internal view of the business process that a use case describes. The process can be shown by the interaction that takes place between the objects that collaborate to support a use case through the use of interaction (sequence and communication) diagrams. It is also possible to show the effect that the set of use cases that make up the system has on the objects in the system through the use of behavioral state machines.

Creating behavioral models is an iterative process that iterates not only over the individual behavioral models [e.g., interaction (sequence and communication) diagrams and behavioral state machines] but also over the functional (see Chapter 4) and structural models (see Chapter 5). As the behavioral models are created, it is not unusual to make changes to the functional and structural models. In this chapter, we describe interaction diagrams, behavioral state machines, and CRUDE analysis and when to use each.

INTERACTION DIAGRAMS

One of the primary differences between class diagrams and interaction diagrams, besides the obvious difference that one describes structure and the other behavior, is that the modeling focus on a class diagram is at the class level, whereas the interaction diagrams focus on the object level. In this section, we review objects, operations, and messages and we cover the two different diagrams (sequence and communication) that can be used to model the interactions that take place between the objects in an information system.

Objects, Operations, and Messages

An object is an instantiation of a *class,* that is, an actual person, place, or thing about which we want to capture information. If we were building an appointment system for a doctor's office, classes might include doctor, patient, and appointment. The specific patients, such as Jim Maloney, Mary Wilson, and Theresa Marks, are considered objects—that is, *instances* of the patient class.

Each object has *attributes* that describe information about the object, such as a patient's name, birth date, address, and phone number. Each object also has *behaviors.* At this point in the development of the evolving system, the behaviors are described by *operations.* An operation is nothing more than an action that an object can perform. For example, an appointment object can probably schedule a new appointment, delete an appointment, and locate the next available appointment. Later on during the development of the evolving system, the behaviors will be implemented as *methods.*

Each object also can send and receive messages. *Messages* are information sent to objects to tell an object to execute one of its behaviors. Essentially, a message is a function or procedure call from one object to another object. For example, if a patient is new to the doctor's office, the system sends an insert message to the application. The patient object receives the instruction (the message) and does what it needs to do to insert the new patient into the system (the behavior).

Sequence Diagrams

Sequence diagrams are one of two types of interaction diagrams. They illustrate the objects that participate in a use case and the messages that pass between them over time for *one* use case. A sequence diagram is a *dynamic model* that shows the explicit sequence of messages that are passed between objects in a defined interaction. Because sequence diagrams emphasize the time-based ordering of the activity that takes place among a set of objects, they are very helpful for understanding real-time specifications and complex use cases.

The sequence diagram can be a *generic sequence diagram* that shows all possible scenarios[1] for a use case, but usually, each analyst develops a set of *instance sequence diagrams,* each of which depicts a single *scenario* within the use case. If you are interested in understanding the flow of control of a scenario by time, you should use a sequence diagram to depict this information. The diagrams are used throughout the analysis and design phases. However, the design diagrams are very implementation specific, often including database objects or specific user interface components as the objects.

Elements of a Sequence Diagram   Figure 6-1 shows an instance sequence diagram that depicts the objects and messages for the Make Old Patient Appt use case, which describes the process by which an existing patient creates a new appointment or cancels or reschedules an

---

[1] Remember that a scenario is a single executable path through a use case.

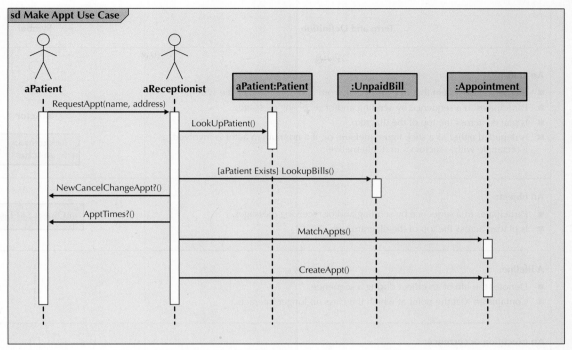

FIGURE 6-1   Example Sequence Diagram

appointment for the doctor's office appointment system. In this specific instance, the Make Old Patient Appt process is portrayed.

Actors and *objects* that participate in the sequence are placed across the top of the diagram using actor symbols from the use-case diagram and object symbols from the object diagram (see Figure 6-2). Notice that the *actors* and objects in Figure 6-1 are aPatient, aReceptionist, aPatient, UnpaidBill, and Appointment.[2] They are not placed in any particular order, although it is nice to organize them in some logical way, such as the order in which they participate in the sequence. For each of the objects, the name of the class of which they are an instance is given after the object's name (e.g., aPatient means that aPatient is an instance of the Patient class).

A dotted line runs vertically below each actor and object to denote the *lifeline* of the actors and objects over time (see Figure 6-1).[3] Sometimes, an object creates a *temporary object;* in this case, an X is placed at the end of the lifeline at the point where the object is destroyed (not shown). For example, think about a shopping cart object for a web commerce application. The shopping cart is used for temporarily capturing line items for an order, but once the order is confirmed, the shopping cart is no longer needed. In this case, an X would be located at the point at which the shopping cart object is destroyed. When objects continue to exist in the system after they are used in the sequence diagram, then the lifeline continues to the bottom of the diagram (this is the case with all of the objects in Figure 6-1).

---

[2] In some versions of the sequence diagram, object symbols are used as surrogates for the actors. However, for clarity, we recommend using actor symbols for actors instead.

[3] Technically speaking, in UML 2.0 the lifeline actually refers to both the object (actor) and the dashed line drawn vertically underneath the object (actor). However, we prefer to use the older terminology because it is more descriptive of what is actually being represented.

Term and Definition	Symbol
An actor: ■ Is a person or system that derives benefit from and is external to the system. ■ Participates in a sequence by sending and/or receiving messages. ■ Is placed across the top of the diagram. ■ Is depicted either as a stick figure (default) or, if a nonhuman actor is involved, as a rectangle with <<actor>> in it (alternative).	anActor <<actor>> anActor
An object: ■ Participates in a sequence by sending and/or receiving messages. ■ Is placed across the top of the diagram.	anObject : aClass
A lifeline: ■ Denotes the life of an object during a sequence. ■ Contains an X at the point at which the class no longer interacts.	
An execution occurrence: ■ Is a long narrow rectangle placed atop a lifeline. ■ Denotes when an object is sending or receiving messages.	
A message: ■ Conveys information from one object to another one. ■ An operation call is labeled with the message being sent and a solid arrow, whereas a return is labeled with the value being returned and shown as a dashed arrow.	aMessage() ReturnValue
A guard condition: ■ Represents a test that must be met for the message to be sent.	[aGuardCondition]:aMessage()
For object destruction: ■ An X is placed at the end of an object's lifeline to show that it is going out of existence.	X
A frame: ■ Indicates the context of the sequence diagram.	Context

FIGURE 6-2   Sequence Diagram Syntax

A thin rectangular box, called the *execution occurrence,* is overlaid onto the lifeline to show when the classes are sending and receiving messages (see Figure 6-2). A message is a communication between objects that conveys information with the expectation that activity will ensue. Many different types of messages can be portrayed on a sequence diagram.

However, in the case of using sequence diagrams to model use cases, two types of messages are typically used: operation call and return. *Operation call messages* passed between objects are shown using solid lines connecting two objects with an arrow on the line showing which way the message is being passed. Argument values for the message are placed in parentheses next to the message's name. The order of messages goes from the top to the bottom of the page, so messages located higher on the diagram represent messages that occur earlier on in the sequence, versus the lower messages that occur later. A *return message* is depicted as a dashed line with an arrow on the end of the line portraying the direction of the return. The information being returned is used to label the arrow. However, because adding return messages tends to clutter the diagram, unless the return messages add a lot of information to the diagram, they can be omitted. For example, in Figure 6-1, no return messages are depicted.[4] In Figure 6-1, LookUpPatient() is a message sent from the actor aReceptionist to the object aPatient to determine whether the aPatient actor is a current patient.

At times, a message is sent only if a *condition* is met. In those cases, the condition is placed between a set of brackets, [ ]—for example, [aPatient Exists] LookupBills(). The condition is placed in front of the message name. However, when using a sequence diagram to model a specific scenario, conditions are typically not shown on any single sequence diagram. Instead, conditions are implied only through the existence of different sequence diagrams.

There are times that a message is repeated. This is designated with an asterisk (*) in front of the message name (e.g., * Request CD). An object can also send a message to itself. This is known as *self-delegation*. Sometimes, an object creates another object. This is shown by the message being sent directly to an object instead of its lifeline.

Figure 6-3 portrays two additional examples of instance-specific sequence diagrams. The first one is related to the Make Lunch use case that was described in the activity diagram portrayed in Figure 4-9. The second one is related to the Place Order use case associated with the activity diagram in Figure 4-8. In both examples, the diagrams simply represent a single scenario. Notice in the Make Lunch sequence diagram there is a message being sent from an actor to itself [CreateSandwich()]. Depending on the complexity of the scenario being modeled, this particular message could have been eliminated. Obviously, both the process of making a lunch and placing an order can be quite a bit more complex. However, from a learning point of view, you should be able to see how the sequence diagrams and the activity diagrams relate to one another.

Guidelines for Creating Sequence Diagrams Ambler[5] provides a set of guidelines when drawing sequence diagrams (see Figure 6-4). In this section, we review six of them.

- Try to have the messages not only in a top-to-bottom order but also, when possible, in a left-to-right order. Given that western cultures tend to read left to right and top to bottom, a sequence diagram is much easier to interpret if the messages are ordered as much as possible in the same way. To accomplish this, order the actors and objects along the top of the diagram in the order that they participate in the scenario of the use case.
- If an actor and an object conceptually represent the same idea, one inside of the software and the other outside, label them with the same name. In fact, this implies that they exist in both the use-case diagram (as an actor) and in the class

---

[4] However, some CASE tools require the return messages to be displayed. Obviously, when you are using these tools, you have to include the return messages on the diagram.

[5] S.W. Ambler, *The Elements of UML 2.0 Style* (Cambridge, England: Cambridge University Press, 2005).

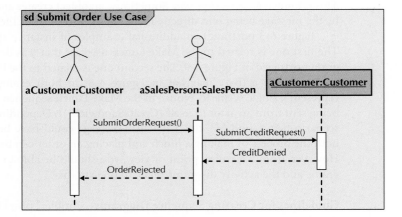

FIGURE 6-3

Additional Sample
Instance-Specific
Sequence Diagrams

diagram (as a class). At first glance, this might seem to lead to confusion. However, if they do indeed represent the same idea, then they should have the same name. For example, a customer actor interacts with the system and the system stores information about the customer. In this case, they do indeed represent the same conceptual idea.

- Strive for left-to-right ordering of messages
- If an actor and object represent the same idea, name them the same.
- Place the initiator of the scenario on the left of diagram.
- When there are multiple objects of the same type, be sure to name them.
- Only show return values when they are not obvious.
- Justify message names and return values near the arrowhead.

FIGURE 6-4

Guidelines for
Creating Sequence
Diagrams

- The initiator of the scenario—actor or object—should be drawn as the farthest left item in the diagram. This guideline is essentially a specialization of the first guideline. In this case, it relates specifically to the actor or object that triggers the scenario.

- When there are multiple objects of the same type, be sure to include a name for the object in addition to the class of the object. For example, in the making a lunch example (see Figure 6-3), there are two objects of type Parent. As such, they should be named. Otherwise, you can simply use the class name. This will simplify the diagram. In this case, the Child object did not have to be named. We could have simply placed a colon in front of the class name instead.

- Show return values only when they are not obvious. Showing all of the returns tends to make a sequence diagram more complex and potentially difficult to comprehend. In many cases, less is more. Only show the returns that actually add information for the reader of the diagram.

- Justify message names and return values near the arrowhead of the message and return arrows, respectively. This makes it much easier to interpret the messages and their return values.

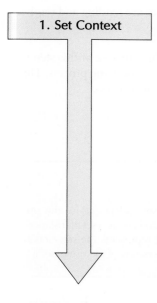

Creating a Sequence Diagram   In this section, we describe a six-step process used to create a sequence diagram.[6] The first step in the process is to determine the context of the sequence diagram. The context of the diagram can be a system, a use case, or a scenario of a use case. The context of the diagram is depicted as a labeled *frame* around the diagram (see Figures 6-1, 6-2, and 6-3). Most commonly, it is one use-case scenario. Figure 6-1 portrays the instance-specific sequence diagram for the scenario from the Make Old Patient Appt use case given in Figure 4-11 for making a new appointment for an existing patient. For each possible scenario for the Make Old Patient Appt use case, a separate instance-specific sequence diagram would be created. On the surface, this seems to be a lot of potentially redundant and useless work. However, at this point in the representation of a system, we are still trying to completely understand the problem. This process of creating instance-specific sequence diagrams for each scenario instead of creating a single generic sequence diagram for the entire use case will enable the developers to attain a more complete understanding of the problem being addressed. Each instance-specific sequence diagram is fairly simple to interpret, whereas a generic sequence diagram can be very complex. The testing of a specific use case is accomplished in a much easier manner by validating and verifying the completeness of the set of instance-specific sequence diagrams instead of trying to work through a single complex generic sequence diagram.

The second step is to identify the actors and objects that participate in the sequence being modeled—that is, the actors and objects that interact with each other during the use-case scenario. The actors were identified during the creation of the functional model, whereas the objects are identified during the development of the structural model. These are the classes on which the objects of the sequence diagram for this scenario will be based. One very useful approach to identifying all of the scenarios associated with a use case is to role-play the *CRC cards* (see Chapter 5). This can help you identify potentially missing operations that are necessary to support the business process, which the use case is representing, in a complete manner. Also, during role-playing, it is likely that new classes, and hence, new objects,

---

[6] The approach described in this section are adapted from Grady Booch, James Rumbaugh, Ivar Jacobson, *The Unified Modeling Language User Guide* (Reading, MA: Addison-Wesley, 1999).

will be uncovered.[7] Don't worry too much about identifying all the objects perfectly; remember that the behavioral modeling process is iterative. Usually, the sequence diagrams are revised multiple times during the behavioral modeling processes.

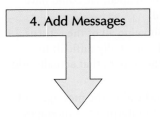

3. Set Lifeline

The third step is to set the lifeline for each object. To do this, you need to draw a vertical dotted line below each class to represent the class's existence during the sequence. An X should be placed below the object at the point on the lifeline where the object goes out of existence.

4. Add Messages

The fourth step is to add the messages to the diagram. This is done by drawing arrows to represent the messages being passed from object to object, with the arrow pointing in the message's transmission direction. The arrows should be placed in order from the first message (at the top) to the last (at the bottom) to show time sequence. Any parameters passed along with the messages should be placed in parentheses next to the message's name. If a message is expected to be returned as a response to a message, then the return message is not explicitly shown on the diagram.

5. Place Execution Occurrence

The fifth step is to place the execution occurrence on each object's lifeline by drawing a narrow rectangle box over the lifelines to represent when the classes are sending and receiving messages.

6. Validate

The sixth and final step is to validate the sequence diagram. The purpose of this step is to guarantee that the sequence diagram completely represents the underlying process. This is done by guaranteeing that the diagram depicts all the steps in the process.[8]

YOUR TURN   6-1 Drawing a Sequence Diagram

In "Your Turn 4-4", you were asked to create a set of use cases and a use-case diagram for the campus housing service that helps students find apartments. In "Your Turn 5-2", you were asked to create a structural model (CRC cards and a class diagram) for those use cases. Select one of the use cases from the use-case diagram and create a set of instance-specific sequence diagrams that represents the interaction among classes in the different scenarios of the use case.

Example   In the previous chapters, we have demonstrated the diagramming and modeling processes using the Borrow Books use case of the Library Book Collection Management System. When considering instance-specific scenario diagrams, we need to draw one sequence diagram per scenario. In the case of the Borrow Books use case in Chapter 4, there are nine different scenarios. Therefore, for this one use case, there would be nine separate diagrams. In this example, we are setting the context of the sequence diagram to only one specific scenario of the Borrow Books use case: Students who have a valid ID and do not have any overdue books or any fines. The other scenarios include Students without a valid ID, Students with a valid ID but who owe fines or have overdue books, and the same three

---

[7] This obviously will cause you to go back and modify the structural model (see Chapter 5).

[8] We describe validation in more detail later in this chapter.

Normal Flow of Events:

    1. The Borrower brings books to the Librarian at the checkout desk.

    2. The Borrower provides Librarian their ID card.

    3. The Librarian checks the validity of the ID Card.

        If the Borrower is a Student Borrower, Validate ID Card against Registrar's Database.

        If the Borrower is a Faculty/Staff Borrower, Validate ID Card against Personnel Database.

        If the Borrower is a Guest Borrower, Validate ID Card against Library's Guest Database.

    4. The Librarian checks whether the Borrower has any overdue books and/or fines.

    5. The Borrower checks out the books.

SubFlows:

Alternate/Exceptional Flows:

    4a. The ID Card is invalid, the book request is rejected.

    5a. The Borrower either has overdue books, fines, or both, the book request is rejected.

FIGURE 6-5   Flow of Events section of the Use-Case Description of the Borrow Books Use Case

scenarios for the other two types of Borrowers: Faculty/Staff and Guest. In this example, we are only drawing the one sequence diagram for the Students with a valid ID scenario. To begin with, we should review the Flow of Events of the use-case description (see Figure 6-5), the activity diagram (see Figure 6-6), and the use-case diagram (see Figure 6-7).

The next step is to identify the actors and objects involved in the scenario. By studying the flow of events and the use-case diagram, we identify students, librarians, and the registrar's database as actors and borrowers, the book collection, and books as the objects. We place the actors and objects across the top of the diagram based on the ordering of

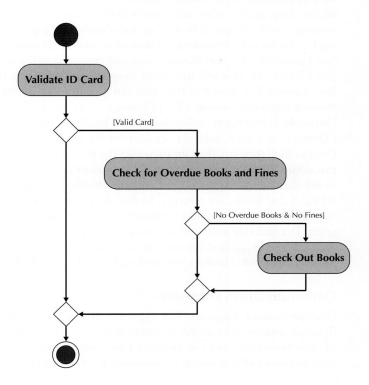

FIGURE 6-6

Activity Diagram of the Borrow Books Use Case (Figure 4-10)

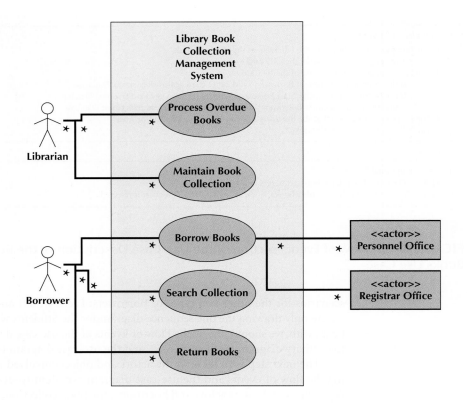

FIGURE 6-7

Use-Case Diagram
for the Library
Book Collection
Management System
(Figure 4-5)

their appearance in the normal flow of events. The next step involves simply drawing the lifelines beneath the actors and objects in the scenario. The fourth step is to add the actual messages to the diagram. To do this, we again review the actual steps taken when executing this scenario by reviewing the flow of events (see Figure 6-5) and the activity diagram (see Figure 6-6). We also should review any results from the role-playing of the CRC cards (see Chapter 5). This will help us to properly portray where the functionality is located. For example, in Figure 6-8, the Librarian executes the CheckOutBooks() procedure (the Student sends the message CheckOutBooks () to ask the Librarian to execute the Check-OutBooks () procedure) when the student hands the librarian the books to check out. The Librarian, in return, asks the Student for the ID card. When the student hands the ID Card to the Librarian, the Librarian asks the Registrar's Database to execute the ValidID() procedure when the Librarian passes the student's ID number over to the database system to ask the database system to validate the student's ID number. This continues until the ID Card and Books are returned to the student. Once we have decided from whom the messages are to be sent and to whom they are sent, we can place the messages on the diagram. The fifth step then is to add the execution occurrence to the diagrams to show when each actor or object is in the process of executing one of its operations. Next, we must validate the diagram. Finally, we should replicate this process for the other eight scenarios.

Communication Diagrams

Communication diagrams, like sequence diagrams, essentially provide a view of the dynamic aspects of an object-oriented system. They can show how the members of a set of objects collaborate to implement a use case or a use-case scenario. They can also be used to model all the interactions among a set of collaborating objects, in other words,

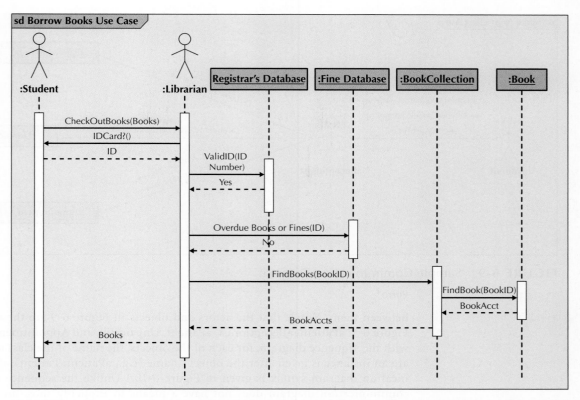

FIGURE 6-8 Sequence Diagram of the Borrow Books Use Case for Students with a Valid ID and No Overdue Books or Fines

a collaboration (see CRC cards in Chapter 5). In this case, a communication diagram can portray how dependent the different objects are on one another.[9] A communication diagram is essentially an object diagram that shows message-passing relationships instead of aggregation or generalization associations. Communication diagrams are very useful to show process patterns (i.e., patterns of activity that occur over a set of collaborating classes).

Communication diagrams are equivalent to sequence diagrams, but they emphasize the flow of messages through a set of objects, whereas the sequence diagrams focus on the time ordering of the messages being passed. Therefore, to understand the flow of control over a set of collaborating objects or to understand which objects collaborate to support business processes, a communication diagram can be used. For time ordering of the messages, a sequence diagram should be used. In some cases, both can be used to more fully understand the dynamic activity of the system.

Elements of a Communication Diagram   Figure 6-9 shows a communication diagram for the Make Old Patient Appt use case. Like the sequence diagram in Figure 6-1, the Make Old Patient Appt process is portrayed.

Actors and objects that collaborate to execute the use case are placed on the communication diagram in a manner to emphasize the message passing that takes place

---

[9] We return to this idea of dependency in Chapters 7 and 8.

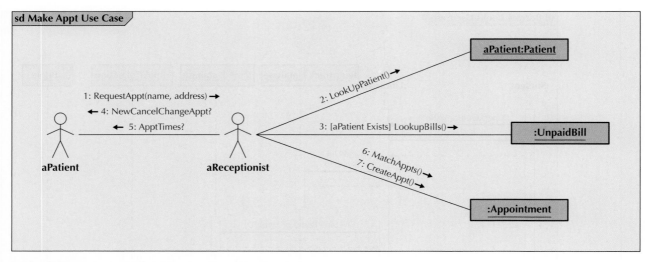

FIGURE 6-9   Sample Communication Diagram

between them. Notice that the actors and objects in Figure 6-9 are the same ones in Figure 6-1: aPatient, aReceptionist, aPatient, UnpaidBill, and Appointment.[10] Again, as with the sequence diagram, for each of the objects, the name of the class of which they are an instance is given after the object's name (e.g., aPatient: Patient). (The communication diagram syntax is given in Figure 6-10.) Unlike the sequence diagram, the communication diagram does not have a means to explicitly show an object being deleted or created. It is assumed that when a delete, destroy, or remove message is sent to an object, it will go out of existence, and a create or new message will cause a new object to come into existence. Another difference between the two interaction diagrams is that the communication diagram never shows returns from message sends, whereas the sequence diagram can optionally show them.

An *association* is shown between actors and objects with an undirected line. For example, an association is shown between the aPatient and aReceptionist actors. Messages are shown as labels on the associations. Included with the labels are lines with arrows showing the direction of the message being sent. For example, in Figure 6-9, the aPatient actor sends the RequestAppt() message to the aReceptionist actor, and the aReceptionist actor sends the NewCancelChangeAppt?() and the ApptTimes?() messages to the aPatient actor. The sequence of the message sends is designated with a sequence number. In Figure 6-9, the RequestAppt() message is the first message sent, whereas the NewCancelChangeAppt?() and the ApptTimes?() messages are the fourth and fifth message sent, respectively.

Like the sequence diagram, the communication diagram can represent conditional messages. For example, in Figure 6-9, the LookupBills() message is sent only if the [aPatient exists] condition is met. If a message is repeatedly sent, an asterisk is placed after the sequence number. Finally, an association that loops onto an object shows self-delegation. The message is shown as the label of the association.

When a communication diagram is fully populated with all the objects, it can become very complex and difficult to understand. When this occurs, it is necessary to simplify the diagram. One approach to simplifying a communication diagram, like use-case diagrams

---

[10] In some versions of the communication diagram, object symbols are used as surrogates for the actors. However, again we recommend using actor symbols for actors instead.

Term and Definition	Symbol
An actor: ■ Is a person or system that derives benefit from and is external to the system. ■ Participates in a collaboration by sending and/or receiving messages. ■ Is depicted either as a stick figure (default) or, if a nonhuman actor is involved, as a rectangle with <<actor>> in it (alternative).	 **anActor** **<<actor>>** **anActor**
An object: ■ Participates in a collaboration by sending and/or receiving messages.	**anObject : aClass**
An association: ■ Shows an association between actors and/or objects. ■ Is used to send messages.	————————
A message: ■ Conveys information from one object to another one. ■ Has direction shown using an arrowhead. ■ Has sequence shown by a sequence number.	SeqNumber: aMessage →
A guard condition: ■ Represents a test that must be met for the message to be sent.	SeqNumber: [aGuardCondition]: aMessage →
A frame: ■ Indicates the context of the communication diagram.	**Context**

FIGURE 6-10   Communication Diagram Syntax

(see Chapter 4) and class diagrams (see Chapter 5), is through the use of *packages* (i.e., logical groups of classes). In the case of communication diagrams, its objects are grouped together based on the messages sent to and received from the other objects.[11]

Figure 6-11 provides two additional examples of communication diagrams. These diagrams are equivalent to the sequence diagrams contained in Figure 6-3. However, when comparing the communication diagrams to the sequence diagrams in these figures, you see that quite a bit of information is lost. For example, the CreateSandwich() message is nowhere to be found. However, the primary purpose of the communication diagram is to show how the different actors and classes interact, and this is exactly the information that is included.

---

[11] For those familiar with structured analysis and design, packages serve a purpose similar to the leveling and balancing processes used in data flow diagramming. Packages and package diagrams are described in Chapter 7.

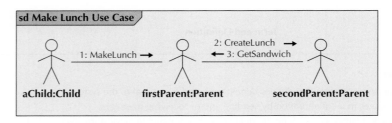

FIGURE 6-11
Additional Sample Communication Diagrams

Guidelines for Creating Communication Diagrams   Ambler[12] provides a set of guidelines when drawing communication diagrams (see Figure 6-12). In this section, in addition to the first four guidelines for drawing sequence diagrams, we consider two more.

- Use the correct diagram for the information you are interested in communicating with the user. In this case, do not use communication diagrams to model process flow. Instead, you should use an activity diagram with swimlanes that represent objects (see Chapter 4). Communication diagrams allow the team to easily identify a set of objects that are intertwined. On the other hand, it would be very difficult to "see" how the objects collaborated in an activity diagram.

- When trying to understand the sequencing of messages, a sequence diagram should be used instead of a communication diagram. As in the previous guideline, this guideline essentially suggests that you should use the diagram that was designed to deal with the issue at hand. Even though communication diagrams can show sequencing of messages, this was never meant to be their primary purpose.

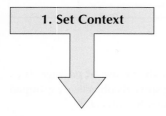

Creating a Communication Diagram[13]   Remember that a communication diagram is basically an object diagram that shows message-passing relationships instead of aggregation or generalization associations. In this section, we describe a five-step process used to build a communication diagram. The first step in the process is to determine the context of the communication diagram. Like a sequence diagram, the context of the diagram can be a system, a use case, or a scenario of a use case. The context of the diagram is depicted as a labeled frame around the diagram (see Figures 6-9, 6-10, and 6-11).

FIGURE 6-12
Guidelines for Creating Communication Diagrams

- Apply sequence diagram guidelines 1 through 4.
- Do not use communication diagrams to model process flow.
- Use a sequence diagram instead of a communication diagram when sequencing is important.

[12] S.W. Ambler, *The Elements of UML 2.0 Style* (Cambridge, England: Cambridge University Press, 2005).

[13] The approach described in this section is adapted from Booch, Rumbaugh, and Jacobson, *The Unified Modeling Language User Guide*.

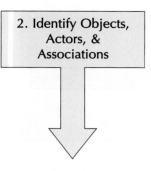

2. Identify Objects, Actors, & Associations

The second step is to identify the objects (actors) and the associations that link the objects (actors) that participate in the collaboration together. Remember, the objects that participate in the collaboration are instances of the classes identified during the development of the structural model (see Chapter 5). Like the sequence-diagramming process, it is likely that additional objects, and hence, classes, will be discovered. Again, this is normal because the underlying development process is iterative and incremental. In addition to the communication diagram being modified, the sequence diagrams and structural model probably also have to be modified. Additional functional requirements might also be uncovered, hence requiring the functional models to be modified as well (see Chapter 4).

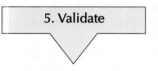

3. Lay Out Diagram

The third step is to lay out the objects (actors) and their associations on the communication diagram by placing them together based on the associations that they have with the other objects in the collaboration. By focusing on the associations between the objects (actors) and minimizing the number of associations that cross over one another, we can increase the understandability of the diagram.

4. Add Messages

The fourth step is to add the messages to the associations between the objects. We do this by adding the name of the message(s) to the association link between the objects and an arrow showing the direction of the message being sent. Each message has a sequence number associated with it to portray the time-based ordering of the message.[14]

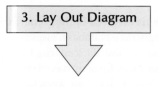

5. Validate

The fifth and final step is to validate the communication diagram. The purpose of this step is to guarantee that the communication diagram faithfully portrays the underlying process(es). This is done by ensuring that all steps in the process are depicted on the diagram.

YOUR TURN 6-2 Drawing a Communication Diagram

In "Your Turn 6-1", you were asked to create a set of instance-specific sequence diagrams for a use case of housing service. Draw a communication diagram for the same situation.

Example   As with the sequence diagramming example, we return to the Borrow Books use case of the Library Book Collection Management System. In this case, to set the context of the diagram, we visit the Student without a valid ID and Student with a Valid ID but owes fines or has overdue books scenarios. We create two communication diagrams, one for each scenario. As with the sequence-diagramming process, we review the Flow of Events of the use-case description (see Figure 6-5), the activity diagram (see Figure 6-6), and the use-case diagram (see Figure 6-7).

---

[14] However, remember the sequence diagram portrays the time-based ordering of the messages in a top-down manner. If your focus is on the time-based ordering of the messages, we recommend that you also use the sequence diagram.

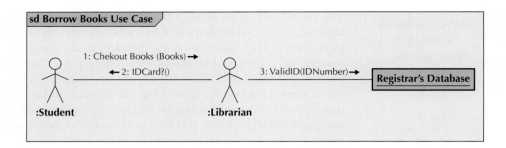

FIGURE 6-13

Communication
Diagram for the
Student Without a
Valid ID

The next step is to identify the actor, objects, and associations involved in the scenario. In both scenarios, the actors are Student, Librarian, and the Registrar's Database. However, because the process is aborted very early in the Student without a valid ID scenario, there are no objects in the scenario. The Student with a Valid ID but owes fines or has overdue books scenario does include one object: Borrower. Both scenarios have an association between the Student and Librarian actors and the Librarian and Registrar's Database actor. The Student with a Valid ID but owes fines or has overdue books scenario also has an association between the Librarian actor and the Borrower object.

The next step is to lay out the diagram. In both cases, because the student initiates the process, we place the Student actor to the far left of the diagram. We then place the other actors on the diagram in the order in which they participate in the process. We also place the :Borrower object to the far bottom right of the diagram that represents the Student with a Valid ID but owes fines or has overdue books scenario, to reflect the left-to-right and top-to-bottom direction of reading for most western cultures.

Now we place the relevant associations between the actors and objects that participate in the scenarios. In this step, we add the messages to the associations. We again review the flow of events (see Figure 6-5) of the use-case description to identify the directionality and content of the messages. Figures 6-13 and 6-14 portray the communication diagrams created.

The last step is to validate the diagrams. As with sequence diagrams, because we are drawing instance-specific versions of the communication diagram, we must also draw the remaining seven diagrams for the other scenarios.

FIGURE 6-14

Communication
Diagram for the
Student With a Valid ID
but Owes Fines or has
Overdue Books

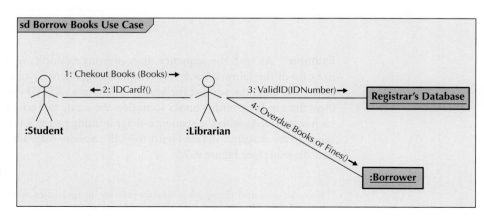

BEHAVIORAL STATE MACHINES

Some of the classes in the *class diagrams* represent a set of objects that are quite dynamic in that they pass through a variety of states over the course of their existence. For example, a patient can change over time from being new to current to former, based on his or her status with the doctor's office. A behavioral state machine is a dynamic model that shows the different states through which a single object passes during its life in response to events, along with its responses and actions. Typically, behavioral state machines are not used for all objects but just to further define complex objects to help simplify the design of algorithms for their methods. The behavioral state machine shows the different states of the object and what events cause the object to change from one state to another. In comparison to the interaction diagrams, behavioral state machines should be used to help understand the dynamic aspects of a single class and how its instances evolve over time,[15] and not with seeing how a particular use case or use-case scenario is executed over a set of classes.

In this section, we describe states, events, transitions, actions, and activities and the use of the behavioral state machine to model the state changes through which complex objects pass. As in the creation of the interaction diagrams, when we create a behavioral state machine for an object, it is possible that we will uncover additional events that need to be included in your functional model (see Chapter 4) and additional operations that need to be included in the structural model (see Chapter 5), so our interaction diagrams might have to be modified again. Because object-oriented development is iterative and incremental, this continuous modification of the evolving models (functional, structural, and behavioral) of the system is to be expected.

States, Events, Transitions, Actions, and Activities

The *state* of an object is defined by the value of its attributes and its relationships with other objects at a particular point in time. For example, a patient might have a state of new, current, or former. The attributes or properties of an object affect the state that it is in; however, not all attributes or attribute changes will make a difference. For example, think about a patient's address. Those attributes make very little difference to changes in a patient's state. However, if states were based on a patient's geographic location (e.g., in-town patients were treated differently than out-of-town patients), changes to the patient's address would influence state changes.

An *event* is something that takes place at a certain point in time and changes a value or values that describe an object, which, in turn, changes the object's state. It can be a designated condition becoming true, the receipt of the call for a method by an object, or the passage of a designated period of time. The state of the object determines exactly what the response will be.

A *transition* is a relationship that represents the movement of an object from one state to another state. Some transitions have a guard condition. A *guard condition* is a Boolean expression that includes attribute values, which allows a transition to occur only if the condition is true. An object typically moves from one state to another based on the outcome of an action triggered by an event. An *action* is an atomic, non-decomposable process that cannot be interrupted. From a practical perspective, actions take zero time, and they are associated with a transition. In contrast, an *activity* is a nonatomic, decomposable process that can be interrupted. Activities take a long period of time to complete, and they can be started and stopped by an action.

---

[15] Some authors refer to this as modeling an object's life cycle.

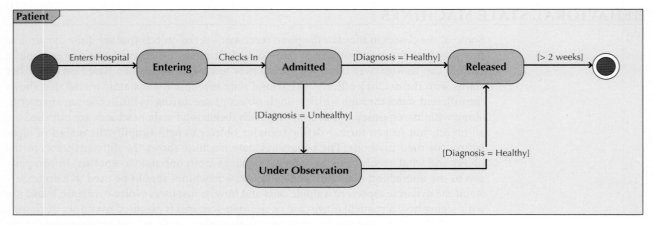

FIGURE 6-15   Sample Behavioral State Machine Diagram

Term and Definition	Symbol
A state: ■ Is shown as a rectangle with rounded corners. ■ Has a name that represents the state of an object.	aState
An initial state: ■ Is shown as a small, filled-in circle. ■ Represents the point at which an object begins to exist.	●
A final state: ■ Is shown as a circle surrounding a small, filled-in circle (bull's-eye). ■ Represents the completion of activity.	◉
An event: ■ Is a noteworthy occurrence that triggers a change in state. ■ Can be a designated condition becoming true, the receipt of an explicit signal from one object to another, or the passage of a designated period of time. ■ Is used to label a transition.	anEvent
A transition: ■ Indicates that an object in the first state will enter the second state. ■ Is triggered by the occurrence of the event labeling the transition. ■ Is shown as a solid arrow from one state to another, labeled by the event name.	→
A frame: ■ Indicates the context of the behavioral state machine.	Context

FIGURE 6-16   Behavioral State Machine Diagram Syntax

Elements of a Behavioral State Machine

Figure 6-15 presents an example of a behavioral state machine representing the patient class in the context of a hospital environment. From this diagram we can tell that a patient enters a hospital and is admitted after checking in. If a doctor finds the patient to be healthy, he or she is released and is no longer considered a patient after two weeks elapses. If a patient is found to be unhealthy, he or she remains under observation until the diagnosis changes.

A *state* is a set of values that describes an object at a specific point in time, and it represents a point in an object's life in which it satisfies some condition, performs some action, or waits for something to happen (see Figure 6-16). In Figure 6-15, states include entering, admitted, released, and under observation. A state is depicted by a *state symbol,* which is a rectangle with rounded corners with a descriptive label that communicates a particular state. There are two exceptions. An *initial state* is shown using a small, filled-in circle, and an object's *final state* is shown as a circle surrounding a small, filled-in circle. These exceptions depict when an object begins and ceases to exist, respectively.

Arrows are used to connect the state symbols, representing the transitions between states. Each arrow is labeled with the appropriate event name and any parameters or conditions that may apply. For example, the two transitions from admitted to released and under observation contain guard conditions. As in the other behavioral diagrams, in many cases it is useful to explicitly show the context of the behavioral state machine using a frame.

Figure 6-17 depicts two additional behavioral state machines. The first one is for the lunch object that was associated with the Make Lunch use-case scenario of Figures 6-3 and 6-11. In this case, there is obviously additional information that has been captured about the lunch object. For example, the scenario of Figures 6-3 and 6-11 did not include information regarding the lunch being taken out of the box or being eaten. This implies additional use cases and/or use-case scenarios that would have to be included in a system that dealt with lunch processing. The second behavioral state machine deals with the life cycle of an order. The order object is associated with the submit order use-case scenario described in Figures 6-3 and 6-11. As in the lunch example, there is quite a bit of additional information contained in this behavioral state machine. For an order-processing system, quite a few additional sequence and communication diagrams would be necessary to completely represent all the processing associated with an order object. Obviously, because behavioral state machines can uncover additional processing requirements, they can be very useful in filling out the complete description of an evolving system.

Sometimes, states and subclasses can be confused. For example, in Figure 6-18, are the classes Freshman, Sophomore, Junior, and Senior subclasses of the class Undergraduate, or are they simply states that an instance of the Undergraduate class goes through during its lifetime? In this case, the latter is the better answer. When trying to identify all potential classes when the structural model is created (see Chapter 5), you might actually identify states of the relevant superclass instead of subclasses. This is another example of how tightly intertwined the functional, structural, and behavioral models can be. From a modeling perspective, even though we had to remove the Freshman, Sophomore, Junior, and Senior subclasses from the structural model, it was better to capture that information as part of the structural model and remove it when we were creating the behavioral models than to omit it and take the chance of missing a crucial piece of information about the problem domain. Remember, object-oriented development is iterative and incremental. As we progress to a correct model of the problem domain, we will make many mistakes.

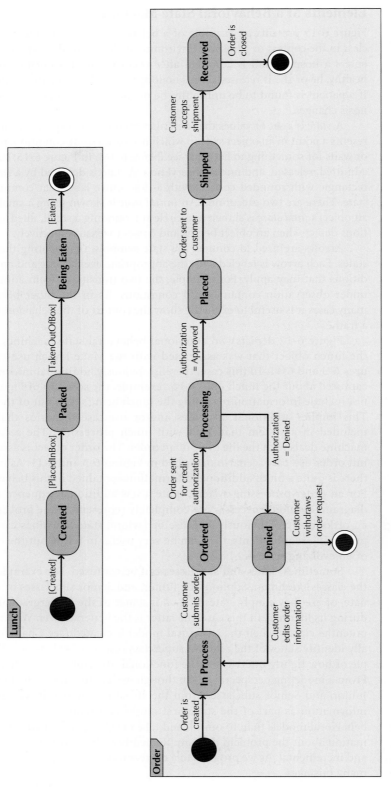

FIGURE 6-17 Additional Behavioral State Machine Diagrams

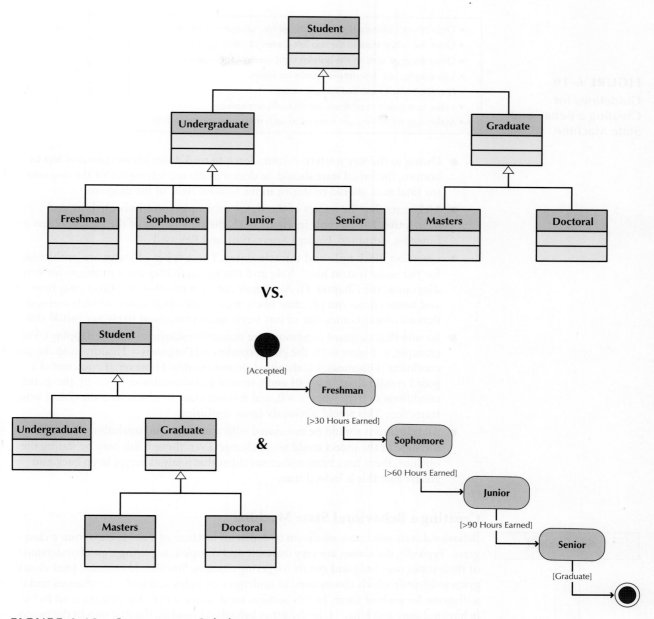

FIGURE 6-18   States versus Subclasses

Guidelines for Creating Behavioral State Machines   As with the sequence and communication diagrams, Amble suggests a set of guidelines when drawing behavior state machines. In this case, we consider seven of his recommendations (see Figure 6-19).[16]

- Create a behavioral state machine for objects whose behavior changes based on the state of the object. In other words, do not create a behavioral state machine for an object whose behavior is always the same regardless of its state. These objects are too simple.

[16] S.W. Ambler, *The Elements of UML 2.0 Style* (Cambridge, England: Cambridge University Press, 2005).

FIGURE 6-19
Guidelines for
Creating a Behavioral
State Machine

- Only create behavioral state machines for "complex" objects.
- Draw the initial state in the top left corner of the diagram.
- Draw the final state in the bottom right corner of the diagram.
- Use simple, but descriptive, names for states.
- Question black hole and miracle states.
- Make sure guard conditions are mutually exclusive.
- Make sure transitions are associated with messages and operations.

■ Owing to the way western cultures learn to read, from left to right and top to bottom, the initial state should be drawn in the top left corner of the diagram and the final state should be drawn in the bottom right of the diagram.

■ Make sure that the names of the states are simple, intuitively obvious, and descriptive. For example in Figure 6-15, the state names of the patient object are Entering, Admitted, Under Observation, and Released.

■ Question black hole and miracle states. These types of states are problematic for the same reason black hole and miracle activities are a problem for activity diagrams (see Chapter 4). *Black hole states*, states that an object goes into and never comes out of, most likely are actually final states. *Miracle states*, states that an object comes out of but never went into, most likely are initial states.

■ Be sure that all guard conditions are mutually exclusive, (not overlapping). For example, in Figure 6-15, the guard condition [Diagnosis = Healthy] and the guard condition [Diagnosis = Unhealthy] do not overlap. However, if you created a guard condition of [x >= 0] and a second guard condition [x <= 0], the guard conditions overlap when x = 0, and it is not clear to which state the object would transition. This would obviously cause confusion.

■ All transitions should be associated with a message and operation. Otherwise, the state of the object could never change. Even though this may be stating the obvious, there have been numerous times that analysts forget to go back and ensure that this is indeed true.

Creating a Behavioral State Machine

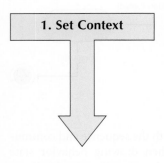

1. Set Context

Behavioral state machines are drawn to depict an instance of a single class from a class diagram. Typically, the classes are very dynamic and complex, requiring a good understanding of their states over time and events triggering changes. You should examine your class diagram to identify which classes need to undergo a complex series of state changes and draw a diagram for each of them. In this section, we describe a five-step process used to build a behavioral state machine.[17] Like the other behavioral models, the first step in the process is to determine the context of the behavioral state machine, which is shown in the label of the frame of the diagram. The context of a behavioral state machine is usually a class. However, it also could be a set of classes, a subsystem, or an entire system.

2. Identify Object States

The second step is to identify the various states that an object will have over its lifetime. This includes establishing the boundaries of the existence of an object by identifying the initial and final states of an object. We also must identify the stable states of an object. The information necessary to perform this is gleaned from reading the use-case descriptions, talking with users, and relying on the requirements-gathering

[17] The approach described in this section is adapted from Booch, Rumbaugh, and Jacobson, *The Unified Modeling Language User Guide.*

techniques that you learned about in Chapter 3. An easy way to identify the states of an object is to write the steps of what happens to an object over time, from start to finish, similar to how the normal flow of events section of a use-case description would be created.

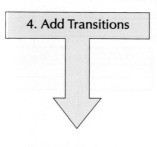

3. Lay Out Diagram

The third step is to determine the sequence of the states that an object will pass through during its lifetime. Using this sequence, the states are placed onto the behavioral state machine in a left-to-right order.

4. Add Transitions

The fourth step is to identify the transitions between the states of the objects and to add the events, actions, and guard conditions associated with the transitions. The events are the *triggers* that cause an object to move from one state to the next state. In other words, an event causes an action to execute that changes the value(s) of an object's attribute(s) in a significant manner. The actions are typically operations contained within the object. Also, guard conditions can model a set of test conditions that must be met for the transition to occur. At this point in the process, the transitions are drawn between the relevant states and labeled with the event, action, or guard condition.

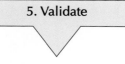

5. Validate

The fifth step is to validate the behavioral state machine by making sure that each state is reachable and that it is possible to leave all states except for final states. Obviously, if an identified state is not reachable, either a transition is missing or the state was identified in error. Only final states can be a dead end from the perspective of an object's life cycle.

YOUR TURN   6-3 Drawing a Behavioral State Machine

*Y*ou have been working with the system for the campus housing service that helps students find apartments. One of the dynamic classes in this system is probably the apartment class. Draw a behavioral state machine to show the various states that an apartment object transitions throughout its lifetime. Can you think of other classes that would make good candidates for a behavioral state machine?

Example   The first step in drawing a behavioral state machine is to set the context. For our purposes, the context typically is an instance of a class that has multiple states and whose behavior depends upon the state in which it currently resides. As suggested earlier, we should review the class diagram (see Figure 6-20) to identify the "interesting" classes. In the case of the Library Book Collection Management System, the obvious class to consider is the Book class.

The next step is to identify the different states through which an instance of the Book class can traverse during its lifetime. Good places to look for possible state changes are the use-case descriptions (see Figure 6-5), the activity diagrams (see Figure 6-6), the sequence diagrams (see Figure 6-8), and the communication diagrams (see Figures 6-13 and 6-14). In the case of a book, even though the states may be similar, you must be careful in identifying the states associated with an instance of the Book class and not the states associated with the physical book itself. In Chapter 5, we observed that there were a number of implied states to consider. These included Checked Out, Overdue, Requested, Available, and Damaged. If the book is damaged, the book could either be repaired and put back into circulation or it could be too damaged to repair and be removed from circulation instead. Even though a Borrower could be fined for an overdue or damaged book, being fined is not a state of a book, it is a state of a borrower.

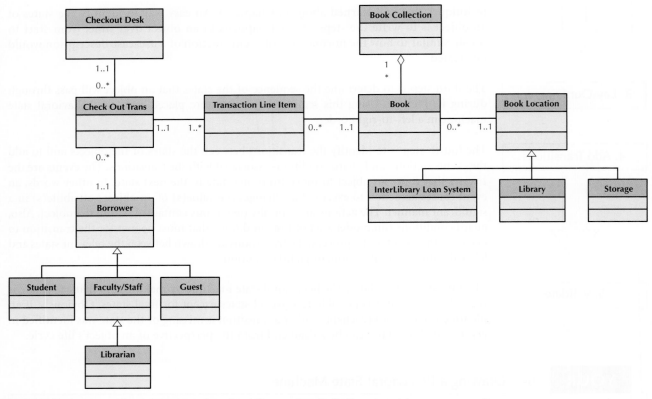

FIGURE 6-20   Class Diagram for the Library Book Collection Management System

Next, we lay out the diagram by ordering the states in a sequential manner based on the life cycle of a book. For example, it probably makes no sense to have a book to go from a repaired state to a damaged state. However, going from a damaged state to a repaired state makes sense. Nor does it make sense for a book to go from an available state directly to an overdue state. However, the converse makes sense. The states we identified for a book object include Available, Checked Out, Overdue, Requested, Damaged, and Being Repaired. Next, we added the transitions between the states and labeled them with the appropriate guard conditions. The behavioral state machine for an instance of the Book class is portrayed in Figure 6-21.

Finally, we validate the diagram by checking for missing states or transitions and ensuring that there are no black hole or miracle states.

CRUDE ANALYSIS

One useful technique for identifying potential collaborations is *CRUDE analysis*.[18] CRUDE analysis uses a *CRUDE matrix*, in which each interaction among objects is labeled with a letter for the type of interaction: C for create, R for read or reference, U for update,

---

[18] CRUD analysis has typically been associated with structured analysis and design [see Alan Dennis, Barbara Haley Wixom and Roberta M. Roth, *Systems Analysis Design*, 3nd ed. (New York: Wiley, 2006)] and information engineering [see James Martin, *Information Engineering, Book II Planning and Analysis* (Englewood Cliffs, NJ: Prentice Hall, 1990)]. In our case, we have simply adapted it to object-oriented systems development. In the case of object orientation, we have added an E to allow us to document the execution of operations that do not create, read, update, or delete but that, instead, simply are executed for possible side-effect purposes. Specific details on collaborations are described in Chapter 7.

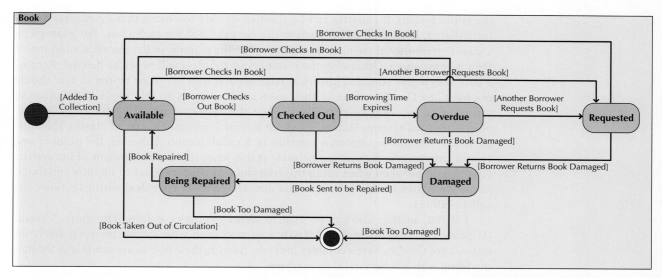

FIGURE 6-21   Behavioral State Machine for an Instance of the Book Class in the Library Book Collection Management System

D for delete, and E for execute. In an object-oriented approach, a class/actor-by-class/actor matrix is used.[19] Each cell in the matrix represents the interaction between instances of the classes. For example, in Figure 6-1, an instance of the Receptionist actor creates an instance of the Appointment class. Assuming a Row:Column ordering, a C is placed in the cell Receptionist:Appointment. Also, in Figure 6-1, an instance of the Receptionist actor references an instance of the Appointments class. In this case, an R is placed in the Receptionist:Appointments cell. Figure 6-22 shows the CRUDE matrix based on the Make Old Patient Appt use case.

Unlike the interaction diagrams and behavioral state machines, a CRUDE matrix is most useful as a system-wide representation. Once a CRUDE matrix is completed for

	Receptionist	PatientList	Patient	UnpaidBills	Appointments	Appointment
Receptionist		RU	CRUD	R	RU	CRUD
PatientList			R			
Patient						
UnpaidBills						
Appointments						R
Appointment						

FIGURE 6-22   CRUDE Matrix for the Make Old Patient Apt Use Case

[19] Another useful but more-detailed form of the CRUDE matrix is a Class/Actor:Operation-by-Class/Actor:Operation matrix. For validation and verification purposes, this more-detailed matrix is more useful. However, for our purposes at this point in our discussion, the Class/Actor-by-Class/Actor matrix is sufficient.

the entire system, the matrix can be scanned quickly to ensure that every class can be instantiated. Each type of interaction can be validated for each class. For example, if a class represents only temporary objects, then the column in the matrix should have a D in it somewhere. Otherwise, the instances of the class will never be deleted. Because a data warehouse contains historical data, objects that are to be stored in one should not have any U or D entries in their associated columns. In this way, CRUDE analysis can be used as a way to partially validate the interactions among the objects in an object-oriented system. Finally, the more interactions among a set of classes, the more likely they should be clustered together in a collaboration. However, the number and type of interactions are only an estimate at this point in the development of the system. Care should be taken when using this technique to cluster classes to identify collaborations. We return to this subject in the next chapter when we deal with partitions and collaborations.

CRUDE analysis also can be used to identify complex objects. The more (C)reate, (U)pdate, or (D)elete entries in the column associated with a class, the more likely the instances of the class have a complex life cycle. As such, these objects are candidates for state modeling with a behavioral state machine.

Example The best way to create a CRUDE matrix is to conceptually merge the sequence and communication diagrams that model all of the scenarios of all of the use cases in a system. The easiest way to accomplish this is simply to create an empty class/actor-by-class/actor matrix. In the case of the Library Book Collection Management System, we have six actors (Student, Faculty/Staff, Guest, Librarian, Personnel Office, and Registrar's Office) and eight classes (Book, Book Collection, Student, Faculty/Staff, Guest, Interlibrary Loan System, Library, and Storage). Once this matrix has been laid out, all you need to do is role-play the scenarios and see what actors and classes interact with each other. Based on the type of interaction, you simply record a C, R, U, D, or E in the appropriate cell of the matrix. You do this repeatedly until all of the scenarios of all of the use cases have been executed. The CRUDE matrix for the Library Book Collection Management System is shown in Figure 6-23. One of the functions that the matrix can serve is to begin the validation process of the entire system. In this case, by quickly reviewing the matrix we can see that absolutely nothing seems to be interacting with the Library and Storage objects. This raises an important question as to whether these objects should exist or not. If nothing calls or uses them and they don't call or use anything, then why are they part of this system? Either they should be removed from the current representation of the system, or we have managed to miss some interaction. Knowing this allows us to go back to the user, in this case, the Librarian, and ask what should be done.

YOUR TURN 6-4 CRUDE Analysis

You have been working with the system for the campus housing service that helps students find apartments. Based on the work completed so far, perform a CRUDE analysis to identify which classes collaborate the most and to perform some validation of the evolving representation of the system.

	Student Actor	Faculty/Staff Actor	Guest Actor	Librarian Actor	Personnel Office Actor	Registrar's Office Actor	Book	Book Collection	Student Class	Faculty/Staff Class	Guest Class	Interlibrary Loan System	Library	Storage
Student Actor				E			R,E	R				E		
Faculty/Staff Actor				E			R,E	R				E		
Guest Actor				E			R,E	R				E		
Librarian Actor					R,E	R,E	C,R,U,D,E	R,U,E	R,U	R,U	C,R,U,D,E	R,E		
Personnel Office Actor														
Registrar's Office Actor														
Book														
Book Collection														
Student Class														
Faculty/Staff Class														
Guest Class														
Interlibrary Loan System														
Library														
Storage														

FIGURE 6-23   CRUDE Matrix for the Library Book Collection Management System

VERIFYING AND VALIDATING THE BEHAVIORAL MODEL[20]

In this chapter, we described three different diagrams (sequence diagram, communication diagram, and behavioral state machine) and CRUDE matrices that could be used to represent the behavioral model. The sequence and communication diagrams modeled the interaction among instances of classes that worked together to support the business processes included in a system, the behavioral state machine described the state changes through which an object traverses during its lifetime, and the CRUDE matrix represented a system-level overview of the interactions among the objects in the system. In this chapter, we combine walkthroughs with CRUDE matrices to more completely verify and validate the behavioral models. Since, in the previous section we covered CRUDE analysis and matrices, we focus only on walkthroughs in this section. We again use the appointment system and focus on Figures 6-1, 6-9, 6-15, and 6-22 to describe a set of rules that can be used to ensure that the behavioral model is internally consistent.

First, every actor and object included on a sequence diagram must be included as an actor and an object on a communication diagram, and vice versa. For example, in Figures 6-1 and 6-5, the aReceptionist actor and the Patients object appear on both diagrams.

Second, if there is a message on the sequence diagram, there must be an association on the communications diagram, and vice versa. For example, Figure 6-1 portrays a message being sent from the aReceptionist actor to the Patient object, and a matching association appears in the corresponding communication diagram (see Figure 6-9).

Third, every message that is included on a sequence diagram must appear as a message on an association in the corresponding communication diagram, and vice versa. For example, the LookUpPatient() message sent by the aReceptionist actor to the Patient object on the sequence diagram (see Figure 6-1) appears as a message on the association between the aReceptionist actor and the Patient object on the communication diagram (see Figure 6-9).

Fourth, if a guard condition appears on a message in the sequence diagram, there must be an equivalent guard condition on the corresponding communication diagram, and vice versa. For example, the message sent from the aReceptionist actor to the UnpaidBills object has a guard condition of [aPatient Exists] (see Figure 6-1). Figure 6-9 shows the matching guard condition included on the communication diagram.

Fifth, the sequence number included as part of a message label in a communications diagram implies the sequential order in which the message will be sent. Therefore, it must correspond to the top-down ordering of the messages being sent on the sequence diagram. For example, the LookUpPatient message sent from the aReceptionist actor to the Patient object on the sequence diagram (see Figure 6-1) is the second from the top of the diagram. The LookUpPatient message sent from the aReceptionist actor to the Patients object on the communications diagram (see Figure 6-9) is labeled with the number 2.[21]

Sixth, all transitions contained in a behavior state machine must be associated with a message being sent on a sequence and communication diagram, and it must be classified as a (C)reate, (U)pdate, or (D)elete message in a CRUDE matrix. For example, in Figure 6-15, the Checks In transition must be associated with a message in the corresponding sequence and communication diagrams. Furthermore, it should be associated with an (U)pdate entry in the CRUDE matrix associated with the hospital patient system.

Seventh, all entries in a CRUDE matrix imply a message being sent from an actor or object to another actor or object. If the entry is a (C)reate, (U)pdate, or (D)elete, then there must be an associated transition in a behavioral state machine that represents the instances of the

[20] The material in this section has been adapted from E. Yourdon, *Modern Structured Analysis* (Englewood Cliffs, NJ: Prentice Hall, 1989).

[21] There are more complicated numbering schemes that could be used. However, for our purposes, a simple sequential number is sufficient.

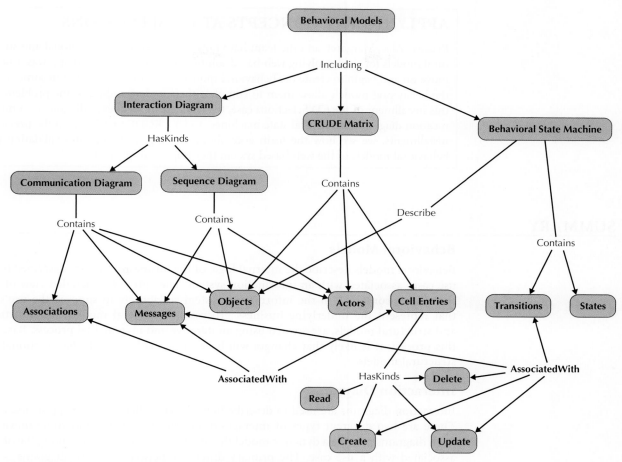

FIGURE 6-24   Interrelationships among Behavioral Models

receiving class. For example, in Figure 6-22, the R and U entries in the Receptionist row and Appointments column imply that instances of the Receptionist actor will read and update instances of the Appointments class. Thus, there should be read and update messages on the sequence and communication diagrams corresponding with the appointments processes. Reviewing Figures 6-1 and 6-9, we see that there is a message, MatchAppts(), from the aReceptionist actor to the Appointments object. However, based on this review, it is unclear whether the MatchAppts() message represents a read, an update, or both. Therefore, additional analysis is required.[22] Because there is an (U)pdate message involved, there must be a transition on a behavioral state machine that portrays the life cycle of an Appointments object.

Finally, many representation-specific rules have been proposed. However, as in the other models, these rules are beyond the scope of this section on verification and validation.[23] Figure 6-24 portrays the associations among the behavioral models.

[22] We have delayed the description of designing operations and methods until Chapter 8. Therefore, the detailed information required to understand a specific message has not been created yet. However, in many cases, enough information will already have been created to validate many of the transitions in behavioral state machines and CRUDE matrices.

[23] A good reference for these types of restrictions is S.W. Ambler, *The Elements of UML 2.0 Style* (Cambridge, England: Cambridge University Press, 2005).

> ## APPLYING THE CONCEPTS AT CD SELECTIONS
>
> Because Alec, Margaret, and the team have now completed rough functional and structural models for their evolving web-based solution, they have decided that it was time to move on and begin to create the behavioral models. Alec understood that in some ways, the behavioral models allow them to complete their understanding of the problem. In this installment of the CD Selections case, the team creates sequence diagrams, communication diagrams, behavioral state machines, and a CRUDE matrix. As in the previous installments, we see how the team goes about creating, verifying, and validating the behavioral models of the web-based system they hope to implement.

SUMMARY

Behavioral Models

Behavioral models describe the internal logic of the business processes described by the use cases associated with an information system. They provide a detailed view of how the objects contained in the information system collaborate to support the use cases that represent the underlying business processes. Behavioral models, like functional and structural models, are created using an iterative and incremental process. Based on this process, it is likely that changes will have to be made to both the functional and structural models.

Interaction Diagrams

Interaction diagrams are used to describe how objects collaborate to support use cases. There are two different types of interaction diagrams: sequence and communication. Both diagrams provide a dynamic model that portrays the interaction among the objects associated with a use case. The primary difference between the two diagrams is that sequence diagrams focus on the time ordering or sequence of the messages being sent between the objects, whereas communication diagrams spotlight the collaborating nature of the objects supporting use cases. A communication diagram is essentially an object diagram (see Chapter 5) that portrays message-sending relationships instead of structural relationships.

Behavioral State Machine

The behavioral state machine shows the different states through which a single class passes during its life in response to events, along with responses and actions. A state is a set of values that describes an object at a specific point in time, and it represents a point in an object's life in which it satisfies some condition, performs some action, or waits for something to happen. An event is something that takes place at a certain point in time and changes a value that describes an object, which, in turn, changes the object's state. As objects move from state to state, they undergo transitions. Typically, behavioral state machines are used to portray the dynamic aspect of a complex class.

CRUDE Analysis

CRUDE analysis is a very useful approach to identifying potential collaborations among classes and to verify and validate a system. In a very concise format (CRUDE matrix) it allows the analyst to see what type of interactions (create, read/reference, update, delete, or execute) the different types of objects have in the system. CRUDE analysis also supports the identification of the more-complex objects that could benefit from state modeling using a behavioral state machine.

Verifying and Validating Behavioral Models

Verifying and validating the behavioral models are very similar to verifying and validating the functional, and structural models. In this case, the sequence diagrams, communication diagrams, behavioral state machines, and CRUDE matrix are carefully checked for consistency. This is done, typically, through the creation of the CRUDE matrix, walkthroughs, or both.

KEY TERMS

Action, 253
Activity, 253
Actor, 239
Association, 248
Attributes, 238
Behavior, 238
Behavior models, 237
Behavioral state machines, 253
Black hole states, 258
Class, 238
Class diagram, 253
Collaboration, 237
Communication diagram, 246
Condition, 241
CRC cards, 243

CRUDE analysis, 260
CRUDE matrix, 260
Dynamic model, 238
Event, 253
Execution occurrence, 240
Final state, 255
Frame, 243
Generic sequence diagram, 238
Guard condition 253
Initial state, 255
Instance, 238
Instance sequence diagram, 238
Lifeline, 239
Message, 238
Method, 238

Miracle states, 258
Object, 239
Operation, 238
Operation call message, 241
Packages, 249
Return message, 241
Scenario, 238
Self-delegation, 241
Sequence diagram, 238
State, 253
State symbol, 255
Temporary object, 239
Transition, 253
Trigger, 259
Use case, 237

QUESTIONS

1. How is behavioral modeling related to structural modeling?
2. How does a use case relate to a sequence diagram? A communication diagram?
3. Contrast the following sets of terms: state, behavior, class, object, action, and activity.
4. Why is iteration important when creating a behavioral model?
5. What are the main building blocks for the sequence diagram? How are they represented on the model?
6. How do you show that a temporary object is to go out of existence on a sequence diagram?

7. Do lifelines always continue down the entire page of a sequence diagram? Explain.
8. Describe the steps used to create a sequence diagram.
9. When drawing a sequence diagram, what guidelines should you follow?
10. Describe the main building blocks for the communication diagram and how they are represented on the model.
11. How do you show the sequence of messages on a communication diagram?
12. How do you show the direction of a message on a communication diagram?

13. Describe the steps used to create a communication diagram.
14. When drawing a communication diagram, what guidelines should you follow?
15. Are states always depicted using rounded rectangles on a behavioral state machine? Explain.
16. What kinds of events can lead to state transitions on a behavioral state machine?
17. What are the steps in building a behavioral state machine?
18. When drawing a behavioral state machine, what guidelines should you follow?
19. How are guard conditions shown on a behavioral state machine?
20. Describe the type of class that is best represented by a behavioral state machine. Give two examples of classes that would be good candidates for a behavioral state machine.
21. What is CRUDE analysis and what is it used for?
22. Identify the models that contain each of the following components: actor, association, class, extends, association, final state, guard condition, initial state, links, message, multiplicity, object, state, transition, and update operation.

EXERCISES

A. Think about sending a first-class letter to an international pen pal. Describe the process that the letter goes through to get from your initial creation of the letter to being read by your friend, from the letter's perspective. Draw a behavioral state machine that depicts the states that the letter moves through.

B. Draw a behavioral state machine that describes the various states that a travel authorization can have through its approval process. A travel authorization form is used in most companies to approve travel expenses for employees. Typically, an employee fills out a blank form and sends it to his or her boss for a signature. If the amount is fairly small (<$300), then the boss signs the form and routes it to accounts payable to be input into the accounting system. The system cuts a check that is sent to the employee for the right amount, and after the check is cashed, the form is filed away with the canceled check. If the check is not cashed within 90 days, the travel form expires. When the amount of the travel voucher is a large amount (>$300), then the boss signs the form and sends it to the CFO, along with a paragraph explaining the purpose of the travel; the CFO signs the form and passes it along to accounts payable. Of course, the boss and the CFO can reject the travel authorization form if they do not feel that the expenses are reasonable. In this case, the employee can change the form to include more explanation or decide to pay the expenses.

C. Think about the system that handles student admissions at your university. The primary function of the system should be able to track a student from the request for information through the admissions process until the student is either admitted to the school or rejected.

1. Write a use-case description that can describe an Admit Student use case.

Assume that applicants who are children of alumni are handled differently from other applicants. Also, assume that a generic Update Student Information use case is available for your system to use.

2. Create a use-case diagram that includes all of the above use cases.

Assume that an admissions form includes the contents of the form, SAT information, and references. Additional information is captured about children of alumni, such as their parent's graduation year, contact information, and college major.

3. Create a class diagram for the use cases identified with questions 1 and 2. Also, be sure to include the above information.

Assume that a temporary student object is used by the system to hold information about people before they send in an admission form. After the form is sent in, these people are considered students.

4. Create sequence diagrams for the scenarios of the above use cases.

5. Create a communication diagram for the scenarios of the above use cases.

6. Create a behavioral state machine to depict a person as he or she moves through the admissions process.

7. Perform a CRUDE analysis to show the interactivity of the objects in the system.

D. For the A Real Estate Inc. problem in Chapters 4 (exercises I, J, and K) and 5 (exercises P and Q):

1. Choose one use case and, for each scenario, create a sequence diagram.

2. Create a communication diagram for each scenario of the use case chosen in Question 1.

3. Create a behavioral state machine to depict one of the classes on the class diagram you created for Chapter 5, exercise P.

4. Perform a CRUDE analysis to show the interactivity of the objects in the system.

5. Perform a verification and validation walkthrough of the problem.

E. For the A Video Store problem in Chapters 4 (exercises L, M, and N) and 5 (exercises R and S):

1. Choose one use case and, for each scenario, create a sequence diagram.

2. Create a communication diagram for each scenario of the use case chosen in Question 1.

3. Create a behavioral state machine to depict one of the classes on the class diagram you created for Chapter 5, exercise R.

4. Perform a CRUDE analysis to show the interactivity of the objects in the system.

5. Perform a verification and validation walkthrough of the problem.

F. For the gym membership problem in Chapters 4 (exercises O, P, and Q) and 5 (exercises T and U):

1. Choose one use case and, for each scenario, create a sequence diagram.

2. Create a communication diagram for each scenario of the use case chosen in Question 1.

3. Create a behavioral state machine to depict one of the classes on the class diagram you created for Chapter 5, exercise T.

4. Perform a CRUDE analysis to show the interactivity of the objects in the system.

5. Perform a verification and validation walkthrough of the problem.

G. For the Picnics R Us problem in Chapters 4 (exercises R, S, and T) and 5 (exercises V and W):

1. Choose one use case and, for each scenario, create a sequence diagram.

2. Create a communication diagram for each scenario of the use case chosen in question 1.

3. Create a behavioral state machine to depict one of the classes on the class diagram you created for Chapter 5, exercise V.

4. Perform a CRUDE analysis to show the interactivity of the objects in the system.

5. Perform a verification and validation walkthrough of the problem.

H. For the Of-the-Month-Club problem in Chapters 4 (exercises U, V, and W) and 5 (exercises X and Y):

1. Choose one use case and, for each scenario, create a sequence diagram.

2. Create a communication diagram for each scenario of the use case chosen in Question 1.

3. Create a behavioral state machine to depict one of the classes on the class diagram you created for Chapter 5, exercise X.

4. Perform a CRUDE analysis to show the interactivity of the objects in the system.

5. Perform a verification and validation walkthrough of the problem.

MINICASES

1. Refer to the functional model (use-case diagram, activity diagrams, and use-case descriptions) you prepared for the Professional and Scientific Staff Management (PSSM) minicase in Chapter 4. Based on your performance, PSSM was so satisfied that they wanted you to develop both the structural and behavioral models so that they could more fully understand both the interaction that would take place between the users and the system and the system itself in greater detail.

a. Create both CRC cards and a class diagram based on the functional models created in Chapter 4.

b. Create a sequence and a communication diagram for each scenario of each use case identified in the functional model.

c. Create a behavioral state machine for each of the complex classes in the class diagram.

d. Perform a CRUDE analysis to show the interactivity of the objects in the system.

e. Perform a verification and validation walkthrough of each model: functional, structural, and behavioral.

2. Having learned from past bad experiences, Professor Takato, president of FRDI (see minicase 2 in Chapter 5),

decided to define very precise and clear rules by which a paper gets published in the conference proceedings. These rules are to be built into his new CMS and can be summarized as follows:

- Paper must be registered in the CMS with title, list of authors, and abstract;
- Paper content must be uploaded into the CMS;
- A paper can be uploaded several times before the review process starts, but uploads are not allowed during the review process;
- Paper must have been reviewed by three independent reviewers and received at least two favorable reviews for acceptance;
- If accepted, the final version must be uploaded;
- At least one of the authors needs to register for the conference and pay the conference fee.

In order to understand how the above rules change the state of the Paper instance objects over time, Professor Takato decided to build a Behavioral State Machine for an instance of the Paper class. Soon, he realized that this task is not exactly trivial, and that there are a couple of key aspects that need more attention. Help Professor Takato build his Behavioral State Machine for a typical Paper object. Point out and comment on the most interesting aspects of the resultant Behavioral State Machine.

PART TWO

DESIGN

Whereas analysis modeling concentrated on the functional requirements of the evolving system, design modeling incorporates the nonfunctional requirements. That is, design modeling focuses on *how* the system will operate. First, the project team verifies and validates the analysis models (functional, structural, and behavioral). Next, a set of factored and partitioned analysis models are created. The class and method designs are illustrated using the class specifications (using CRC cards and class diagrams), contracts, and method specifications. Next, the data management layer is addressed by designing the actual database or file structure to be used for object persistence, and a set of classes that will map the class specifications into the object persistence format chosen. Concurrently, the team produces the user interface layer design using use scenarios, windows navigation diagrams, real use cases, interface templates, storyboards, Windows layout diagrams, and user interface prototypes. The physical architecture layer design is created using deployment diagrams and hardware/software specifications. This collection of deliverables represents the system specification that is handed to the programming team for implementation.

CHAPTER 7
SYSTEM DESIGN

Factored/Partitioned
  Functional Models
Factored/Partitioned
  Structural Models
Factored/Partitioned
  Behavioral Models
Package Diagrams
Alternative Matrix

CHAPTER 8
CLASS AND METHOD DESIGN

Class Specifications
Contracts
Method Specifications

CHAPTER 9
DATA BASE DESIGN

Object Persistence Design
Data Access & Manipulation
  Class Design

CHAPTER 10
USER INTERFACE DESIGN

Use Scenarios
Windows Navigation diagrams
Real Use Cases
Interface Templates
Storyboards
Windows Layout diagrams
User Interface Prototypes

CHAPTER 11
ARCHITECTURE

Deployment Diagrams
Hardware/software
  Specifications

Factored/Partitioned Functional Models

Factored/Partitioned Structural Models

Factored/Partitioned Behavioral Models

Package Diagrams

Alternative Matrix

Class Specifications

Contracts

Method Specifications

Object Persistence Design

Data Access & Manipulation Class

User Interface Design

Deployment Diagrams

Hardware/Software Specifications

CHAPTER 7

SYSTEM DESIGN

Object-oriented system development uses the requirements that were gathered during analysis to create a blueprint for the future system. A successful object-oriented design builds upon what was learned in earlier phases and leads to a smooth implementation by creating a clear, accurate plan of what needs to be done. This chapter describes the initial transition from analysis to design and presents three ways to approach the design for the new system.

OBJECTIVES

- Understand the verification and validation of the analysis models
- Understand the transition from analysis to design
- Understand the use of factoring, partitions, and layers
- Be able to create package diagrams
- Be familiar with the custom, packaged, and outsource design alternatives
- Be able to create an alternative matrix.

CHAPTER OUTLINE

INTRODUCTION

The purpose of analysis is to figure out what the business needs are. The purpose of design is to decide how to build the system. The major activity that takes place during *design* is evolving the set of analysis representations into design representations.

Throughout design, the project team carefully considers the new system with respect to the current environment and systems that exist within the organization as a whole. Major considerations of the *how* of a system are environmental factors, such as integrating with existing systems, converting data from legacy systems, and leveraging skills that exist in-house. Although the planning and analysis are undertaken to develop a possible system, the goal of design is to create a blueprint for a system that makes sense to implement.

An important initial part of design is to examine several design strategies and decide which will be used to build the system. Systems can be built from scratch, purchased and customized, or outsourced to others, and the project team needs to investigate the viability of each alternative. This decision influences the tasks that are to be accomplished during design.

At the same time, detailed design of the individual classes and methods that are used to map out the nuts and bolts of the system and how they are to be stored must still be completed. Techniques such as CRC cards, class diagrams, contract specification, method specification, and database design provide the final design details in preparation for the implementation phase, and they ensure that programmers have sufficient information to build the right system efficiently. These topics are covered in Chapters 8 and 9.

Design also includes activities such as designing the user interface, system inputs, and system outputs, which involve the ways that the user interacts with the system. Chapter 10 describes these three activities in detail, along with techniques such as storyboarding and prototyping, which help the project team design a system that meets the needs of its users and is satisfying to use.

Finally, physical architecture decisions are made regarding the hardware and software that will be purchased to support the new system and the way that the processing of the system will be organized. For example, the system can be organized so that its processing is centralized at one location, distributed, or both centralized and distributed, and each solution offers unique benefits and challenges to the project team. Because global issues and security influence the implementation plans that are made, they need to be considered along with the system's technical architecture. Physical architecture, security, and global issues are described in Chapter 11.

The many steps of design are highly interrelated and, as with the steps in analysis, the analysts often go back and forth among them. For example, prototyping in the interface design step often uncovers additional information that is needed in the system. Alternatively, a system that is being designed for an organization that has centralized systems might require substantial hardware and software investments if the project team decides to change to a system in which all the processing is distributed.

In this chapter, we overview the processes that are used to evolve the analysis models into design models. But before we move on into design, we really need to be sure that the current analysis models are consistent. Thus, we next discuss how to verify and validate the analysis models. Afterward, we describe different higher-level constructs that can be used to evolve the analysis models into design models. Then we introduce the use of packages and package diagrams as a means of representing the higher-level constructs used to evolve the models. Finally, we examine the three fundamental approaches to developing new systems: make, buy, or outsource.

PRACTICAL   Avoiding Classic Design

    TIP

*I*n Chapter 2, we discussed several classic mistakes and how to avoid them. Here, we summarize four classic mistakes in design and discuss how to avoid them.

1. *Reducing design time:* If time is short, there is a temptation to reduce the time spent in "unproductive" activities such as design, so that the team can jump into "productive" programming. This results in missing important details that have to be investigated later at a much higher cost in terms of time and money (usually at least ten times higher).

   Solution: If time pressure is intense, use timeboxing to eliminate functionality or move it into future versions.

2. *Feature creep:* Even if you are successful at avoiding scope creep, about 25 percent of system requirements will still change. And, changes—big and small—can significantly increase time and cost.

   Solution: Ensure that all changes are vital and that the users are aware of the impact on cost and time. Try to move proposed changes into future versions.

3. *Silver bullet syndrome:* Analysts sometimes believe the marketing claims for some design tools that claim to solve all problems and magically reduce time and costs. No *one* tool or technique can eliminate overall time or costs by more than 25 percent (although some can reduce individual steps by this much).

   Solution: If a design tool has claims that appear too good to be true, just say no.

4. *Switching tools midproject:* Sometimes analysts switch to what appears to be a better tool during design in the hopes of saving time or costs. Usually, any benefits are outweighed by the need to learn the new tool. This also applies even to minor upgrades to current tools.

   Solution: Don't switch or upgrade unless there is a compelling need for *specific* features in the new tool, and then explicitly *increase* the schedule to include learning time.

Adapted from Steve McConnell, *Rapid Development* (Redmond, WA: Microsoft Press, 1966).

VERIFYING AND VALIDATING THE ANALYSIS MODELS[1]

Before we evolve our analysis representations into design representations, we need to verify and validate the current set of analysis models to ensure that they faithfully represent the problem domain under consideration. This includes testing the fidelity of each model; for example, we must be sure that the activity diagram(s), use-case descriptions, and use-case diagrams all describe the same functional requirements. It also involves testing the fidelity between the models; for instance, transitions on a behavioral state machine are associated with operations contained in a class diagram. In Chapters 4, 5, and 6, we focused on verifying and validating the individual models: function, structural, and behavioral. In this chapter, we center our attention on ensuring that the different models are consistent. Figure 7-1 portrays the fact that the object-oriented analysis models are highly interrelated. For example, do the functional and structural models agree? What about the functional and behavioral models? And finally, are the structural and behavioral models trustworthy? In this section, we describe a set of rules that are useful to verify and validate the intersections of the analysis models. Depending on the specific constructs of each actual model, different interrelationships are relevant. The process of ensuring the consistency among them is known as *balancing the models.*

[1] The material in this section has been adapted from E. Yourdon, *Modern Structured Analysis* (Englewood Cliffs, NJ: Prentice Hall, 1989). Verifying and validating are a type of testing. We also describe testing in Chapter 12.

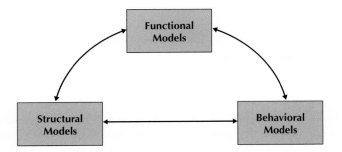

FIGURE 7-1
Object-Oriented
Analysis Models

Balancing Functional and Structural Models

To balance the functional and structural models, we must ensure that the two sets of models are consistent with each other. That is, the activity diagrams, use-case descriptions, and use-case diagrams must agree with the CRC cards and class diagrams that represent the evolving model of the problem domain. Figure 7-2 shows the interrelationships between the functional and structural models. By reviewing this figure, we uncover four sets of associations between the models. This gives us a place to begin balancing the functional and structural models.[2]

First, every class on a class diagram and every CRC card must be associated with at least one use case, and vice versa. For example, the CRC card portrayed in Figure 7-3 and its related class contained in the class diagram (see Figure 7-4) are associated with the Make Old Patient Appt use case described in Figure 7-5.

Second, every activity or action contained in an activity diagram (see Figure 7-6) and every event contained in a use-case description (see Figure 7-5) should be related to one or more responsibilities on a CRC card and one or more operations in a class on a class diagram, and vice versa. For example, the Get Patient Information activity on the example activity diagram (see Figure 7-6) and the first two events on the use-case description (see Figure 7-5) are associated with the make appointment responsibility on the CRC card (see Figure 7-3) and the makeAppointment() operation in the Patient class on the class diagram (see Figure 7-4).

Third, every object node on an activity diagram must be associated with an instance of a class on a class diagram (i.e., an object) and a CRC card or an attribute contained in a class and on a CRC card. However, in Figure 7-6, there is an object node, Appt Request Info, that does not seem to be related to any class in the class diagram portrayed in Figure 7-4. Thus, either the activity or class diagram is in error or the object node must represent an attribute. In this case, it does not seem to represent an attribute. We could add a class to the class diagram that creates temporary objects associated with the object node on the activity diagram. However, it is unclear what operations, if any, would be associated with these temporary objects. Therefore, a better solution would be to delete the Appt Request Info object nodes from the activity diagram. In reality, this object node represented only a set of bundled attribute values, that is, data that would be used in the appointment system process (see Figure 7-7).

Fourth, every attribute and association/aggregation relationships contained on a CRC card (and connected to a class on a class diagram) should be related to the subject or object of an event in a use-case description. For example, in Figure 7-5, the second event states: The Patient provides the Receptionist with his or her name and address. By reviewing the CRC

---

[2] Role-playing the CRC cards (see Chapter 5) also can be very useful in verifying and validating the relationships among the functional and structural models.

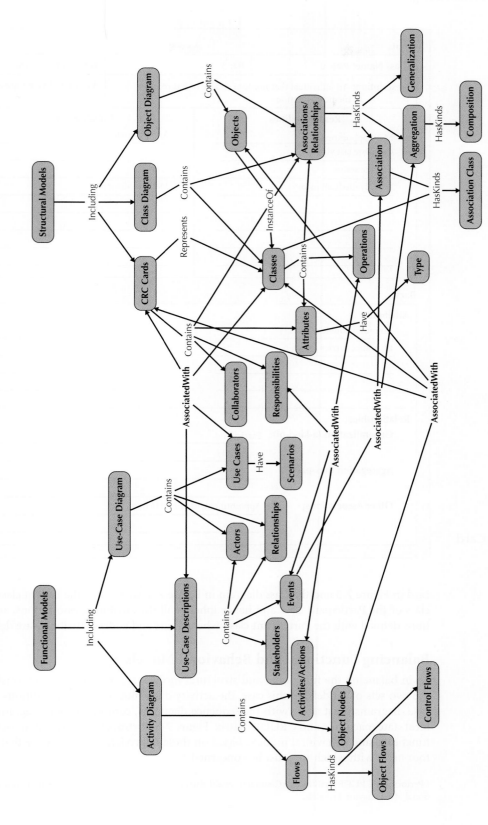

FIGURE 7-2 Relationships among Functional and Structural Models

Front:			
Class Name: Patient		**ID:** 3	**Type:** Concrete, Domain
Description: An individual that needs to recieve or has received medical attention			**Associated Use Cases:** 2

Responsibilities	Collaborators
Make appointment	Appointment
Calculate last visit	
Change status	
Provide medical history	Medical history

Back:

Attributes:

Amount (double)

Insurance carrier (text)

Relationships:

Generalization (a-kind-of): Participant

Aggregation (has-parts):

Other Associations: Appointment, Medical History

FIGURE 7-3
Old Patient CRC Card
(Figure 5-23)

card in Figure 7-3 and the class diagram in Figure 7-4, we see that the Patient class is a subclass of the Participant class and hence inherits all the attributes, associations, and operations defined with the Participant class, where name and address attributes are defined.

Balancing Functional and Behavioral Models

As in balancing the functional and structural models, we must ensure the consistency of the two sets of models. In this case, the activity diagrams, use-case descriptions, and use-case diagrams must agree with the sequence diagrams, communication diagrams, behavioral state machines, and CRUDE matrix. Figure 7-8 portrays the relationships between the functional and behavioral models. Based on these interrelationships, we see that there are four areas with which we must be concerned.[3]

[3] Performing CRUDE analysis (see Chapter 6) could also be useful in reviewing the intersections among the functional and behavioral models.

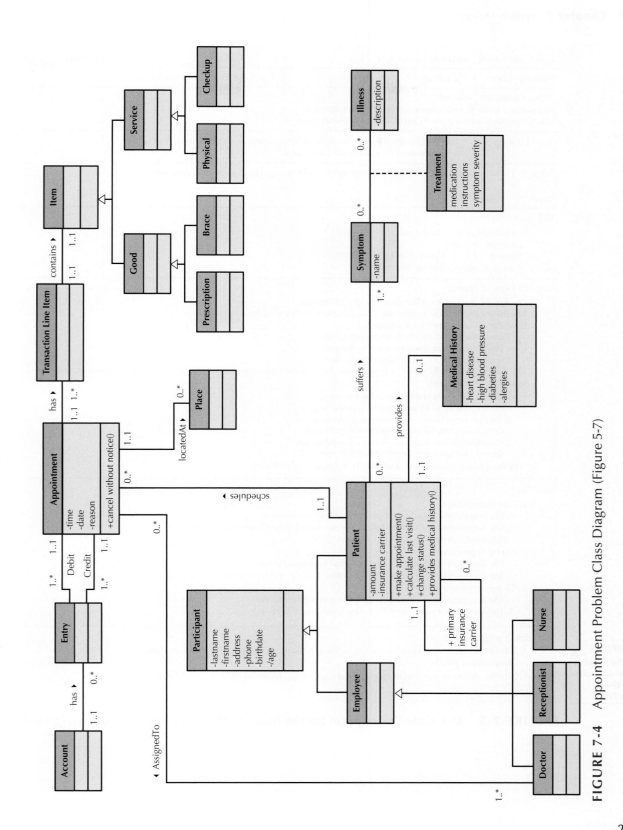

FIGURE 7-4 Appointment Problem Class Diagram (Figure 5-7)

Use Case Name: Make Old Patient Appt		ID: 2	Importance Level: High
Primary Actor: Old Patient		Use Case Type: Detail, Essential	

Stakeholders and Interests:
Old patient - wants to make, change, or cancel an appointment
Doctor - wants to ensure patient's needs are met in a timely manner

Brief Description: This use case describes how we make an appointment as well as changing or canceling
an appointment for a previously seen patient.

Trigger: Patient calls and asks for a new appointment or asks to cancel or change an existing appointment

Type: External

Relationships:
   Association: Old Patient
   Include:
   Extend: Update Patient Information
   Generalization: Manage Appointments

Normal Flow of Events:
   1. The Patient contacts the office regarding an appointment.
   2. The Patient provides the Receptionist with his or her name and address.
   3. If the Patient's information has changed
         Execute the Update Patient Information use case.
   4. If the Patient's payment arrangements has changed
         Execute the Make Payments Arrangements use case.
   5. The Receptionist asks Patient if he or she would like to make a new appointment, cancel an existing appointment, or change
      an existing appointment.

      If the patient wants to make a new appointment,
         the S-1: new appointment subflow is performed.
      If the patient wants to cancel an existing appointment,
         the S-2: cancel appointment subflow is performed.
      If the patient wants to change an existing appointment,
         the S-3: change appointment subflow is performed.
   6. The Receptionist provides the results of the transaction to the Patient.

SubFlows:
   S-1: New Appointment
         1. The Receptionist asks the Patient for possible appointment times.
         2. The Receptionist matches the Patient's desired appointment times with available dates and
            times and schedules the new appointment.
   S-2: Cancel Appointment
         1. The Receptionist asks the Patient for the old appointment time.
         2. The Receptionist finds the current appointment in the appointment file and cancels it.
   S-3: Change Appointment
         1. The Receptionist performs the S-2: cancel appointment subflow.
         2. The Receptionist performs the S-1: new appointment subflow.

Alternate/Exceptional Flows:
   S-1, 2a1: The Receptionist proposes some alternative appointment times based on what is available in the
         appointment schedule.
   S-1, 2a2: The Patient chooses one of the proposed times or decides not to make an appointment.

TEMPLATE
can be found at
www.wiley.com
/go/global/
tegarden

FIGURE 7-5   Use-Case Description for the Make Old Patient Appt Use Case (Figure 4-11)

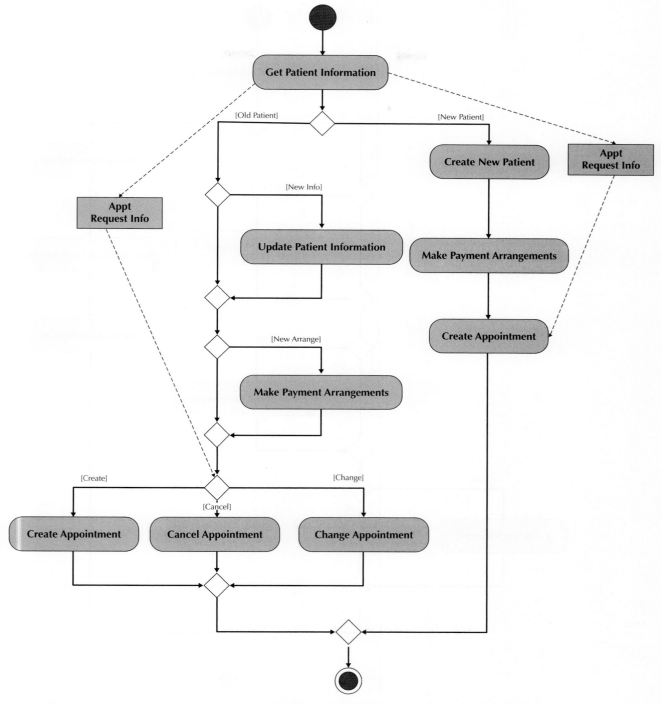

FIGURE 7-6 Activity Diagram for the Manage Appointments Use Case (Figure 4-7)

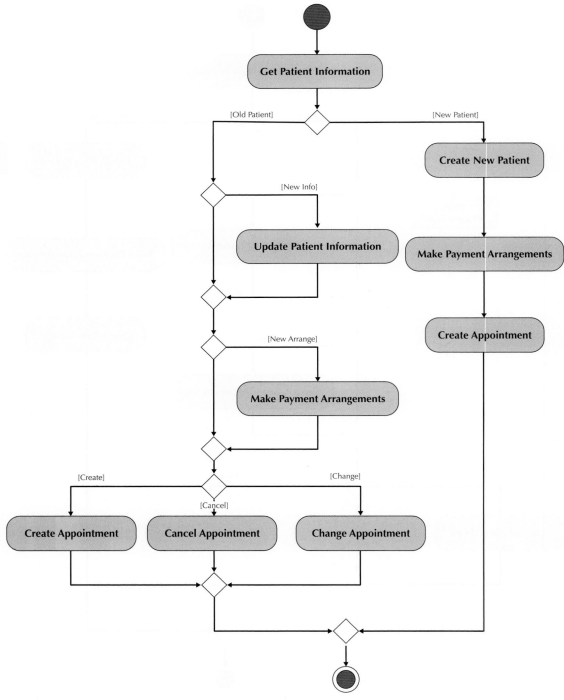

FIGURE 7-7 Corrected Activity Diagram for the Manage Appointments Use Case

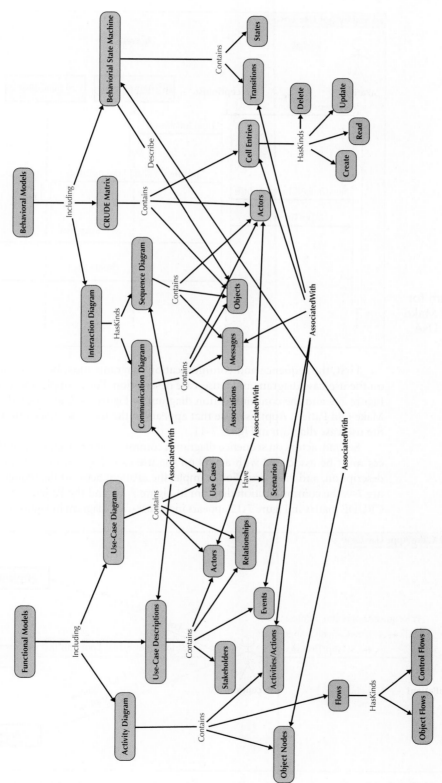

FIGURE 7-8   Relationships between Functional and Behavioral Models

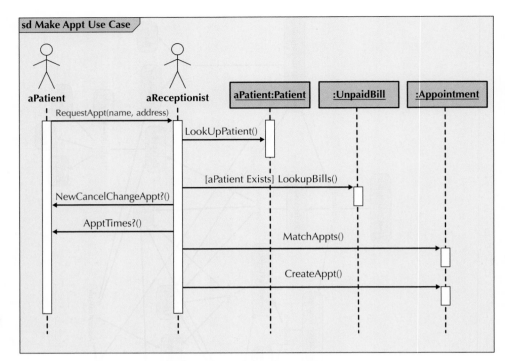

FIGURE 7-9
Sequence Diagram for a Scenario of the Make Old Patient Appt Use Case (Figure 6-1)

First, the sequence and communication diagrams must be associated with a use case on the use-case diagram and a use-case description. For example, the sequence diagram in Figure 7-9 and the communication diagram in Figure 7-10 are related to scenarios of the Make Old Patient Appt use case that appears in the use-case description in Figure 7-5 and the use-case diagram in Figure 7-11.

Second, actors on sequence diagrams, communication diagrams, and/or CRUDE matrices must be associated with actors on the use-case diagram or referenced in the use-case description, and vice versa. For example, the aPatient actor in the sequence diagram in Figure 7-9, the communication diagram in Figure 7-10, and the Patient row and column in the CRUDE matrix in Figure 7-12 appears in the use-case diagram in Figure 7-11 and the use-case

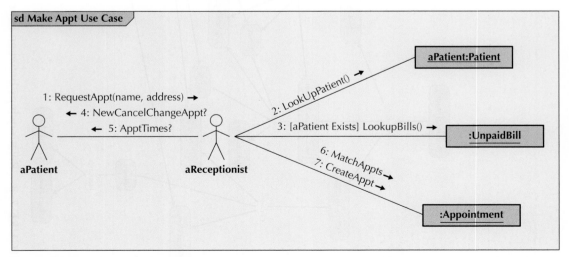

FIGURE 7-10 Communication Diagram for a Scenario of the Make Old Patient Appt Use Case (Figure 6-9)

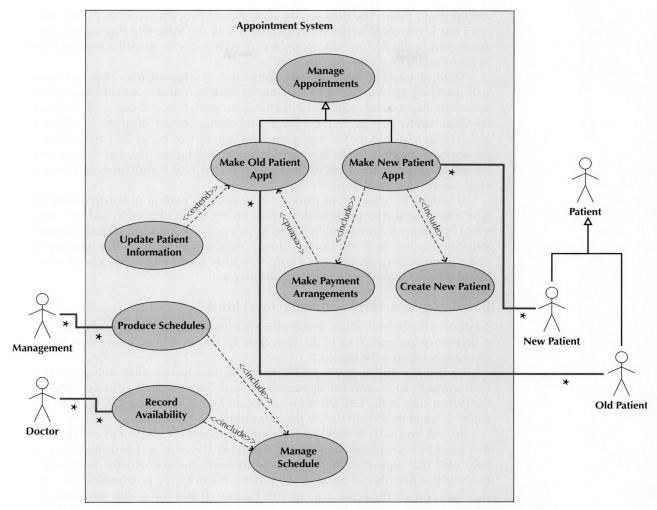

FIGURE 7-11   Modified Use-Case Diagram for the Appointment System (Figure 4-15)

	Receptionist	PatientList	Patient	UnpaidBills	Appointments	Appointment
Receptionist		RU	CRUD	R	RU	CRUD
PatientList			R			
Patient						
UnpaidBills						
Appointments						R
Appointment						

FIGURE 7-12   CRUDE Matrix for the Make Old Patient Apt Use Case (Figure 6-22)

description in Figure 7-5. However, the aReceptionist does not appear in the use-case diagram but is referenced in the events associated with the Make Old Patient Appt use-case description. In this case, the aReceptionist actor is obviously an internal actor, which cannot be portrayed on UML's use-case diagram.

Third, messages on sequence and communication diagrams, transitions on behavioral state machines, and entries in a CRUDE matrix must be related to activities and actions on an activity diagram and events listed in a use-case description, and vice versa. For example, the CreateAppt() message on the sequence and communication diagrams (see Figures 7-9 and 7-10) is related to the CreateAppointment activity (see Figure 7-7) and the S-1: New Appointment subflow on the use-case description (see Figure 7-5). The C entry into the Receptionist/Appointment cell of the CRUDE matrix is also associated with these messages, activity, and subflow.

Fourth, all complex objects represented by an object node in an activity diagram must have a behavioral state machine that represents the object's life cycle, and vice versa. As stated in Chapter 6, complex objects tend to be very dynamic and pass through a variety of states during their lifetimes. However, in this case, because we no longer have any object nodes in the activity diagram (see Figure 7-7), there is no necessity for a behavioral state machine to be created based on the activity diagram.

Balancing Structural and Behavioral Models

To discover the relationships between the structural and behavioral models, we use the concept map in Figure 7-13. In this case, there are five areas in which we must ensure the consistency between the models.[4]

First, objects that appear in a CRUDE matrix must be associated with classes that are represented by CRC cards and appear on the class diagram, and vice versa. For example, the Patient class in the CRUDE matrix in Figure 7-12 is associated with the CRC card in Figure 7-3, and the Patient class in the class diagram in Figure 7-4.

Second, because behavioral state machines represent the life cycle of complex objects, they must be associated with instances (objects) of classes on a class diagram and with a CRC card that represents the class of the instance. For example, the behavioral state machine that describes an instance of a Patient class in Figure 7-14 implies that a Patient class exists on a related class diagram (see Figure 7-4) and that a CRC card exists for the related class (see Figure 7-3).

Third, communication and sequence diagrams contain objects that must be an instantiation of a class that is represented by a CRC card and is located on a class diagram. For example, Figure 7-9 and Figure 7-10 have an anAppt object that is an instantiation of the Appointment class. Therefore, the Appointment class must exist in the class diagram (see Figure 7-4) and a CRC card should exist that describes it. However, there is an object on the communication and sequence diagrams associated with a class that did not exist on the class diagram: UnpaidBill. At this point, the analyst must decide to either modify the class diagram by adding these classes or rethink the communication and sequence diagrams. In this case, it is better to add the class to the class diagram (see Figure 7-15).

Fourth, messages contained on the sequence and communication diagrams, transitions on behavioral state machines, and cell entries on a CRUDE matrix must be associated with responsibilities and associations on CRC cards and operations in classes and associations connected to the classes on class diagrams. For example, the CreateAppt()

---

[4] Role-playing (see Chapter 5) and CRUDE analysis (see Chapter 6) also can be very useful in this undertaking.

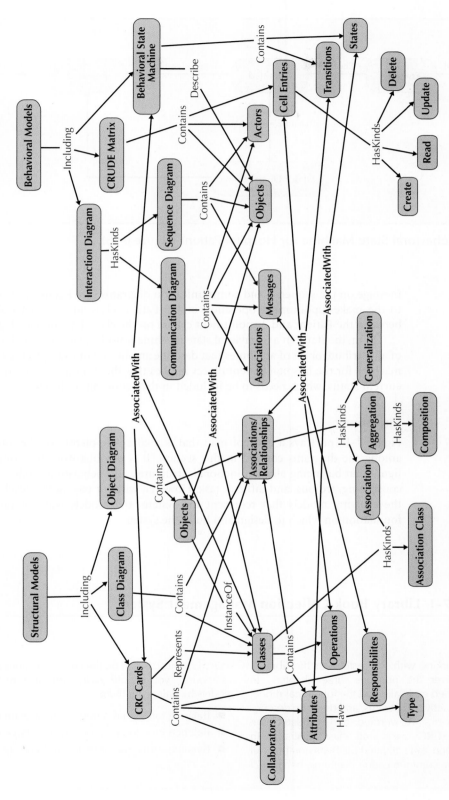

FIGURE 7-13 Relationships between Structural and Behavioral Models

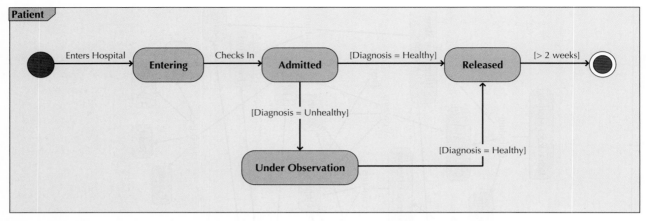

FIGURE 7-14 Behavioral State Machine for Hospital Patient (Figure 6-15)

message on the sequence and communication diagrams (see Figures 7-9 and 7-10) relate to the makeAppointment operation of the Patient class and the schedules association between the Patient and Appointment classes on the class diagram (see Figure 7-15).

Fifth, the states in a behavioral state machine must be associated with different values of an attribute or set of attributes that describe an object. For example, the behavioral state machine for the hospital patient object implies that there should be an attribute, possibly current status, which needs to be included in the definition of the class.

Summary

Figure 7-16 portrays a concept map that is a complete picture of the interrelationships among the diagrams covered in this section. It is obvious from the complexity of this figure that balancing all the functional, structural, and behavioral models is a very time-consuming, tedious, and difficult task. However, without paying this level of attention to the evolving models that represent the system, the models will not provide a sound foundation on which to design and build the system.

YOUR
TURN

7-1 Library Book Collection Management System

We have been working with the system for the library's book collection over the previous three chapters. In Chapter 4, we verified and validated the functional model (use-case diagram, activity diagrams, and use-case descriptions). In Chapter 5, we verified and validated the functional model (CRC cards and class diagram). In Chapter 6, we verified and validated the behavioral model (sequence diagrams, communication diagrams, behavioral state machines). Based on the current versions of the analysis models, you should verify and validate the analysis models by balancing them.

- Balance functional models with structural models.
- Balance functional models with behavioral models.
- Balance structural models with behavioral models.

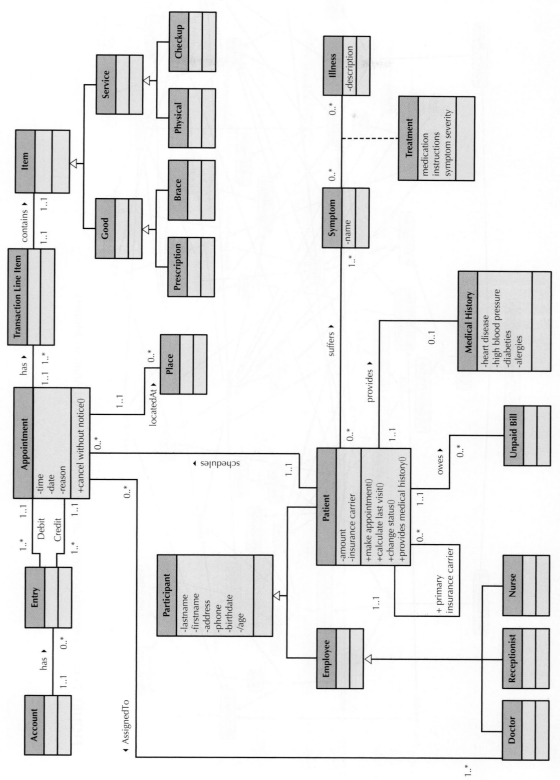

FIGURE 7-15 Corrected Appointment System Class Diagram

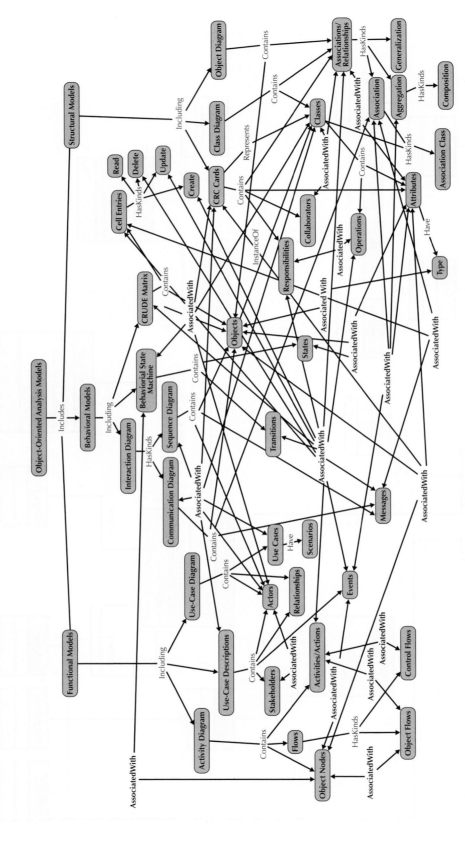

FIGURE 7-16 Interrelationships among Object-Oriented Analysis Models

EVOLVING THE ANALYSIS MODELS INTO DESIGN MODELS

Now that we have successfully verified and validated our analysis models, we need to begin evolving them into appropriate design models. The purpose of the analysis models was to represent the underlying business problem domain as a set of collaborating objects. In other words, the analysis activities defined the functional requirements. To achieve this, the analysis activities ignored nonfunctional requirements such as performance and the system environment issues (e.g., distributed or centralized processing, user-interface issues, and database issues). In contrast, the primary purpose of the design models is to increase the likelihood of successfully delivering a system that implements the functional requirements in a manner that is affordable and easily maintainable. Therefore, in systems design, we address both the functional and nonfunctional requirements.

From an object-oriented perspective, system design models simply refine the system analysis models by adding system environment (or solution domain) details to them and refining the problem domain information already contained in the analysis models. When evolving the analysis model into the design model, you should first carefully review the use cases and the current set of classes (their operations and attributes and the relationships between them). Are all the classes necessary? Are there any missing classes? Are the classes fully defined? Are any attributes or methods missing? Do the classes have any unnecessary attributes and methods? Is the current representation of the evolving system optimal? Obviously, if we have already verified and validated the analysis models, quite a bit of this has already taken place. Yet, object-oriented systems development is both incremental and iterative. Therefore, we must review the analysis models again. However, this time we begin looking at the models of the problem domain through a design lens. In this step, we make modifications to the problem domain models that will enhance the efficiency and effectiveness of the evolving system.

In the following sections, we introduce factoring, partitions and collaborations, and layers as a way to evolve problem domain-oriented analysis models into optimal solution domain-oriented design models. From an enhanced Unified Process perspective (see Figure 1-18), we are moving from the analysis workflow to the design workflow, and we are moving further into the Elaboration phase and partially into the Construction phase.

Factoring

Factoring is the process of separating out a *module* into a stand-alone module. The new module can be a new *class* or a new *method*. For example, when reviewing a set of classes, it may be discovered that they have a similar set of attributes and methods. Thus, it might make sense to factor out the similarities into a separate class. Depending on whether the new class should be in a superclass relationship to the existing classes or not, the new class can be related to the existing classes through a *generalization (a-kind-of)* or possibly through an *aggregation (has-parts)* relationship. Using the appointment system example, if the Employee class had not been identified, we could possibly identify it at this stage by factoring out the similar methods and attributes from the Nurse, Receptionist, and Doctor classes. In this case, we would relate the new class (Employee) to the existing classes using the generalization (a-kind-of) relationship. Obviously, by extension we also could have created the Participant class if it had not been previously identified.

Abstraction and *refinement* are two processes closely related to factoring. Abstraction deals with the creation of a higher-level idea from a set of ideas. Identifying the Employee class is an example of abstracting from a set of lower classes to a higher one. In some cases, the abstraction process identifies *abstract classes,* whereas in other situations, it identifies additional *concrete classes.*[5] The refinement process is the opposite of the abstraction process. In the appointment system example, we could identify additional subclasses of the Employee class, such as Secretary and Bookkeeper. Of course, we would add the new classes only if there were sufficient differences among them. Otherwise, the more general class, Employee, would suffice.

Partitions and Collaborations

Based on all the factoring, refining, and abstracting that can take place to the evolving system, the sheer size of the system representation can overload the user and the developer. At this point in the evolution of the system, it might make sense to split the representation into a set of *partitions*. A partition is the object-oriented equivalent of a subsystem,[6] where a subsystem is a decomposition of a larger system into its component systems (e.g., an accounting information system could be functionally decomposed into an accounts-payable system, an accounts-receivable system, a payroll system, etc.). From an object-oriented perspective, partitions are based on the pattern of activity (messages sent) among the objects in an object-oriented system. We describe an easy approach to model partitions and collaborations later in this chapter: packages and package diagrams.

A good place to look for potential partitions is the *collaborations* modeled in UML's communication diagrams (see Chapter 6). If you recall, one useful way to identify collaborations is to create a communication diagram for each use case. However, because an individual class can support multiple use cases, an individual class can participate in multiple use-case–based collaborations. In cases where classes are supporting multiple use cases, the collaborations should be merged. The class diagram should be reviewed to see how the different classes are related to one another. For example, if attributes of a class have complex object types, such as Person, Address, or Department, and these object types were not modeled as associations in the class diagram, we need to recognize these implied associations. Creating a diagram that combines the class diagram with the communication diagrams can be very useful to show to what degree the classes are coupled.[7] The greater the coupling between classes, the more likely the classes should be grouped together in a collaboration or partition. By looking at a CRUDE matrix, we can use CRUDE analysis (see Chapter 6) to identify potential classes on which to merge collaborations.

One of the easiest techniques to identify the classes that could be grouped to form a collaboration is through the use of cluster analysis or multiple dimensional scaling. These statistical techniques enable the team to objectively group classes together based on their affinity for each other. The affinity can be based on semantic relationships, different types of messages being sent between them (e.g., create, read, update, delete, or execute), or some weighted combination of both. There are many different similarity measures and many different algorithms on which the clusters can be based, so one must be careful when using these techniques. Always make sure that the collaborations identified using these techniques make sense from the problem domain perspective. Just because a mathematical algorithm suggests that the classes belong together does not make it so. However, this is a good approach to create a first-cut set of collaborations.

Depending on the complexity of the merged collaboration, it may be useful in decomposing the collaboration into multiple partitions. In this case, in addition to having collaborations

---

[5] See Chapter 5 for the differences between abstract and concrete classes.

[6] Some authors refer to partitions as subsystems [e.g., see R. Wirfs-Brock, B. Wilkerson, and L. Weiner, *Designing Object-Oriented Software* (Englewood Cliffs, NJ: Prentice Hall, 1990)], whereas others refer to them as layers [e.g., see I. Graham, *Migrating to Object Technology* (Reading, MA: Addison-Wesley, 1994)]. However, we have chosen to use the term *partition* [C. Larman, *Applying UML and Patterns: An Introduction to Object-Oriented Analysis and Design* (Englewood Cliffs, NJ: Prentice Hall, 1998)] to minimize confusion between subsystems in a traditional systems development approach and layers associated with Rational's Unified Approach.

[7] We describe the concept of coupling in Chapter 8.

YOUR

TURN

7-2 Campus Housing

You have been working with the system for the campus housing service over the previous three chapters. In Chapter 4, you were asked to create a set of use cases and to create a use-case diagram (Your Turn 4-2). In Chapter 5, you created a structural model (CRC cards and class diagram) for the same situation (Your Turn 5-2). In Chapter 6, you created a sequence diagram (Your Turn 6-1) and a communication diagram (Your Turn 6-2) for one of the use cases you had identified in Chapter 4. Finally, you also created a behavioral state machine for the apartment class in Chapter 6 (Your Turn 6-3).

Based on the current set of functional, structural, and behavioral models portrayed in these diagrams, apply the abstraction and refinement processes to identify additional abstract and concrete classes that would be useful to include in the evolving system.

between objects, it is possible to have collaborations among partitions. The general rule is the more messages sent between objects, the more likely the objects belong in the same partition. The fewer messages sent, the less likely the two objects belong together.

Another useful approach to identifying potential partitions is to model each collaboration between objects in terms of clients, servers, and contracts. A *client* is an instance of a *class* that sends a *message* to an instance of another class for a *method* to be executed; a *server* is the instance of a class that receives the message; and a *contract* is the specification that formalizes the interactions between the client and server objects (see Chapters 5 and 8). This approach allows the developer to build up potential partitions by looking at the contracts that have been specified between objects. In this case, the more contracts there are between objects, the more likely that the objects belong in the same partition. The fewer contracts, the less chance there is that the two classes belong in the same partition.

Remember, the primary purpose of identifying collaborations and partitions is to determine which classes should be grouped together in design.

Layers

Until this point in the development of our system, we have focused only on the problem domain; we have totally ignored the system environment (data management, user interface, and physical architecture). To successfully evolve the analysis model of the system into a design model of the system, we must add the system environment information. One useful way to do this, without overloading the developer, is to use *layers*. A layer represents an element of the software architecture of the evolving system. We have focused only on one layer in the evolving software architecture: the problem domain layer. There should be a layer for each of the different elements of the system environment (e.g., data management, user interface, physical architecture). Like partitions and collaborations, layers also can be portrayed using packages and package diagrams (see the next section of this chapter).

The idea of separating the different elements of the architecture into separate layers can be traced back to the MVC architecture of *Smalltalk*.[8] When Smalltalk was first created,[9] the authors decided to separate the application logic from the logic of the user interface. In this manner, it was possible to easily develop different user interfaces that worked with the same application. To accomplish this, they created the *Model–View–Controller*

---

[8] See S. Lewis, *The Art and Science of Smalltalk: An Introduction to Object-Oriented Programming Using Visual-Works* (Englewood Cliffs, NJ: Prentice Hall, 1995).

[9] Smalltalk was invented in the early 1970s by a software-development research team at Xerox PARC. It introduced many new ideas into the area of programming languages (e.g., object orientation, windows-based user interfaces, reusable class library, and the development environment). In many ways, Smalltalk is the parent of all object-based and object-oriented languages, such as Visual Basic, C++, and Java.

Layers	Examples	Relevant Chapters
Foundation	Date, Enumeration	7, 8
Problem Domain	Employee, Customer	4, 5, 6, 7, 8
Data Management	DataInputStream, FileInputStream	8, 9
Human–Computer Interaction	Button, Panel	8, 10
Physical Architecture	ServerSocket, URLConnection	8, 11

FIGURE 7-17
Layers and
Sample Classes

(MVC) architecture, where *Models* implemented the application logic (problem domain) and *Views* and *Controllers* implemented the logic for the user interface. Views handled the output and Controllers handled the input. Because graphical user interfaces were first developed in the Smalltalk language, the MVC architecture served as the foundation for virtually all graphical user interfaces that have been developed today (including the Mac interfaces, the Windows family, and the various Unix-based GUI environments).

Based on Smalltalk's innovative MVC architecture, many different software layers have been proposed.[10] Based on these proposals, we suggest the following layers on which to base software architecture: foundation, problem domain, data management, human–computer interaction, and physical architecture (see Figure 7-17). Each layer limits the types of classes that can exist on it (e.g., only user interface classes may exist on the human–computer interaction layer).

Foundation   The *foundation layer* is, in many ways, a very uninteresting layer. It contains classes that are necessary for any object-oriented application to exist. They include classes that represent fundamental data types (e.g., integers, real numbers, characters, strings), classes that represent fundamental data structures, sometimes referred to as *container classes* (e.g., lists, trees, graphs, sets, stacks, queues), and classes that represent useful abstractions, sometimes referred to as *utility classes* (e.g., date, time, money). Today, the classes found on this layer are typically included with the object-oriented development environments.

Problem Domain   The *problem-domain layer* is what we have focused our attention on up until now. At this stage of the development of our system, we need to further detail the classes so that we can implement them in an effective and efficient manner. Many issues need to be addressed when designing classes, no matter on which layer they appear. For example, there are issues related to factoring, cohesion and coupling, connascence, encapsulation, proper use of inheritance and polymorphism, constraints, contract specification, and detailed method design. These issues are discussed in Chapter 8.

Data Management   The *data management layer* addresses the issues involving the persistence of the objects contained in the system. The types of classes that appear in this layer deal with

[10] For example, Problem Domain, Human Interaction, Task Management, and Data Management (P. Coad and E. Yourdon, *Object-Oriented Design* [Englewood Cliffs, NJ: Yourdon Press, 1991)]; Domain, Application, and Interface (I. Graham, *Migrating to Object Technology* [Reading, MA: Addison-Wesley, 1994)]; Domain, Service, and Presentation [C. Larman, *Applying UML and Patterns: An Introduction to Object-Oriented Analysis and Design* (Englewood Cliffs, NJ: Prentice Hall, 1998)]; Business, View, and Access [A. Bahrami, *Object-Oriented Systems Development using the Unified Modeling Language* (New York: McGraw-Hill, 1999)]; Application-Specific, Application-General, Middleware, System-Software [I. Jacobson, G. Booch, and J. Rumbaugh, *The Unified Software Development Process* (Reading, MA: Addison-Wesley, 1999)]; and Foundation, Architecture, Business, and Application [M. Page-Jones, *Fundamentals of Object-Oriented Design in UML* (Reading, MA: Addison-Wesley, 2000)].

how objects can be stored and retrieved. The classes contained in this layer allow the problem domain classes to be independent of the storage used and, hence, increase the portability of the evolving system. Some of the issues related to this layer include choice of the storage format and optimization. In today's world, there is a plethora of different options in which to choose to store objects. These include sequential files, random access files, relational databases, object/relational databases, object-oriented databases, and NoSQL data stores. Each of these options has been optimized to provide solutions for different access and storage problems. Today, from a practical perspective, there is no single solution that optimally serves all applications. The correct solution is most likely some combination of the different storage options. A complete description of all the issues related to the *data management layer* is well beyond the scope of this book.[11] However, we do present the fundamentals in Chapter 9.

Human–Computer Interaction   The *human–computer interaction layer* contains classes associated with the View and Controller idea from Smalltalk. The primary purpose of this layer is to keep the specific user-interface implementation separate from the problem domain classes. This increases the portability of the evolving system. Typical classes found on this layer include classes that can be used to represent buttons, windows, text fields, scroll bars, check boxes, drop-down lists, and many other classes that represent user-interface elements.

When it comes to designing the user interface for an application, many issues must be addressed: How important is consistency across different user interfaces? What about differing levels of user experience? How is the user expected to be able to navigate through the system? What about help systems and online manuals? What types of input elements should be included (e.g., text box, radio buttons, check boxes, sliders, drop-down list boxes)? What types of output elements should be included (e.g., text, tables, graphs)? Other questions that must be addressed are related to the platform on which the software will be deployed. For example, is the application going to run on a stand-alone computer, is it going to be distributed, or is the application going mobile? If it is expected to run on mobile devices, what type of platform: notebooks, tablets, or phones? Will it be deployed using web technology, which runs on multiple devices, or will it be created using apps that are based on Android from Google, Blackberry OS from RIM, iOS from Apple, or Windows Phone from Microsoft? Depending on the answer to these questions, different types of user interfaces are possible.

With the advent of social networking platforms, such as Facebook, Twitter, blogs, YouTube, LinkedIn, and Second Life, the implications for the user interface can be mind boggling. Depending on the application, different social networking platforms may be appropriate for different aspects of the application. Furthermore, each of the different social networking platforms enables (or prevents) consideration of different types of user interfaces. Finally, with the potential audience of your application being global, many different cultural issues will arise in the design and development of culturally aware user interfaces (such as multilingual requirements). Obviously, a complete description of all the issues related to human–computer interaction is beyond the scope of this book.[12]

---

[11] There are many good database design books that are relevant to this layer; see, for example, M. Gillenson, *Fundamentals of Database Management Systems* (Hoboken, NJ: John Wiley & Sons, 2005); F. R. McFadden, J. A. Hoffer, Mary B. Prescott, *Modern Database Management*, 4th ed. (Reading, MA: Addison-Wesley, 1998); M. Blaha and W. Premerlani, *Object-Oriented Modeling and Design for Database Applications* (Englewood Cliffs, NJ: Prentice Hall, 1998); and R. J. Muller, *Database Design for Smarties: Using UML for Data Modeling* (San Francisco: Morgan Kaufmann, 1999).

[12] Books on user interface design that address these issues include B. Schheiderman, *Designing the User Interface: Strategies for Effective Human Computer Interaction*, 3rd ed. (Reading, MA: Addison-Wesley, 1998); J. Tidwell, *Designing Interfaces: Patterns for Effective Interaction Design,* 2nd ed. (Sebastopol, CA: O'Reilly Media, 2010); S. Krug, *Don't Make Me Think: A Common Sense Approach to Web Usability* (Berkeley, CA: New Riders Publishing, 2006); and N. Singh and A. Pereira, *The Culturally Customized Web Site: Customizing Web Sites for the Global Marketplace* (Oxford, UK: Elsevier, 2005).

However, from the user's perspective, the user interface is the system. We present the basic issues in user interface design in Chapter 10.

Physical Architecture   The *physical architecture layer* addresses how the software will execute on specific computers and networks. This layer includes classes that deal with communication between the software and the computer's operating system and the network. For example, classes that address how to interact with the various ports on a specific computer are included in this layer.

Unlike in the foundation layer, many design issues must be addressed before choosing the appropriate set of classes for this layer. These design issues include the choice of a computing or network architecture (such as the various client-server architectures), the actual design of a network, hardware and server software specification, and security issues. Other issues that must be addressed with the design of this layer include computer hardware and software configuration (choice of operating systems such as Linux, Mac OS, and Windows, processor types and speeds, amount of memory, data storage, and input/output technology), standardization, virtualization, grid computing, distributed computing, and web services. This then leads us to one of the proverbial gorillas on the corner. What do you do with the cloud? The *cloud* is essentially a form of distributed computing. In this case, the cloud allows you to treat the platform, infrastructure, software, and even business processes as remote services that can be managed by another firm. In many ways, the cloud allows much of IT to be outsourced (see the discussion of outsourcing later in this chapter). Also, as brought up with the human–computer interaction layer, the whole issue of mobile computing is very relevant to this layer. In particular, the different devices, such as phones and tablets, are relevant and the way they will communicate with each other, such as through cellular networks or Wi-Fi, is also important.

Finally, given the amount of power that IT requires today, the whole topic of Green IT must be addressed. Topics that need to be addressed related to Green IT are the location of the data center, data center cooling, alternative power sources, reduction of consumables, the idea of a paperless office, Energy Star compliance, and the potential impact of virtualization, the cloud, and mobile computing. Like the data management and human–computer interaction layers, a complete description of all the issues related to the physical architecture is beyond the scope of this book.[13] However, we do present the basic issues in Chapter 11.

PACKAGES AND PACKAGE DIAGRAMS

In UML, collaborations, partitions, and layers can be represented by a higher-level construct: a package.[14] In fact, a package serves the same purpose as a folder on your computer. When packages are used in programming languages such as Java, packages are actually implemented as folders. A *package* is a general construct that can be applied to any of the elements in UML models. In Chapter 4, we introduced the idea of packages as a way to group use cases together to make the use-case diagrams easier to read and to keep the models at a reasonable level of complexity. In Chapters 5 and 6, we did the same thing for class and communication diagrams, respectively. In this section, we describe a *package diagram*: a diagram composed only of packages. A package diagram is effectively a class diagram that only shows packages.

The symbol for a package is similar to a tabbed folder (see Figure 7-18). Depending on where a package is used, packages can participate in different types of relationships. For

---

[13] Some books that cover these topics include S.D. Burd, *Systems Architecture*, 6th ed. (Boston: Course Technology, 2011); I. Englander, *The Architecture of Computer Hardware, Systems Software, & Networking: An Information Technology Approach* (Hoboken, NJ: Wiley, 2009); and K.K. Hausman and S. Cook, *IT Architecture for Dummies* (Hoboken, NJ: Wiley Publishing, 2011).

[14] This discussion is based on material in Chapter 7 of M. Fowler with K. Scott, *UML Distilled: A Brief Guide to the Standard Object Modeling Language,* 3rd ed. (Reading, MA: Addison-Wesley, 2004).

Packages and Package Diagrams 297

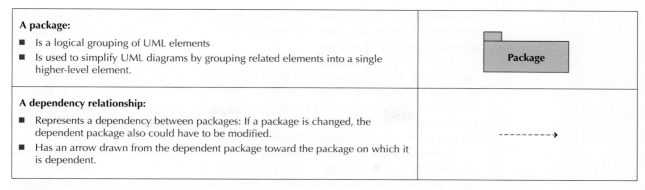

A package: ■ Is a logical grouping of UML elements ■ Is used to simplify UML diagrams by grouping related elements into a single higher-level element.	
A dependency relationship: ■ Represents a dependency between packages: If a package is changed, the dependent package also could have to be modified. ■ Has an arrow drawn from the dependent package toward the package on which it is dependent.	

FIGURE 7-18   Syntax for Package Diagram

example, in a class diagram, packages represent groupings of classes. Therefore, aggregation and association relationships are possible.

In a package diagram, it is useful to depict a new relationship, the *dependency relationship*. A dependency relationship is portrayed by a dashed arrow (see Figure 7-18). A dependency relationship represents the fact that a modification dependency exists between two packages. That is, it is possible that a change in one package could cause a change to be required in another package. Figure 7-19 portrays the dependencies among the different layers (foundation, problem domain, data management, human–computer interaction, and physical architecture). For example, if a change occurs in the problem domain layer, it most likely will cause changes to occur in the human–computer interaction, physical architecture, and data management layers. Notice that these layers point to the problem domain layer and therefore are dependent on it. However, the reverse is not true.[15]

At the class level, there could be many causes for dependencies among classes. For example, if the protocol for a method is changed, then this causes the interface for all objects of this class to change. Therefore, all classes that have objects that send messages to

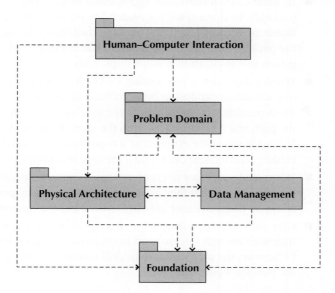

FIGURE 7-19

Package Diagram of Dependency Relationships among Layers

[15] A useful side effect of the dependencies among the layers is that the project manager can divide the project team up into separate subteams: one for each design layer. This is possible because each of the design layers is dependent on the problem domain layer, which has been the focus of analysis. In design, the team can gain some productivity-based efficiency by working on the different layer designs in parallel.

the instances of the modified class might have to be modified. Capturing dependency relationships among the classes and packages helps the organization in maintaining object-oriented information systems.

Collaborations, partitions, and layers are modeled as packages in UML. Collaborations are normally factored into a set of partitions, which are typically placed on a layer. Partitions can be composed of other partitions. Also, it is possible to have classes in partitions, which are contained in another partition, which is placed on a layer. All these groupings are represented using packages in UML. Remember that a package is simply a generic grouping construct used to simplify UML models through the use of composition.[16]

A simple package diagram, based on the appointment system example from the previous chapters, is shown in Figure 7-20. This diagram portrays only a very small portion of the entire system. In this case, we see that the Patient UI, Patient-DAM, and Patient Table classes depend on the Patient class. Furthermore, the Patient-DAM class depends on the Patient Table class. The same can be seen with the classes dealing with the actual appointments. By isolating the Problem Domain classes (such as the Patient and Appt classes) from the actual object-persistence classes (such as the Patient Table and Appt Table classes) through the use of the intermediate Data Management classes (Patient-DAM and Appt-DAM classes), we isolate the Problem Domain classes from the actual storage medium.[17] This greatly simplifies the maintenance and increases the reusability of the Problem Domain classes. Of course, in a complete description of a real system, there would be many more dependencies.

Guidelines for Creating Package Diagrams

As with the UML diagrams described in the earlier chapters, we provide a set of guidelines that we have adapted from Ambler to create package diagrams.[18] In this case, we offer six guidelines (see Figure 7-21).

- Use package diagrams to logically organize designs. Specifically, use packages to group classes together when there is an inheritance, aggregation, or composition relationship between them or when the classes form a collaboration.
- In some cases, inheritance, aggregation, or association relationships exist between packages. In those cases, for readability purposes, try to support inheritance relationships vertically, with the package containing the superclass being placed above the package containing the subclass. Use horizontal placement to support aggregation and association relationships, with the packages being placed side by side.
- When a dependency relationship exists on a diagram, it implies that there is at least one semantic relationship between elements of the two packages. The direction of the dependency is typically from the subclass to the superclass, from the whole to the part, and with contracts, from the client to the server. In other words, a subclass is dependent on the existence of a superclass, a whole is dependent upon its parts existing, and a client can't send a message to a nonexistent server.
- When using packages to group use cases together, be sure to include the actors and the associations that they have with the use cases grouped in the package. This will allow the diagram's user to better understand the context of the diagram.
- Give each package a simple but descriptive name, to provide the package diagram user with enough information to understand what the package encapsulates. Otherwise, the user will have to drill-down or open up the package to understand the package's purpose.

---

[16] For those familiar with traditional approaches, such as structured analysis and design, packages serve a similar purpose as the leveling and balancing processes used in data flow diagramming.

[17] These issues are described in more detail in Chapter 9.

[18] S.W. Ambler, *The Elements of UML 2.0 Style* (Cambridge, UK: Cambridge University Press, 2005).

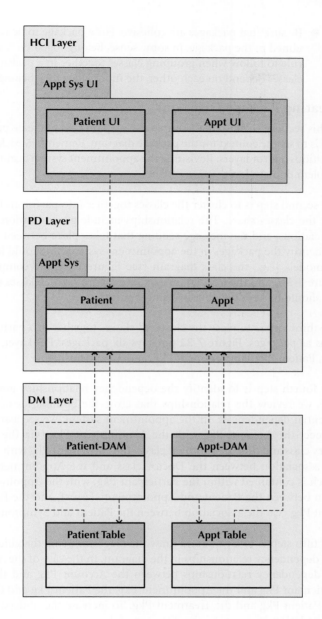

FIGURE 7-20
Partial Package
Diagram of the
Appointment System

- Create package diagrams to logically organize designs.
- Organize package diagrams based on semantic relationships. Use vertical positioning to support inheritance; use horizontal positioning to support aggregation and association.
- Dependency relationships between packages should reflect the existence of semantic relationships between elements of one package with elements of another package.
- In the case of use-case–based package diagrams, be sure to include the actors to portray use-case usage.
- Give packages simple but descriptive names.
- Make packages cohesive.

FIGURE 7-21
Guidelines for Creating
a Package Diagram

■ Be sure that packages are cohesive. For a package to be cohesive, the classes contained in the package, in some sense, belong together. A simple, but not perfect, rule to follow when grouping classes together in a package is that the more the classes depend on each other, the more likely they belong together in a package.

Creating Package Diagrams

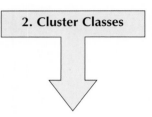

1. Set Context

In this section, we describe a simple five-step process to create package diagrams. The first step is to set the context for the package diagram. Remember, packages can be used to model partitions and/or layers. Revisiting the appointment system again, let's set the context as the problem domain layer.

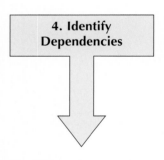

2. Cluster Classes

The second step is to cluster the classes together into partitions based on the relationships that the classes share. The relationships include generalization, aggregation, the various associations, and the message sending that takes place between the objects in the system. To identify the packages in the appointment system, we should look at the different analysis models [e.g., the class diagram (see Figure 6-2), the communication diagrams (see Figure 7-5)], and the CRUDE matrix (see Figure 7-14). Classes in a generalization hierarchy should be kept together in a single partition.

3. Create Packages

The third step is to place the clustered classes together in a partition and model the partitions as packages. Figure 7-22 portrays six packages: PD Layer, Account Pkg, Participant Pkg, Patient Pkg, Appointment Pkg, and Treatment Pkg.

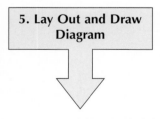

4. Identify Dependencies

The fourth step is to identify the dependency relationships among the packages. In this case, we review the relationships that cross the boundaries of the packages to uncover potential dependencies. In the appointment system, we see association relationships that connect the Account Pkg with the Appointment Pkg (via the associations between the Entry class and the Appointment class), the Participant Pkg with the Appointment Pkg (via the association between the Doctor class and the Appointment class), the Patient Pkg, which is contained within the Participant Pkg, with the Appointment Pkg (via the association between the Patient and Appointment classes), and the Patient Pkg with the Treatment Pkg (via the association between the Patient and Symptom classes).

5. Lay Out and Draw Diagram

The fifth step is to lay out and draw the diagram. Using the guidelines, place the packages and dependency relationships in the diagram. In the case of the Appointment system, there are dependency relationships between the Account Pkg and the Appointment Pkg, the Participant Pkg and the Appointment Pkg, the Patient Pkg and the Appointment Pkg, and the Patient Pkg and the Treatment Pkg. To increase the understandability of the dependency relationships among the different packages, a pure package diagram that shows only the dependency relationships among the packages can be created (see Figure 7-23).

YOUR TURN   **7-3 Campus Housing**

*B*ased on the factoring of the evolving system in "Your Turn 7-2", identify a set of partitions for the problem domain layer and model them in a package diagram.

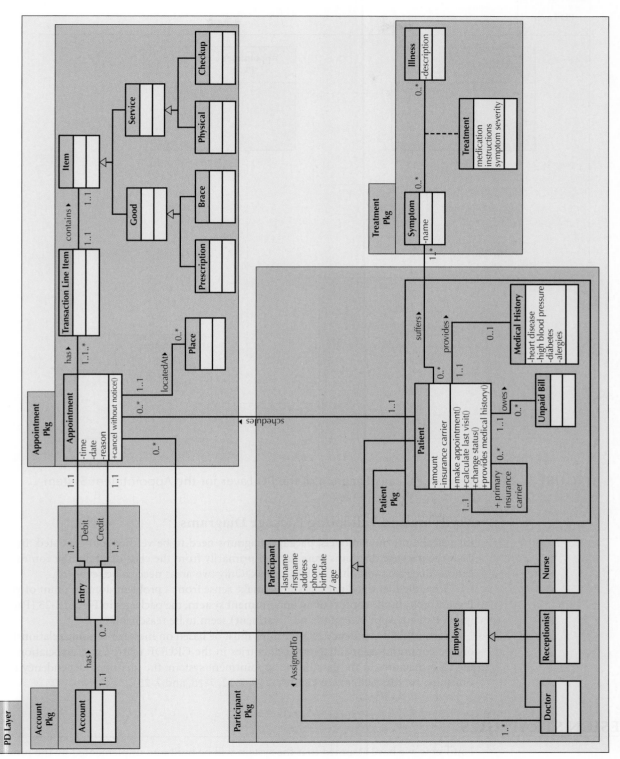

FIGURE 7-22 Package Diagram of the PD Layer of the Appointment Problem

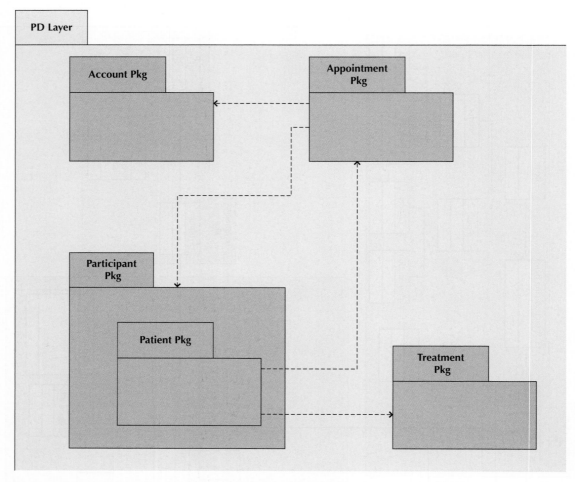

FIGURE 7-23 Overview Package Diagram of the PD Layer for the Appointment System

Verifying and Validating Package Diagrams

Like all the previous models, package diagrams need to be verified and validated. In this case, the package diagrams were derived primarily from the class diagram, the communications diagrams, and the CRUDE matrix. Only two areas need to be reviewed.

First, the identified packages must make sense from a problem domain point of view. For example, in the context of an appointment system, the packages in Figure 7-23 (Participant, Patient, Appt, Account, and Treatment) seem to be reasonable.

Second, all dependency relationships must be based on message-sending relationships on the communications diagram, cell entries in the CRUDE matrix, and associations on the class diagram. In the case of the appointment system, the identified dependency relationships are reasonable (see Figures 7-10, 7-12, 7-15, and 7-23).

DESIGN STRATEGIES

Until now, we have assumed that the system will be built and implemented by the project team; however, there are actually three ways to approach the creation of a new system: developing a custom application in-house, buying a packaged system and customizing it, and relying on an external vendor, developer, or service provider to build the system. Each of these choices has its strengths and weaknesses, and each is more appropriate in different scenarios. The following sections describe each design choice in turn, and then we present criteria that you can use to select one of the three approaches for your project.

Custom Development

Many project teams assume that *custom development,* or building a new system from scratch, is the best way to create a system. For one thing, teams have complete control over the way the system looks and functions. Custom development also allows developers to be flexible and creative in the way they solve business problems. Additionally, a custom application would be easier to change to include components that take advantage of current technologies that can support such strategic efforts.

Building a system in-house also builds technical skills and functional knowledge within the company. As developers work with business users, their understanding of the business grows and they become better able to align IS with strategies and needs. These same developers climb the technology learning curve so that future projects applying similar technology require much less effort.

Custom application development, however, also includes a dedicated effort that involves long hours and hard work. Many companies have a development staff that already is overcommitted to filling huge backlogs of systems requests, and they just do not have time for another project. Also, a variety of skills—technical, interpersonal, functional, project management, and modeling—all have to be in place for the project to move ahead smoothly. IS professionals, especially highly skilled individuals, are quite difficult to hire and retain.

The risks associated with building a system from the ground up can be quite high, and there is no guarantee that the project will succeed. Developers could be pulled away to work on other projects, technical obstacles could cause unexpected delays, and the business users could become impatient with a growing timeline.

CONCEPTS

IN ACTION

7-A Building a Custom System—with Some Help

I worked with a large financial institution in the southeast that suffered serious financial losses several years ago. A new CEO was brought in to change the strategy of the organization to being more "customer focused." The new direction was quite innovative, and it was determined that custom systems, including a data warehouse, would have to be built to support the new strategic efforts. The problem was that the company did not have the in-house skills for these kinds of custom projects.

The company now has one of the most successful data-warehouse implementations because of its willingness to use outside skills and because of its focus on project management. To supplement skills within the company, eight sets of external consultants, including hardware vendors, systems integrators, and business strategists, were hired to take part and transfer critical skills to internal employees. An in-house project manager coordinated the data-warehouse implementation full-time, and her primary goals were to set clear expectations

and define responsibilities and to communicate the interdependencies that existed among the team members.

This company shows that successful custom development can be achieved, even when the company does not start off with the right skills in-house. But this kind of project is not easy to pull off—it takes a talented project manager to keep the project moving along and to transition the skills to the right people over time.

—*Barbara Wixom*

Questions

1. What are the risks in building a custom system without the right technical expertise?
2. Why did the company select a project manager from within the organization?
3. Would it have been better to hire an external professional project manager to coordinate the project?

Packaged Software

Many business needs are not unique, and because it makes little sense to reinvent the wheel, many organizations buy *packaged software* that has already been written rather than developing their own custom solution. In fact, there are thousands of commercially available software programs that have already been written to serve a multitude of purposes. Think about your own need for a word processor—did you ever consider writing your own word processing software? That would be very silly considering the number of good software packages available that are relatively inexpensive.

Similarly, most companies have needs that can be met quite well by packaged software, such as payroll or accounts receivable. It can be much more efficient to buy programs that have already been created, tested, and proved, and a packaged system can be bought and installed in a relatively short time when compared with a custom system. Plus, packaged systems incorporate the expertise and experience of the vendor who created the software.

Packaged software can range from reusable components to small, single-function tools to huge, all-encompassing systems such as *enterprise resource planning (ERP)* applications that are installed to automate an entire business. Implementing ERP systems is a process in which large organizations spend millions of dollars installing packages by companies such as SAP, PeopleSoft, Oracle, and Baan and then change their businesses accordingly. Installing ERP software is much more difficult than installing small application packages because benefits can be harder to realize and problems are much more serious.

One problem is that companies buying packaged systems must accept the functionality that is provided by the system, and rarely is there a perfect fit. If the packaged system is large in scope, its implementation could mean a substantial change in the way the company does business. Letting technology drive the business can be a dangerous way to go.

Most packaged applications allow *customization,* or the manipulation of system parameters to change the way certain features work. For example, the package might have a way to accept information about your company or the company logo that would then appear on input screens. Or an accounting software package could offer a choice of various ways to handle cash flow or inventory control so that it can support the accounting practices in different organizations. If the amount of customization is not enough and the software package has a few features that don't quite work the way the company needs it to work, the project team can create workarounds.

A *workaround* is a custom-built add-on program that interfaces with the packaged application to handle special needs. It can be a nice way to create needed functionality that does not exist in the software package. But workarounds should be a last resort for several reasons. First, workarounds are not supported by the vendor who supplied the packaged software, so upgrades to the main system might make the workaround ineffective. Also, if problems arise, vendors have a tendency to blame the workaround as the culprit and refuse to provide support.

Although choosing a packaged software system is simpler than custom development, it, too, can benefit from following a formal methodology, just as if a custom application were being built.

Systems integration refers to the process of building new systems by combining packaged software, existing legacy systems, and new software written to integrate these. Many consulting firms specialize in systems integration, so it is not uncommon for companies to select the packaged software option and then outsource the integration of a variety of packages to a consulting firm. (Outsourcing is discussed in the next section).

The key challenge in systems integration is finding ways to integrate the data produced by the different packages and legacy systems. Integration often hinges on taking data produced by one package or system and reformatting it for use in another package or system. The project team starts by examining the data produced by and needed by the different packages or systems and identifying the transformations that must occur to move the data from one to the other. In many cases, this involves fooling the different packages or systems into thinking that the data were produced by an existing program module that the package or system expects to produce the data, rather than the new package or system that is being integrated.

A third approach is through the use of an *object wrapper*.[19] An object wrapper is essentially an object that "wraps around" a legacy system, enabling an object-oriented system to send messages to the legacy system. Effectively, object wrappers create an application program interface (API) to the legacy system. The creation of an object wrapper protects the corporation's investment in the legacy system.

Outsourcing

The design choice that requires the least amount of in-house resources is *outsourcing*—hiring an external vendor, developer, or service provider to create the system. Outsourcing has become quite popular in recent years. Some estimate that as many as 50 percent of companies with IT budgets of more than $5 million are currently outsourcing or evaluating the approach.

With outsourcing, the decision making and/or management control of a business function is transferred to an outside supplier. This transfer requires two-way coordination, exchange of information, and trust between the supplier and the business. From an IT perspective, IT outsourcing can include hiring consultants to solve a specific problem, hiring contract programmers to implement a solution, hiring a firm to manage the IT function and assets of a company, or actually outsourcing the entire IT function to a separate firm. Today, through the use of application service providers (ASPs), web services technology, and cloud services, it is possible to use a pay-as-you-go approach for a software package.[20] Essentially, IT outsourcing involves hiring a third party to perform some IT function that traditionally would be performed in-house.

There can be great benefit to having someone else develop a company's system. The outside company may be more experienced in the technology or have more resources, such as experienced programmers. Many companies embark upon outsourcing deals to reduce costs, whereas others see it as an opportunity to add value to the business.

For whatever reason, outsourcing can be a good alternative for a new system. However, it does not come without costs. If you decide to leave the creation of a new system in the hands of someone else, you could compromise confidential information or lose control over future development. In-house professionals are not benefiting from the skills that could be learned from the project; instead, the expertise is transferred to the outside organization. Ultimately, important skills can walk right out the door at the end of the contract. Furthermore, when offshore outsourcing is being considered, we must also be cognizant of language issues, time-zone differences, and cultural differences (for example, acceptable business practices as understood in one country that may be unacceptable in another). All these concerns, if not dealt with properly, can prevail over any advantage that outsourcing or offshore outsourcing could realize.

---

[19] Ian Graham, *Object-Oriented Methods: Principles & Practice*, 3rd ed. (Reading, MA: Addison-Wesley, 2001).

[20] For an economic explanation of how this could work, see H. Baetjer, *Software as Capital: An Economic Perspective on Software Engineering* (Los Alamitos, CA: IEEE Computer Society Press, 1997).

CONCEPTS

7-B Collecting Data for the Long Term

IN ACTION

*P*harmaceutical companies are generally heavily regulated. It can take years for a new drug to make it to the market because of time for the development phase, highly monitored testing, and final approval by the Food and Drug Administration (FDA). Once the drug is on the market, other companies can try to produce generic drugs that seem to be compatible with the name-brand drug.

Occasionally, an approved drug, some years into its life span, gets scrutiny for higher-than-expected side effects. For example, a drug that is effective in lowering cholesterol might also cause the side effect of an increased chance for cataract growth that was not discovered during the initial testing and approval cycle. Data are collected on all aspects of clinical trials and from the marketplace, but some relationships are just harder to find.

Questions

1. Is there a particular systems approach to being able to collect and analyze data from a mountain of data?

2. If you were building a strategic planning system for tracking a drug from proposal through development and testing and into the marketplace, how would you approach it?

3. What requirements might be necessary to build such a system?

Most risks can be addressed if a company decides to outsource, but two are particularly important. First, the company must thoroughly assess the requirements for the project—a company should never outsource what is not understood. If rigorous planning and analysis has occurred, then the company should be well aware of its needs. Second, the company should carefully choose a vendor, developer, or service with a proven track record with the type of system and technology that its system needs.

Three primary types of contracts can be drawn to control the outsourcing deal. A *time-and-arrangements contract* is very flexible because a company agrees to pay for whatever time and expenses are needed to get the job done. Of course, this agreement could result in a large bill that exceeds initial estimates. This works best when the company and the outsourcer are unclear about what it is going to take to finish the job.

A company will pay no more than expected with a *fixed-price contract* because if the outsourcer exceeds the agreed-upon price, it will have to absorb the costs. Outsourcers are much more careful about defining requirements clearly up front, and there is little flexibility for change.

The type of contract gaining in popularity is the *value-added contract,* whereby the outsourcer reaps some percentage of the completed system's benefits. The company has very little risk in this case, but it must expect to share the wealth once the system is in place.

Creating fair contracts is an art because flexibility must be carefully balanced with clearly defined terms. Often, needs change over time. Therefore, the contract should not be so specific and rigid that alterations cannot be made. Think about how quickly technology like the World Wide Web changes. It is difficult to foresee how a project might evolve over a long period of time. Short-term contracts help leave room for reassessment if needs change or if relationships are not working out the way both parties expected. In all cases, the relationship with the outsourcer should be viewed as a partnership where both parties benefit and communicate openly.

Managing the outsourcing relationship is a full-time job. Thus, someone needs to be assigned full time to manage the outsourcer, and the level of that person should be appropriate for the size of the job (a multimillion dollar outsourcing engagement should be handled by a high-level executive). Throughout the relationship, progress should be tracked

FIGURE 7-24
Outsourcing
Guidelines

Outsourcing
• Keep the lines of communication open between you and your outsourcer.
• Define and stabilize requirements before signing a contact.
• View the outsourcing relationship as a partnership.
• Select the vendor, developer, or service provider carefully.
• Assign a person to managing the relationship.
• Don't outsource what you don't understand.
• Emphasize flexible requirements, long-term relationships, and short-term contracts.

and measured against predetermined goals. If a company does embark upon an outsourcing design strategy, it should be sure to get adequate information. Many books have been written that provide much more detailed information on the topic.[21] Figure 7-24 summarizes some guidelines for outsourcing.

CONCEPTS
IN ACTION

7-C EDS Value-Added Contract

Value-added contracts can be quite rare—and very dramatic. They exist when a vendor is paid a percentage of revenue generated by the new system, which reduces the up-front fee, sometimes to zero. The City of Chicago and EDS (a large consulting and systems integration firm) agreed to reengineer the process by which the city collects the fines on 3.6 million parking tickets per year and thus signed a landmark deal of this type in 1992. At the time, because of clogged courts and administrative problems, the city collected on only about 25 percent of all tickets issued. It had a $60 million backlog of uncollected tickets.

Dallas-based EDS invested an estimated $25 million in consulting and new systems in exchange for the right to up to 26 percent of the uncollected fines, a base

processing fee for new tickets, and software rights. As of 1995, EDS had taken in well over $50 million on the deal, analysts say. The deal came under some fire from various quarters as an example of an organization giving away too much in a risk-reward-sharing deal. City officials, however, counter that the city has pulled in about $45 million in previously uncollected fines and has improved its collection rate to 65 percent with little up-front investment.

Source: Jeff Moad. "Outsourcing? Go out on the limb together." pp. 58–61, Vol. 41, No. 2. *Datamation,* February 1, 1995.

Questions

Do you think the City of Chicago got a good deal from this arrangement? Why or why not?

Selecting a Design Strategy

Each of the design strategies just discussed has its strengths and weaknesses, and no one strategy is inherently better than the others. Thus, it is important to understand the strengths and weaknesses of each strategy and when to use each. Figure 7-25 summarizes the characteristics of each strategy.

[21] For more information on outsourcing, we recommend M. Lacity and R. Hirschheim, *Information Systems Outsourcing: Myths, Metaphors, and Realities* (New York, NY: Wiley, 1993); L. Willcocks and G. Fitzgerald, *A Business Guide to Outsourcing Information Technology* (London: Business Intelligence, 1994); E. Carmel, *Offshoring Information Technology: Sourcing and Outsourcing to a Global Workforce* (Cambridge, England: Cambridge University Press, 2005); J. K Halvey and B. M. Melby, *Information Technology Outsourcing Transactions: Process, Strategies, and Contracts,* 2nd Ed. (Hoboken, NJ: Wiley, 2005); and T. L. Friedman, *The World is Flat: A Brief History of the Twenty-First Century, Updated and Expanded Edition* (New York: Farrar, Straus, and Giroux, 2006).

	Use Custom Development When…	Use a Packaged System When…	Use Outsourcing When…
Business Need	The business need is unique.	The business need is common.	The business need is not core to the business.
In-house Experience	In-house functional and technical experience exists.	In-house functional experience exists.	In-house functional or technical experience does not exist.
Project Skills	There is a desire to build in-house skills.	The skills are not strategic.	The decision to outsource is a strategic decision.
Project Management	The project has a highly skilled project manager and a proven methodology.	The project has a project manager who can coordinate the vendor's efforts.	The project has a highly skilled project manager at the level of the organization that matches the scope of the outsourcing deal.
Time frame	The time frame is flexible.	The time frame is short.	The time frame is short or flexible.

FIGURE 7-25   Selecting a Design Strategy

Business Need   If the business need for the system is common and technical solutions already exist that can meet the business need of the system, it makes little sense to build a custom application. Packaged systems are good alternatives for common business needs. A custom alternative must be explored when the business need is unique or has special requirements. Usually, if the business need is not critical to the company, then outsourcing is the best choice—someone outside of the organization can be responsible for the application development.

In-house Experience   If in-house experience exists for all the functional and technical needs of the system, it will be easier to build a custom application than if these skills do not exist. A packaged system may be a better alternative for companies that do not have the technical skills to build the desired system. For example, a project team that does not have web commerce technology skills might want to acquire a web commerce package that can be installed without many changes. Outsourcing is a good way to bring in outside experience that is missing in-house so that skilled people are in charge of building a system.

Project Skills   The skills that are applied during projects are either technical (e.g., Java, SQL) or functional (e.g., electronic commerce), and different design alternatives are more viable, depending on how important the skills are to the company's strategy. For example, if certain functional and technical expertise that relate to Internet sales applications and web commerce application development are important to an organization because it expects the Internet to play an important role in its sales over time, then it makes sense for the company to develop web commerce applications in-house, using company employees, so that the skills can be developed and improved. On the other hand, some skills, such as network security, may be beyond the technical expertise of employees or not of interest to the company's strategists—it is just an operational issue that needs to be addressed. In this case, packaged systems or outsourcing should be considered so internal employees can focus on other business-critical applications and skills.

Project Management   Custom applications require excellent project management and a proven methodology. So many things, such as funding obstacles, staffing holdups, and overly demanding business users, can push a project off-track. Therefore, the project team should choose to develop a custom application only if it is certain that the underlying coordination and control mechanisms will be in place. Packaged and outsourcing alternatives also need to be managed; however, they are more shielded from internal obstacles because the external parties have their own objectives and priorities (e.g., it may be easier for an outside contractor to say no to a user than it is for a person within the company). The latter alternatives typically have their own methodologies, which can benefit companies that do not have an appropriate methodology to use.

Time Frame   When time is a factor, the project team should probably start looking for a system that is already built and tested. In this way, the company will have a good idea of how long the package will take to put in place and what the final result will contain. The time frame for custom applications is hard to pin down, especially when you consider how many projects end up missing important deadlines. If a company must choose the custom development alternative and the time frame is very short, it should consider using techniques such as timeboxing to manage this problem. The time to produce a system using outsourcing really depends on the system and the outsourcer's resources. If a service provider has services in place that can be used to support the company's needs, then a business need could be implemented quickly. Otherwise, an outsourcing solution could take as long as a custom development initiative.

YOUR

TURN

7-4 Choose a Design Strategy

*S*uppose that your university is interested in creating a new course-registration system that can support mobile-based registration. What should the university consider when determining whether to invest in a custom, packaged, or outsourced system solution?

DEVELOPING THE ACTUAL DESIGN

Once the project team has a good understanding of how well each design strategy fits with the project's needs, it must begin to understand exactly *how* to implement these strategies. For example, what tools and technology would be used if a custom alternative were selected? What vendors make packaged systems that address the project's needs? What service providers would be able to build this system if the application were outsourced? This information can be obtained from people working in the IS department and from recommendations by business users. Alternatively, the project team can contact other companies with similar needs and investigate the types of systems that they have put in place. Vendors and consultants usually are willing to provide information about various tools and solutions in the form of brochures, product demonstrations, and information seminars. However, a company should be sure to validate the information it receives from vendors and consultants. After all, they are trying to make a sale. Therefore, they may stretch the capabilities of their tool by focusing on only the positive aspects of the tool while omitting the tool's drawbacks.

It is likely that the project team will identify several ways that a system could be constructed after weighing the specific design options. For example, the project team might have found three vendors that make packaged systems that potentially could meet the project's needs. Or the team may be debating over whether to develop a system using Java as a development tool and the database management system from Oracle or to outsource the development effort to a consulting firm such as Accenture or American Management Systems. Each alternative has pros and cons associated with it that need to be considered, and only one solution can be selected in the end.

Alternative Matrix

An *alternative matrix* can be used to organize the pros and cons of the design alternatives so that the best solution will be chosen in the end. This matrix is created using the same steps as the feasibility analysis, which was presented in Chapter 2. The only difference is that the alternative matrix combines several feasibility analyses into one matrix so that the alternatives can easily be compared. An alternative matrix is a grid that contains the technical, budget, and organizational feasibilities for each system candidate, pros and cons associated with adopting each solution, and other information that is helpful when making comparisons. Sometimes weights are provided for different parts of the matrix to show when some criteria are more important to the final decision.

To create the alternative matrix, draw a grid with the alternatives across the top and different criteria (e.g., feasibilities, pros, cons, and other miscellaneous criteria) along the side. Next, fill in the grid with detailed descriptions about each alternative. This becomes a useful document for discussion because it clearly presents the alternatives being reviewed and comparable characteristics for each one.

Suppose that a company is thinking about implementing a packaged financial system such as Oracle E-Business Financials or Microsoft's Dynamics GP, but there is not enough expertise in-house to be able to create a thorough alternative matrix. This situation is quite common—often the alternatives for a project are unfamiliar to the project team, so outside expertise is needed to provide information about the alternatives' criteria.

One helpful tool is the *request for proposals (RFP)*. An RFP is a document that solicits proposals to provide the alternative solutions from a vendor, developer, or service provider. Basically, an RFP explains the system that a company is trying to build and the criteria that it will use to select a system. Vendors then respond by describing what it would mean for them to be a part of the solution. They communicate the time, the cost, and exactly how their product or services will address the needs of the project.

There is no formal way to write an RFP, but it should include basic information such as the description of the desired system, any special technical needs or circumstances, evaluation criteria, instructions for how to respond, and the desired schedule. An RFP can be a very large document (i.e., hundreds of pages) because companies try to include as much detail as possible about their needs so that the respondent can be just as detailed in the solution that would be provided. Thus, RFPs typically are used for large projects rather than small ones because they take a lot of time to create; even more time and effort are needed for vendors, developers, and service providers to develop high-quality responses— only a project with a fairly large price tag would be worth the time and cost to develop a response for an RFP.

A less effort-intensive tool is a *request for information (RFI)* that includes the same format as the RFP. The RFI is shorter and contains less detailed information about a company's needs, and it requires general information from respondents that communicates the basic services that they can provide.

The final step, of course, is to decide which solution to design and implement. The decision should be made by a combination of business users and technical professionals after the issues involved with the different alternatives are well understood. Once the decision is finalized, design can continue as needed, based on the selected alternative.

YOUR **7-5 Alternative Matrix**
TURN

*S*uppose you have been assigned the task of selecting a CASE tool for your class to use for a semester project. Using the web or other reference resources, select three CASE tools (e.g., ArgoUML, IBM's Rational Rose, or Visual Paradigm). Create an alternative matrix that can be used to compare the three software products so that a selection decision can be made.

APPLYING THE CONCEPTS AT CD SELECTIONS

In the previous installments of the CD Selections case, we saw how Alec, Margaret, and the team had identified the functional and nonfunctional requirements and had completed the functional, structural, and behavioral models of their evolving web-based solution. However, before they can move into design, they realize that they needed to logically partition the model of the problem domain. To do this, they have decided to create a package diagram that will represent an overview of the analysis models for their evolving system. As in the previous installments, we see how the team goes about creating, verifying, and validating the analysis models of the web-based system they hope to implement.

SUMMARY

Design contains many steps that guide the project team through planning out exactly how a system needs to be constructed. The requirements that were identified and the models that were created during analysis serve as the primary inputs for the design activities. In object-oriented design, the primary activity is to evolve the analysis models into design models by optimizing the problem domain information already contained in the analysis models and adding system environment details to them.

Verifying and Validating the Analysis Models

Before actually adding system environment details to the analysis models, the various representations need to be verified and validated. One very useful approach to test the fidelity of the representations is to perform a walkthrough in which developers walk through the representations by presenting the different models to members of the analysis team, members of the design team, and representatives of the client. The walkthrough must validate each model to be sure that the different representations within the model all agree with one another. That is, the different models (functional, structural, and behavioral) must also be consistent.

Evolving the Analysis Models into Design Models

When evolving the analysis models into design models, the analysis models—activity diagrams, use-case descriptions, use-case diagrams, CRC cards, class and object diagrams, sequence diagrams, communication diagrams, and behavioral state machines—should first be carefully reviewed. During this review, factoring, refinement, and abstraction processes can be used to polish the current models. During this polishing, the analysis models can become overly complex. If this occurs, then the models should be partitioned based on the interactivity (message sending) and relationships (generalization, aggregation, and association) shared among the classes. The more a class has in common with another class (i.e., the more relationships shared), the more likely they belong in the same partition.

The second thing to do to evolve the analysis model is to add the system environment (physical architecture, user interface, and data access and management) information to the problem domain information already contained in the model. To accomplish this and to control the complexity of the models, layers are used. A layer represents an element of the software architecture of the system. We recommend five different layers: foundation, physical architecture, human–computer interaction, data access and management, and problem domain. Each layer supports only certain types of classes (e.g., database manipulation classes would be allowed only on the data access and management layer).

Packages and Package Diagrams

A package is a general UML construct used to represent collaborations, partitions, and layers. Its primary purpose is to support the logical grouping of other UML constructs (e.g., use cases and classes by the developer and user to simplify and increase the understandability of a UML diagram). There are instances in which a diagram that contains only packages is useful. A package diagram contains packages and dependency relationships. A dependency relationship represents the possibility of a modification dependency existing between two packages (i.e., changes in one package could cause changes in the dependent package).

Identifying packages and creating a package diagram is accomplished using a five-step process. The five steps can be summed up as setting the context, clustering similar classes, placing the clustered classes into a package, identifying dependency relationships among the packages, and placing the dependency relationship on the package diagram.

Design Strategies

During design, the project team also needs to consider three approaches to creating the new system, including developing a custom application in-house, buying a packaged system and customizing it, and relying on an external vendor, developer, or system provider to build and/or support the system.

Custom development allows developers to be flexible and creative in the way they solve business problems and it builds technical and functional knowledge within the organization. However, many companies have a development staff that is already overcommitted to filling huge backlogs of systems requests and they just don't have time to devote to a project where a system is built from scratch. It can be much more efficient to buy programs that have been created, tested, and proved. A packaged system can be bought and installed in a relatively short time as compared with a custom solution. Workarounds can be used to meet the needs that are not addressed by the packaged application.

The third design strategy is to outsource the project and pay an external vendor, developer, or service provider to create the system. It can be a good alternative to approaching the new system; however, it does not come without costs. If a company decides to leave the creation of a new system in the hands of someone else, the organization could compromise confidential information or lose control over future development.

Each of these design strategies has its strengths and weaknesses, and no one strategy is inherently better than the others. Thus, it is important to consider such issues as the uniqueness of business need for the system, the amount of in-house experience that is available to build the system, and the importance of the project skills to the company. Also, the existence of good project management and the amount of time available to develop the application play a role in the selection process.

Developing the Actual Design

Ultimately, the decision must be made regarding the specific type of system that needs to be designed. An alternative matrix can help make this decision by presenting feasibility information for several candidate solutions in a way they can be compared easily. The request for proposal and the request for information are two ways to gather accurate information regarding the alternatives.

KEY TERMS

A-kind-of, 291
Abstract classes, 291
Abstraction, 291
Aggregation, 291
Alternative matrix, 310
Balancing the models, 275
Class, 291
Client, 293
Collaboration, 292
Concrete classes, 291
Contract, 293
Controller, 294
Custom development, 303
Customization, 304
Data management layer, 294
Dependency relationship, 297
Enterprise resource
   systems (ERP), 304

Factoring, 291
Fixed-price contract, 306
Foundation layer, 294
Generalization, 291
Has-parts, 291
Human–computer
   interaction layer, 295
Layer, 293
Message, 293
Method, 293
Model, 294
Model-View-Controller
   (MVC), 293
Module, 291
Object wrapper, 305
Outsourcing, 305
Package, 296
Package diagram, 296

Packaged software, 304
Partition, 292
Physical architecture
   layer, 296
Problem domain layer, 294
Refinement, 291
Request for information (RFI), 310
Request for proposals (RFP), 310
Server, 293
Smalltalk, 293
Systems integration, 304
Time-and-arrangements
   contract, 306
Validation, 275
Value-added contract, 306
Verification, 275
View, 294
Workaround, 304

QUESTIONS

1. What is the primary difference between an analysis model and a design model?
2. What is meant by balancing the models?
3. What are the interrelationships among the functional, structural, and the behavioral models that need to be tested?

4. What does factoring mean? How is it related to abstraction and refinement?
5. What is a partition? How does a partition relate to a collaboration?
6. What is a layer? Name the different layers.
7. What is the purpose of the different layers?
8. Describe the different types of classes that can appear on each of the layers.
9. What issues or questions arise on each of the different layers?
10. What is a package? How are packages related to partitions and layers?
11. What is a dependency relationship? How do you identify them?
12. What are the five steps for identifying packages and creating package diagrams?
13. What needs to be verified and validated in package diagrams?
14. When drawing package diagrams, what guidelines should you follow?
15. What situations are most appropriate for a custom development design strategy?
16. What are some problems with using a packaged software approach to building a new system? How can these problems be addressed?
17. Why do companies invest in ERP systems?
18. What are the pros and cons of using a workaround?
19. When is outsourcing considered a good design strategy? When is it not appropriate?
20. What are the differences between the time-and-arrangements, fixed-price, and value-added contracts for outsourcing?
21. How are the alternative matrix and feasibility analysis related?
22. What is an RFP? How is this different from an RFI?

EXERCISES

A. For the A Real Estate Inc. problem in Chapters 4 (exercises I, J, and K), 5 (exercises P and Q), and 6 (exercise D):

1. Perform a verification and validation walkthrough of the functional, structural, and behavioral models to ensure that all between-model issues have been resolved.
2. Using the communication diagrams and the CRUDE matrix, create a package diagram of the problem domain layer.
3. Perform a verification and validation walkthrough of the package diagram.
4. Based on the analysis models that have been created and your current understanding of the firm's position, what design strategy would you recommend? Why?

B. For the A Video Store problem in Chapters 4 (exercises L, M, and N), 5 (exercises R and S), and 6 (exercise E):

1. Perform a verification and validation walkthrough of the functional, structural, and behavioral models to ensure that all between-model issues have been resolved.
2. Using the communication diagrams and the CRUDE matrix, create a package diagram of the problem domain layer.
3. Perform a verification and validation walkthrough of the package diagram.

4. Based on the analysis models that have been created and your current understanding of the firm's position, what design strategy would you recommend? Why?

C. For the health club membership problem in Chapters 4 (exercises O, P, and Q), 5 (exercises T and U), and 6 (exercise F):

1. Perform a verification and validation walkthrough of the functional, structural, and behavioral models to ensure that all between-model issues have been resolved.
2. Using the communication diagrams and the CRUDE matrix, create a package diagram of the problem domain layer.
3. Perform a verification and validation walkthrough of the package diagram.
4. Based on the analysis models that have been created and your current understanding of the firm's position, what design strategy would you recommend? Why?

D. For the Picnics R Us problem in Chapters 4 (exercises R, S, and T), 5 (exercises V and W), and 6 (exercise G):

1. Perform a verification and validation walkthrough of the functional, structural, and behavioral models to ensure that all between-model issues have been resolved.

2. Using the communication diagrams and the CRUDE matrix, create a package diagram of the problem domain layer.

3. Perform a verification and validation walkthrough of the package diagram.

4. Based on the analysis models that have been created and your current understanding of the firm's position, what design strategy would you recommend? Why?

E. For the Of-the-Month-Club problem in Chapters 4 (exercises U, V, and W), 5 (exercises X and Y), and 6 (exercise H):

1. Perform a verification and validation walkthrough of the functional, structural, and behavioral models to ensure that all between-model issues have been resolved.

2. Using the communication diagrams and the CRUDE matrix, create a package diagram of the problem domain layer.

3. Perform a verification and validation walkthrough of the package diagram.

4. Based on the analysis models that have been created and your current understanding of the firm's position, what design strategy would you recommend? Why?

F. Suppose you are leading a project that will implement a new course-enrollment system for your university. You are thinking about either using a packaged course-enrollment application or outsourcing the job to an external consultant. Create an outline for an RFP to which interested vendors and consultants could respond.

G. Suppose you and your friends are starting a small business painting houses in the summertime. You need to buy a software package that handles the financial transactions of the business. Create an alternative matrix that compares three packaged systems (e.g., Quicken, MS Money, Quickbooks). Which alternative appears to be the best choice?

MINICASES

1. Ting Ting Lim is the general manager of the Hong Kong office of Davies International, a British multinational human resources firm. The firm began in the UK in the early 1920s, and gradually opened offices in many British colonies around the world. Its information systems were largely developed at its UK head office. Over the last decade, there has been significant growth in business outside of the UK, and the Hong Kong office is now the firm's largest office. Offices in other countries such as India and Singapore have also grown dramatically.

Ms. Lim is reflecting on the client management software system that Davies International purchased five years ago. At that time, the firm had just gone through a growth spurt, and the combination of automated and manual procedures that managed client accounts had become unwieldy. Ms. Lim was in the committee that researched and selected the package currently used. She had learned of the software at a seminar. Initially, it worked fairly well for the firm. Some office procedures had to be changed to fit the package, but the office staff had expected this and were prepared for it.

Since that time, the Hong Kong office has continued to grow, not only through an expanding client base, but also from the acquisition of several smaller,

employment-related businesses in Hong Kong and an expansion onto the Chinese mainland. In a move to diversify its human resource management services, the firm's support staff has also expanded. Ms. Lim is particularly proud of the IS department the Hong Kong office has built up over the years. Using strong ties with a local university, an attractive compensation package, and a good working environment, the IS department is well-staffed with competent, innovative people, plus a steady stream of college interns that keeps the department vibrant. One of the IS teams pioneered the use of the Internet to offer services to a whole new Pankajet segment, a move that has proven to be successful and spread to Davies International offices overseas.

It seems clear that a major change is needed in the client management software, and Ms. Lim has already begun to budget for such a project. This software is a central part of Davies International operations, and Ms. Lim wants to be sure that a quality system is obtained this time. She knows that the vendor of their current system has made some revisions and additions to its product line. There are other software vendors who offer similar products. Ms. Lim is also considering the idea of appointing the Hong Kong office's IS department to develop a custom software application.

    a. Outline the issues that Ms. Lim should consider that would support the development of a custom software application in-house.

    b. Outline the issues that Ms. Lim should consider which would support the purchase of a software package.

    c. Within the context of a systems development project, when should the decision of "make-versus-buy" be made? How should Ms. Lim proceed? Explain your answer.

2. Refer to the Singapore Importers Pte Ltd minicase in Chapter 5. After completing all the analysis models (both the as-is and to-be models), the general manager finally understood why it was important to understand the as-is system before delving into the development of the to-be system. However, you now tell him that the to-be models are only the problem domain portion of the design. To say the least, he is now very confused. After explaining to him the advantages of using a layered approach to developing the system, he says, "I don't care about reusability or maintenance. I only want the system to be implemented as soon as possible. You IS people are always trying to pull a fast one on the users. Just get the system done!"

    What is your response to the general manager? Do you jump into the implementation he had insisted on? What do you do next?

3. Refer to the analysis models that you created for professional and scientific staff management (PSSM) for minicase 2 in Chapter 4 and for minicase 1 in Chapter 6.

    a. Perform a verification and validation walkthrough of the functional, structural, and behavioral models to ensure that all between-model issues have been resolved.

    b. Using the communication diagrams and the CRUDE matrix, create a package diagram of the problem domain layer.

    c. Perform a verification and validation walkthrough of the package diagram.

    d. Based on the analysis models that have been created and your current understanding of the firm's position, what design strategy would you recommend? Why?

4. Professor Takato, president of FRDI (see minicase 2, Chapter 5, and minicase 2, Chapter 6) is considering the choice of a design strategy for his new CMS. He wants this decision to be well-founded as he plans to use his CMS for several years and for several conferences. He has a limited budget for development and, aside from his own software development skills, which he is willing to put to work in his limited time available, he can only count on a few enthusiastic students who volunteered to help. Professor Takato identified the following packages as part of his to-be CMS:

    a. General Presentation and Marketing — presents the conference, conference topics, venue, traveling and accommodation information, program committees, important dates, and other.

    b. Communication Automation — supports generation and submission of possibly personalized emails or other types of messages to various categories of recipients, such as potential participants drawn from a database of previous conference participants to advertise the upcoming conference, submitting authors (e.g., for confirmation of paper acceptance), reviewers (e.g., for sending list of papers to review), others.

    c. Submission System — supports submission of papers.

    d. Review System — supports management of the paper review process.

    e. Registration System — supports conference registration and payment of conference fees.

    f. Post-conference Survey System — supports the collection and evaluation of post-conference online feedback from participants.

Given the above considerations, comment on the pros and cons Professor Takato would have to consider for each of the three design strategies: Custom Development, Packaged Software, and Outsourcing.

CHAPTER 8

CLASS AND METHOD DESIGN

The most important step of the design phase is designing the individual classes and methods. Object-oriented systems can be quite complex, so analysts need to create instructions and guidelines for programmers that clearly describe what the system must do. This chapter presents a set of criteria, activities, and techniques used to design classes and methods. Together, they are used to ensure the object-oriented design communicates how the system needs to be coded.

OBJECTIVES

- Become familiar with coupling, cohesion, and connascence
- Be able to specify, restructure, and optimize object designs
- Be able to identify the reuse of predefined classes, libraries, frameworks, and components
- Be able to specify constraints and contracts
- Be able to create a method specification

CHAPTER OUTLINE

INTRODUCTION

WARNING: *This material may be hazardous to your mental stability*. Not really, but now that we have your attention, you must realize that this material is fairly technical in nature and that it is extremely important in today's "flat" world. Today, much of the actual implementation will be done in a different geographic location than where the analysis and design are performed. We must ensure that the design is specified in a "correct" manner and that there is no, or at least minimal, ambiguity in the design specification.

In today's flat world, the common language spoken among developers is very likely to be UML and some object-oriented language, such as Java, and not English. English has always been and always will be ambiguous. Furthermore, to what variety of English do we refer? As both Oscar Wilde and George Bernard Shaw independently pointed out, the United States and England are divided by a common language. A simple, but relevant, example is the number of zeros in one billion. In American English, there are nine, but in British English there are twelve. Obviously, this could lead to problems when one is building financial information systems.

Practically speaking, Class and Method design is where all the work actually gets done during design. No matter which layer you are focusing on, the classes, which will be used to create the system objects, must be designed. Some people believe that with reusable class libraries and off-the-shelf components, this type of low-level, or detailed, design is a waste of time and that we should jump immediately into the "real" work: coding the system. However, if the past shows us anything, it shows that low-level, or detailed, design is critical despite the use of libraries and components. Detailed design is still very important for three reasons. First, with today's modern CASE tools, quite a bit of the actual code can be generated by the tool from the detailed design. Second, even preexisting classes and components need to be understood, organized, and pieced together. Third, it is still common for the project team to have to write some code and produce original classes that support the application logic of the system.

Jumping right into coding will guarantee results that can be disastrous. For example, even though the use of layers can simplify the individual classes, they can increase the complexity of the interactions between them. If the classes are not designed carefully, the resulting system can be very inefficient. Or worse, the instances of the classes (i.e., the objects) will not be capable of communicating with each other, which will result in the system's not working properly.

In an object-oriented system, changes can take place at different levels of abstraction. These levels include variable, method, class/object, package,[1] library, and/or application/system levels (see Figure 8-1). The changes that take place at one level can affect other levels (e.g., changes to a class can affect the package level, which can affect both the system level and the library level, which, in turn, can cause changes back down at the class level). Finally, changes can occur at different levels at the same time.

The good news is that the detailed design of the individual classes and methods is fairly straightforward, and the interactions among the objects on the problem domain layer have been designed, in some detail, during analysis (see Chapters 4 through 6). The other layers (data management, human–computer interaction, and physical architecture) are highly dependent on the problem domain layer. Therefore, if the problem domain classes are designed correctly, the design of the classes on the other layers will fall into place, relatively speaking.

That being said, it has been our experience that many project teams are much too quick at jumping into writing code for the classes without first designing them. Some of this has been caused by the fact that object-oriented systems analysis and design has evolved from object-oriented programming. Until recently, there has been a general lack of accepted guidelines on how to design and develop effective object-oriented systems.

---

[1] A package is a group of collaborating objects. Other names for a package include cluster, partition, pattern, subject, and subsystem.

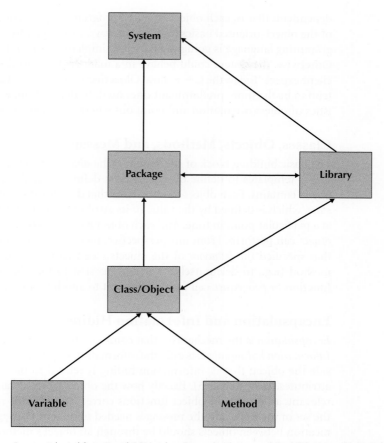

FIGURE 8-1
Levels of Abstraction
in Object-Oriented
Systems

Source: Adapted from David P. Tegarden, Steven D. Sheetz, and David E. Monarchi, "A Software Complexity Model of Object-Oriented Systems," *Decision Support Systems* 13 (March 1995): 241–262.

However, with the acceptance of UML as a standard object notation, standardized approaches based on the work of many object methodologists have begun to emerge.[2]

In this chapter, we begin by reviewing the basic characteristics of object orientation. Next, we present a set of useful design criteria and activities that are applicable across any layer for class and method design. Finally, we present a set of techniques that are useful for designing methods: contracts and method specifications.

REVIEW OF THE BASIC CHARACTERISTICS
OF OBJECT ORIENTATION

Object-oriented systems can be traced back to the Simula and the Smalltalk programming languages. However, until the increase in processor power and the decrease in processor cost that occurred in the 1980s, object-oriented approaches were not practical. Many of the specific details concerning the basic characteristics of object orientation are language

---

[2] For example, OPEN [I. Graham, B. Henderson-Seller, and H. Yanoussi, *The Open Process Specification* (Reading, MA: Addison-Wesley, 1997)], RUP [P. Kruchten, *The Rational Unified Process: An Introduction*, 2nd ed. (Reading, MA: Addison-Wesley, 2000)], and the Enhanced Unified Process (see Chapter 1).

dependent; that is, each object-oriented programming language tends to implement some of the object-oriented basics in different ways. Consequently, we need to know which programming language is going to be used to implement the different aspects of the solution. Otherwise, the system could behave in a manner different than the analyst, designer, and client expect. Today, the C++, Java, Objective-C, and Visual Basic programming languages tend to be the more predominant ones used. In this section, we review the basic characteristics of object orientation and point out where the language-specific issues emerge.

Classes, Objects, Methods, and Messages

The basic building block of the system is the *object*. Objects are *instances* of *classes* that we use as templates to define objects. A class defines both the data and processes that each object contains. Each object has *attributes* that describe data about the object. Objects have *state*, which is defined by the value of its attributes and its relationships with other objects at a particular point in time. And each object has *methods*, which specify what processes the object can perform. From our perspective, methods are used to implement the *operations* that specified the *behavior* of the objects (see Chapter 5). To get an object to perform a method (e.g., to delete itself), a *message* is sent to the object. A message is essentially a function or procedure call from one object to another object.

Encapsulation and Information Hiding

Encapsulation is the mechanism that combines the processes and data into a single object. *Information hiding* suggests only the information required to use an object be available outside the object; that is, information hiding is related to the *visibility* of the methods and attributes (see Chapter 5). Exactly how the object stores data or performs methods is not relevant, as long as the object functions correctly. All that is required to use an object are the set of methods and the messages needed to be sent to trigger them. The only communication between objects should be through an object's methods. The fact that we can use an object by sending a message that calls methods is the key to reusability, because it shields the internal workings of the object from changes in the outside system, and it keeps the system from being affected when changes are made to an object.

Polymorphism and Dynamic Binding

Polymorphism means having the ability to take several forms. By supporting polymorphism, object-oriented systems can send the same message to a set of objects, which can be interpreted differently by different classes of objects. Based on encapsulation and information hiding, an object does not have to be concerned with *how* something is done when using other objects. It simply sends a message to an object and that object determines how to interpret the message. This is accomplished through the use of dynamic binding.

Dynamic binding refers to the ability of object-oriented systems to defer the data typing of objects to run time. For example, imagine that you have an array of type employee that contains instances of hourly employees and salaried employees (see Figure 8-2). Both these types of employees implement a compute pay method. An object can send the message to each instance contained in the array to compute the pay for that individual instance. Depending on whether the instance is an hourly employee or a salaried employee, a different method will be executed. The specific method is chosen at run time. With this ability, individual classes are easier to understand. However, the specific level of support for polymorphism and dynamic binding is language specific. Most object-oriented programming languages support dynamic binding of methods, and some support dynamic binding of attributes.

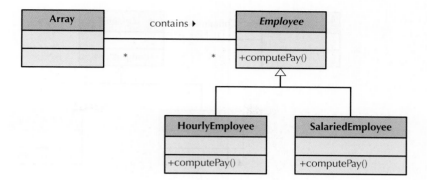

FIGURE 8-2
Example of
Polymorphism

But polymorphism can be a double-edged sword. Through the use of dynamic binding, there is no way to know before run time which specific object will be asked to execute its method. In effect, there is a decision made by the system that is not coded anywhere.[3] Because all these decisions are made at run time, it is possible to send a message to an object that it does not understand (i.e., the object does not have a corresponding method). This can cause a run-time error that, if the system is not programmed to handle it correctly, can cause the system to abort.[4]

Finally, if the methods are not semantically consistent, the developer cannot assume that all methods with the same name will perform the same generic operation. For example, imagine that you have an array of type person that contains instances of employees and customers (see Figure 8-3). These both implement a compute pay method. An object can send the message to each instance contained in the array to execute the compute pay method for that individual instance. In the case of an instance of employee, the compute pay method computes the amount that the employee is owed by the firm, whereas the compute pay method associated with an instance of a customer computes the amount owed the firm by the customer. Depending on whether the instance is an employee or a customer, a different meaning is associated with the method. Therefore, the semantics of each method must be determined individually. This substantially increases the difficulty of understanding individual objects. The key to controlling the difficulty of understanding object-oriented systems when using polymorphism is to ensure that all methods with the same name implement that same generic operation (i.e., they are semantically consistent).

Inheritance

Inheritance allows developers to define classes incrementally by reusing classes defined previously as the basis for new classes. Although we could define each class separately, it might be simpler to define one general superclass that contains the data and methods needed by the subclasses, and then have these classes inherit the properties of the superclass. Subclasses inherit the attributes and methods from the superclasses above them. Inheritance makes it simpler to define classes.

---

[3] From a practical perspective, there is an implied case statement. The system chooses the method based on the type of object being asked to execute it and the parameters passed as arguments to the method. This is typically done through message dispatch tables that are hidden from the programmer.

[4] In most object-oriented programming languages, these errors are referred to as exceptions that the system "throws" and must "catch." In other words, the programmer must correctly program the throw and catch or the systems will abort. Again, each programming language can handle these situations in a unique manner.

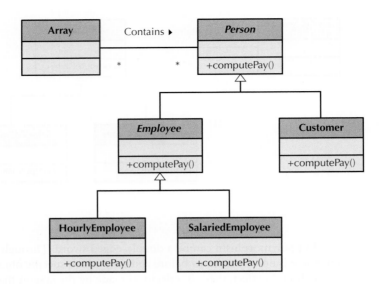

FIGURE 8-3

Example of
Polymorphism Misuse

There have been many different types of inheritance mechanisms associated with object-oriented systems.[5] The most common inheritance mechanisms include different forms of single and multiple inheritance. *Single inheritance* allows a subclass to have only a single parent class. Currently, all object-oriented methodologies, databases, and programming languages permit extending the definition of the superclass through single inheritance.

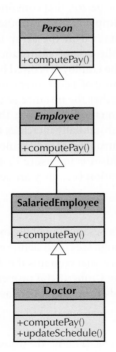

Some object-oriented methodologies, databases, and programming languages allow a subclass to redefine some or all the attributes and/or methods of its superclass. With *redefinition* capabilities, it is possible to introduce an *inheritance conflict* [i.e., an attribute (or method) of a subclass with the same name as an attribute (or method) of a super-class]. For example in Figure 8-4, Doctor is a subclass of Employee. Both have methods named ComputePay(). This causes an inheritance conflict. Furthermore, when the definition of a superclass is modified, all its subclasses are affected. This can introduce additional inheritance conflicts in one (or more) of the superclass's subclasses. For example in Figure 8-4, Employee could be modified to include an additional method, UpdateSchedule(). This would add another inheritance conflict between Employee and Doctor. Therefore, developers must be aware of the effects of the modification not only in the superclass but also in each subclass that inherits the modification.

Finally, through redefinition capabilities, it is possible for a programmer to arbitrarily cancel the inheritance of methods by placing stubs[6] in the subclass that will override the definition of the inherited method. If the cancellation of methods is necessary for the correct definition of the subclass, then it is likely that the subclass has been misclassified (i.e., it is inheriting from the wrong superclass).

FIGURE 8-4

Example of
Redefinition and
Inheritance Conflict

---

[5] See, for example, M. Lenzerini, D. Nardi, and M. Simi, *Inheritance Hierarchies in Knowledge Representation and Programming Languages* (New York: Wiley, 1991).

[6] In this case, a stub is simply the minimal definition of a method to prevent syntax errors from occurring.

As you can see, from a design perspective, inheritance conflicts and redefinition can cause all kinds of problems with interpreting the final design and implementation.[7] However, most inheritance conflicts are due to poor classification of the subclass in the inheritance hierarchy (the generalization a-kind-of semantics are violated), or the actual inheritance mechanism violates the encapsulation and information hiding principle (i.e., subclasses are capable of directly addressing the attributes or methods of a superclass). To address these issues, Jim Rumbaugh and his colleagues suggested the following guidelines:[8]

- Do not redefine query operations.
- Methods that redefine inherited ones should restrict only the semantics of the inherited ones.
- The underlying semantics of the inherited method should never be changed.
- The signature (argument list) of the inherited method should never be changed.

However, many existing object-oriented programming languages violate these guidelines. When it comes to implementing the design, different object-oriented programming languages address inheritance conflicts differently. Therefore, it is important at this point in the development of the system to know what the chosen programming language supports. We must be sure that the design can be implemented as intended. Otherwise, the design needs to be modified before it is turned over to remotely located programmers.

When considering the interaction of inheritance with polymorphism and dynamic binding, object-oriented systems provide the developer with a very powerful, but dangerous, set of tools. Depending on the object-oriented programming language used, this interaction can allow the same object to be associated with different classes at different times. For example, an instance of Doctor can be treated as an instance of Employee or any of its direct and indirect superclasses, such as SalariedEmployee and Person, respectively (see Figure 8-4). Therefore, depending on whether static or dynamic binding is supported, the same object may execute different implementations of the same method at different times. Or, if the method is defined only with the SalariedEmployee class and it is currently treated as an instance of the Employee class, the instance could cause a run-time error to occur.[9] It is important to know what object-oriented programming language is going to be used so that these kinds of issues can be solved with the design, instead of the implementation, of the class.

With *multiple inheritance*, a subclass may inherit from more than one superclass. In this situation, the types of inheritance conflicts are multiplied. In addition to the possibility of having an inheritance conflict between the subclass and one (or more) of its superclasses, it is now possible to have conflicts between two (or more) superclasses. In this latter case, three different types of additional inheritance conflicts can occur:

- Two inherited attributes (or methods) have the same name (spelling) and semantics.
- Two inherited attributes (or methods) have different names but identical semantics (i.e., they are *synonyms*).
- Two inherited attributes (or methods) have the same name but different semantics (i.e., they are *heteronyms, homographs,* or *homonyms*). This also violates the proper use of polymorphism.

---

[7] For more information, see Ronald J. Brachman, "I Lied about the Trees Or, Defaults and Definitions in Knowledge Representation," *AI Magazine* 5, no. 3 (Fall 1985): 80–93.

[8] J. Rumbaugh, M. Blaha, W. Premerlani, F. Eddy, and W. Lorensen, *Object-Oriented Modeling and Design* (Englewood Cliffs, NJ: Prentice Hall, 1991).

[9] This happens with novices quite regularly when using C++.

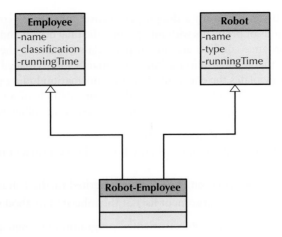

FIGURE 8-5
Additional Inheritance
Conflicts with Multiple
Inheritance

For example, in Figure 8-5, Robot-Employee is a subclass of both Employee and Robot. In this case, Employee and Robot conflict with the attribute name. Which one should Robot-Employee inherit? Because they are the same, semantically speaking, does it really matter? It is also possible that Employee and Robot could have a semantic conflict on the classification and type attributes if they have the same semantics. Practically speaking, the only way to prevent this situation is for the developer to catch it during the design of the subclass. Finally, what if the runningTime attributes have different semantics? In the case of Employee objects, the runningTime attribute stores the employee's time running a mile, whereas the runningTime attribute for Robot objects stores the average time between checkups. Should Robot-Employee inherit both of them? It really depends on whether the robot employees can run the mile or not. With the potential for these additional types of conflicts, there is a risk of decreasing the understandability in an object-oriented system instead of increasing it through the use of multiple inheritance. Our advice is to use great care when using multiple inheritance.

CONCEPTS
IN ACTION

8-A Inheritance Abuses

Meilir Page-Jones, through his consulting company, identified a set of abuses of inheritance. In some cases, these abuses led to lengthy and bloody disputes and gruesome implementations; in one case, it led to the destruction of the development team. In all cases, the error was in not enforcing a generalization (a-kind-of) semantics. In one case, the inheritance hierarchy was inverted: Board-Member was a superclass of Manager, which was a superclass of Employee. However, in this case, an Employee is *not* a-kind-of Manager, which is *not* a-kind-of Board-Member. In fact, the opposite was true. However, if you think of an Organization Chart, a BoardMember is superior to a Manager, which is superior to an Employee. In

another example, the client's firm attempted to use inheritance to model a membership idea (e.g., Student is a member of a club). However, the club should have had an attribute that contained the student members. In the other examples, inheritance was used to implement an association relationship and an aggregation relationship.

Source: Meilir Page-Jones, *Fundamentals of Object-Oriented Design in UML* (Reading, MA: Addison-Wesley, 2000).

Question

As an analyst, how can you attempt to avoid these types of inheritance abuses?

DESIGN CRITERIA

When considering the design of an object-oriented system, a set of criteria exists that can be used to determine whether the design is a good one or a bad one. According to Coad and Yourdon,[10] "A good design is one that balances trade-offs to minimize the total cost of the system over its entire lifetime." These criteria include coupling, cohesion, and connascence.

Coupling

Coupling refers to how interdependent or interrelated the modules (classes, objects, and methods) are in a system. The higher the interdependency, the more likely changes in part of a design can cause changes to be required in other parts of the design. For object-oriented systems, Coad and Yourdon[11] identified two types of coupling to consider: interaction and inheritance.

Interaction coupling deals with the coupling among methods and objects through message passing. Lieberherr and Holland put forth the *law of Demeter* as a guideline to minimize this type of coupling.[12] Essentially, the law minimizes the number of objects that can receive messages from a given object. The law states that an object should send messages only to one of the following:

- Itself. (For example, in Figure 8-6a, Object1 can send Message1 to itself. In other words, a method associated with Object1 can use other methods associated with Object1.[13])
- An object that is contained in an attribute of the object or one of its superclasses. (For example, in Figure 8-6b, PO1 should be able to send messages using both its Customer and Date attributes.)
- An object that is passed as a parameter to the method. (For example, in Figure 8-6c, the aPatient instance sends the message RequestAppt(name, address) to the aReceptionist instance, which is allowed to send messages to the instances contained in the name and address parameters.)
- An object that is created by the method. (For example, in Figure 8-6c, the method RequestAppt associated with the aReceptionist instance creates an instance of the Appointment class. The RequestAppt method is allowed to send messages to that instance.)
- An object that is stored in a global variable.[14]

In each case, interaction coupling is increased. For example, the coupling increases between the objects if the calling method passes attributes to the called method or if the calling method depends on the value being returned by the called method.

---

[10] Peter Coad and Edward Yourdon, *Object-Oriented Design* (Englewood Cliffs, NJ: Yourdon Press, 1991), p. 128.

[11] Ibid.

[12] Karl J. Lieberherr and Ian M. Holland, "Assuring Good Style for Object-Oriented Programs," *IEEE Software*, 6, no. 5 (September, 1989): 38–48; and Karl J. Lieberherr, *Adaptive Object-Oriented Software: The Demeter Method with Propagation Patterns* (Boston, MA: PWS Publishing, 1996).

[13] Obviously, this is stating what is expected.

[14] From a design perspective, global variables should be avoided. Most pure object-oriented programming languages do not explicitly support global variables, and we do not address them any further.

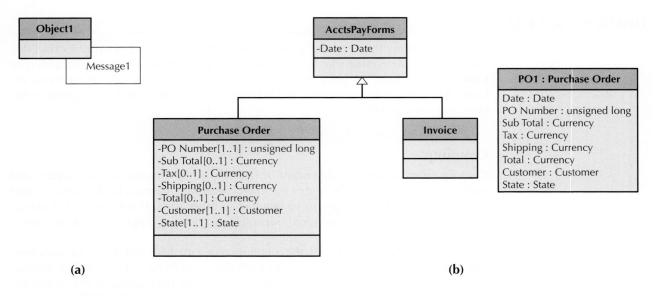

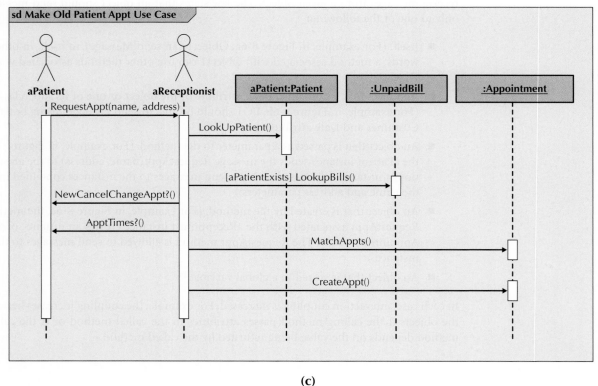

FIGURE 8-6   Examples of Interaction Coupling

There are six types of interaction coupling, each falling on different parts of a good-to-bad continuum. They range from no direct coupling to content coupling. Figure 8-7 presents the different types of interaction coupling. In general, interaction coupling should be minimized. The one possible exception is that non–problem-domain classes must be coupled to their corresponding problem domain classes. For example, a report object (on

Level	Type	Description
Good	No Direct Coupling	The methods do not relate to one another; that is, they do not call one another.
	Data	The calling method passes a variable to the called method. If the variable is composite (i.e., an object), the entire object is used by the called method to perform its function.
	Stamp	The calling method passes a composite variable (i.e., an object) to the called method, but the called method only uses a portion of the object to perform its function.
	Control	The calling method passes a control variable whose value will control the execution of the called method.
	Common or Global	The methods refer to a "global data area" that is outside the individual objects.
Bad	Content or Pathological	A method of one object refers to the inside (hidden parts) of another object. This violates the principles of encapsulation and information hiding. However, C++ allows this to take place through the use of "friends."

Source: These types were adapted from Meilir Page-Jones, *The Practical Guide to Structured Systems Design,* 2nd ed. (Englewood Cliffs, NJ: Yardon Press, 1988); and Glenford Myers, *Composite/Structured Design (*New York: Van Nostrand Reinhold, 1978).

FIGURE 8-7
Types of Interaction Coupling

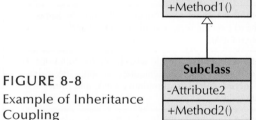

FIGURE 8-8
Example of Inheritance Coupling

the human–computer interaction layer) that displays the contents of an employee object (on the problem domain layer) will be dependent on the employee object. In this case, for optimization purposes, the report class may be even content or pathologically coupled to the employee class. However, problem domain classes should never be coupled to non–problem-domain classes.

Inheritance coupling, as its name implies, deals with how tightly coupled the classes are in an inheritance hierarchy. Most authors tend to say simply that this type of coupling is desirable. However, depending on the issues raised previously with inheritance—inheritance conflicts, redefinition capabilities, and dynamic binding—a high level of inheritance coupling might not be a good thing. For example, in Figure 8-8, should Method2() defined in Subclass be allowed to call Method1() defined in Superclass? Or, should Method2() defined in Subclass refer to Attribute1 defined in Superclass? Or, even more confusing, assuming that Superclass is an abstract class, can a Method1() call Method2() or use Attribute2 defined in Subclass? Obviously, the first two examples have some intuitive sense. Using the properties of a superclass is the primary purpose of inheriting from it in the first place. On the other hand, the third example is somewhat counterintuitive. However, owing to the way that different object-oriented programming languages support dynamic binding, polymorphism, and inheritance, all these examples could be possible.

As Snyder has pointed out, most problems with inheritance involve the ability within the object-oriented programming languages to violate the encapsulation and information-hiding principles.[15] From a design perspective, the developer needs to optimize the trade-offs

[15] Alan Snyder, "Encapsulation and Inheritance in Object-Oriented Programming Languages," in N. Meyrowitz, ed., *OOPSLA '86 Conference Proceedings, ACM SigPlan Notices,* 21, no. 11 (November 1986); and Alan Snyder, "Inheritance and the Development of Encapsulated Software Components," in B. Shriver and P. Wegner, eds., *Research Directions in Object-Oriented Programming* (Cambridge, MA: MIT Press, 1987).

of violating the encapsulation and information-hiding principles and increasing the desirable coupling between subclasses and its superclasses. The best way to solve this conundrum is to ensure that inheritance is used only to support generalization/specialization (a-kind-of) semantics and the principle of substitutability (see Chapter 5). All other uses should be avoided.

Cohesion

Cohesion refers to how single-minded a module (class, object, or method) is within a system. A class or object should represent only one thing, and a method should solve only a single task. Three general types of cohesion have been identified by Coad and Yourdon for object-oriented systems: method, class, and generalization/specialization.[16]

Method cohesion addresses the cohesion within an individual method (i.e., how single-minded a method is). Methods should do one and only one thing. A method that actually performs multiple functions is more difficult to understand—and, therefore, to implement and maintain—than one that performs only a single function. Seven types of method cohesion have been identified (see Figure 8-9). They range from functional cohesion (good) down to coincidental cohesion (bad). In general, method cohesion should be maximized.

Level	Type	Description
Good	Functional	A method performs a single problem-related task (e.g., calculate current GPA).
	Sequential	The method combines two functions in which the output from the first one is used as the input to the second one (e.g., format and validate current GPA).
	Communicational	The method combines two functions that use the same attributes to execute (e.g., calculate current and cumulative GPA).
	Procedural	The method supports multiple weakly related functions. For example, the method could calculate student GPA, print student record, calculate cumulative GPA, and print cumulative GPA.
	Temporal or Classical	The method supports multiple related functions in time (e.g., initialize all attributes).
	Logical	The method supports multiple related functions, but the choice of the specific function is chosen based on a control variable that is passed into the method. For example, the called method could open a checking account, open a savings account, or calculate a loan, depending on the message that is sent by its calling method.
Bad	Coincidental	The purpose of the method cannot be defined or it performs multiple functions that are unrelated to one another. For example, the method could update customer records, calculate loan payments, print exception reports, and analyze competitor pricing structure.

Source: These types were adapted from Page-Jones, *The Practical Guide to Structured Systems*, and Myers, *Composite/Structured Design.*

FIGURE 8-9
Types of Method Cohesion

[16] Coad and Yourdon, *Object-Oriented Design.*

Class cohesion is the level of cohesion among the attributes and methods of a class (i.e., how single-minded a class is). A class should represent only one thing, such as an employee, a department, or an order. All attributes and methods contained in a class should be required for the class to represent the thing. For example, an employee class should have attributes that deal with a social security number, last name, first name, middle initial, addresses, and benefits, but it should not have attributes such as door, engine, or hood. Furthermore, there should be no attributes or methods that are never used. In other words, a class should have only the attributes and methods necessary to fully define instances for the problem at hand. In this case, we have *ideal class cohesion*. Glenford Meyers suggested that a cohesive class[17] should have these attributes:

- It should contain multiple methods that are visible outside the class (i.e., a single-method class rarely makes sense).
- Each visible method performs only a single function (i.e., it has functional cohesion; see Figure 8-9).
- All methods reference only attributes or other methods defined within the class or one of its superclasses (i.e., if a method is going to send a message to another object, the remote object must be the value of one of the local object's attributes).[18]
- It should not have any control couplings between its visible methods (see Figure 8-7).

Page-Jones[19] has identified three less-than-desirable types of class cohesion: mixed-instance, mixed-domain, and mixed-role (see Figure 8-10). An individual class can have a mixture of any of the three types.

Level	Type	Description
Good	Ideal	The class has none of the mixed cohesions.
	Mixed-Role	The class has one or more attributes that relate objects of the class to other objects on the same layer (e.g., the problem domain layer), but the attribute(s) have nothing to do with the underlying semantics of the class.
	Mixed-Domain	The class has one or more attributes that relate objects of the class to other objects on a different layer. As such, they have nothing to do with the underlying semantics of the thing that the class represents. In these cases, the offending attribute(s) belongs in another class located on one of the other layers. For example, a port attribute located in a problem domain class should be in a system architecture class that is related to the problem domain class.
Worse	Mixed-Instance	The class represents two different types of objects. The class should be decomposed into two separate classes. Typically, different instances only use a portion of the full definition of the class.
		Source: Page-Jones, Fundamentals of Object-Oriented Design in UML.

FIGURE 8-10
Types of Class
Cohesion

[17] We have adapted his informational-strength module criteria from structured design to object-oriented design. [see Glenford J. Myers, *Composite/Structured Design* (New York, NY: Van Nostrand Reinhold, 1978)].

[18] This restricts messages passing to only the first, second, and fourth conditions supported by the law of Demeter. For example, in Figure 8-6c, aReceptionist must have attributes associated with it that contains objects for Patients, Unpaid Bills, and Appointments. Furthermore, once an instance of Appointment is created, aReceptionist must have an attribute with the instance as its value to send any additional messages.

[19] See Meilir Page-Jones, *Fundamentals of Object-Oriented Design in UML* (Reading, MA: Addison-Wesley, 2000).

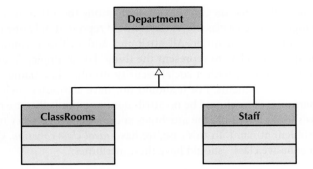

FIGURE 8-11
Generalization/
Specialization vs.
Inheritance Abuse

Generalization/specialization cohesion addresses the sensibility of the inheritance hierarchy. How are the classes in the inheritance hierarchy related? Are the classes related through a generalization/specialization (a-kind-of) semantics? Or, are they related via some association, aggregation, or membership type of relationship that was created for simple reuse purposes? Recall all the issues raised previously on the use of inheritance. For example, in Figure 8-11, the subclasses ClassRooms and Staff inherit from the superclass Department. Obviously, instances of the ClassRooms and Staff classes are not a-kind-of Department. However, in the early days of object-oriented programming, this use of inheritance was quite common. When a programmer saw that there were some common properties that a set of classes shared, the programmer would create an artificial abstraction that defined the commonalities. This was potentially useful in a reuse sense, but it turned out to cause many maintenance nightmares. In this case, instances of the ClassRooms and Staff classes are associated with or a-part-of an instance of Department. Today, we know that highly cohesive inheritance hierarchies should support only the semantics of generalization and specialization (a-kind-of) and the principle of substitutability.

Connascence

Connascence[20] generalizes the ideas of cohesion and coupling, and it combines them with the arguments for encapsulation. To accomplish this, three levels of encapsulation have been identified. Level-0 encapsulation refers to the amount of encapsulation realized in an individual line of code, level-1 encapsulation is the level of encapsulation attained by combining lines of code into a method, and level-2 encapsulation is achieved by creating classes that contain both methods and attributes. Method cohesion and interaction coupling address primarily level-1 encapsulation. Class cohesion, generalization/specialization cohesion, and inheritance coupling address only level-2 encapsulation. Connascence, as a generalization of cohesion and coupling, addresses both level-1 and level-2 encapsulation.

But what exactly is connascence? Connascence literally means to be born together. From an object-oriented design perspective, it really means that two modules (classes or methods) are so intertwined that if you make a change in one, it is likely that a change in the other will be required. On the surface, this is very similar to coupling and, as such, should be minimized. However, when you combine it with the encapsulation levels, it is not quite as simple as that. In this case, we want to minimize overall connascence by eliminating any unnecessary connascence throughout the system; minimize connascence across any

---

[20] See Meilir Page-Jones, "Comparing Techniques by Means of Encapsulation and Connascence," *Communications of the ACM* 35, no. 9 (September 1992): 147–151.

Type	Description
Name	If a method refers to an attribute, it is tied to the name of the attribute. If the attribute's name changes, the content of the method will have to change.
Type or Class	If a class has an attribute of type A, it is tied to the type of the attribute. If the type of the attribute changes, the attribute declaration will have to change.
Convention	A class has an attribute in which a range of values has a semantic meaning (e.g., account numbers whose values range from 1,000 to 1,999 are assets). If the range would change, then every method that used the attribute would have to be modified.
Algorithm	Two different methods of a class are dependent on the same algorithm to execute correctly (e.g., insert an element into an array and find an element in the same array). If the underlying algorithm would change, then the insert and find methods would also have to change.
Position	The order of the code in a method or the order of the arguments to a method is critical for the method to execute correctly. If either is wrong, then the method will, at least, not function correctly.

Source: Meilir Page-Jones, "Comparing Techniques by Means of Encapsulation and Connascence" and Meilir Page-Jones, *Fundamentals of Object-Oriented Design in UML.*

FIGURE 8-12
Types of Connascence

encapsulation boundaries, such as method boundaries and class boundaries; and maximize connascence within any encapsulation boundary.

Based on these guidelines, a subclass should never directly access any hidden attribute or method of a superclass [i.e., a subclass should not have special rights to the properties of its superclass(es)]. If direct access to the nonvisible attributes and methods of a superclass by its subclass is allowed—and is permitted in most object-oriented programming languages—and a modification to the superclass is made, then owing to the connascence between the subclass and its superclass, it is likely that a modification to the subclass also is required.[21] In other words, the subclass has access to something across an encapsulation boundary (the class boundary between the subclass and the superclass). Practically speaking, you should maximize the cohesion (connascence) within an encapsulation boundary and minimize the coupling (connascence) between the encapsulation boundaries. There are many possible types of connascence. Figure 8-12 describes five of the types.

OBJECT DESIGN ACTIVITIES

The design activities for classes and methods are really an extension of the analysis and evolution activities presented previously (see Chapters 4 through 7). In this case, we expand the descriptions of the partitions, layers, and classes. Practically speaking, the expanded descriptions are created through the activities that take place during the detailed design of the classes and methods. The activities used to design classes and methods include additional specification of the current model, identifying opportunities for reuse, restructuring

---

[21] Based on these guidelines, the use of the protected visibility, as supported in Java and C++, should be minimized, if not avoided. "Friends" as defined in C++ also should be minimized or avoided. Owing to the level of dependencies these language features create, any convenience afforded to a programmer is more than offset in potential design, understandability, and maintenance problems. These features must be used with great caution and must be fully documented.

the design, optimizing the design, and, finally, mapping the problem domain classes to an implementation language. Of course, any changes made to a class on one layer can cause the classes on the other layers that are coupled to it to be modified as well. The object design activities are described in this section.

Adding Specifications

At this point in the development of the system, it is crucial to review the current set of functional, structural, and behavioral models. First, we should ensure that the classes on the problem domain layer are both necessary and sufficient to solve the underlying problem. To do this, we need to be sure that there are no missing attributes or methods and no extra or unused attributes or methods in each class. Furthermore, are there any missing or extra classes? If we have done our job well during analysis, there will be few, if any, attributes, methods, or classes to add to the models. And it is unlikely that we have any extra attributes, methods, or classes to delete from the models. However, we still need to ensure that we have factored, abstracted, and refined the evolving models and created the relevant partitions and collaborations (see Chapter 7). We have mentioned this before, but we cannot overemphasize the importance of constantly reviewing the evolving system. Remember, it is always better to be safe than sorry.

Second, we need to finalize the visibility (hidden or visible) of the attributes and methods in each class. Depending on the object-oriented programming language used, this could be predetermined. [For example, in Smalltalk, attributes are hidden and methods are visible. Other languages allow the programmer to set the visibility of each attribute or method. For example, in C++ and Java, you can set the visibility to private (hidden), public (visible), or protected (visible to subclasses, but not to other classes)].[22] By default, most object-oriented analysis and design approaches assume Smalltalk's approach.

Third, we need to decide on the signature of every method in every class. The *signature* of a method comprises three parts: the name of the method, the parameters or arguments that must be passed to the method, including their object type, and the type of value that the method will return to the calling method. The signature of a method is related to the method's *contract*.[23]

Fourth, we need to define any constraints that must be preserved by the objects (e.g., an attribute of an object that can have values only in a certain range). There are three different types of constraints: preconditions, postconditions, and invariants.[24] These are captured in the form of contracts (described later in this chapter) and assertions added to the CRC cards and class diagrams. We also must decide how to handle a violation of a constraint. Should the system simply abort? Should the system automatically undo the change that caused the violation? Should the system let the end user determine the approach to correct the violation? In other words, the designer must design the errors that the system is expected to handle. It is best not to leave these types of design decisions for the programmer to solve. Violations of a constraint are known as *exceptions* in languages such as C++ and Java.

Even though we have described these activities in the context of the problem domain layer, they are also applicable to the other layers: data management (Chapter 9), human–computer interaction (Chapter 10), and physical architecture (Chapter 11).

---

[22] It is also possible to control visibility through packages and friends (see Footnote 21).

[23] Contracts were introduced in Chapter 5 and they are described in detail later in this chapter.

[24] Constraints are described in more detail later in this chapter.

Identifying Opportunities for Reuse

Previously, we looked at possibly employing reuse in our models in analysis through the use of *patterns* (see Chapter 5). In design, in addition to using analysis patterns, there are opportunities for using design patterns, frameworks, libraries, and components. The opportunities vary depending on which layer is being reviewed. For example, it is doubtful that a class library will be of much help on the problem domain layer, but a class library could be of great help on the foundation layer. In this section, we describe the use of design patterns, frameworks, libraries, and components.

Like analysis patterns, design patterns are simply useful groupings of collaborating classes that provide a solution to a commonly occurring problem. The primary difference between analysis and design patterns is that design patterns are useful in solving "a general design problem in a particular context,"[25] whereas analysis patterns tended to aid in filling out a problem domain representation. For example, a useful pattern is the Whole-Part pattern (see Figure 8-13a). The Whole-Part pattern explicitly supports the Aggregation and Composition relationships within the UML. Another useful design pattern is the Iterator pattern (see Figure 8-13b). The primary purpose of the Iterator pattern is to provide the designer with a standard approach to support traversing different types of collections. By using this pattern, regardless of the collection type (ConcreteAggregate), the designer knows that the collection will need to create an iterator (ConcreteIterator) that customizes the standard operations used to traverse the collection: first(), next(), isDone(), and currentItem(). Given the number of collections typically found in business applications, this pattern is one of the more useful ones. For example in Figure 8-14a, we replicate a portion of both the Appointment and Library problems discussed in previous chapters and in Figure 8-14b, we show how the Iterator pattern can be applied to those sections of their evolving designs. Finally, some of the design patterns support different physical architectures (see Chapter 11). For example, the Forwarder–Receiver pattern (see Figure 8-13c) supports a peer-to-peer architecture. Many design patterns are available in C++ or Java source code.

A *framework* is composed of a set of implemented classes that can be used as a basis for implementing an application. Most frameworks allow us to create subclasses to inherit from classes in the framework. There are object-persistence frameworks that can be purchased and used to add persistence to the problem domain classes, which would be helpful on the data management layer. Of course, when inheriting from classes in a framework, we are creating a dependency (i.e., increasing the inheritance coupling from the subclass to the superclass). Therefore, if we use a framework and the vendor makes changes to the framework, we will have to at least recompile the system when we upgrade to the new version of the framework.

A *class library* is similar to a framework in that it typically has a set of implemented classes that were designed for reuse. However, frameworks tend to be more domain specific. In fact, frameworks may be built using a class library. A typical class library could be purchased to support numerical or statistical processing, file management (data management layer), or user interface development (human–computer interaction layer). In some cases, instances of classes contained in the class library can be created, and in other cases, classes in the class library can be extended by creating subclasses based on them. As with frameworks, if we use inheritance to reuse the classes in the class library, we will run into all the issues dealing with inheritance coupling and connascence. If we directly instantiate classes in the class library, we will create a

[25] Erich Gamma, Richard Helm, Ralph Johnson, and John Vlissides, *Design Patterns: Elements of Reusable Object-Oriented Software* (Reading, MA: Addison-Wesley, 1995).

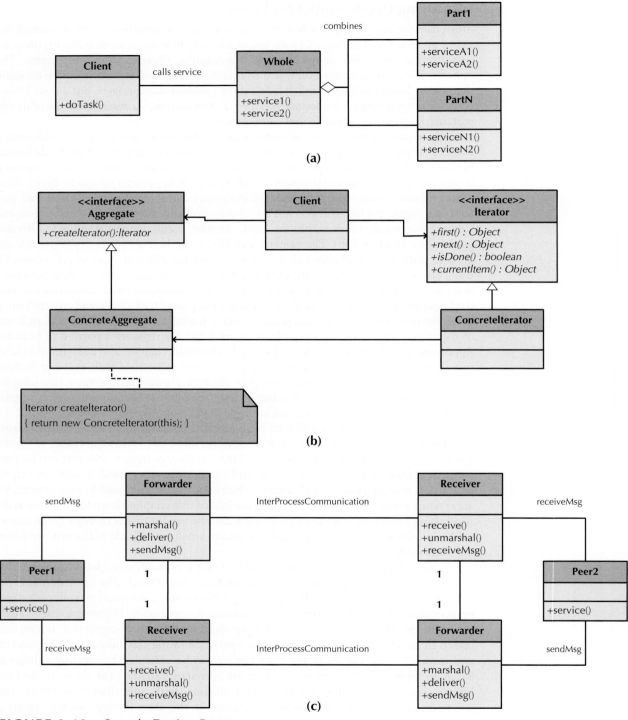

FIGURE 8-13   Sample Design Patterns

Source: F. Buschmann, R. Meunier, H. Rohnert, P. Sommerlad, and M. Stal, *Pattern-Oriented Software Architecture: A System of Patterns* (Chichester, UK: Wiley, 1996), E. Gamma, R. Helm, R. Johnson, and J. Vlissides, *Design Patterns: Elements of Reusable Object-Oriented Software* (Reading, MA: Addison-Wesley, 1995).

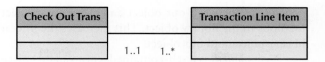

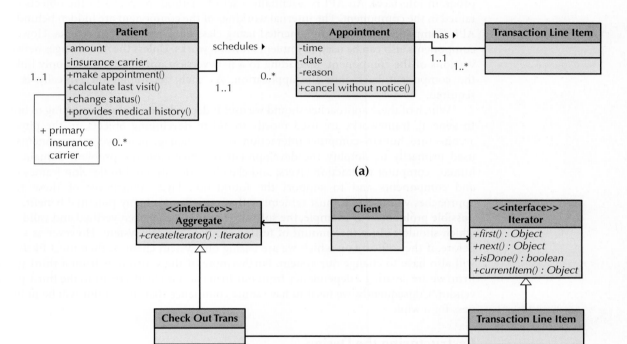

(a)

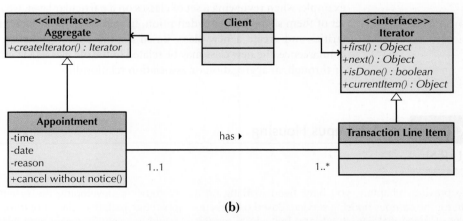

(b)

FIGURE 8-14   Iterator Design Pattern Applied to Library and Appointment Problems

dependency between our object and the library object based on the signatures of the methods in the library object. This increases the interaction coupling between the class library object and our object.

A *component* is a self-contained, encapsulated piece of software that can be plugged into a system to provide a specific set of required functionalities. Today, there are many components available for purchase. A component has a well-defined *API* (application program interface). An API is essentially a set of method interfaces to the objects contained in the component. The internal workings of the component are hidden behind the API. Components can be implemented using class libraries and frameworks. However, components also can be used to implement frameworks. Unless the API changes between versions of the component, upgrading to a new version normally requires only linking the component back into the application. As such, recompilation typically is not required.

Which of these approaches should we use? It depends on what we are trying to build. In general, frameworks are used mostly to aid in developing objects on the physical architecture, human–computer interaction, or data management layers; components are used primarily to simplify the development of objects on the problem domain and human–computer interaction layers; and class libraries are used to develop frameworks and components and to support the foundation layer. Whichever of these reuse approaches you use, you must remember that reuse brings many potential benefits and possible problems. For example, the software has previously been verified and validated, which should reduce the amount of testing required for our system. However as stated before, if the software on which we are basing our system changes, then most likely, we will also have to change our system. Furthermore, if the software is from a third-party firm, we are creating a dependency from our firm (or our client's firm) to the third-party vendor. Consequently, we need to have some confidence that the vendor will be in business for a while.

Restructuring the Design

Once the individual classes and methods have been specified and the class libraries, frameworks, and components have been incorporated into the evolving design, we should use factoring to restructure the design. *Factoring* (Chapter 7) is the process of separating out aspects of a method or class into a new method or class to simplify the overall design. For example, when reviewing a set of classes on a particular layer, we might discover that a subset of them shares a similar definition. In that case, it may be useful to factor out the similarities and create a new class. Based on the issues related to cohesion, coupling, and connascence, the new class may be related to the old classes via inheritance (generalization) or through an aggregation or association relationship.

YOUR TURN   8-1 Campus Housing

In the previous chapters, you have been working on a system for the campus housing service. Based on the current set of functional, structural, and behavioral models that you have developed, are there potential opportunities of reuse in developing the system? Search the web for potential patterns, class libraries, and components that could be useful in developing this system.

Another process that is useful for restructuring the evolving design is *normalization*. Normalization is described in Chapter 9 in relation to relational databases. However, normalization can be useful at times to identify potential classes that are missing from the design. Also related to normalization is the requirement to implement the actual association and aggregation relationships as attributes. Virtually no object-oriented programming language differentiates between attributes and association and aggregation relationships. Therefore, all association and aggregation relationships must be converted to attributes in the classes. For example, in Figure 8-15a, the Customer and State classes are associated with the Order class. Furthermore, the Product-Order association class is associated with both the Order and Product classes. One of the first things that must be done is to convert the Product Order Association class to a normal class. Notice the multiplicity values for the new associations between the Order and the Product Order classes and the Product Order and Product classes (see Figure 8-15b). Next, we need to convert all associations to attributes that represent the relationships between the affected classes. In this case, the Customer class must have an Orders attribute added to represent the set of orders that an instance of the Customer class may possess; the Order class must add attributes to reference instances of the Customer, State, and Product Order classes; the State class must have an attribute added to it to reference all of the instances of the Order class that is associated with that particular state; the new Product Order class must have attributes that allow an instance of the Product Order class to reference which instance of the Order class and which instance of the Product class is relevant to it; and, finally, the Product class must add an attribute that references the relevant instances of the Product Order class (see Figure 8-15c). As you can see, even in this very small example, many changes need to be made to ready the design for implementation.

Finally, all inheritance relationships should be challenged to ensure that they support only a generalization/specialization (a-kind-of) semantics. Otherwise, all the problems mentioned previously with inheritance coupling, class cohesion, and generalization/specialization cohesion will come to pass.

Optimizing the Design[26]

Up until now, we have focused our energy on developing an understandable design. With all the classes, patterns, collaborations, partitions, and layers designed, and with all the class libraries, frameworks, and components included in the design, understandability has been, as it should have been, our primary focus. However, increasing the understandability of a design typically creates an inefficient design. Conversely, focusing on efficiency issues will deliver a design that is more difficult to understand. A good practical design manages the inevitable trade-offs that must occur to create an acceptable system. In this section, we describe a set of simple optimizations that can be used to create a more efficient design.[27]

The first optimization to consider is to review the access paths between objects. In some cases, a message from one object to another has a long path to traverse (i.e., it goes through many objects). If the path is long and the message is sent frequently, a redundant path should be considered. Adding an attribute to the calling object that will store a direct connection to the object at the end of the path can accomplish this.

[26] The material contained in this section is based on James Rumbaugh, Michael Blaha, William Premerlani, Frederick Eddy, and William Lorensen, *Object-Oriented Modeling and Design* (Englewood Cliffs, NJ: Prentice Hall, 1991); and Bernd Brugge and Allen H. Dutoit, *Object-Oriented Software Engineering: Conquering Complex and Changing Systems* (Englewood Cliffs, NJ: Prentice Hall, 2000).

[27] The optimizations described here are only suggestions. In all cases, the decision to implement one or more of these optimizations really depends on the problem domain of the system and the environment on which the system will reside, that is, the data management layer (see Chapter 9), the human–computer interaction layer (see Chapter 10), and the physical architecture layer (see Chapter 11).

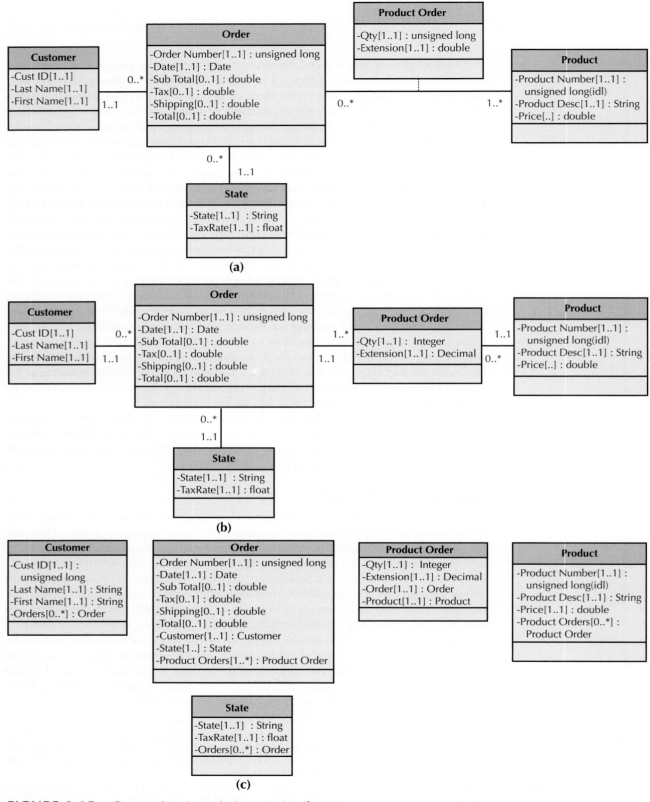

FIGURE 8-15   Converting Associations to Attributes

A second optimization is to review each attribute of each class. Which methods use the attributes and which objects use the methods should be determined. If the only methods that use an attribute are read and update methods and only instances of a single class send messages to read and update the attribute, then the attribute may belong with the calling class instead of the called class. Moving the attribute to the calling class will substantially speed up the system.

A third optimization is to review the direct and indirect fan-out of each method. *Fan-out* refers to the number of messages sent by a method. The direct fan-out is the number of messages sent by the method itself, whereas the indirect fan-out also includes the number of messages sent by the methods called by the other methods in a message tree. If the fan-out of a method is high relative to the other methods in the system, the method should be optimized. One way to do this is to consider adding an index to the attributes used to send the messages to the objects in the message tree.

A fourth optimization is to look at the execution order of the statements in often-used methods. In some cases, it is possible to rearrange some of the statements to be more efficient. For example, if it is known, based on the objects in the system, that a search routine can be narrowed by searching on one attribute before another one, then the search algorithm should be optimized by forcing it to always search in a predefined order.

A fifth optimization is to avoid recomputation by creating a *derived attribute* (or *active value*) (e.g., a total that stores the value of the computation). This is also known as *caching computational results*. It can be accomplished by adding a *trigger* to the attributes contained in the computation (i.e., attributes on which the derived attribute is dependent). This would require a recomputation to take place only when one of the attributes that go into the computation is changed. Another approach is to simply mark the derived attribute for recomputation and delay the recomputation until the next time the derived attribute is accessed. This last approach delays the recomputation as long as possible. In this manner, a computation does not occur unless it must occur. Otherwise, every time a derived attribute needs to be accessed, a computation will be required.

A sixth optimization that should be considered deals with objects that participate in a one-to-one association; that is, they both must exist for either to exist. In this case, it might make sense, for efficiency purposes, to collapse the two defining classes into a single class. However, this optimization might need to be reconsidered when storing the "fatter" object in a database. Depending on the type of object persistence used (see Chapter 9), it can actually be more efficient to keep the two classes separate. Alternatively, it could make more sense for the two classes to be combined on the problem domain layer but kept separate on the data management layer.

YOUR TURN   8-2 Campus Housing

Assume that you are the project leader for the campus housing system that you have been developing over the previous chapters and that you just modified in Your Turn 8-1. However, as you review the current set of models, you realize that even though the models provide a rather complete description of the problem domain layer, the evolving models have begun to become unmanageable.

As project leader, you also need to guarantee that the design will be efficient. Create a set of discussion points that you will use to explain to your development team the importance of optimizing the design before jumping into coding. Be sure to include an example of each optimization technique that can be used in the current set of models for the campus housing system.

Mapping Problem Domain Classes to Implementation Languages[28]

Up until this point in the development of the system, it has been assumed that the classes and methods in the models would be implemented directly in an object-oriented programming language. However, now it is important to map the current design to the capabilities of the programming language used. For example, if we have used multiple inheritance in our design but we are implementing in a language that supports only single inheritance, then the multiple inheritance must be factored out of the design. If the implementation is to be done in an object-based language, one that does not support inheritance,[29] or a non–object-based language, such as C, we must map the problem-domain objects to programming constructs that can be implemented using the chosen implementation environment. In this section, we describe a set of rules that can be used to do the necessary mapping.

Implementing Problem Domain Classes in a Single-Inheritance Language The only issue associated with implementing problem domain objects is the factoring out of any multiple inheritance—that is, the use of more than one superclass—used in the evolving design. For example, if you were to implement the solution in Java, Smalltalk, or Visual Basic.net, you must factor out any multiple inheritance. The easiest way to do this is to use the following rule:

> **RULE 1a:** Convert the additional inheritance relationships to association relationships. The multiplicity of the new association from the subclass to the superclass should be 1..1. If the additional superclasses are concrete, that is, they can be instantiated themselves, then the multiplicity from the superclass to the subclass is 0..1. Otherwise, it is 1..1. Furthermore, an exclusive-or (XOR) constraint must be added between the associations. Finally, you must add appropriate methods to ensure that all information is still available to the original class.

or

> **RULE 1b:** Flatten the inheritance hierarchy by copying the attributes and methods of the additional superclass(es) down to all of the subclasses and remove the additional superclass from the design.[30]

Figure 8-16 demonstrates the application of these rules. Figure 8-16a portrays a simple example of multiple inheritance where Flying Car inherits from both Airplane and Car, and Amphibious Car inherits from both Car and Boat. Assuming that Car is concrete, we apply Rule 1a to part a, and we end up with the diagram in part b, where we have added the association between Flying Car and Car and the association between Amphibious Car and Boat. The multiplicities have been added correctly, and the XOR constraint has been applied. If we apply Rule 1b to part a, we end up with the diagram in part c, where all the attributes of Car have been copied down into Flying Car and Amphibious Car. In this latter case, you might have to deal with the effects of inheritance conflicts (see earlier in the chapter).

The advantage of Rule 1a is that all problem domain classes identified during analysis are preserved. This allows maximum flexibility of maintenance of the design of the problem

---

[28] The mapping rules presented in this section are based on material in Coad and Yourdon, *Object-Oriented Design*.

[29] In this case, we are talking about implementation inheritance, not the interface inheritance. Interface inheritance supported by Visual Basic and Java supports only inheriting the requirements to implement certain methods, not any implementation. Java and Visual Basic.net also support single inheritance as described in this text, whereas Visual Basic 6 supports only interface inheritance.

[30] It is also a good idea to document this modification in the design so that in the future, modifications to the design can be maintained easily.

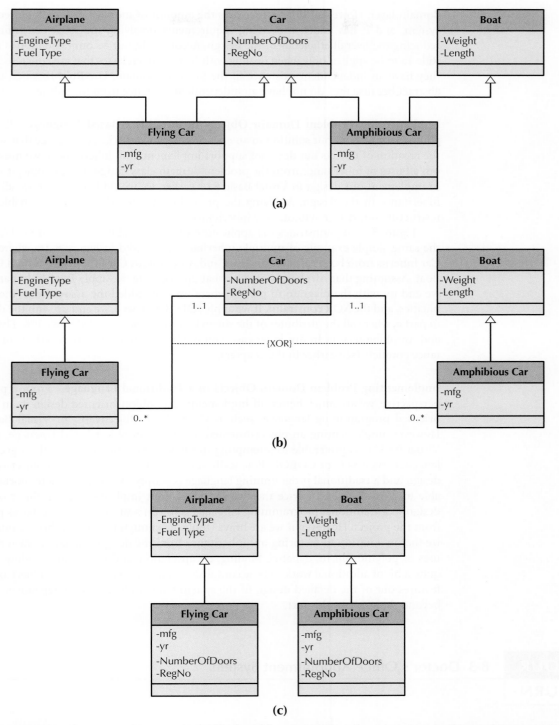

FIGURE 8-16   Factoring Out Multiple-Inheritance Effect for a Single-Inheritance Language

domain layer. However, Rule 1a increases the amount of message passing required in the system, and it has added processing requirements involving the XOR constraint, thus reducing the overall efficiency of the design. Accordingly, our recommendation is to limit Rule 1a to be applied only when dealing with "extra" superclasses that are concrete because they have an independent existence in the problem domain. Use Rule 1b when they are abstract because they do not have an independent existence from the subclass.

Implementing Problem Domain Objects in an Object-Based Language   If we are going to implement our solution in an *object-based language* (i.e., a language that supports the creation of objects but does not support implementation inheritance), we must factor out all uses of inheritance from the problem domain class design. For example, if we were to implement our design in Visual Basic 6 or earlier, we would have to remove all uses of inheritance in the design. Applying the preceding rule to all superclasses enables us to restructure our design without any inheritance.

Figure 8-17 demonstrates the application of the preceding rules. Figure 8-17a shows the same simple example of multiple inheritance portrayed in Figure 8-16, where Flying Car inherits from both Airplane and Car, and Amphibious Car inherits from both Car and Boat. Assuming that Airplane, Car, and Boat are concrete, we apply Rule 1a to part a and we end up with the diagram in part b, where we have added the associations, the multiplicities, and the XOR constraint. If we apply Rule 1b to part a, we end up with the diagram in part c, where all the attributes of the superclasses have been copied down into Flying Car and Amphibious Car. In this latter case, you might have to deal with the effects of inheritance conflicts (see earlier in the chapter).

Implementing Problem Domain Objects in a Traditional Language   From a practical perspective, we are much better off implementing an object-oriented design in an object-oriented programming language, such as C++, Java, Objective-C, or Visual Basic.net. However, implementing an object-oriented design in an object-based language, such as Visual Basic 6, is preferable to attempting to implement it in a traditional programming language, such as C or COBOL. Practically speaking, the gulf between an object-oriented design and a traditional programming language is simply too great for mere mortals to be able to cross. The best advice that we can give about implementing an object-oriented design in a traditional programming language is to run away as fast and as far as possible from the project. However, if we are brave (foolish?) enough to attempt this, we must realize that in addition to factoring out inheritance from the design, we have to factor out all uses of polymorphism, dynamic binding, encapsulation, and information hiding. This is quite a bit of additional work to be accomplished. The way we factor these object-oriented features out of the detailed design of the system tends to be language dependent. This is beyond the scope of this text.

YOUR TURN

8-3 Doctor's Office Appointment System

In the previous chapters, we have been using a doctor's office appointment system as an example. Assume that you now know that the system must be implemented in Visual Basic 6, which does not support implementation inheritance. Redraw the class diagram, factoring out the use of inheritance in the design by applying the above rules.

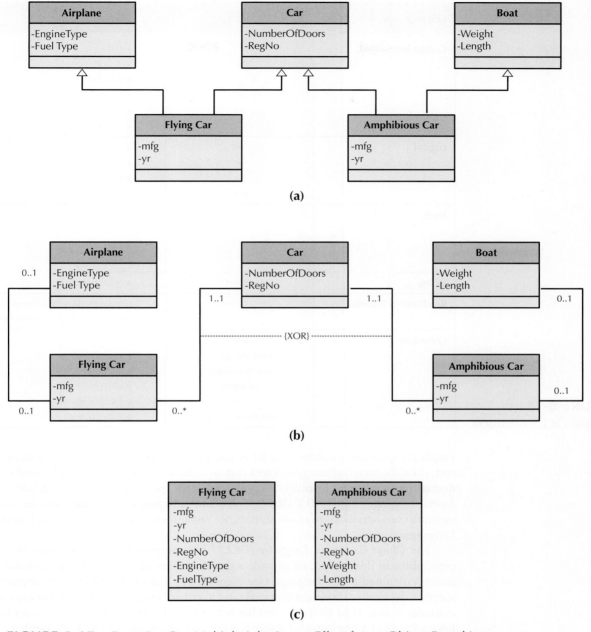

FIGURE 8-17 Factoring Out Multiple-Inheritance Effect for an Object-Based Language

CONSTRAINTS AND CONTRACTS

Contracts were introduced in Chapter 5 in association with collaborations. A contract formalizes the interactions between the client and server objects, where a *client* (*consumer*) object is an instance of a class that sends a message to a *server* (*supplier*) object that executes one of its methods in response to the request. Contracts are modeled on the legal notion of a contract, where both parties, client and server objects, have obligations and rights.

Operator Type	Operator	Example
Comparison	=	a = 5
	<	a < 100
	<=	a <= 100
	>	a > 100
	>=	a >= 100
	<>	a <> 100
Logical	and	a and b
	or	a or b
	xor	a xor b
	not	not a
Math	+	a + b
	−	a − b
	*	a * b
	/	a / b
String	concat	a = b.concat(c)
Relationship Traversal	.	relationshipAttributeName.b
	::	superclassName::propertyName
Collection	size	a.size
	count(object)	a.count(b)
	includes(object)	a.includes(b)
	isEmpty	a.isEmpty
	sum()	a.sum(b,c,d)
	select(expression)	a.select(b > d)

FIGURE 8-18
Sample OCL Constructs

Practically speaking, a contract is a set of constraints and guarantees. If the constraints are met, then the server object guarantees certain behavior.[31] Constraints can be written in a natural language (e.g., English), a semiformal language (e.g., *Structured English*[32]), or a formal language (e.g., UML's Object Constraint Language). Given the need for precise, unambiguous specification of constraints, we recommend using UML's Object Constraint Language.

The *Object Constraint Language (OCL)*[33] is a complete language designed to specify constraints. In this section, we provide a short overview of some of the more useful constructs contained in the language (see Figure 8-18). Essentially, all OCL expressions are simply a declarative statement that evaluates to either being true or false. If the expression evaluates to true, then the constraint has been satisfied. For example, if a customer had to have a less than a one hundred dollar balance owed to be allowed to place another credit order, the OCL expression would be:

balance owed <= 100.00

---

[31] The idea of using contracts in design evolved from the "Design by Contract" technique developed by Bertrand Meyer. See Bertrand Meyer, *Object-Oriented Software Construction* (Englewood Cliffs, NJ: Prentice Hall, 1988).

[32] We describe Structured English with Method Specification later in this chapter.

[33] For a complete description of the object constraint language, see Jos Warmer and Anneke Kleppe, *The Object Constraint Language: Precise Modeling with UML* (Reading, MA: Addison-Wesley, 1999).

OCL also has the ability to traverse relationships between objects, e.g., if the amount on a purchase order is required to be the sum of the values of the individual purchase order lines, this can be modeled as:

amount = OrderLine.sum(getPrice())

OCL also provides the ability to model more complex constraints with a set of logical operators: and, or, xor, and not. For example, if customers were to be given a discount only if they were a senior citizen or a "prime" customer, OCL could be used to model the constraint as:

age > 65 or customerType = "prime"

OCL provides many other constructs that can be used to build unique constraints. These include math-oriented operators, string operators, and relationship traversal operators. For example, if the printed name on a customer order should be the concatenation of the customer's first name and last name, then OCL could represent this constraint as:

printedName = firstName.concat(lastName)

We already have seen an example of the '.' operator being used to traverse a relationship from Order to OrderLine above. The '::' operator allows the modeling of traversing inheritance relationships.

OCL also provides a set of operations that are used to support constraints over a collection of objects. For example, we demonstrated the use of the sum() operator above where we wanted to guarantee that the amount was equal to the summation of all of the prices of the items in the collection. The size operation returns the number of items in the collection. The count operation returns the number of occurrences in the collection of the specific object passed as its argument. The includes operation tests whether the object passed to it is already included in the collection. The isEmpty operation determines whether the collection is empty or not. The select operation provides support to model the identification of a subset of the collection based on the expression that is passed as its argument. Obviously, OCL provides a rich set of operators and operations in which to model constraints.

Types of Constraints

Three different types of constraints are typically captured in object-oriented design: preconditions, postconditions, and invariants.

Contracts are used primarily to establish the preconditions and postconditions for a method to be able to execute properly. A *precondition* is a constraint that must be met for a method to execute. For example, the parameters passed to a method must be valid for the method to execute. Otherwise, an exception should be raised. A *postcondition* is a constraint that must be met after the method executes, or the effect of the method execution must be undone. For example, the method cannot make any of the attributes of the object take on an invalid value. In this case, an exception should be raised, and the effect of the method's execution should be undone.

Whereas preconditions and postconditions model the constraints on an individual method, *invariants* model constraints that must always be true for all instances of a class. Examples of invariants include domains or types of attributes, multiplicity of attributes, and the valid values of attributes. This includes the attributes that model association and aggregation relationships. For example, if an association relationship is required, an invariant should be created that will enforce it to have a valid value for the instance to exist. Invariants are normally attached to the class. We can attach invariants to the CRC cards or class diagram by adding a set of assertions to them.

In Figure 8-19, the back of the CRC card constrains the attributes of an Order to specific types. For example, Order Number must be an unsigned long, and Customer must be

Front:

Class Name: Order	ID: 2	Type: Concrete, Domain

Description: An Individual that needs to receive or has received medical attention	Associated Use Cases: 3

Responsibilities	Collaborators
Calculate subtotal	
Calculate tax	
Calculate shipping	
Calculate total	

(a)

Back:

Attributes:

Order Number	(1..1)	(unsigned long)	
Date	(1..1)	(Date)	
Sub Total	(0..1)	(double)	{Sub Total = ProductOrder. sum(GetExtension())}
Tax	(0..1)	(double)	(Tax = State.GetTaxRate() * Sub Total)
Shipping	(0..1)	(double)	
Total	(0..1)	(double)	
Customer	(1..1)	(Customer)	
Cust ID	(1..1)	(unsigned long)	{Cust ID = Customer. GetCustID()}
State	(1..1)	(State)	
StateName	(1..1)	(String)	{State Name = State. GetState()}

Relationships:

Generalization (a-kind-of):

Aggregation (has-parts):

Other Associations: Customer {1..1} State {1..1}Product {1..*}

FIGURE 8-19

Invariants on a CRC Card

(b)

an instance of the Customer class. Furthermore, additional invariants were added to four of the attributes. For example, Cust ID must not only be an unsigned long, but it also must have one and only one value [i.e., a multiplicity of (1..1)], and it must have the same value as the result of the GetCustID() message sent to the instance of Customer stored in the Customer attribute. Also shown is the constraint for an instance to exist, an instance of the Customer class, an instance of the State class, and at least one instance of the Product class must be associated with the Order object (see the Relationships section of the CRC card where the multiplicities are 1..1, 1..1, and 1..*, respectively). Figure 8-20 portrays the same set of invariants on a class diagram. However, if all invariants are placed on a class diagram, the diagram becomes very difficult to understand. Consequently, we recommend either extending the CRC card to document the invariants instead of attaching them all to the class diagram or creating a separate text document that contains them (see Figure 8-21).

YOUR	**8-4 Campus Housing**
TURN	

*I*n Your Turn 5-2, you created a set of CRC cards and a class diagram. Add invariants to the class diagram and to the set of CRC cards.

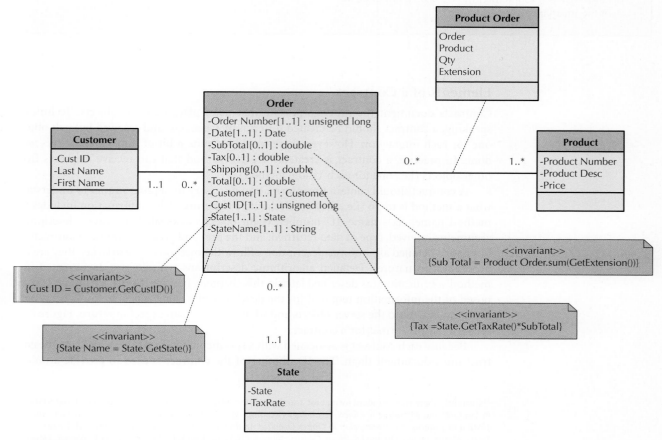

FIGURE 8-20   Invariants on a Class Diagram

FIGURE 8-21
Invariants in a Text File

> **Order class invariants:**
> Cust ID = Customer.GetCustID()
> State Name = Sate.GetState()
> Sub Total = ProductOrder.sum(GetExtension())
> Tax = State.GetTaxRate() * Sub Total1

YOUR TURN

8-5 Invariants

*U*sing the CRC card in Figure 8-19 and the class diagram in Figure 8-20 as guides, add invariants for the Customer class, the State class, the Product class, and the Product-Order Association to their respective CRC cards and the class diagram.

Questions

1. How easy is it to interpret the class diagram once all of the invariants were added?

2. Look at the class diagram in Figure 7-15 and the package diagram in Figure 7-22. What would they look like if the invariants were all attached to the diagrams? What would you recommend doing to avoid this situation?

Elements of a Contract

Contracts document the message passing that takes place between objects. Technically speaking, a contract should be created for each message sent and received by each object, one for each interaction. However, there would be quite a bit of duplication if this were done. In practice, a contract is created for each method that can receive messages from other objects (i.e., one for each visible method).

A contract should contain the information necessary for a programmer to understand what a method is to do (i.e., they are declarative in nature). This information includes the method name, class name, ID number, client objects, associated use cases, description, arguments received, type of data returned, and the pre- and postconditions.[34] Contracts do not have a detailed algorithmic description of how the method is to work (i.e., they are not procedural in nature). Detailed algorithmic descriptions typically are documented in a method specification (as described later in this chapter). In other words, a contract is composed of the information required for the developer of a client object to know what messages can be sent to the server objects and what the client can expect in return. Figure 8-22 shows a sample format for a contract.

Because each contract is associated with a specific method and a specific class, the contract must document them. The ID number of the contract is used to provide a unique

---

[34] Currently, there is no standard format for a contract. The contract in Figure 8-22 is based on material contained in Ian Graham, *Migrating to Object Technology* (Reading, MA: Addison-Wesley, 1995); Craig Larman, *Applying UML and Patterns: An Introduction to Object-Oriented Analysis and Design* (Englewood Cliffs, NJ: Prentice Hall, 1998); Meyer, *Object-Oriented Software Construction;* and R. Wirfs-Brock, B. Wilkerson, and L. Wiener, *Designing Object-Oriented Software* (Englewood Cliffs, NJ: Prentice Hall, 1990).

Method Name:	Class Name:		ID:
Clients (Consumers):			
Associated Use Cases:			
Description of Responsibilities:			
Arguments Received:			
Type of Value Returned:			
Preconditions:			
Postconditions:			

FIGURE 8-22
Sample Contract Form

identifier for every contract. The Clients (Consumers) element of a contract is a list of classes and methods that send a message to this specific method. This list is determined by reviewing the sequence diagrams associated with the server class. The Associated Use Cases element is a list of use cases in which this method is used to realize the implementation of the use case. The use cases listed here can be found by reviewing the server class's CRC card and the associated sequence diagrams.

The Description of Responsibilities provides an informal description of what the method is to perform, not how it is to do it. The arguments received are the data types of the parameters passed to the method, and the value returned is the data type of the value that the method returns to its clients. Together with the method name, they form the signature of the method.

The precondition and postcondition elements are where the pre- and postconditions for the method are recorded. Recall that pre- and postconditions can be written in a natural language, a semiformal language, or a formal language. However, the more precisely they are written, the less likely it is that the programmer will misunderstand them. As with invariants, we recommend that you use UML's Object Constraint Language.[35]

Example   In this example, we return to the order example shown in Figures 8-15, 8-19, 8-20, and 8-21. In this case, we limit the discussion to the design of the addOrder method for the Customer class. The first decision we must make is how to specify the design of the relationship from Customer to Order. By reviewing Figures 8-15, 8-19, and 8-20, we see that the relationships have a multiplicity of 0..*, which means that an instance of customer

---

[35] See Warmer and Kleppe, *The Object Constraint Language: Precise Modeling with UML.*

may exist without having any orders or an instance of customer could have many orders. As shown in Figure 8-15c, the relationship has been converted to an attribute that can contain many instances of the Order class.

However, an important question that would not typically come up during analysis is whether the order objects should be kept in sorted order or not. Another question that is necessary to have answered for design purposes is how many orders could be expected by a customer. The answers to these two questions will determine how we should organize the orders from the customer object's perspective. If the number of orders is going to be relatively small and the orders don't have to be kept in sorted order, then using a built-in programming language construct such as a vector is sufficient. However, if the number of orders is going to be large or the orders must be kept in sorted order, then some form of a sorted data structure, such as a linked list, is necessary. For example purposes, we assume that a customer's orders will need to be kept in sorted order and that there will be a large number of them. Therefore, instead of using a vector to contain the orders, we use a sorted singly linked list.

To keep the design of the Customer class as close to the problem domain representation as possible, the design of the Customer class is based on the Iterator pattern in Figure 8-13. For simplicity purposes, we assume that an order is created before it is associated with the specific customer. Otherwise, given the additional constraints of the instance of State class and the instance of the Product Order class existing before an instance of Order can be created would also have to be taken into consideration. This assumption allows us to ignore the fact that an instance of State can have many orders, an instance of Order can have many instances of Product Order associated with it, and an instance of Product can have many instances of Product Order associated with it, which would require us to design many additional containers (vectors or other data structures).

Based on all of the above, a new class diagram fragment was created that represents a linked list-based relationship between instances of the Customer class and instances of the Order class (see Figure 8-23). By carefully comparing Figures 8-15 and 8-23, we see that the Iterator pattern idea has been included between the Customer and Order classes. The domain of the Orders relationship-based attribute of the Customer class has been replaced with OrderList to show that the list of orders will be contained in a list data structure. Figure 8-24 portrays an object-diagram-based representation of how the relationship between a customer instance and a set of order instances is stored in a sorted singly linked list data structure. In this case, we see that a Customer object has an OrderList object associated with it, each OrderList object could have N OrderNode objects, and each OrderNode object will have an Order object. We see that each Order object is associated with a single Customer object. By comparing Figures 8-15 and 8-24, we see that the intention of the multiplicity constraints of the Orders attribute of Customer, where a customer can have many orders, and the multiplicity constraints of the Customer attribute of Orders is being modeled correctly. Finally, notice that one of the operations contained in the OrderList class is a private method. We will return to this specific point in the next section that addresses method specification.

Using Figures 8-22, 8-23, and 8-24, contracts for the addOrder method of the Customer class and the insertOrder method for the OrderList class can be specified (see Figure 8-25). In the case of the addOrder method of the Customer class, we see that only instances of the Order class use the method (see Clients section), that the method only implements part of the logic that supports the addCustomerOrder use case (see Associated Use Cases section), and that the contract includes a short description of the methods responsibilities. We also see that the method receives a single argument of type Order and that it does not return anything (void). Finally, we see that both a precondition and a postcondition were

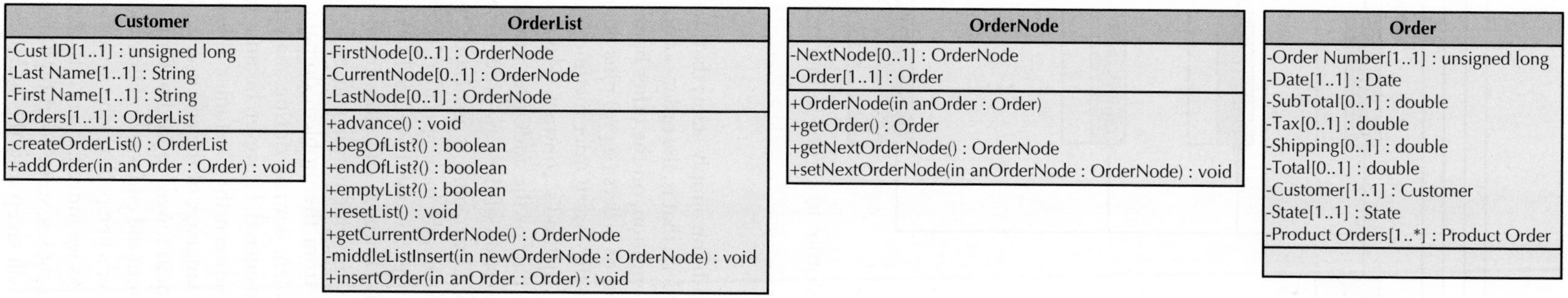

Customer

-Cust ID[1..1] : unsigned long
-Last Name[1..1] : String
-First Name[1..1] : String
-Orders[1..1] : OrderList

-createOrderList() : OrderList
+addOrder(in anOrder : Order) : void

OrderList

-FirstNode[0..1] : OrderNode
-CurrentNode[0..1] : OrderNode
-LastNode[0..1] : OrderNode

+advance() : void
+begOfList?() : boolean
+endOfList?() : boolean
+emptyList?() : boolean
+resetList() : void
+getCurrentOrderNode() : OrderNode
-middleListInsert(in newOrderNode : OrderNode) : void
+insertOrder(in anOrder : Order) : void

OrderNode

-NextNode[0..1] : OrderNode
-Order[1..1] : Order

+OrderNode(in anOrder : Order)
+getOrder() : Order
+getNextOrderNode() : OrderNode
+setNextOrderNode(in anOrderNode : OrderNode) : void

Order

-Order Number[1..1] : unsigned long
-Date[1..1] : Date
-SubTotal[0..1] : double
-Tax[0..1] : double
-Shipping[0..1] : double
-Total[0..1] : double
-Customer[1..1] : Customer
-State[1..1] : State
-Product Orders[1..*] : Product Order

FIGURE 8-23 Class Diagram Fragment of the Customer to Order Relationship Modeled as a Sorted Singly Linked List

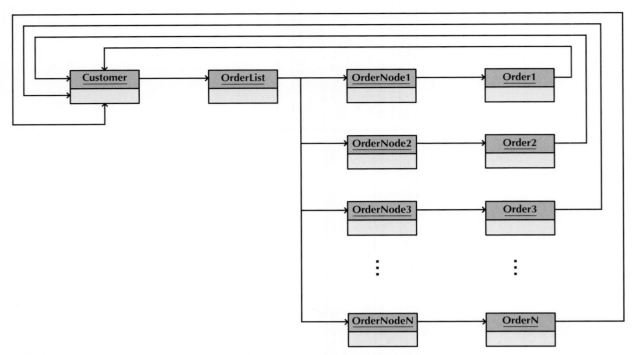

FIGURE 8-24 Object Diagram of the Customer to Order Relationship Modeled as a Sorted Singly Linked List

specified. The precondition simply states that the new Order object cannot be included in the current list of Orders; that is, the order cannot have previously been associated with this customer. The postcondition, on the other hand, specifies that the new list of orders must be equal to the old list of orders (@pre) plus the new order object (including).

The contract for the insertOrder method for the OrderList class is somewhat simpler than the addOrder method's contract. From a practical perspective, the insertOrder method implements part of the addOrder method's logic. Specifically speaking, it implements that actual insertion of the new order object into the specific data structure chosen to manage the list of Order objects associated with the specific Customer object. Consequently, because we already have specified the precondition and postcondition for the addOrder method, we do not have to further specify the same constraints for the insertOrder method. However, this does implicitly increase the dependence of the Customer objects on the implementation chosen for the list of customer orders. This is a good example of moving from the problem domain to the solution domain. While we were focusing on the problem domain during analysis, the actual implementation of the list of orders was never considered. However, because we now are designing the implementation of the relationship between the Customer objects and the Order objects, we have had to move away from the language of the end user and toward the language of the programmer. During design, the focus moves toward optimizing the code to run faster on the computer and not worrying about the end user's ability to understand the inner workings of the system; from an end user's perspective, the system should become more of a black box with which they interact. As we move farther into the detailed design of the implementation of the problem domain classes, some solution domain classes, such as the approach to implement relationships, will creep into the specification of the problem domain layer. In this

Method Name:	addOrder	Class Name:	Customer	ID:	36

Clients (consumers):	Order

Associated Use Cases:

        addCustomerOrder

Description of Responsibilities:

    Implement the necessary behavior to add a new order to an existing customer keeping
the orders in sorted order by the order's order number.

Arguments Received:

        anOrder:Order

Type of Value Returned: void

Preconditions:

        not orders.includes(anOrder)

Postconditions:

        Orders = Orders@pre.including(anOrder)

Method Name:	insertOrder	Class Name:	OrderList	ID:	123

Clients (consumers):	Customer

Associated Use Cases:

        addCustomerOrder

Description of Responsibilities:
    Implement inserting an Order object into an OrderNode object and manage the
insertion of the OrderNode object into the current location in the sorted singly
linked list of orders.

Arguments Received:

        anOrder:Order

Type of Value Returned: void

Preconditions:

       None.

Postconditions:

       None.

FIGURE 8-25

Sample Contract for the
addOrder Method of the
Customer class and the
insertOrder method of
the OrderList class

particular example, the OrderList and OrderNode classes also could be used to implement the relationships from State objects to Order objects, from Order objects to Product Order objects, and from Product objects to Product Order objects (see Figure 8-15). Given our simple example, one can clearly see that specifying the design of the problem domain layer could include many additional solution domain classes to be specified on the problem domain layer.

<table>
<tr><td>YOUR
TURN</td><td>8-6 Contract</td></tr>
</table>

Using the CRC card in Figure 8-19, the class diagram in Figure 8-20, and the contract forms in Figure 8-25 as guides, create contracts for the calculate subtotal, calculate tax, calculate shipping, and calculate total methods.

METHOD SPECIFICATION

Once the analyst has communicated the big picture of how the system needs to be put together, he or she needs to describe the individual classes and methods in enough detail so that programmers can take over and begin writing code. Methods on the CRC cards, class diagram, and contracts are described using *method specifications*. Method specifications are written documents that include explicit instructions on how to write the code to implement the method. Typically, project team members write a specification for each method and then pass them all along to programmers, who write the code during implementation of the project. Specifications need to be very clear and easy to understand, or programmers will be slowed down trying to decipher vague or incomplete instructions.

There is no formal syntax for a method specification, so every organization uses its own format, often using a form like the one in Figure 8-26. Typical method specification forms contain four components that convey the information that programmers will need for writing the appropriate code: general information, events, message passing, and algorithm specification.

General Information

The top of the form in Figure 8-26 contains general information, such as the name of the method, name of the class in which this implementation of the method will reside, ID number, Contract ID (which identifies the contract associated with this method implementation), programmer assigned, the date due, and the target programming language. This information is used to help manage the programming effort.

Events

The second section of the form is used to list the events that trigger the method. An *event* is a thing that happens or takes place. Clicking the mouse generates a mouse event, pressing a key generates a keystroke event—in fact, almost everything the user does generates an event.

In the past, programmers used procedural programming languages (e.g., COBOL, C) that contained instructions that were implemented in a predefined order, as determined by the computer system, and users were not allowed to deviate from the order. Many

Method Name:		Class Name:		ID:
Contract ID:		Programmer:		Date Due:

Programming Language:

☐ Visual Basic   ☐ Smalltalk   ☐ C++   ☐ Java

Triggers/Events:

Arguments Received: Data Type:	Notes:

Messages Sent & Arguments Passed: ClassName.MethodName:	Data Type:	Notes:

Arguments Returned: Data Type:	Notes:

Algorithm Specification:

Misc. Notes:

FIGURE 8-26

Method Specification
Form

programs today are *event driven* (e.g., programs written in languages such as Visual Basic, Objective C, C++, or Java), and event-driven programs include methods that are executed in response to an event initiated by the user, system, or another method. After initialization, the system waits for an event to occur. When it does, a method is fired that carries out the appropriate task, and then the system waits once again.

We have found that many programmers still use method specifications when programming in event-driven languages, and they include the event section on the form to capture when the method will be invoked. Other programmers have switched to other design tools that capture event-driven programming instructions, such as the behavioral state machine described in Chapter 6.

Message Passing

The next sections of the method specification describe the message passing to and from the method, which are identified on the sequence and collaboration diagrams. Programmers need to understand what arguments are being passed into, passed from, and returned by the method because the arguments ultimately translate into attributes and data structures within the actual method.

Algorithm Specifications

Algorithm specifications can be written in Structured English or some type of formal language.[36] Structured English is simply a formal way of writing instructions that describe the steps of a process. Because it is the first step toward the implementation of the method, it looks much like a simple programming language. Structured English uses short sentences that clearly describe exactly what work is performed on what data. There are many versions of Structured English because there are no formal standards; each organization has its own type of Structured English. Figure 8-27 shows some examples of commonly used Structured English statements.

Common Statements	Example
Action Statement	Profits = Revenues – Expenses Generate Inventory-Report
If Statement	IF Customer Not in the Customer Object Store THEN Add Customer record to Customer Object Store ELSE Add Current-Sale to Customer's Total-Sales Update Customer record in Customer Object Store

For Statement	FOR all Customers in Customer Object Store DO
	    Generate a new line in the Customer-Report
	    Add Customer's Total-Sales to Report-Total

Case Statement	CASE
	    IF Income < 10,000: Marginal-tax-rate = 10 percent
	    IF Income < 20,000: Marginal-tax-rate = 20 percent
	    IF Income < 30,000: Marginal-tax-rate = 31 percent
	    IF Income < 40,000: Marginal-tax-rate = 35 percent
	    ELSE Marginal-Tax-Rate = 38 percent
	ENDCASE

FIGURE 8-27
Structured English

[36] For our purposes, Structured English will suffice. However, there has been some work with the Catalysis, Fusion, and Syntropy methodologies to include formal languages, such as VDM and Z, into specifying object-oriented systems.

Action statements are simple statements that perform some action. An If statement controls actions that are performed under different conditions, and a For statement (or a While statement) performs some actions until some condition is reached. A Case statement is an advanced form of an If statement that has several mutually exclusive branches.

If the algorithm of a method is complex, a tool that can be useful for algorithm specification is UML's *activity diagram* (see Figure 8-28 and Chapter 4). Recall that activity diagrams can be used to specify any type of process. Obviously, an algorithm specification represents a process. However, owing to the nature of object orientation, processes tend to be highly distributed over many little methods over many objects. Needing to use an activity diagram to specify the algorithm of a method can, in fact, hint at a problem in the design. For example, the method should be further decomposed or there could be missing classes.

The last section of the method specification provides space for other information that needs to be communicated to the programmer, such as calculations, special business rules, calls to subroutines or libraries, and other relevant issues. This also can point out changes or improvements that will be made to any of the other design documentation based on problems that the analyst detected during the specification process.[37]

Example

This example continues the addition of a new order for a customer described in the previous section (see Figure 8-29). Even though in most cases, because there are libraries of data structure classes available that you could simply reuse and therefore would not need to specify the algorithm to insert into a sorted singly linked list, we use it as an example of how method specification can be accomplished. The general information section of the specification documents the method's name, its class, its unique ID number, the ID number of its associated contract, the programmer assigned, the date that its implementation is due, and the programming language to be used. Second, the trigger/event that caused this method to be executed is identified. Third, the data type of the argument passed to this method is documented (Order). Fourth, owing to the overall complexity of inserting a new node into the list, we have factored out one specific aspect of the algorithm into a separate private method (middleListInsert()) and we have specified that this method will be sending messages to instances of the OrderNode class and the Order class. Fifth, we specify the type of return value that insertOrder will produce. In this case, the insertOrder method will not return anything (void). Finally, we specify the actual algorithm. In this example, for the sake of completeness, we provide both a Structured English–based (see Figure 8-30) and an activity diagram–based algorithm specification (see Figure 8-31). Previously, we stated that we had factored out the logic of inserting into the middle of the list into a separate private method: middleListInsert(). Figure 8-32 shows the logic of this method. Imagine collapsing this logic back into the logic of the insertOrder method, i.e., replace the middleListInsert(newOrderNode) activity in Figure 8-31 with the contents of Figure 8-32. Obviously, the insertOrder method would be more complex.

---

[37] Remember that the development process is very incremental and iterative. Therefore, changes could be cascaded back to any point in the development process (e.g., to use-case descriptions, use-case diagrams, CRC cards, class diagrams, object diagrams, sequence diagrams, communication diagrams, behavioral state machines, and package diagrams).

An action: ■ Is a simple, non-decomposable piece of behavior. ■ Is labeled by its name.	Action
An activity: ■ Is used to represent a set of actions. ■ Is labeled by its name.	Activity
An object node: ■ Is used to represent an object that is connected to a set of object flows. ■ Is labeled by its class name.	Class Name
A control flow: ■ Shows the sequence of execution.	→
An object flow: ■ Shows the flow of an object from one activity (or action) to another activity (or action).	------→
An initial node: ■ Portrays the beginning of a set of actions or activities.	●
A final-activity node: ■ Is used to stop all control flows and object flows in an activity (or action).	⊗
A final-flow node: ■ Is used to stop a specific control flow or object flow.	◉
A decision node: ■ Is used to represent a test condition to ensure that the control flow or object flow only goes down one path. ■ Is labeled with the decision criteria to continue down the specific path.	[Decision Criteria] [Decision Criteria]
A merge node: ■ Is used to bring back together different decision paths that were created using a decision node.	
A Fork node: ■ Is used to split behavior into a set of parallel or concurrent flows of activities (or actions).	
A Join node: ■ Is used to bring back together a set of parallel or concurrent flows of activities (or actions).	
A Swimlane: ■ Is used break up an activity diagram into rows and columns to assign the individual activities (or actions) to the individuals or objects that are responsible for executing the activity (or action). ■ Is labeled with the name of the individual or object responsible.	Swimlane

FIGURE 8-28   Syntax for an Activity Diagram (Figure 4-6)

Method Name: insertOrder	Class Name: OrderList	ID: 100
Contract ID: 123	Programmer: J. Doe	Date Due: 1/1/12

Programming Language:

❏ **Visual Basic**   ❏ **Smalltalk**   ❏ **C++**   ❏ **Java**

Triggers/Events:

Customer places an order

Arguments Received:

Data Type:	Notes:
Order	The new customer's new order.

Messages Sent & Arguments Passed:

ClassName.MethodName:	Data Type:	Notes:
OrderNode.new()	Order	
OrderNode.getOrder()		
Order.getOrderNumber()		
OrderNode.setNextNode()	OrderNode	
self.middleListInsert()	OrderNode	

Arguments Returned:

Data Type:	Notes:
void	

Algorithm Specification:

See Figures 8-30 and 8-31.

Misc. Notes:

None.

FIGURE 8-29   Method Specification for the insertOrder method

FIGURE 8-30

Structured English-based Algorithm Specification for the insertOrder Method

```
Create new OrderNode with the new Order
IF emplyList?()
   FirstNode = LastNode = CurrentNode = newOrderNode
ELSE IF newOrderNode.getOrder().getOrderNumber() < FirstNode.getOrder().getOrderNumber()
   newOrderNode.setNextNode(FirstNode)
   FirstNode = newOrderNode
ELSE IF newOrderNode.getOrder().getOrderNumber() > LastNode.getOrder().getOrderNumber()
   LastNode.setNextNode(newOrderNode)
   LastNode = newOrderNode
ELSE
   middleListInsert(newOrderNode)
```

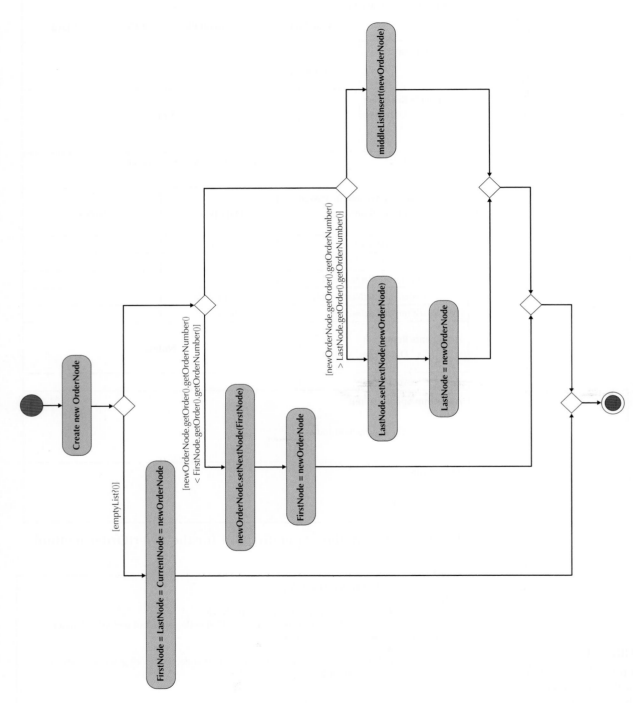

FIGURE 8-31   Activity-Diagram-based Algorithm Specification for the insertOrder Method

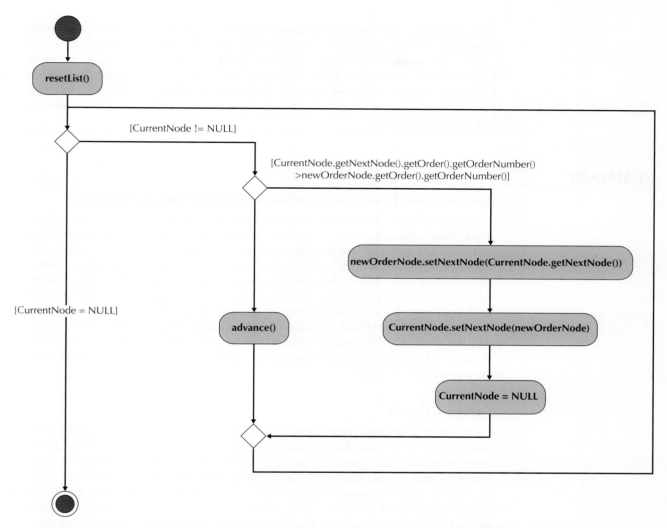

FIGURE 8-32   Activity Diagram–based Algorithm Specification for the middleListInsert Method

YOUR	**8-7 Method Specification**
TURN	

*U*sing the CRC card in Figure 8-19, the class diagram in Figure 8-20, and the contracts created in Your Turn 8-6 as guides, create method specifications for the calculate subtotal, calculate tax, calculate shipping, and calculate total methods.

APPLYING THE CONCEPTS AT CD SELECTIONS

Up until now, Alec, Margaret, and the development team members have been focusing on being sure that they captured the underlying behavior and structure of the evolving system. During this installment, Alec instructs the team members to make sure that the connascence is minimized at all levels of the design, to identify any opportunities for reuse, to consider restructuring and optimizing the evolving specification. Furthermore, he instructed them to identify any and all constraints that need to be modeled. He also suggested that they define the invariants in a separate text file and to define the preconditions and postconditions for all public methods using contracts. Finally, he instructed the team to specify every method using the method specification form.

SUMMARY

Review of the Basic Characteristics of Object-Oriented Systems

A class is a template on which objects can be instantiated. An object is a person, place, or thing about which we want to capture information. Each object has attributes and methods. The methods are executed by objects sending messages that trigger them. Encapsulation and information hiding allows an object to conceal its inner processes and data from the other objects. Polymorphism and dynamic binding allow a message to be interpreted differently by different kinds of objects. However, if polymorphism is not used in a semantically consistent manner, it can make an object design incomprehensible. Classes can be arranged in a hierarchical fashion in which subclasses inherit attributes and methods from superclasses to reduce the redundancy in development. However, through redefinition capabilities or multiple inheritance, inheritance conflicts can be introduced into the design.

Design Criteria

Coupling, cohesion, and connascence have been put forth as a set of criteria by which a design of an object-oriented system can be evaluated. Two types of coupling, interaction and inheritance, and three types of cohesion, method, class, and generalization/specialization, were described. Interaction coupling deals with the communication that takes place among the objects, and inheritance coupling deals with the innate dependencies in the use of inheritance in object-oriented systems. Method cohesion addresses how single-minded a method is. The fewer things a method does, the more cohesive it is. Class cohesion does the same for classes. A class should be a representation of one and only one thing. Generalization/specialization cohesion deals with the quality of an inheritance hierarchy. Good inheritance hierarchies support only generalization and specialization (a-kind-of) semantics. Connascence generalizes coupling and cohesion and then combines them with the different levels of encapsulation. The general rule is to maximize the cohesion (connascence) within an encapsulation boundary and minimize the coupling (connascence) between the encapsulation boundaries.

Object Design Activities

There are five basic object design activities. First, additional specification is possible by carefully reviewing the models, deciding on the proper visibility of the attributes and methods, setting the signature for each method, and identifying any constraints associated with the classes or the classes' methods. Second, look for opportunities for reuse by reviewing the model and looking at possible patterns, class libraries, frameworks, and components that could be used to enhance the system. Third, restructure the model through the use of factoring and normalization. Be sure to take the programming language into consideration. It may be necessary to map the current design into the restricted capabilities of the language (e.g., the language supports only single inheritance). Also, be certain that the inheritance in your model supports only generalization/specialization (a-kind-of) semantics. Fourth, optimize the design. However, be careful in the process of optimizing the design. Optimizations typically decrease the understandability of the model. Fifth, map the problem domain classes to an implementation language.

Constraints and Contracts

Three types of *constraints* are associated with object-oriented design: invariants, preconditions, and postconditions. Invariants capture constraints that must always be true for all instances of a class (e.g., domains and values of attributes or multiplicity of relationships). Typically, invariants are attached to class diagrams and CRC cards. However, for clarity purposes, we suggest placing them on the CRC cards and/or in a separate text file.

Contracts formalize the interaction between objects (i.e., the message passing). Thus, they include the pre- and postconditions that must be enforced for a method to execute properly. Contracts provide an approach to modeling the rights and obligations when client and server objects interact. From a practical perspective, all interactions that can take place between all possible client and server objects are not modeled on separate contracts. Instead, a single contract is drawn up for using each visible method of a server object.

Method Specification

A method specification is a written document that provides clear and explicit instructions on how a method is to behave. Without clear and unambiguous method specifications, critical design decisions have to be made by programmers instead of by designers. Even though there is no standard format for a method specification, typically, four types of information are captured. First, there is general information such as the name of the method, name of the class, contract ID, programmer assigned, the date due, and the target programming language. Second, owing to the rise in popularity of GUI-based and event-driven systems, events also are captured. Third, the information that deals with the signature of the method, data received, data passed on to other methods, and data returned by the method is recorded. Finally, an unambiguous specification of the algorithm is given. The algorithm typically is modeled using Structured English, an activity diagram, or a formal language.

KEY TERMS

QUESTIONS

1. What are the basic characteristics of object-oriented systems?
2. What is dynamic binding?
3. Define polymorphism. Give one example of a good use of polymorphism and one example of a bad use of polymorphism.
4. What is an inheritance conflict? How does an inheritance conflict affect the design?
5. Why is cancellation of methods a bad thing?
6. Give the guidelines to avoid problems with inheritance conflicts.
7. How important is it to know which object-oriented programming language is going to be used to implement the system?
8. What additional types of inheritance conflicts are there when using multiple inheritance?
9. What is the law of Demeter?
10. What are the six types of interaction coupling? Give one example of good interaction coupling and one example of bad interaction coupling.
11. What are the seven types of method cohesion? Give one example of good method cohesion and one example of bad method cohesion.
12. What are the four types of class cohesion? Give one example of each type.
13. What are the five types of connascence described in your text? Give one example of each type.
14. When designing a specific class, what types of additional specification for a class could be necessary?
15. What are exceptions?
16. What are constraints? What are the three different types of constraints?

17. What are patterns, frameworks, class libraries, and components? How are they used to enhance the evolving design of the system?
18. How are factoring and normalization used in designing an object system?
19. What are the different ways to optimize an object system?
20. What is the typical downside of system optimization?
21. What is the purpose of a contract? How are contracts used?
22. What is the Object Constraint Language? What is its purpose?
23. What is Structured English? What is its purpose?
24. What is an invariant? How are invariants modeled in a design of a class? Give an example of an invariant for an hourly employee class using the Object Constraint Language.
25. Create a contract for a compute pay method associated with an hourly employee class. Specify the preconditions and postconditions using the Object Constraint Language.
26. How do you specify a method's algorithm? Give an example of an algorithm specification for a compute pay method associated with an hourly employee class using Structured English.
27. How do you specify a method's algorithm? Give an example of an algorithm specification for a compute pay method associated with an hourly employee class using an activity diagram.
28. How are methods specified? Give an example of a method specification for a compute pay method associated with an hourly employee class.

EXERCISE

A. For the A Real Estate Inc. problem in Chapters 4 (exercises I, J, and K), 5 (exercises P and Q), 6 (exercise D), and 7 (exercise A):

1. Choose one of the classes and create a set of invariants for attributes and relationships, and add them to the CRC card for the class.

2. Choose one of the methods in the class that you chose and create a contract and a method specification for it. Use OCL to specify any pre- or postcondition and use both Structured English and an activity diagram to specify the algorithm.

B. For the A Video Store problem in Chapters 4 (exercises L, M, N K), 5 (exercises R and S), 6 (exercise E), and 7 (exercise B):

1. Choose one of the classes and create a set of invariants for attributes and relationships, and add them to the CRC card for the class.

2. Choose one of the methods in the class that you chose and create a contract and a method specification for it. Use OCL to specify any pre- or postcondition and use both Structured English and an activity diagram to specify the algorithm.

C. For the gym membership problem in Chapters 4 (exercises O, P, and Q), 5 (exercises T and U), 6 (exercise F), and 7 (exercise C):

1. Choose one of the classes and create a set of invariants for attributes and relationships, and add them to the CRC card for the class.

2. Choose one of the methods in the class that you chose and create a contract and a method specification for it. Use OCL to specify any pre- or postcondition and use both Structured English and an activity diagram to specify the algorithm.

D. For the Picnics R Us problem in Chapters 4 (exercises R, S, and T), 5 (exercises V and W), 6 (exercise G), and 7 (exercise D):

1. Choose one of the classes and create a set of invariants for attributes and relationships, and add them to the CRC card for the class.

2. Choose one of the methods in the class that you chose and create a contract and a method specification for it. Use OCL to specify any pre- or postcondition and use both Structured English and an activity diagram to specify the algorithm.

E. For the Of-the-Month-Club problem in Chapters 4 (exercises U, V, and W), 5 (exercises X and Y), 6 (exercise H), and 7 (exercise E):

1. Choose one of the classes and create a set of invariants for attributes and relationships, and add them to the CRC card for the class.

2. Choose one of the methods in the class that you chose and create a contract and a method specification for it. Use OCL to specify any pre- or postcondition and use both Structured English and an activity diagram to specify the algorithm.

F. Describe the difference in meaning between the following two class diagrams. Which is a better model? Why?

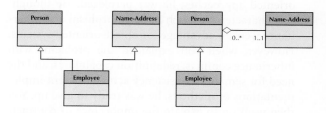

G. From a cohesion, coupling, and connascence perspective, is the following class diagram a good model? Why or why not?

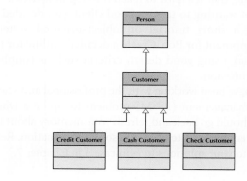

H. From a cohesion, coupling, and connascence perspective, are the following class diagrams good models? Why or why not?

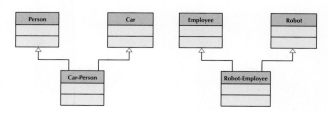

I. Create a set of inheritance conflicts for the two inheritance structures in the class diagrams of exercise H.

MINICASES

1. You are a team leader with a major IS outsourcing firm based in India. Bob Smith is the project manager of a joint major development project between your firm and a large US company. Bob has been in the software industry for 30 years and has always prided himself on his expertise in developing software. He is familiar with structured analysis and design as well as information engineering of the 1980s and 1990s. He is also well-versed with the advantage of rapid application development. But the other day, when you and Bob were talking about the advantages of object-oriented approaches, he was perplexed. He thought that characteristics such as polymorphism and inheritance were an advantage for object-oriented systems. However, when you explained the problems with inheritance conflicts, redefinition capabilities, and the need for semantic consistency across different implementations of methods, he was ready to give up. You then went on to explain the importance of contracts in maintaining the system. At this point in the conversation, he basically threw in the towel. As he walked off, you heard him say something like "I guess it's true, that it's hard to teach old dog new tricks."

   Not wanting to upset a good client, you decided to write a short tutorial on object-oriented systems development for Bob. Create a detailed outline for the tutorial, using good design criteria such as coupling and cohesion.

2. You have been working with the professional and scientific management (PSSM) problem for quite a while. You should go back and refresh your memory about the problem before attempting to solve this situation. Refer back to your solutions to Minicase 3 in Chapter 7.

   a. For each class in the structural model, using OCL, create a set of invariants for attributes and relationships and add them to the CRC cards for the classes.
   b. Choose one of the classes in the structural model. Create a contract for each method in that class. Be sure to use OCL to specify the preconditions and the postconditions. Be as complete as possible.
   c. Create a method specification for each method in the class you chose for question b. Use both Structured English and activity diagrams for the algorithm specification.

3. Refer to the Behavioral State Machine you provided as solution to describe the evolution of the state of a Paper instance object in Professor Takato's CMS (minicase 2, Chapter 6). Part of this Behavioral State Machine describes the state transitions of the paper object through the review process and is to be implemented by the Review_Paper() method of the paper object. Provide a Structured-English-based algorithm specification for the Review_Paper() method. Specify the preconditions, postconditions, and parameters of this method. What are the paper class's attributes used in this method? Assume, according to Minicase 2, Chapter 5, that paper authors and reviewers are represented by instances of two different classes: Author and Reviewer. Given that Professor Takato insists that his CMS prevents authors from reviewing their own papers, explain what are the issues arising from the current design in enforcing this condition. What if a class Person is added and Author and Reviewer are defined as sub-classes of it? Would this approach make it easier and safer to enforce the condition above?

CHAPTER 9

DATA BASE DESIGN

A project team designs the data management layer of a system using a four-step process: selecting the format of the storage, mapping the problem domain classes to the selected format, optimizing the storage to perform efficiently, and then designing the necessary data access and manipulation classes. This chapter describes the different ways objects can be stored and several important characteristics that should be considered when choosing among object persistence formats. It describes a problem domain class to object persistence format mapping process for the most important object persistence formats. Because the most popular storage format today is the relational database, the chapter focuses on the optimization of relational databases from both storage and access perspectives. We describe the effect that nonfunctional requirements have on the data management layer. Last, the chapter describes how to design data access and manipulation classes.

OBJECTIVES

- Become familiar with several object persistence formats
- Be able to map problem domain objects to different object persistence formats
- Be able to apply the steps of normalization to a relational database
- Be able to optimize a relational database for object storage and access
- Become familiar with indexes for relational databases
- Be able to estimate the size of a relational database
- Understand the affect of nonfunctional requirements on the data management layer
- Be able to design the data access and manipulation classes

CHAPTER OUTLINE

INTRODUCTION

As explained in Chapter 7, the work done by any application can be divided into a set of layers. This chapter focuses on the *data management layer,* which includes both data access and manipulation logic, along with the actual design of the storage. The data storage component of the data management layer manages how data are stored and handled by the programs that run the system. This chapter describes how a project team designs the storage for objects (*object persistence*) using a four-step approach: selecting the format of the storage, mapping the problem domain objects to the object persistence format, optimizing the object persistence format, and designing the data access and manipulation classes necessary to handle the communication between the system and the database.

Applications are of little use without the data that they support. How useful is a multimedia application that can't support images or sound? Why would someone log into a system to find information if it took him or her less time to locate the information manually? Design includes four steps to object persistence design that decrease the chances of ending up with inefficient systems, long system response times, and users who cannot get to the information that they need in the way that they need it—all of which can affect the success of the project.

The first part of this chapter describes a variety of storage formats and explains how to select the appropriate one for your application. From a practical perspective, there are five basic types of formats that can be used to store objects for application systems: files (sequential and random), object-oriented databases, object-relational databases, relational databases, or NoSQL datastores.[1] Each type has certain characteristics that make it more appropriate for some types of systems over others.

Once the object persistence format is selected to support the system, the problem domain objects need to drive the design of the actual object storage. Then the object storage needs to be designed to optimize its processing efficiency, which is the focus of the next part of the chapter. One of the leading complaints by end users is that the final system is too slow, so to avoid such complaints, project team members must allow time during design to carefully make sure that the file or database performs as fast as possible. At the same time, the team must keep hardware costs down by minimizing the storage space that the application will require. The goals of maximizing access to the objects and minimizing the amount of space taken to store objects can conflict, and designing object persistence efficiency usually requires trade-offs.

Finally, it is necessary to design a set of data access and manipulation classes to ensure the independence of the problem domain classes from the storage format. The data access and manipulation classes handle all communication with the database. In this manner, the problem domain is decoupled from the object storage, allowing the object storage to be changed without affecting the problem domain classes.

OBJECT PERSISTENCE FORMATS

There are five main types of object persistence formats: files (sequential and random access), object-oriented databases, object-relational databases, relational databases, and NoSQL datastores. *Files* are electronic lists of data that have been optimized to perform a particular transaction. For example, Figure 9-1 shows a customer order file with information about

---

[1] There are other types of files, such as relative, indexed sequential, and multi-indexed sequential, and databases, such as hierarchical, network, and multidimensional. However, these formats typically are not used for object persistence.

Order Number	Date	Cust ID	Last Name	First Name	Amount	Tax	Total	Prior Customer	Payment Type
234	11/23/00	2242	DeBerry	Ann	$ 90.00	$5.85	$ 95.85	Y	MC
235	11/23/00	9500	Chin	April	$ 12.00	$0.60	$ 12.60	Y	VISA
236	11/23/00	1556	Fracken	Chris	$ 50.00	$2.50	$ 52.50	N	VISA
237	11/23/00	2242	DeBerry	Ann	$ 75.00	$4.88	$ 79.88	Y	AMEX
238	11/23/00	2242	DeBerry	Ann	$ 60.00	$3.90	$ 63.90	Y	MC
239	11/23/00	1035	Black	John	$ 90.00	$4.50	$ 94.50	Y	AMEX
240	11/23/00	9501	Kaplan	Bruce	$ 50.00	$2.50	$ 52.50	N	VISA
241	11/23/00	1123	Williams	Mary	$120.00	$9.60	$129.60	N	MC
242	11/24/00	9500	Chin	April	$ 60.00	$3.00	$ 63.00	Y	VISA
243	11/24/00	4254	Bailey	Ryan	$ 90.00	$4.50	$ 94.50	Y	VISA
244	11/24/00	9500	Chin	April	$ 24.00	$1.20	$ 25.20	Y	VISA
245	11/24/00	2242	DeBerry	Ann	$ 12.00	$0.78	$ 12.78	Y	AMEX
246	11/24/00	4254	Bailey	Ryan	$ 20.00	$1.00	$ 21.00	Y	MC
247	11/24/00	2241	Jones	Chris	$ 50.00	$2.50	$ 52.50	N	VISA
248	11/24/00	4254	Bailey	Ryan	$ 12.00	$0.60	$ 12.60	Y	AMEX
249	11/24/00	5927	Lee	Diane	$ 50.00	$2.50	$ 52.50	N	AMEX
250	11/24/00	2242	DeBerry	Ann	$ 12.00	$0.78	$ 12.78	Y	MC
251	11/24/00	9500	Chin	April	$ 15.00	$0.75	$ 15.75	Y	MC
252	11/24/00	2242	DeBerry	Ann	$132.00	$8.58	$140.58	Y	MC
253	11/24/00	2242	DeBerry	Ann	$ 72.00	$4.68	$ 76.68	Y	AMEX

FIGURE 9-1   Customer Order File

customers' orders, in the form in which it is used, so that the information can be accessed and processed quickly by the system.

A *database* is a collection of groupings of information, each of which is related to each other in some way (e.g., through common fields). Logical groupings of information could include such categories as customer data, information about an order, product information, and so on. A *database management system (DBMS)* is software that creates and manipulates these databases (see Figure 9-2 for a relational database example). Such *end user DBMSs* as Microsoft Access support small-scale databases that are used to enhance personal productivity, whereas *enterprise DBMSs,* such as DB2, Versant, and Oracle, can manage huge volumes of data and support applications that run an entire company. An end-user DBMS is significantly less expensive and easier for novice users to use than its enterprise counterpart, but it does not have the features or capabilities that are necessary to support mission-critical or large-scale systems.

In the next sections, we describe sequential and random-access files, relational databases, object-relational databases, and object-oriented databases that can be used to handle a system's object persistence requirements. We also describe a new exciting technology that shows much promise for object persistence: NoSQL datastores. Finally, we describe a set of characteristics on which the different formats can be compared.

Sequential and Random Access Files

From a practical perspective, most object-oriented programming languages support sequential and random access files as part of the language.[2] In this section, we describe

---

[2] For example, see the FileInputStream, FileOutputStream, and RandomAccessFile classes in the java.io package.

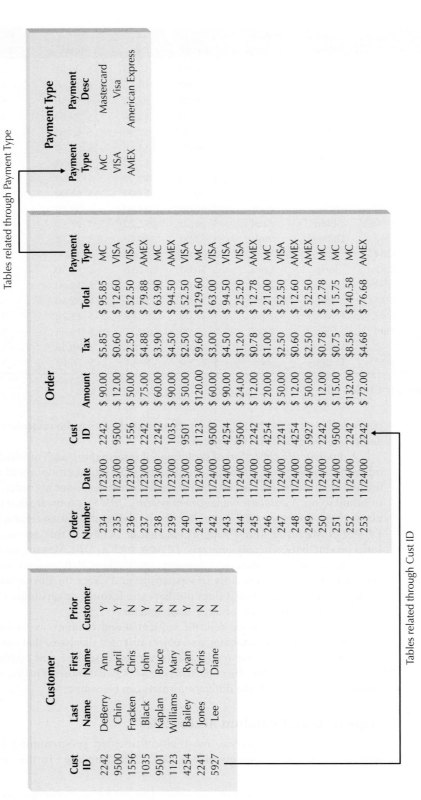

Customer

Cust ID	Last Name	First Name	Prior Customer
2242	DeBerry	Ann	Y
9500	Chin	April	Y
1556	Fracken	Chris	N
1035	Black	John	Y
9501	Kaplan	Bruce	N
1123	Williams	Mary	N
4254	Bailey	Ryan	Y
2241	Jones	Chris	N
5927	Lee	Diane	N

Order

Order Number	Date	Cust ID	Amount	Tax	Total	Payment Type
234	11/23/00	2242	$ 90.00	$5.85	$ 95.85	MC
235	11/23/00	9500	$ 12.00	$0.60	$ 12.60	VISA
236	11/23/00	1556	$ 50.00	$2.50	$ 52.50	VISA
237	11/23/00	2242	$ 75.00	$4.88	$ 79.88	AMEX
238	11/23/00	2242	$ 60.00	$3.90	$ 63.90	MC
239	11/23/00	1035	$ 90.00	$4.50	$ 94.50	AMEX
240	11/23/00	9501	$ 50.00	$2.50	$ 52.50	VISA
241	11/23/00	1123	$120.00	$9.60	$129.60	MC
242	11/24/00	9500	$ 60.00	$3.00	$ 63.00	VISA
243	11/24/00	4254	$ 90.00	$4.50	$ 94.50	VISA
244	11/24/00	9500	$ 24.00	$1.20	$ 25.20	VISA
245	11/24/00	2242	$ 12.00	$0.78	$ 12.78	AMEX
246	11/24/00	4254	$ 20.00	$1.00	$ 21.00	MC
247	11/24/00	2241	$ 50.00	$2.50	$ 52.50	VISA
248	11/24/00	4254	$ 12.00	$0.60	$ 12.60	AMEX
249	11/24/00	5927	$ 50.00	$2.50	$ 52.50	AMEX
250	11/24/00	2242	$ 12.00	$0.78	$ 12.78	MC
251	11/24/00	9500	$ 15.00	$0.75	$ 15.75	MC
252	11/24/00	2242	$132.00	$8.58	$140.58	MC
253	11/24/00	2242	$ 72.00	$4.68	$ 76.68	AMEX

Payment Type

Payment Type	Payment Desc
MC	Mastercard
VISA	Visa
AMEX	American Express

Tables related through Payment Type

Tables related through Cust ID

FIGURE 9-2   Customer Order Database

what sequential access and random access files are.[3] We also describe how sequential access and random access files are used to support an application. For example, they can be used to support master files, look-up files, transaction files, audit files, and history files.

Sequential access files allow only sequential file operations to be performed (e.g., read, write, and search). Sequential access files are very efficient for sequential operations, such as report writing. However, for random operations, such as finding or updating a specific object, they are very inefficient. On the average, 50 percent of the contents of a sequential access file will have to be searched through before finding the specific object of interest in the file. They come in two flavors: ordered and unordered.

An *unordered sequential access file* is basically an electronic list of information stored on disk. Unordered files are organized serially (i.e., the order of the file is the order in which the objects are written to the file). Typically, new objects simply are added to the file's end.

Ordered sequential access files are placed into a specific sorted order (e.g., in ascending order by customer number). However, there is overhead associated with keeping files in a particular sorted order. The file designer can keep the file in sorted order by always creating a new file each time a delete or addition occurs, or he or she can keep track of the sorted order via the use of a *pointer,* which is information about the location of the related record. A pointer is placed at the end of each record, and it "points" to the next record in a series or set. The underlying data/file structure in this case is the *linked list*[4] data structure demonstrated in the previous chapter.

Random access files allow only random or direct file operations to be performed. This type of file is optimized for random operations, such as finding and updating a specific object. Random access files typically give a faster response time to find and update operations than any other type of file. However, because they do not support sequential processing, applications such as report writing are very inefficient. The various methods to implement random access files are beyond the scope of this book.[5]

There are times when it is necessary to be able to process files in both a sequential and random manner. One simple way to do this is to use a sequential file that contains a list of the keys (the field in which the file is to be kept in sorted order) and a random access file for the actual objects. This minimizes the cost of additions and deletions to a sequential file while allowing the random file to be processed sequentially by simply passing the key to the random file to retrieve each object in sequential order. It also allows fast random processing to occur by using only the random access file, thus optimizing the overall cost of file processing. However, if a file of objects needs to be processed in both a random and sequential manner, the developer should consider using a database (relational, object-relational, or object-oriented) instead.

There are many different application types of files—for example, master files, lookup files, transaction files, audit files, and history files. *Master files* store core information that is important to the business and, more specifically, to the application, such as order information or customer mailing information. They usually are kept for long periods of time, and new records are appended to the end of the file as new orders or new customers are captured by the system. If changes need to be made to existing records, programs must be written to update the old information.

---

[3] For a more complete coverage of issues related to the design of files, see Owen Hanson, *Design of Computer Data Files* (Rockville, MD: Computer Science Press, 1982).

[4] For more information on various data structures, see Ellis Horowitz and Sartaj Sahni, *Fundamentals of Data Structures* (Rockville, MD: Computer Science Press, 1982); and Michael T. Goodrich and Roberto Tamassia, *Data Structures and Algorithms in Java* (New York: Wiley, 1998).

[5] For a more detailed look at the underlying data and file structures of the different types of files, see Mary E. S. Loomis, *Data Management and File Structures,* 2nd ed. (Englewood Cliffs, NJ: Prentice Hall, 1989); and Michael J. Folk and Bill Zoeellick, *File Structures: A Conceptual Toolkit* (Reading, MA: Addison-Wesley, 1987).

Lookup files contain static values, such as a list of valid ZIP codes or the names of the U.S. states. Typically, the list is used for validation. For example, if a customer's mailing address is entered into a master file, the state name is validated against a lookup file that contains U.S. states to make sure that the operator entered the value correctly.

A *transaction file* holds information that can be used to update a master file. The transaction file can be destroyed after changes are added, or the file may be saved in case the transactions need to be accessed again in the future. Customer address changes, for one, would be stored in a transaction file until a program is run that updates the customer address master file with the new information.

For control purposes, a company might need to store information about how data change over time. For example, as human resources clerks change employee salaries in a human resources system, the system should record the person who made the changes to the salary amount, the date, and the actual change that was made. An *audit file* records before and after images of data as they are altered so that an audit can be performed if the integrity of the data is questioned.

Sometimes files become so large that they are unwieldy, and much of the information in the file is no longer used. The *history file* (or archive file) stores past transactions (e.g., old customers, past orders) that are no longer needed by system users. Typically, the file is stored off-line, yet it can be accessed on an as-needed basis. Other files, such as master files, can then be streamlined to include only active or very recent information.

YOUR TURN

9-1 Student Admissions System

*S*uppose you are building a web-based system for the admissions office at your university that will be used to accept electronic applications from students. All the data for the system will be stored in a variety of files.

Question

Give an example using this system for each of the following file types: master, lookup, transaction, audit, and history. What kind of information would each file contain, and how would the file be used?

Relational Databases

A relational database is the most popular kind of database for application development today. A relational database is based on collections of tables with each table having a *primary key*—a field or fields whose values are unique for every row of the table. The tables are related to one another by placing the primary key from one table into the related table as a *foreign key* (see Figure 9-3). Most *relational database management systems (RDBMS)* support *referential integrity,* or the idea of ensuring that values linking the tables together through the primary and foreign keys are valid and correctly synchronized. For example, if an order-entry clerk using the tables in Figure 9-3 attempted to add order 254 for customer number 1111, he or she would have made a mistake because no customer exists in the Customer table with that number. If the RDBMS supported referential integrity, it would check the customer numbers in the Customer table; discover that the number 1111 is invalid; and return an error to the entry clerk. The clerk would then go back to the original order form and recheck the customer information. Can you imagine the problems that would occur if the RDBMS let the entry clerk add the order with the wrong information? There would be no way to track down the name of the customer for order 254.

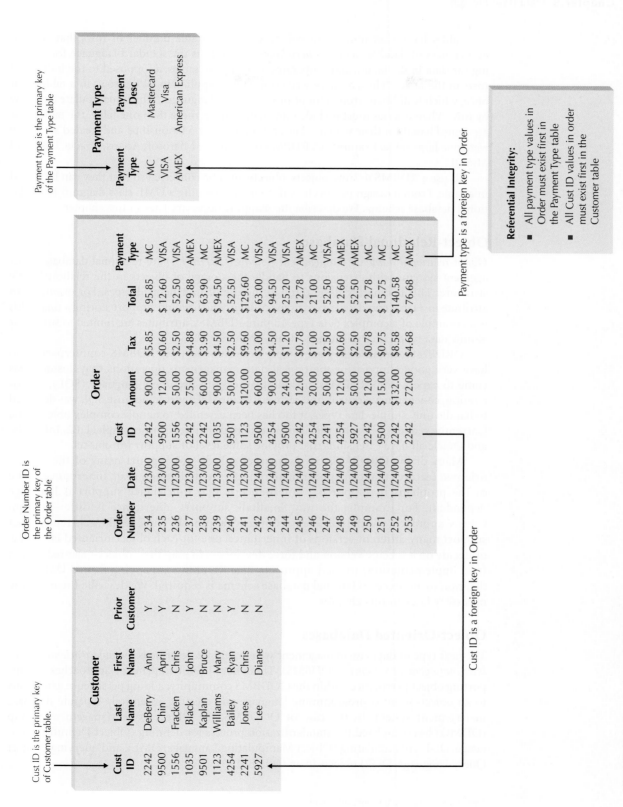

Cust ID is the primary key of Customer table.

Customer

Cust ID	Last Name	First Name	Prior Customer
2242	DeBerry	Ann	Y
9500	Chin	April	Y
1556	Fracken	Chris	N
1035	Black	John	Y
9501	Kaplan	Bruce	N
1123	Williams	Mary	N
4254	Bailey	Ryan	Y
2241	Jones	Chris	N
5927	Lee	Diane	N

Order Number ID is the primary key of the Order table

Order

Order Number	Date	Cust ID	Amount	Tax	Total	Payment Type
234	11/23/00	2242	$ 90.00	$5.85	$ 95.85	MC
235	11/23/00	9500	$ 12.00	$0.60	$ 12.60	VISA
236	11/23/00	1556	$ 50.00	$2.50	$ 52.50	VISA
237	11/23/00	2242	$ 75.00	$4.88	$ 79.88	AMEX
238	11/23/00	2242	$ 60.00	$3.90	$ 63.90	MC
239	11/23/00	1035	$ 90.00	$4.50	$ 94.50	AMEX
240	11/23/00	9501	$ 50.00	$2.50	$ 52.50	VISA
241	11/23/00	1123	$120.00	$9.60	$129.60	MC
242	11/24/00	9500	$ 60.00	$3.00	$ 63.00	VISA
243	11/24/00	4254	$ 90.00	$4.50	$ 94.50	VISA
244	11/24/00	9500	$ 24.00	$1.20	$ 25.20	VISA
245	11/24/00	2242	$ 12.00	$0.78	$ 12.78	AMEX
246	11/24/00	4254	$ 20.00	$1.00	$ 21.00	MC
247	11/24/00	2241	$ 50.00	$2.50	$ 52.50	VISA
248	11/24/00	4254	$ 12.00	$0.60	$ 12.60	AMEX
249	11/24/00	5927	$ 50.00	$2.50	$ 52.50	AMEX
250	11/24/00	2242	$ 12.00	$0.78	$ 12.78	MC
251	11/24/00	9500	$ 15.00	$0.75	$ 15.75	MC
252	11/24/00	2242	$132.00	$8.58	$140.58	MC
253	11/24/00	2242	$ 72.00	$4.68	$ 76.68	AMEX

Cust ID is a foreign key in Order

Payment type is the primary key of the Payment Type table

Payment Type

Payment Type	Payment Desc
MC	Mastercard
VISA	Visa
AMEX	American Express

Payment type is a foreign key in Order

Referential Integrity:

- All payment type values in Order must exist first in the Payment Type table
- All Cust ID values in order must exist first in the Customer table

FIGURE 9-3   Relational Database

Tables have a set number of columns and a variable number of rows that contain occurrences of data. *Structured query language (SQL)* is the standard language for accessing the data in the tables, and it operates on complete tables, as opposed to the individual rows in the tables. Thus, a query written in SQL is applied to all the rows in a table all at once, which is different from a lot of programming languages, which manipulate data row by row. When queries must include information from more than one table, the tables first are *joined* based on their primary key and foreign key relationships and treated as if they were one large table. Examples of RDBMS software are Microsoft Access, Oracle, DB2, and MySQL.

To use a RDBMS to store objects, objects must be converted so that they can be stored in a table. From a design perspective, this entails mapping a UML class diagram to a relational database schema. We describe the mapping necessary later in this chapter.

Object-Relational Databases

Object-relational database management systems (ORDBMSs) are relational database management systems with extensions to handle the storage of objects in the relational table structure. This is typically done through the use of user-defined types. For example, an attribute in a table could have a data type of *map,* which would support storing a map. This is an example of a complex data type. In pure RDBMSs, attributes are limited to simple or atomic data types, such as integers, floats, or chars.

ORDBMSs, because they are simply extensions to their RDBMS counterparts, also have very good support for the typical data management operations that business has come to expect from RDBMSs, including an easy-to-use query language (SQL), authorization, concurrency-control, and recovery facilities. However, because SQL was designed to handle only simple data types, it too has been extended to handle complex object data. Currently, vendors deal with this issue in different manners. For example, DB2, Informix, and Oracle all have extensions that provide some level of support for objects.

Many of the ORDBMSs on the market still do not support many of the object-oriented features that can appear in an object-oriented design (e.g., inheritance). One of the problems in supporting inheritance is that inheritance support is language dependent. For example, the way Smalltalk supports inheritance is different from C++'s approach, which is different from Java's approach. Thus, vendors currently must support many different versions of inheritance, one for each object-oriented language, or decide on a specific version and force developers to map their object-oriented design (and implementation) to their approach. Like RDBMSs, a mapping from a UML class diagram to an object-relational database schema is required. We describe the mapping necessary later in this chapter.

Object-Oriented Databases

The next type of database management system that we describe is the *object-oriented database management systems (OODBMS).* There have been two primary approaches to supporting object persistence within the OODBMS community: adding persistence extensions to an object-oriented programming language and creating an entirely separate database management system. In the case of OODBMS, the Object Data Management Group (ODMG) has completed the standardization process for defining (Object Definition Language, ODL), manipulating (Object Manipulating Language, OML), and querying (Object Query Language, OQL) objects in an OODBMS.[6]

---

[6] See www.odbms.org for more information.

With an OODBMS, collections of objects are associated with an extent. An *extent* is simply the set of instances associated with a particular class (i.e., it is the equivalent of a table in a RDBMS). Technically speaking, each instance of a class has a unique identifier assigned to it by the OODBMS: the *Object ID*. However, from a practical point of view, it is still a good idea to have a semantically meaningful primary key (even though from an OODBMS perspective, this is unnecessary). Referential integrity is still very important. In an OODBMS, from the user's perspective, it looks as if the object is contained within the other object. However, the OODBMS actually keeps track of these relationships through the use of the Object ID, and therefore foreign keys are not necessary.[7]

OODBMSs provide support for some form of inheritance. However, as already discussed, inheritance tends to be language dependent. Currently, most OODBMSs are tied closely to either a particular *object-oriented programming language (OOPL)* or a set of OOPLs. Most OODBMSs originally supported either Smalltalk or C++. Today, many of the commercially available OODBMSs provide support for C++, Java, and Smalltalk.

OODBMSs also support the idea of *repeating groups (fields)* or *multivalued attributes*. These are supported through the use of *attribute sets* and *relationships sets*. RDBMSs do not explicitly allow multivalued attributes or repeating groups. This is considered to be a violation of the first normal form (discussed later in this chapter) for relational databases. Some ORDBMSs do support repeating groups and multivalued attributes.

Up until recently, OODBMSs have mainly been used to support multimedia applications or systems that involve complex data (e.g., graphics, video, sound). Application areas, such as computer-aided design and manufacturing (CAD/CAM), financial services, geographic information systems, health care, telecommunications, and transportation, have been the most receptive to OODBMSs. They are also becoming popular technologies for supporting electronic commerce, online catalogs, and large, web multimedia applications. Examples of pure OODBMSs include Gemstone, Objectivity, db4o, and Versant.

Although pure OODBMS exist, most organizations currently invest in *ORDBMS* technology. The market for OODBMS is expected to grow, but its ORDBMS and RDBMS counterparts dwarf it. One reason for this situation is that there are many more experienced developers and tools in the RDBMS arena. Furthermore, relational users find that using an OODBMS comes with a fairly steep learning curve.

NoSQL Data Stores

NoSQL data stores are the newest type of object persistence available. Depending on whom you talk to, NoSQL either stands for No SQL or Not Only SQL. Regardless, the data stores that are described as NoSQL typically do not support SQL. Currently, there is no standard for NoSQL data stores. Most NoSQL data stores were created to address problems associated with storing large amounts of distributed data in RDBMSs. They tend to support very fast queries. However, when it comes to updating, they normally do not support a locking mechanism, and consequently, all copies of a piece of data are not required to be consistent at all times. Instead, they tend to support an eventually consistent based model. So it is technically possible to have different values for different copies of the same object stored in different locations in a distributed system. Depending on the application, this could cause problems for decision makers. Therefore, their applicability is limited and they are not applicable to traditional business transaction processing systems. Some of the

---

[7] Depending on the storage and updating requirements, it usually is a good idea to use a foreign key in addition to the Object ID. The Object ID has no semantic meaning. Therefore, in the case of needing to rebuild relationships between objects, Object IDs are difficult to validate. Foreign keys, by contrast, should have some meaning outside of the DBMS.

better known NoSQL data stores include Google's Big Table, Amazon's Dynamo, Apache's HBase, Apache's CouchDB, and Apache/Facebook's Cassandra. There are many different types of NoSQL data stores, including key-value stores, document stores, column-oriented stores, and object databases. Besides object databases, which are either ORDBMSs or OODBMSs, we describe each below.

Key-value data stores essentially provide a distributed index (primary key) to where a BLOB (binary large object) is stored. A BLOB treats a set of attributes as one large object. A good example of this type of NoSQL data store is Amazon's Dynamo. Dynamo provides support for many of the core services for Amazon. Obviously, being one of the largest e-commerce sites in the world, Amazon had to have a solution for object persistence that was scalable, distributable, and reliable. Typical RDBMS-based solutions would not work for some of these applications. Typical applications that use key-value data stores are web-based shopping carts, product catalogs, and bestseller lists. These types of applications do not require updating the underlying data. For example, you do not update the title of a book in your shopping cart when you are making a purchase at Amazon. Given the scale and distributed nature of this type of system, there are bound to be many failures across the system. Being fault tolerant and temporarily sacrificing some consistency across all copies of an object is a reasonable trade-off.

Document data stores, as the name suggests, are built around the idea of documents. The idea of document databases has been around for a long time. One of the early systems that used this approach was Lotus Notes. These types of stores are considered to be schema free. By that we mean there is no detailed design of the database. A good example of an application that would benefit from this type of approach is a business card database. In a relational database, multiple tables would need to be designed. In a document data store, the design is done more in a "just in time" manner. As new business cards are input into the system, attributes not previously included are simply added to the evolving design. Previously entered business cards would simply not have those attributes associated with them. One major difference between key-value data stores and document data stores is that the "document" has structure and can be easily searched based on the non-key attributes contained in the document, whereas the key-value data store simply treats the "value" as one big monolithic object. Apache's CouchDB is a good example of this type of data store.

Columnar data stores organize the data into columns instead of rows. However, there seems to be some confusion as to what this actually implies. In the first approach to columnar data stores, the rows represent the attributes and the columns represent the objects. This is in comparison to a relational database, where columns represent the attributes and the objects are represented by rows. These types of data stores are very effective in business intelligence, data mining, and data warehousing applications where the data are fairly static and many computations are performed over a single, or a small, subset of the available attributes. In comparison to a relational database where you would have to select a set of attributes from all rows; with this type of data store, you would simply have to select a set of rows. This should be a lot faster than with a relational database. A few good examples of this type of columnar data store include Oracle's Retail Predictive Application Server, HP's Vertica, and SAP's Sybase IQ. The second approach to columnar data stores, which includes Apache's HBase, Apache/Facebook's Cassandra, and Google's BigTable, is designed to handle very large data sets (petabytes of data) that can be accessed as if the data are stored in columns. However, in this case, the data are actually stored in a three-dimensional map composed of object ID, attribute name, timestamp, and value instead of using columns and rows. This approach is highly scalable and distributable. These types of data stores support social applications such as Twitter and Facebook, and support search applications such as Google Maps, Earth, and Analytics.

Given the popularity of social computing, business intelligence, data mining, data warehousing, e-commerce, and their need for highly scalable, distributable, and reliable data storage, NoSQL data stores is an area that should be considered as part of an object persistence solution. However, at this time, NoSQL data stores lack the maturity to be considered for most business applications. Given the overall diversity and complexity of NoSQL data stores and their limited applicability to traditional business applications, we do not consider them any further in this text.

Selecting an Object Persistence Format

Each of the file and database storage formats that have been presented has its strengths and weaknesses, and no one format is inherently better than the others. In fact, sometimes a project team chooses multiple formats (e.g., a relational database for one, a file for another, and an object-oriented database for a third). Thus, it is important to understand the strengths and weaknesses of each format and when to use each one. Figure 9-4 presents a summary of the characteristics of each and the characteristics that can help identify when each type of format is more appropriate.

Major Strengths and Weaknesses   The major strengths of files include that some support for sequential and random access files is normally part of an OOPL, files can be designed to be very efficient, and they are a good alternative for temporary or short-term storage. However, all file manipulation must be done through the OOPL. Files do not have

	Sequential and Random Access Files	Relational DBMS	Object Relational DBMS	Object-Oriented DBMS
Major Strengths	Usually part of an object-oriented programming language Files can be designed for fast performance Good for short-term data storage	Leader in the database market Can handle diverse data needs	Based on established, proven technology, e.g., SQL Able to handle complex data	Able to handle complex data Direct support for object orientation
Major Weaknesses	Redundant data Data must be updated using programs, i.e., no manipulation or query language No access control	Cannot handle complex data No support for object orientation Impedance mismatch between tables and objects	Limited support for object orientation Impedance mismatch between tables and objects	Technology is still maturing Skills are hard to find
Data Types Supported	Simple and Complex	Simple	Simple and Complex	Simple and Complex
Types of Application Systems Supported	Transaction processing	Transaction processing and decision making	Transaction processing and decision making	Transaction processing and decision making
Existing Storage Formats	Organization dependent	Organization dependent	Organization dependent	Organization dependent
Future Needs	Poor future prospects	Good future prospects	Good future prospects	Good future prospects

FIGURE 9-4   Comparison of Object Persistence Formats

any form of access control beyond that of the underlying operating system. Finally, in most cases, if files are used for permanent storage, redundant data most likely will result. This can cause many update anomalies.

RDBMSs bring with them proven commercial technology. They are the leaders in the DBMS market. Furthermore, they can handle very diverse data needs. However, they cannot handle complex data types, such as images. Therefore, all objects must be converted to a form that can be stored in tables composed of atomic or simple data. They provide no support for object orientation. This lack of support causes an *impedance mismatch* between the objects contained in the OOPL and the data stored in the tables. An impedance mismatch refers to the amount of work done by both the developer and DBMS, and the potential information loss that can occur when converting objects to a form that can be stored in tables.

Because ORDBMSs are typically object-oriented extensions to RDBMSs, they inherit the strengths of RDBMSs. They are based on established technologies, such as SQL, and unlike their predecessors, they can handle complex data types. However, they provide only limited support for object orientation. The level of support varies among the vendors; therefore, ORDBMSs also suffer from the impedance mismatch problem.

OODBMSs support complex data types and have the advantage of directly supporting object orientation. Therefore, they do not suffer from the impedance mismatch that the previous DBMSs do. Even though the ODMG has released version 3.0 of its set of standards, the OODBMS community is still maturing. Therefore, this technology might still be too risky for some firms. The other major problems with OODBMS are the lack of skilled labor and the perceived steep learning curve of the RDBMS community.

Data Types Supported   The first issue is the type of data that will need to be stored in the system. Most applications need to store simple data types, such as text, dates, and numbers, and all files and DBMSs are equipped to handle this kind of data. The best choice for simple data storage, however, is usually the RDBMS because the technology has matured over time and has continuously improved to handle simple data very effectively.

Increasingly, applications are incorporating complex data, such as video, images, or audio. ORDBMSs or OODBMSs are best able to handle data of this type. Complex data stored as objects can be manipulated much faster than with other storage formats.

Type of Application System   There are many different kinds of application systems that can be developed. *Transaction-processing systems* are designed to accept and process many simultaneous requests (e.g., order entry, distribution, payroll). In transaction-processing systems, the data are continuously updated by a large number of users, and the queries that are asked of the systems typically are predefined or targeted at a small subset of records (e.g., List the orders that were backordered today or What products did customer #1234 order on May 12, 2001?).

Another set of application systems is the set designed to support decision making, such as *decision support systems (DSS), management information systems (MIS), executive information systems (EIS),* and *expert systems (ES).* These decision-making support systems are built to support users who need to examine large amounts of read-only historical data. The questions that they ask are often ad hoc, and they include hundreds or thousands of records at a time (e.g., List all customers in the West region who purchased a product costing more than $500 at least three times, or What products had increased sales in the summer months that have not been classified as summer merchandise?).

Transaction-processing systems and DSSs, thus, have very different data storage needs. Transaction-processing systems need data storage formats that are tuned for a lot of data updates and fast retrieval of predefined, specific questions. Files, relational databases, object-relational databases, and object-oriented databases can all support these kinds of requirements. By contrast, systems to support decision making are usually only reading data (not updating it), often in ad hoc ways. The best choices for these systems usually are RDBMSs because these formats can be configured specially for needs that may be unclear and less apt to change the data.

Existing Storage Formats   The storage format should be selected primarily on the basis of the kind of data and application system being developed. However, project teams should consider the existing storage formats in the organization when making design decisions. In this way, they can better understand the technical skills that already exist and how steep the learning curve will be when the storage format is adopted. For example, a company that is familiar with RDBMS will have little problem adopting a relational database for the project, whereas an OODBMS might require substantial developer training. In the latter situation, the project team might have to plan for more time to integrate the object-oriented database with the company's relational systems, or possibly consider moving toward an ORDBMS solution.

Future Needs   Not only should a project team consider the storage technology within the company, but it should also be aware of current trends and technologies that are being used by other organizations. A large number of installations of a specific type of storage format suggest that skills and products are available to support the format. Therefore, the selection of that format is safe. For example, it would probably be easier and less expensive to find RDBMS expertise when implementing a system than to find help with an OODBMS.

Other Miscellaneous Criteria   Other criteria that should be considered include cost, licensing issues, concurrency control, ease of use, security and access controls, version management, storage management, lock management, query management, language bindings, and APIs. We also should consider performance issues, such as cache management, insertion, deletion, retrieval, and updating of complex objects. Finally, the level of support for object orientation (such as objects, single inheritance, multiple inheritance, polymorphism, encapsulation and information hiding, methods, multivalued attributes, repeating groups) is critical.

YOUR TURN   **9-2 Donation Tracking System**

A major public university graduates approximately 10,000 students per year, and the development office has decided to build a web-based system that solicits and tracks donations from the university's large alumni body. Ultimately, the development officers hope to use the information in the system to better understand the alumni giving patterns so that they can improve giving rates.

Question

1. What kind of system is this? Does it have characteristics of more than one?
2. What different kinds of data will this system use?
3. On the basis of your answers, what kind of data storage format(s) do you recommend for this system?

MAPPING PROBLEM DOMAIN OBJECTS TO OBJECT PERSISTENCE FORMATS[8]

As described in the previous section, there are many different formats from which to choose to support object persistence. Each of the different formats can have some conversion requirements. Regardless of the object persistence format chosen, we suggest supporting primary keys and foreign keys by adding them to the problem domain classes at this point. However, this does imply some additional processing required. The developer has to set the value for the foreign key when adding the relationship to an object. In some cases, this overhead may be too costly. In those cases, this suggestion should be ignored. In the remainder of this section, we describe how to map the problem domain classes to the different object persistence formats. From a practical perspective, file formats are used mostly for temporary storage. Thus, we do not consider them further.

We also recommend that the data management functionality specifics, such as retrieval and updating of data from the object storage, be included only in classes contained in the data management layer. This will ensure that the data management classes are dependent on the problem domain classes and not the other way around. Furthermore, this allows the design of problem domain classes to be independent of any specific object persistence environment, thus increasing their portability and their potential for reuse. Like our previous recommendation, this one also implies additional processing. However, the increased portability and potential for reuse realized should more than compensate for the additional processing required.

Mapping Problem Domain Objects to an OODBMS Format

If we support object persistence with an OODBMS, the mappings between the problem domain objects and the OODBMS tend to be fairly straightforward. As a starting point, we suggest that each concrete problem domain class should have a corresponding object persistence class in the OODBMS. There will also be a data access and manipulation (DAM) class that contains the functionality required to manage the interaction between the object persistence class and the problem domain layer. For example, using the appointment system example from the previous chapters, the Patient class is associated with an OODBMS class (see Figure 9-5). The Patient class essentially will be unchanged from analysis. The Patient-OODBMS class will be a new class that is dependent on the Patient class, whereas the Patient-DAM class will be a new class that depends on both the Patient class and the Patient-OODBMS class. The Patient-DAM class must be able to read from and write to the OODBMS. Otherwise, it will not be able to store and retrieve instances of the Patient class. Even though this does add overhead to the installation of the system, it allows the problem domain class to be independent of the OODBMS being used. If, at a later time, another OODBMS or object persistence format is adopted, only the DAM classes will have to be modified. This approach increases both the portability and the potential for reuse of the problem domain classes.

Even though we are implementing the DAM layer using an OODBMS, a mapping from the problem domain layer to the OODBMS classes in the data access and management layer may be required, depending on the level of support of inheritance in the OODBMS and the level of inheritance used in the problem domain classes. If multiple inheritance is used in the problem domain but not supported by the OODBMS, then the multiple inheritance must be factored out of the OODBMS classes. For each case of multiple inheritance (i.e., more than one superclass), the following rules can be used to

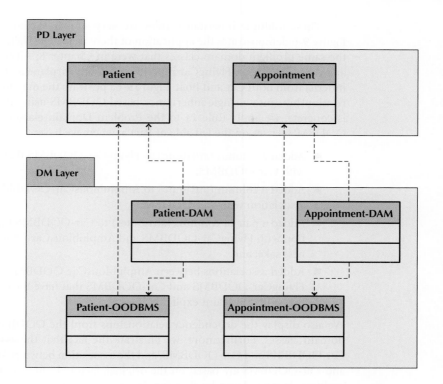

FIGURE 9-5
Appointment System
Problem Domain and
DM Layers

factor out the multiple inheritance effects in the design of the OODBMS classes.[8]

Rule 1a: Add a column(s) to the OODBMS class(es) that represents the subclass(es) that will contain an Object ID of the instance stored in the OODBMS class that represents the "additional" superclass(es). This is similar in concept to a foreign key in an RDBMS. The multiplicity of this new association from the subclass to the "superclass" should be 1..1. Add a column(s) to the OODBMS class(es) that represents the superclass(es) that will contain an Object ID of the instance stored in the OODBMS class that represents the subclass(es). If the superclasses are concrete, that is, they can be instantiated themselves, then the multiplicity from the superclass to the subclass is 0..1, otherwise, it is 1..1. An exclusive-or (XOR) constraint must be added between the associations. Do this for each "additional" superclass.

or

Rule 1b: Flatten the inheritance hierarchy of the OODBMS classes by copying the attributes and methods of the additional OODBMS superclass(es) down to all of the OODBMS subclasses, and remove the additional superclass from the design.[9]

---

[8] The rules presented in this section are based on material in Ali Bahrami, *Object-Oriented Systems Development using the Unified Modeling Language* (New York: McGraw-Hill, 1999); Michael Blaha and William Premerlani, *Object-Oriented Modeling and Design for Database Applications* (Upper Saddle River, NJ: Prentice Hall, 1998); Akmal B. Chaudri and Roberto Zicari, *Succeeding with Object Databases: A Practical Look at Today's Implementations with Java and XML* (New York: Wiley, 2001); Peter Coad and Edward Yourdon, *Object-Oriented Design* (Upper Saddle River, NJ: Yourdon Press, 1991); and Paul R. Read, Jr., *Developing Applications with Java and UML* (Boston: Addison-Wesley, 2002).

[9] It is also a good idea to document this modification in the design so that in the future, modifications to the design can be easily maintained.

These multiple inheritance rules are very similar to those described in Chapter 8. Figure 9-6 demonstrates the application of these rules. The right side of the figure portrays the same problem domain classes that were in Chapter 8: Airplane, Car, Boat, FlyingCar, and AmphibiousCar. FlyingCar inherits from both Airplane and Car, and AmphibiousCar inherits from both Car and Boat. Figure 9-6a portrays the mapping of multiple inheritance relationships into a single inheritance-based OODBMS using Rule 1a. Assuming that Car is concrete, we apply Rule 1a to the Problem Domain classes, and we end up with the OODBMS classes on the left side of Part a, where we have:

- Added a column (attribute) to FlyingCar-OODBMS that represents an association with Car-OODBMS;
- Added a column (attribute) to AmphibiousCar-OODBMS that represents an association with Car-OODBMS;
- Added a pair of columns (attributes) to Car-OODBMS that represents an association with FlyingCar-OODBMS and AmphibiousCar-OODBMS, and for completeness sake; and
- Added associations between AmphibiousCar-OODBMS and Car-OODBMS and FlyingCar-OODBMS and Car-OODBMS that have the correct multiplicities and the XOR constraint explicitly shown.

We also display the dependency relationships from the OODBMS classes to the problem domain classes. Furthermore, we illustrate the fact that the association between Flying-Car-OODBMS and Car-OODBMS, and the association between AmphibiousCar-OODBMS and Car-OODBMS are based on the original, factored-out inheritance relationships in the problem domain classes by showing dependency relationships from the associations to the inheritance relationships.

On the other hand, if we apply Rule 1b to map the Problem Domain classes to a single-inheritance-based OODBMS, we end up with the mapping in Figure 9-6b, where all the attributes of Car have been copied into the FlyingCar-OODBMS and AmphibiousCar-OODBMS classes. In this latter case, you may have to deal with the effects of inheritance conflicts (see Chapter 8).

The advantage of Rule 1a is that all problem domain classes identified during analysis are preserved in the database. This allows maximum flexibility of maintenance of the design of the data management layer. However, Rule 1a increases the amount of message passing required in the system, and it has added processing requirements involving the XOR constraint, thus reducing the overall efficiency of the design. Our recommendation is to limit Rule 1a to be applied only when dealing with "extra" superclasses that are concrete because they have an independent existence in the problem domain. Use Rule 1b when they are abstract because they do not have an independent existence from the subclass.

YOUR TURN

9-3 Doctor's Office Appointment System

In the previous chapters, we have been using a doctor's office appointment system as an example. Assume that you now know that the OODBMS that will be used to support the system will support only single inheritance. Using a class diagram, draw the design for the database.

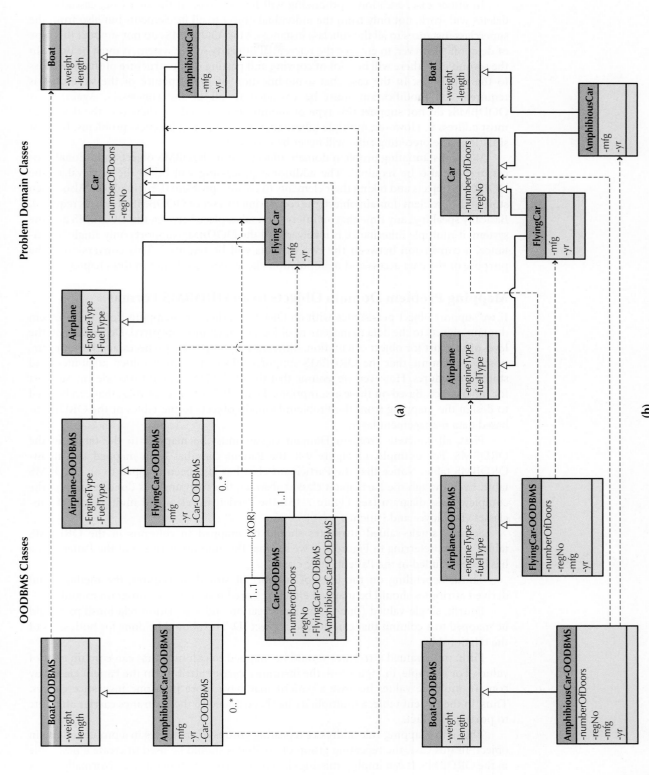

Problem Domain Classes

OODBMS Classes

(a)

(b)

FIGURE 9-6   Mapping Problem Domain Objects to a Single Inheritance–Based OODBMS

383

In either case, additional processing will be required. In the first case, cascading of deletes will work, not only from the individual object to all its elements but also from the superclass instances to all the subclass instances. Most OODBMSs do not support this type of deletion. However, to enforce the referential integrity of the system, it must be done. In the second case, there will be a lot of copying and pasting of the structure of the superclass to the subclasses. In the case that a modification of the structure of the superclass is required, the modification must be cascaded to all of the subclasses. Again, most OODBMSs do not support this type of modification cascading; therefore, the developer must address it. However, multiple inheritance is rare in most business problems. In most situations, the preceding rules will never be necessary.

When instantiating problem domain objects from OODBMS objects, additional processing will also be required. The additional processing will be in the retrieval of the OODBMS objects and taking their elements to create a problem domain object. Also, when storing the problem domain object, the conversion to a set of OODBMS objects is required. Basically speaking, any time that an interaction takes place between the OODBMS and the system, if multiple inheritance is involved and the OODBMS supports only single inheritance, a conversion between the two formats will be required. This conversion is the purpose of the data access and manipulation classes described later in this chapter.

Mapping Problem Domain Objects to an ORDBMS Format

If we support object persistence with an ORDBMS, then the mapping from the problem domain objects to the data management objects is much more involved. Depending on the level of support for object orientation, different mapping rules are necessary. For our purposes, we assume that the ORDBMS supports Object IDs, multivalued attributes, and stored procedures. However, we assume that the ORDBMS does not provide any support for inheritance. Based on these assumptions, Figure 9-7 lists a set of rules that can be used to design the mapping from the Problem Domain objects to the tables of the ORDBMS-based data management layer.

First, all concrete Problem Domain classes must be mapped to the tables in the ORDBMS. For example, in Figure 9-8, the Patient class has been mapped to Patient-ORDBMS table. Notice that the Participant class has also been mapped to an ORDBMS table. Even though the Participant class is abstract, this mapping was done because in the complete class diagram (see Figure 7-15), the Participant class had multiple direct subclasses (Employee and Patient).

Second, single-valued attributes should be mapped to columns in the ORDBMS tables. Again, referring to Figure 9-8, we see that the amount attribute of the Patient class has been included in the Patient Table class.

Third, depending on the level of support of stored procedures, the methods and derived attributes should be mapped either to stored procedures or program modules.

Fourth, single-valued (one-to-one) aggregation and association relationships should be mapped to a column that can store an Object ID. This should be done for both sides of the relationship.

Fifth, multivalued attributes should be mapped to columns that can contain a set of values. For example, in Figure 9-8, the insurance carrier attribute in the Patient class may contain multiple values because a patient may have more than one insurance carrier. Thus, in the Patient table, a multiplicity has been added to the insurance carrier attribute to portray this fact.

The sixth mapping rule addresses repeating groups of attributes in a problem domain object. In this case, the repeating group of attributes should be used to create a new table in the ORDBMS. It can imply a missing class in the problem domain layer. Normally, when

Rule 1: Map all concrete Problem Domain classes to the ORDBMS tables. Also, if an abstract problem domain class has multiple direct subclasses, map the abstract class to an ORDBMS table.

Rule 2: Map single-valued attributes to columns of the ORDBMS tables.

Rule 3: Map methods and derived attributes to stored procedures or to program modules.

Rule 4: Map single-valued aggregation and association relationships to a column that can store an Object ID. Do this for both sides of the relationship.

Rule 5: Map multivalued attributes to a column that can contain a set of values.

Rule 6: Map repeating groups of attributes to a new table and create a one-to-many association from the original table to the new one.

Rule 7: Map multivalued aggregation and association relationships to a column that can store a set of Object IDs. Do this for both sides of the relationship.

Rule 8: For aggregation and association relationships of mixed type (one-to-many or many-to-one), on the single-valued side (1..1 or 0..1) of the relationship, add a column that can store a set of Object IDs. The values contained in this new column will be the Object IDs from the instances of the class on the multivalued side. On the multivalued side (1..* or 0..*), add a column that can store a single Object ID that will contain the value of the instance of the class on the single-valued side.

For generalization/inheritance relationships:

Rule 9a: Add a column(s) to the table(s) that represents the subclass(es) that will contain an Object ID of the instance stored in the table that represents the superclass. This is similar in concept to a foreign key in an RDBMS. The multiplicity of this new association from the subclass to the "superclass" should be 1..1. Add a column(s) to the table(s) that represents the superclass(es) that will contain an Object ID of the instance stored in the table that represents the subclass(es). If the superclasses are concrete, that is, they can be instantiated themselves, then the multiplicity from the superclass to the subclass is 0..1; otherwise, it is 1..1. An exclusive-or (XOR) constraint must be added between the associations. Do this for each superclass.

or

Rule 9b: Flatten the inheritance hierarchy by copying the superclass attributes down to all of the subclasses and remove the superclass from the design.*

*It is also a good idea to document this modification in the design so that in the future, modifications to the design can be maintained easily.

FIGURE 9-7   Schema for Mapping Problem Domain Objects to ORDBMS

a set of attributes repeats together as a group, it implies a new class. Finally, we must create a one-to-many association from the original table to the new one.

The seventh rule supports mapping multivalued (many-to-many) aggregation and association relationships to columns that can store a set of Object IDs. Basically, this is a combination of the fourth and fifth rules. Like the fourth rule, this should be done for both sides of the relationships. For example, in Figure 9-8, the Symptom table has a multivalued attribute (Patients) that can contain multiple Object IDs to Patient Table objects, and Patient table has a multivalued attribute (Symptoms) that can contain multiple Object IDs to Symptom Table objects.

The eighth rule combines the intentions of Rules 4 and 7. In this case, the rule maps one-to-many and many-to-one relationships. On the single-valued side (1..1 or 0..1) of the relationship, a column that can store a set of Object IDs from the table on the multivalued side (1..* or 0..*) of the relationship should be added. On the multivalued side, a column should be added to the table that can store an Object ID from an instance stored in the table on the single-valued side of the relationship. For example, in Figure 9-8, the Patient table has a multivalued attribute (Appts) that can contain multiple Object IDs to Appointment Table objects, whereas the Appointment table has a single-valued attribute (Patient) that can contain an Object ID to a Patient Table object.

The ninth, and final, rule deals with the lack of support for generalization and inheritance. In this case, there are two different approaches. These approaches are virtually identical to the rules described with the preceding OODBMS object persistence formats. For example, in Figure 9-8, the Patient table contains an attribute (Participant) that can contain an Object ID for a Participant Table object, and the Participant table contains an attribute

ORDBMS Tables

Problem Domain Classes

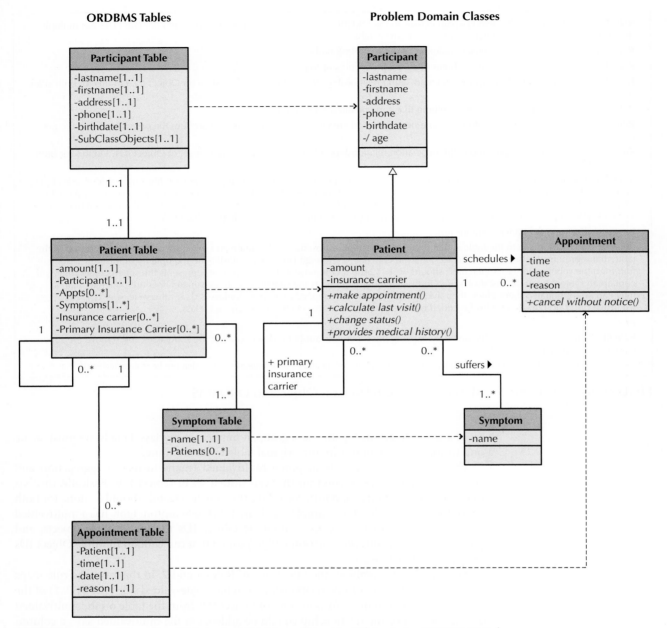

FIGURE 9-8 Example of Mapping Problem Domain Objects to ORDBMS Schema

(SubClassObjects) that contain an Object ID for an object, in this case, stored in the Patient table. In the other case, the inheritance hierarchy is flattened.

Of course, additional processing is required any time an interaction takes place between the database and the system. Every time an object must be created or retrieved from the database, updated, or deleted, the ORDBMS object(s) must be converted to the problem domain object, or vice versa. Again, this is the purpose of the data access and manipulation classes. The only other choice is to modify the problem domain objects. However, such a modification can cause problems between the problem domain layer and the physical architecture and human–computer interface layers. Generally speaking, the cost of conversion between the ORDBMS and the problem domain layer will be more than offset by the savings in development time associated with the interaction between the problem domain and physical architecture and human–computer interaction layers, and the ease of maintenance of a semantically clean problem domain layer. In the long run, owing to the conversions necessary, the development and production cost of using an OODBMS may be less than the development and production cost implementing the object persistence in an ORDBMS.

YOUR

TURN

9-4 Doctor's Office Appointment System

*I*n Your Turn 9-3, you created a design for the database assuming that the database would be implemented using an OODBMS that supported only single inheritance. In this case, assume that you now know that an ORDBMS will be used and that it does not support any inheritance. However, it does support Object IDs, multivalued attributes, and stored procedures. Using a class diagram, draw the design for the database.

Mapping Problem Domain Objects to a RDBMS Format

If we support object persistence with an RDBMS, then the mapping from the problem domain objects to the RDBMS tables is similar to the mapping to an ORDBMS. However, the assumptions made for an ORDBMS are no longer valid. Figure 9-9 lists a set of rules that can be used to design the mapping from the problem domain objects to the RDBMS-based data management layer tables.

Rule 1: Map all concrete problem domain classes to the RDBMS tables. Also, if an abstract Problem Domain class has multiple direct subclasses, map the abstract class to a RDBMS table.

Rule 2: Map single-valued attributes to columns of the tables.

Rule 3: Map methods to stored procedures or to program modules.

Rule 4: Map single-valued aggregation and association relationships to a column that can store the key of the related table, i.e., add a foreign key to the table. Do this for both sides of the relationship.

Rule 5: Map multivalued attributes and repeating groups to new tables and create a one-to-many association from the original table to the new ones.

Rule 6: Map multivalued aggregation and association relationships to a new associative table that relates the two original tables together. Copy the primary key from both original tables to the new associative table, i.e., add foreign keys to the table.

Rule 7: For aggregation and association relationships of mixed type, copy the primary key from the single-valued side (1..1 or 0..1) of the relationship to a new column in the table on the multivalued side (1..* or 0..*) of the relationship that can store the key of the related table, i.e., add a foreign key to the table on the multivalued side of the relationship.

For generalization/inheritance relationships:

Rule 8a: Ensure that the primary key of the subclass instance is the same as the primary key of the superclass. The multiplicity of this new association from the subclass to the "superclass" should be 1..1. If the superclasses are concrete, that is, they can be instantiated themselves, then the multiplicity from the superclass to the subclass is 0..1; otherwise, it is 1..1. Furthermore, an exclusive-or (XOR) constraint must be added between the associations. Do this for each superclass.

OR

Rule 8b: Flatten the inheritance hierarchy by copying the superclass attributes down to all of the subclasses and remove the superclass from the design.*

* It is also a good idea to document this modification in the design so that in the future, modifications to the design can be maintained easily.

FIGURE 9-9   Schema for Mapping Problem Domain Objects to RDBMS

The first four rules are basically the same set of rules used to map problem domain objects to ORDBMS-based data management objects. First, all concrete problem domain classes must be mapped to tables in the RDBMS. Second, single-valued attributes should be mapped to columns in the RDBMS table. Third, methods should be mapped to either stored procedures or program modules, depending on the complexity of the method. Fourth, single-valued (one-to-one) aggregation and association relationships are mapped to columns that can store the foreign keys of the related tables. This should be done for both sides of the relationship. For example in Figure 9-10, we needed to include tables in the RDBMS for the Participant, Patient, Symptom, and Appointment classes.

The fifth rule addresses multivalued attributes and repeating groups of attributes in a problem domain object. In these cases, the attributes should be used to create new tables in the RDBMS. As in the ORDBMS mappings, repeating groups of attributes can imply missing classes in the Problem Domain layer. In that case, a new problem domain class may be required. Finally, we should create a one-to-many or zero-to-many association from the original table to the new one. For example, in Figure 9-10, we needed to create a new table for insurance carrier because it was possible for a patient to have more than one insurance carrier.

The sixth rule supports mapping multivalued (many-to-many) aggregation and association relationships to a new table that relates the two original tables. In this case, the new table should contain foreign keys back to the original tables. For example, in Figure 9-10, we needed to create a new table that represents the suffer association between the Patient and Symptom problem domain classes.

The seventh rule addresses one-to-many and many-to-one relationships. With these types of relationships, the multivalued side (0..* or 1..*) should be mapped to a column in its table that can store a foreign key back to the single-valued side (0..1 or 1..1). It is possible that we have already taken care of this situation because we earlier recommended inclusion of both primary and foreign key attributes in the problem domain classes. In the case of Figure 9-10, we had already added the primary key from the Patient class to the Appointment class as a foreign key (see participantNumber). However, in the case of the reflexive relationship, primary insurance carrier, associated with the Patient class, we need to add a new attribute (primaryInsuranceCarrier) to be able to store the relationship.

The eighth, and final, rule deals with the lack of support for generalization and inheritance. As in the case of an ORDBMS, there are two different approaches. These approaches are virtually identical to the rules described with OODBMS and ORDBMS object persistence formats given earlier. The first approach is to add a column to each table that represents a subclass for each of the concrete superclasses of the subclass. Essentially, this ensures that the primary key of the subclass is the same as the primary key for the superclass. If we had previously added the primary and foreign keys to the problem domain objects, as we recommended, then we do not have to do anything else. The primary keys of the tables will be used to rejoin the instances stored in the tables that represent each of the pieces of the problem domain object. Conversely, the inheritance hierarchy can be flattened and the rules (Rules 1 through 7) can be reapplied.

As in the case of the ORDBMS approach, additional processing will be required any time that an interaction takes place between the database and the system. Every time an object must be created, retrieved from the database, updated, or deleted, the mapping between the problem domain and the RDBMS must be used to convert between the two different formats. In this case, a great deal of additional processing will be required. However, from a practical point of view, it is more likely that you will use a RDBMS for storage of objects than the other approaches because RDBMSs are by far the most popular format in the marketplace. We will focus on how to optimize the RDBMS format for object persistence in the next section of this chapter.

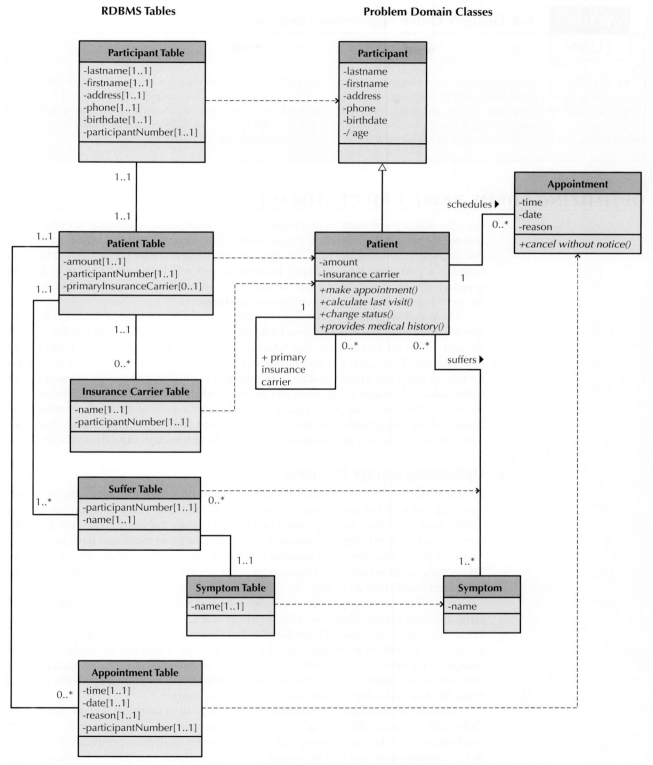

FIGURE 9-10   Example of Mapping Problem Domain Objects to RDBMS Schema

| YOUR | 9-5 Doctor's Office Appointment System |
| TURN | |

*I*n "Your Turn 9-3", you created a design for the database assuming that the database would be implemented using an OODBMS that supported only single inheritance. And, in Your Turn 9-4, you created a design for the database assuming that the database would be implemented using an ORDBMS that did not support any inheritance but did support Object IDs, multivalued attributes, and stored procedures. In this case, you should assume that the system will be supported by an RDBMS. Using a class diagram, draw the design for the database.

OPTIMIZING RDBMS-BASED OBJECT STORAGE

Once the object persistence format is selected, the second step is to optimize the object persistence for processing efficiency. The methods of optimization vary based on the format that you select; however, the basic concepts remain the same. Once you understand how to optimize a particular type of object persistence, you will have some idea as to how to approach the optimization of other formats. This section focuses on the optimization of the most popular storage format: relational databases.

There are two primary dimensions in which to optimize a relational database: for storage efficiency and for speed of access. Unfortunately, these two goals often conflict because the best design for access speed may take up a great deal of storage space as compared to other, less speedy designs. The first section describes how to optimize the object persistence for storage efficiency using a process called *normalization*. The next section presents design techniques, such as denormalization and indexing, which can speed up the performance of the system. Ultimately, the project team will go through a series of trade-offs until the ideal balance of the two optimization dimensions is reached. Finally, the project team must estimate the size of the data storage needed to ensure there is enough capacity on the server(s).

Optimizing Storage Efficiency

The most efficient tables in a relational database in terms of storage space have no redundant data and very few null values. The presence of null values suggests that space is being wasted (and more data to store means higher data storage hardware costs). For example, the table in Figure 9-11 repeats customer information, such as name and state, each time a customer places an order, and it contains many null values in the product-related columns. These nulls occur whenever a customer places an order for fewer than three items (the maximum number on an order).

In addition to wasting space, redundancy and null values also allow more room for error and increase the likelihood that problems will arise with the integrity of the data. What if customer 1035 moved from Maryland to Georgia? In the case of Figure 9-11, a program must be written to ensure that all instances of that customer are updated to show Georgia as the new state of residence. If some of the instances are overlooked, then the table will contain an *update anomaly*, whereby some of the records contain the correctly updated value for state and other records contain the old information.

Nulls threaten data integrity because they are difficult to interpret. A blank value in the Order table's product fields could mean the customer did not want more than one or two products on his or her order, the operator forgot to enter in all three products on the order, or the customer canceled part of the order and the products were deleted by the operator. It is impossible to be sure of the actual meaning of the nulls.

Sample Records:

Order

- Order Number : unsigned long
- Date : Date
- Cust ID : unsigned long
- Last Name : String
- First Name : String
- State : String
- Tax Rate : float
- Product 1 Number : unsigned long
- Product 1 Desc. : String
- Product 1 Price : double
- Product 1 Qty. : unsigned long
- Product 2 Number : unsigned long
- Product 2 Desc. : String
- Product 2 Price : double
- Product 2 Qty. : unsigned long
- Product 3 Number : unsigned long
- Product 3 Desc. : String
- Product 3 Price : double
- Product 3 Qty. : unsigned long

Redundant Data

Null Cells

Order Number	Date	Cust ID	Last Name	First Name	State	Tax Rate	Prod. 1 Number	Prod. 1 Desc.	Prod. 1 Price	Prod. 1 Qty.	Prod. 2 Number	Prod. 2 Desc.	Prod. 2 Price	Prod. 2 Qty.	Prod. 3 Number	Prod. 3 Desc.	Prod. 3 Price	Prod. 3 Qty.
239	11/23/00	1035	Black	John	MD	0.05	555	Cheese Tray	$45.00	2								
260	11/24/00	1035	Black	John	MD	0.05	444	Wine Gift Pack	$60.00	1								
273	11/27/00	1035	Black	John	MD	0.05	222	Bottle Opener	$12.00	2								
241	11/23/00	1123	Williams	Mary	CA	0.08	444	Wine Gift Pack	$60.00	2								
262	11/24/00	1123	Williams	Mary	CA	0.08	222	Bottle Opener	$12.00	2								
287	11/27/00	1123	Williams	Mary	CA	0.08	222	Bottle Opener	$12.00	2								
290	11/30/00	1123	Williams	Mary	CA	0.08	555	Cheese Tray	$45.00	3								
234	11/23/00	2242	DeBerry	Ann	DC	0.065	555	Cheese Tray	$45.00	2								
237	11/23/00	2242	DeBerry	Ann	DC	0.065	111	Wine Guide	$15.00	1	444	Wine Gift Pack	$60.00	1				
238	11/24/00	2242	DeBerry	Ann	DC	0.065	444	Wine Gift Pack	$60.00	1								
245	11/24/00	2242	DeBerry	Ann	DC	0.065	222	Bottle Opener	$12.00	1								
250	11/24/00	2242	DeBerry	Ann	DC	0.065	222	Bottle Opener	$12.00	1								
252	11/24/00	2242	DeBerry	Ann	DC	0.065	222	Bottle Opener	$12.00	1	444	Wine Gift Pack	$60.00	2				
253	11/24/00	2242	DeBerry	Ann	DC	0.065	222	Bottle Opener	$12.00	1	444	Wine Gift Pack	$60.00	1				
297	11/30/00	2242	DeBerry	Ann	DC	0.065	333	Jams & Jellies	$20.00	2								
243	11/24/00	4254	Bailey	Ryan	MD	0.05	555	Cheese Tray	$45.00	2								
246	11/24/00	4254	Bailey	Ryan	MD	0.05	333	Jams & Jellies	$20.00	3								
248	11/24/00	4254	Bailey	Ryan	MD	0.05	222	Bottle Opener	$12.00	1	333	Jams & Jellies	$20.00	2				
235	11/23/00	9500	Chin	April	KS	0.05	222	Bottle Opener	$12.00	2								
242	11/23/00	9500	Chin	April	KS	0.05	333	Jams & Jellies	$20.00	3								
244	11/24/00	9500	Chin	April	KS	0.05	222	Bottle Opener	$12.00	2								
251	11/24/00	9500	Chin	April	KS	0.05	111	Wine Guide	$15.00	2	333	Jams & Jellies	$20.00	2	111	Wine Guide	$15.00	1

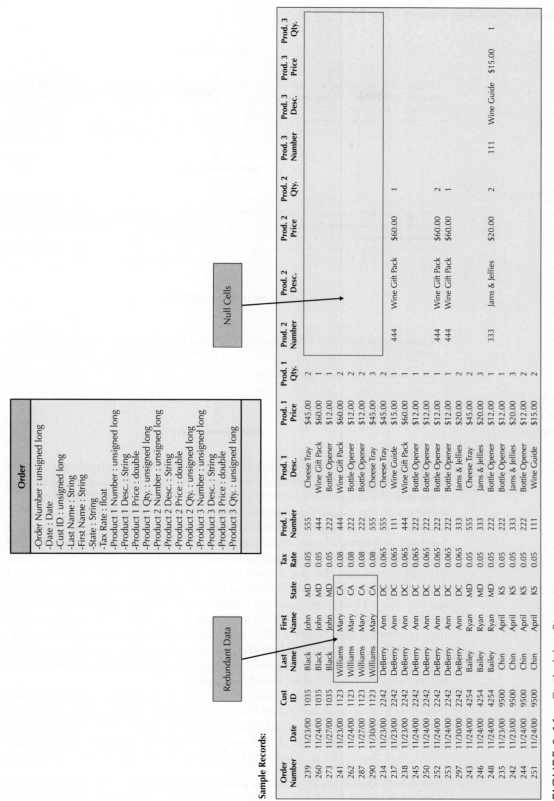

FIGURE 9-11 Optimizing Storage

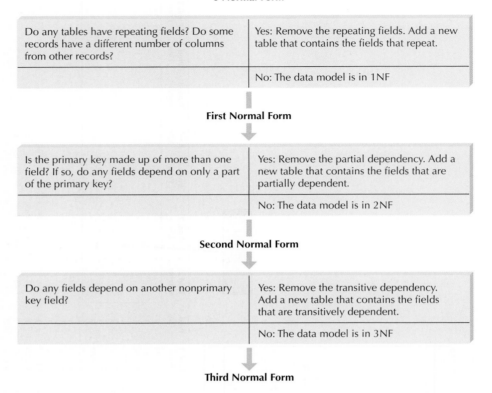

FIGURE 9-12
The Steps of
Normalization

For both these reasons—wasted storage space and data integrity threats—project teams should remove redundancy and nulls from the table. During design, the class diagram is used to examine the design of the RDBMS tables (e.g., see Figure 9-10) and optimize it for storage efficiency. If you follow the modeling instructions and guidelines that were presented in Chapter 5, you will have little trouble creating a design that is highly optimized in this way because a well-formed logical data model does not contain redundancy or many null values.

Sometimes, however, a project team needs to start with a model that was poorly constructed or with one that was created for files or a nonrelational type of format. In these cases, the project team should follow a series of steps that serve to check the model for storage efficiency. These steps make up a process called normalization.[10] *Normalization* is a process whereby a series of rules are applied to the RDBMS tables to determine how well they are formed (see Figure 9-12). These rules help analysts identify tables that are not represented correctly. Here, we describe three normalization rules that are applied regularly in practice. Figure 9-11 shows a model in 0 Normal Form, which is an unnormalized model before the normalization rules have been applied.

A model is in *first normal form (1NF)* if it does not lead to multivalued fields, fields that allow a set of values to be stored, or repeating fields, which are fields that repeat within a table to capture multiple values. The rule for 1NF says that all tables must contain the same number of columns (i.e., fields) and that all the columns must contain a single value. Notice that the model in Figure 9-11 violates 1NF because it causes product number,

---

[10] Normalization also can be performed on the problem domain layer. However, the normalization process should be used on the problem domain layer only to uncover missing classes. Otherwise, optimizations that have nothing to do with the semantics of the problem domain can creep into the problem domain layer.

description, price, and quantity to repeat three times for each order in the table. The resulting table has many records that contain nulls in the product-related columns, and orders are limited to three products because there is no room to store information for more.

A much more efficient design (and one that conforms to 1NF) leads to a separate table to hold the repeating information; to do this, we create a separate table on the model to capture product order information. A zero-to-many relationship would then exist between the two tables. As shown in Figure 9-13, the new design eliminates nulls from the Order table and supports an unlimited number of products that can be associated with an order.

Second normal form (2NF) requires first that the data model is in 1NF and second that the data model leads to tables containing fields that depend on a *whole* primary key. This means that the primary key value for each record can determine the value for all the other fields in the record. Sometimes fields depend on only part of the primary key (i.e., *partial dependency*), and these fields belong in another table.

For example, in the new Product Order table that was created in Figure 9-13, the primary key is a combination of the order number and product number, but the product description and price attributes are dependent only upon product number. In other words, by knowing product number, we can identify the product description and price. However, knowledge of the order number and product number is required to identify the quantity. To rectify this violation of 2NF, a table is created to store product information, and the description and price attributes are moved into the new table. Now, product description is stored only once for each instance of a product number as opposed to many times (every time a product is placed on an order).

A second violation of 2NF occurs in the Order table: customer first name and last name depend only upon the customer ID, not the whole key (Cust ID and Order number). As a result, every time the customer ID appears in the Order table, the names also appear. A much more economical way of storing the data is to create a Customer table with the Customer ID as the primary key and the other customer-related fields (i.e., last name and first name) listed only once within the appropriate record. Figure 9-14 illustrates how the model would look when placed in 2NF.

Third normal form (3NF) occurs when a model is in both 1NF and 2NF and, in the resulting tables, none of the fields depend on nonprimary key fields (i.e., *transitive dependency*). Figure 9-14 contains a violation of 3NF: the tax rate on the order depends upon the state to which the order is being sent. The solution involves creating another table that contains state abbreviations serving as the primary key and the tax rate as a regular field. Figure 9-15 presents the end results of applying the steps of normalization to the original model from Figure 9-11.

YOUR TURN

9-6 Normalizing a Student Activity File

Suppose that you have been asked to build a system that tracks student involvement in activities around campus. You have been given a file with information that needs to be imported into the system, and the file contains the following fields:

student social security number	student first name	activity 2 description
student last name	student advisor name	activity 2 start date
	student advisor phone	activity 3 code
activity 1 start date	activity 1 code	activity 3 description
activity 2 code	activity 1 description	activity 3 start date

Normalize the file. Show how the logical data model would change at each step.

Revised Model:

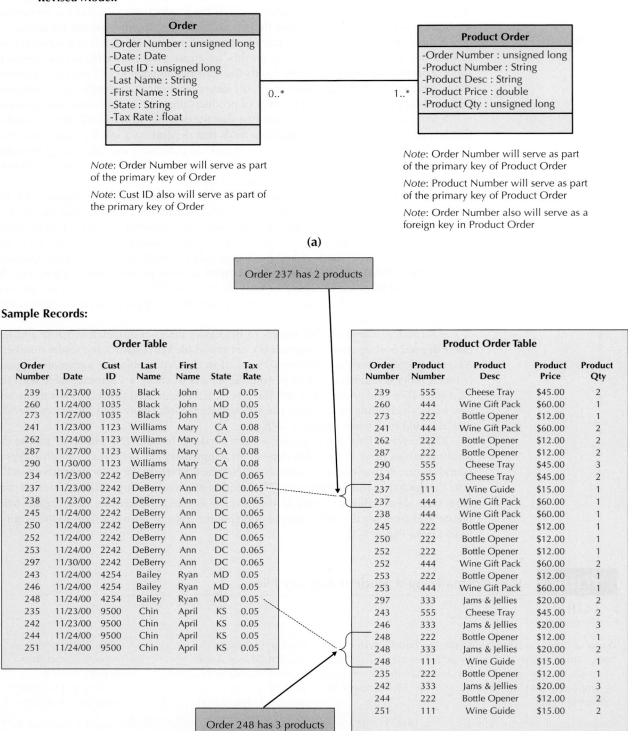

Order

-Order Number : unsigned long
-Date : Date
-Cust ID : unsigned long
-Last Name : String
-First Name : String
-State : String
-Tax Rate : float

0..*   1..*

Product Order

-Order Number : unsigned long
-Product Number : String
-Product Desc : String
-Product Price : double
-Product Qty : unsigned long

Note: Order Number will serve as part of the primary key of Order

Note: Cust ID also will serve as part of the primary key of Order

Note: Order Number will serve as part of the primary key of Product Order

Note: Product Number will serve as part of the primary key of Product Order

Note: Order Number also will serve as a foreign key in Product Order

(a)

Order 237 has 2 products

Sample Records:

Order Table

Order Number	Date	Cust ID	Last Name	First Name	State	Tax Rate
239	11/23/00	1035	Black	John	MD	0.05
260	11/24/00	1035	Black	John	MD	0.05
273	11/27/00	1035	Black	John	MD	0.05
241	11/23/00	1123	Williams	Mary	CA	0.08
262	11/24/00	1123	Williams	Mary	CA	0.08
287	11/27/00	1123	Williams	Mary	CA	0.08
290	11/30/00	1123	Williams	Mary	CA	0.08
234	11/23/00	2242	DeBerry	Ann	DC	0.065
237	11/23/00	2242	DeBerry	Ann	DC	0.065
238	11/23/00	2242	DeBerry	Ann	DC	0.065
245	11/24/00	2242	DeBerry	Ann	DC	0.065
250	11/24/00	2242	DeBerry	Ann	DC	0.065
252	11/24/00	2242	DeBerry	Ann	DC	0.065
253	11/24/00	2242	DeBerry	Ann	DC	0.065
297	11/30/00	2242	DeBerry	Ann	DC	0.065
243	11/24/00	4254	Bailey	Ryan	MD	0.05
246	11/24/00	4254	Bailey	Ryan	MD	0.05
248	11/24/00	4254	Bailey	Ryan	MD	0.05
235	11/23/00	9500	Chin	April	KS	0.05
242	11/23/00	9500	Chin	April	KS	0.05
244	11/24/00	9500	Chin	April	KS	0.05
251	11/24/00	9500	Chin	April	KS	0.05

Product Order Table

Order Number	Product Number	Product Desc	Product Price	Product Qty
239	555	Cheese Tray	$45.00	2
260	444	Wine Gift Pack	$60.00	1
273	222	Bottle Opener	$12.00	1
241	444	Wine Gift Pack	$60.00	2
262	222	Bottle Opener	$12.00	2
287	222	Bottle Opener	$12.00	2
290	555	Cheese Tray	$45.00	3
234	555	Cheese Tray	$45.00	2
237	111	Wine Guide	$15.00	1
237	444	Wine Gift Pack	$60.00	1
238	444	Wine Gift Pack	$60.00	1
245	222	Bottle Opener	$12.00	1
250	222	Bottle Opener	$12.00	1
252	222	Bottle Opener	$12.00	1
252	444	Wine Gift Pack	$60.00	2
253	222	Bottle Opener	$12.00	1
253	444	Wine Gift Pack	$60.00	1
297	333	Jams & Jellies	$20.00	2
243	555	Cheese Tray	$45.00	2
246	333	Jams & Jellies	$20.00	3
248	222	Bottle Opener	$12.00	1
248	333	Jams & Jellies	$20.00	2
248	111	Wine Guide	$15.00	1
235	222	Bottle Opener	$12.00	1
242	333	Jams & Jellies	$20.00	3
244	222	Bottle Opener	$12.00	2
251	111	Wine Guide	$15.00	2

Order 248 has 3 products

(b)

FIGURE 9-13 1NF: Remove Repeating Fields

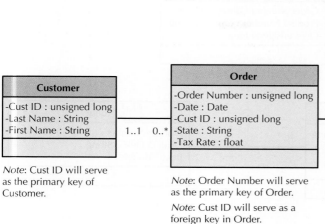

Product Order
-Order Number : unsigned long
-Product Number : unsigned long
-Qty : unsigned long

Customer
-Cust ID : unsigned long
-Last Name : String
-First Name : String

1..1   0..*

Order
-Order Number : unsigned long
-Date : Date
-Cust ID : unsigned long
-State : String
-Tax Rate : float

0..*   1..*

Product
-Product Number : unsigned long
-Product Desc : String
-Price : double

Note: Cust ID will serve as the primary key of Customer.

Note: Order Number will serve as the primary key of Order.

Note: Cust ID will serve as a foreign key in Order.

Note: Order Number will serve as part of the primary key of Product Order.

Note: Order Number also will serve as a foreign key in Product Order.

Note: Product Number will serve as part of the primary key in Product Order.

Note: Product Number also will serve as a foreign key in Product Order.

Note: Product Number will serve as part of the primary key of Product Order.

Sample Records:

Customer Table

Cust ID	Last Name	First Name
1035	Black	John
1123	Williams	Mary
2242	DeBerry	Ann
4254	Bailey	Ryan
9500	Chin	April

Last Name and First Name was moved to the Customer table to eliminate redundancy

Order Table

Order Number	Date	Cust ID	State	Tax Rate
239	11/23/00	1035	MD	0.05
260	11/24/00	1035	MD	0.05
273	11/27/00	1035	MD	0.05
241	11/23/00	1123	CA	0.08
262	11/24/00	1123	CA	0.08
287	11/27/00	1123	CA	0.08
290	11/30/00	1123	CA	0.08
234	11/23/00	2242	DC	0.065
237	11/23/00	2242	DC	0.065
238	11/23/00	2242	DC	0.065
245	11/24/00	2242	DC	0.065
250	11/24/00	2242	DC	0.065
252	11/24/00	2242	DC	0.065
253	11/24/00	2242	DC	0.065
297	11/30/00	2242	DC	0.065
243	11/23/00	4254	MD	0.05
246	11/24/00	4254	MD	0.05
248	11/24/00	4254	MD	0.05
235	11/23/00	9500	KS	0.05
242	11/23/00	9500	KS	0.05
244	11/24/00	9500	KS	0.05
251	11/24/00	9500	KS	0.05

Product Order Table

Order Number	Product Number	Product Qty
239	555	2
260	444	1
273	222	1
241	444	2
262	222	2
287	222	2
290	555	3
234	555	2
237	111	1
237	444	1
238	444	1
245	222	1
250	222	1
252	222	1
252	444	2
253	222	1
253	444	1
297	333	2
243	555	2
246	333	3
248	222	1
248	333	2
248	111	1
235	222	1
242	333	3
244	222	2
251	111	2

Product Table

Product Number	Product Desc	Product Price
111	Wine Guide	$15.00
222	Bottle Opener	$12.00
333	Jams & Jellies	$20.00
444	Wine Gift Pack	$60.00
555	Cheese Tray	$45.00

Product Desc and Price was moved to the Product table to eliminate redundancy

FIGURE 9-14   2NF Partial Dependencies Removed

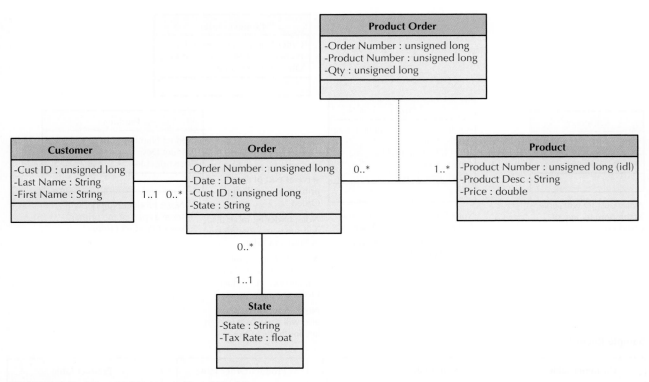

FIGURE 9-15   3NF Normalized Field

Optimizing Data Access Speed

After you have optimized the design of the object storage for efficiency, the end result is that data are spread out across a number of tables. When data from multiple tables need to be accessed or queried, the tables must be first joined. For example, before a user can print out a list of the customer names associated with orders, first the Customer and Order tables need to be joined, based on the customer number field (see Figure 9-15). Only then can both the order and customer information be included in the query's output. Joins can take a lot of time, especially if the tables are large or if many tables are involved.

Consider a system that stores information about 10,000 different products, 25,000 customers, and 100,000 orders, each averaging three products per order. If an analyst wanted to investigate whether there were regional differences in music preferences, he or she would need to combine all the tables to be able to look at products that have been ordered while knowing the state of the customers placing the orders. A query of this information would result in a huge table with 300,000 rows (i.e., the number of products that have been ordered) and 11 columns (the total number of columns from all of the tables combined).

The project team can use several techniques to try to speed up access to the data, including denormalization, clustering, and indexing.

Denormalization   After the object storage is optimized, the project team may decide to denormalize, or add redundancy back into the design. *Denormalization* reduces the number of joins that need to be performed in a query, thus speeding up access. Figure 9-16 shows a denormalized model for customer orders. The customer last name was added back into the Order table because the project team learned during analysis that queries about orders usually require the customer last name field. Instead of joining the Order table

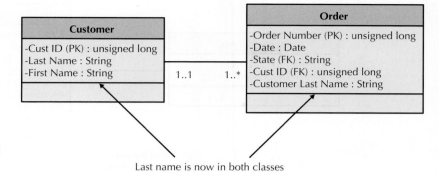

FIGURE 9-16
Denormalized Physical
Data Model

Last name is now in both classes

repeatedly to the Customer table, the system now needs to access only the Order table because it contains all of the relevant information.

Denormalization should be applied sparingly for the reasons described in the previous section, but it is ideal in situations in which information is queried frequently but updated rarely. There are three cases in which you may rely upon denormalization to reduce joins and improve performance. First, denormalization can be applied in the case of look-up tables, which are tables that contain descriptions of values (e.g., a table of product descriptions or a table of payment types). Because descriptions of codes rarely change, it may be more efficient to include the description along with its respective code in the main table to eliminate the need to join the look-up table each time a query is performed (see Figure 9-17a).

Second, one-to-one relationships are good candidates for denormalization. Although logically two tables should be separated, from a practical standpoint the information from both tables may regularly be accessed together. Think about an order and its shipping information. Logically, it might make sense to separate the attributes related to shipping into a separate table, but as a result the queries regarding shipping will probably always need a join to the Order table. If the project team finds that certain shipping information, such as state and shipping method, are needed when orders are accessed, they may decide to combine the tables or include some shipping attributes in the Order table (see Figure 9-17b).

Third, at times it is more efficient to include a parent entity's attributes in its child entity on the physical data model. For example, consider the Customer and Order tables in Figure 9-16, which share a one-to-many relationship, with Customer as the parent and Order as the child. If queries regarding orders continuously require customer information, the most popular customer fields can be placed in Order to reduce the required joins to the Customer table, as was done with Customer Last Name.

YOUR TURN

9-7 Denormalizing a Student Activity File

Consider the logical data model that you created for Your Turn 9-3. Examine the model and describe possible opportunities for denormalization. How would you change the physical data model for this file, and what are the benefits of your changes?

Clustering   Speed of access also is influenced by the way that the data are retrieved. Think about shopping in a grocery store. If you have a list of items to buy but you are unfamiliar with the store's layout, you need to walk down every aisle to make sure that you don't miss

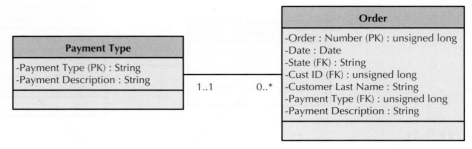

Notice that the payment description field
appears in both Payment Type and Order.

(a)

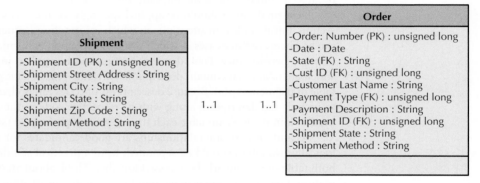

FIGURE 9-17

Denormalization
Situations (FK, foreign
key; PK, primary key)

Notice that the shipment state and shipment method
are included in both Shipment and Order.

(b)

anything from your list. Likewise, if records are arranged on a hard disk in no particular order (or in an order that is irrelevant to your data needs), then any query of the records results in a *table scan* in which the DBMS has to access every row in the table before retrieving the result set. Table scans are the most inefficient of data retrieval methods.

One way to improve access speed is to reduce the number of times that the storage medium needs to be accessed during a transaction. One method is to *cluster* records together physically so that similar records are stored close together. With *intrafile clustering*, like records in the table are stored together in some way, such as in order by primary key or, in the case of a grocery store, by item type. Thus, whenever a query looks for records, it can go directly to the right spot on the hard disk (or other storage medium) because it knows in what order the records are stored, just as we can walk directly to the bread aisle to pick up a loaf of bread. *Interfile clustering* combines records from more than one table that typically are retrieved together. For example, if customer information is usually accessed with the related order information, then the records from the two tables may be physically stored in a way that preserves the customer-order relationship. Returning to the grocery store scenario, an interfile cluster would be similar to storing peanut butter, jelly, and bread next to each other in the same aisle because they are usually purchased together, not because they are similar types of items. Of course, each table can have only one clustering strategy because the records can be arranged physically in only one way.

Indexing   A familiar time saver is an index located in the back of a textbook, which points directly to the page or pages that contain a topic of interest. Think of how long it would take to find all the times that relational database appears in this textbook without the index to rely on! An *index* in data storage is like an index in the back of a textbook; it is a minitable that contains values from one or more columns in a table and the location of the values within the table. Instead of paging through the entire textbook, we can move directly to the right pages and get the information we need. Indexes are one of the most important ways to improve database performance. Whenever there are performance problems, the first place to look is an index.

A query can use an index to find the locations of only those records that are included in the query answer, and a table can have an unlimited number of indexes. Figure 9-18 shows an index that orders records by payment type. A query that searches for all the customers who used American Express can use this index to find the locations of the records that contain American Express as the payment type without having to scan the entire Order table.

Project teams can make indexes perform even faster by placing them into the main memory of the data storage hardware. Retrieving information from memory is much faster than from another storage medium, such as a hard disk—think about how much faster it is to retrieve a memorized phone number versus one that must be looked up in a phone book. Similarly, when a database has an index in memory, it can locate records very, very quickly.

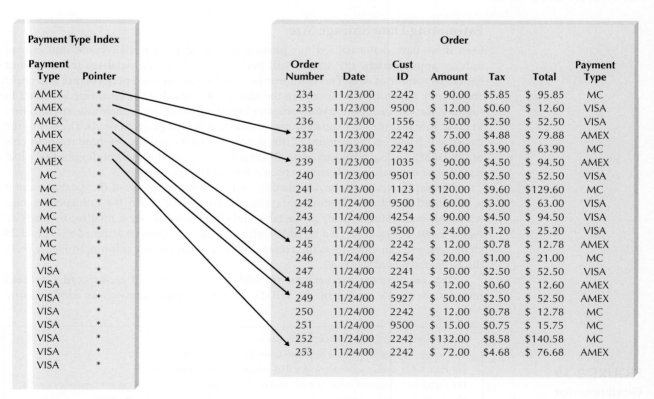

FIGURE 9-18   Payment Type Index

Of course, indexes require overhead in that they take up space on the storage medium. Also, they need to be updated as records in tables are inserted, deleted, or changed. Thus, although queries lead to faster access to the data, they slow down the update process. In general, we should create indexes sparingly for transaction systems or systems that require a lot of updates, but we should apply indexes generously when designing systems for decision support (see Figure 9-19).

CONCEPTS

IN ACTION

9-A Mail-Order Index

A Virginia-based mail-order company sends out approximately 25 million catalogs each year using a Customer table with 10 million names. Although the primary key of the Customer table is customer identification number, the table also contains an index of Customer Last Name. Most people who call to place orders remember their last name, not their customer identification number, so this index is used frequently.

An employee of the company explained that indexes are critical to reasonable response times. A fairly complicated query was written to locate customers by the state in which they lived, and it took more than three weeks to return an answer. A customer state index was created, and that same query provided a response in twenty minutes: that's 1,512 times faster!

Question

As an analyst, how can you make sure that the proper indexes have been put in place so that users are not waiting for weeks to receive the answers to their questions?

Estimating Data Storage Size

Even if we have denormalized our physical data model, clustered records, and created indexes appropriately, the system will perform poorly if the database server cannot handle its volume of data. Therefore, one last way to plan for good performance is to apply *volumetrics,* which means estimating the amount of data that the hardware will need to support. You can incorporate your estimates into the database server hardware specification to make sure that the database hardware is sufficient for the project's needs. The size of the database is based on the amount of *raw data* in the tables and the *overhead* requirements of the DBMS. To estimate size, you will need to have a good understanding of the initial size of your database as well as its expected growth rate over time.

Raw data refers to all the data that are stored within the tables of the database, and it is calculated based on a bottom-up approach. First, write down the estimated average width for each column (field) in the table and sum the values for a total record size (see Figure 9-20). For example, if a variable-width Last Name column is assigned a width of 20 characters, you can enter 13 as the average character width of the column. In Figure 9-20, the estimated record size is 49.

Next, calculate the overhead for the table as a percentage of each record. Overhead includes the room needed by the DBMS to support such functions as administrative

FIGURE 9-19

Guidelines for Creating Indexes

Use indexes sparingly for transaction systems.

Use many indexes to increase response times in decision support systems.

For each table, create a unique index that is based on the primary key.

For each table, create an index that is based on the foreign key to improve the performance of joins.

Create an index for fields that are used frequently for grouping, sorting, or criteria.

Field	Average Size
Order Number	8
Date	7
Cust ID	4
Last Name	13
First Name	9
State	2
Amount	4
Tax Rate	2
Record Size	49
Overhead	30%
Total Record Size	63.7
Initial Table Size	50,000
Initial Table Volume	3,185,000
Growth Rate/Month	1,000
Table Volume @ 3 years	5,478,200

FIGURE 9-20
Calculating
Volumetrics

actions and indexes, and it should be assigned based on past experience, recommendations from technology vendors, or parameters that are built into software that was written to calculate volumetrics. For example, your DBMS vendor might recommend that you allocate 30 percent of the records' raw data size for overhead storage space, creating a total record size of 63.7 in the Figure 9-20 example.

Finally, record the number of initial records that will be loaded into the table, as well as the expected growth per month. This information should have been collected during analysis. According to Figure 9-20, the initial space required by the first table is 3,185,000, and future sizes can be projected based on the growth figure. These steps are repeated for each table to get a total size for the entire database.

Many CASE tools provide you with database-size information based on how you set up the object persistence, and they calculate volumetrics estimates automatically. Ultimately, the size of the database needs to be shared with the design team so that the proper technology can be put in place to support the system's data and potential performance problems can be addressed long before they affect the success of the system.

CONCEPTS
IN ACTION

9-B Return on Investment from Virtualization—A Hard Factor to Determine

Many companies are undergoing server virtualization. This is the concept of putting multiple virtual servers onto one physical device. The payoffs can be significant: fewer servers, less electricity, less generated heat, less air conditioning, less infrastructure and administration costs; increased flexibility; less physical presence (i.e., smaller server rooms), faster maintenance of servers, and more. There are (of course) costs—such as licensing the virtualization software, labor costs in establishing the virtual servers onto a physical device, and labor costs in updating tables and access. But, determining the Return on Investment can be a challenge. Some companies have lost money on server virtualization, but most would say they have gained a positive return on investment with virtualization but have not really quantified the results.

Questions

1. How might a company really determine the return on investment for server virtualization?

2. Would server virtualization impact the amount of data storage required? Why or why not?

3. Is this a project that a systems analyst might be involved in? Why or why not?

DESIGNING DATA ACCESS AND MANIPULATION CLASSES

The final step in developing the data management layer is to design the *data access and manipulation classes* that act as a translator between the object persistence and the problem domain objects. Thus, they should always be capable of at least reading and writing both the object persistence and problem domain objects. As described earlier and in Chapter 8, the object persistence classes are derived from the concrete problem domain classes, whereas the data access and manipulation classes depend on both the object persistence and problem domain classes.

Depending on the application, a simple rule to follow is that there should be one data access and manipulation class for each concrete problem domain class. In some cases, it might make sense to create data access and manipulation classes associated with the human–computer interaction classes (see Chapter 10). However, this creates a dependency from the data management layer to the human–computer interaction layer. Adding this additional complexity to the design of the system normally is not recommended.

Returning to the ORDBMS solution for the Appointment system example (see Figure 9-8), we see that we have four problem domain classes and four ORDBMS tables. Following the previous rule, the DAM classes are rather simple. They have to support only a one-to-one translation between the concrete problem domain classes and the ORDBMS tables (see Figure 9-21). Because the Participant problem domain class is an abstract class, only three data access and manipulation classes are required: Patient-DAM, Symptom-DAM, and Appointment-DAM. However, the process to create an instance of the Patient problem domain class can be fairly complicated. The Patient-DAM class might have to be able to retrieve information from all four ORDBMS tables. To accomplish this, the Patient-DAM class retrieves the information from the Patient table. Using the Object-IDs stored in the attribute values associated with the Participant, Appts, and Symptoms attributes, the remaining information required to create an instance of Patient is easily retrieved by the Patient-DAM class.

In the case of using an RDBMS to provide persistence, the data access and manipulation classes tend to become more complex. For example, in the Appointment system, there are still four problem domain classes, but, owing to the limitations of RDBMSs, we have to support six RDBMS tables (see Figure 9-10). The data access and manipulation class for the Appointment problem domain class and the Appointment RDBMS table is no different from those supported for the ORDBMS solution (see Figures 9-21 and 9-22). However, owing to the multivalued attributes and relationships associated with the Patient and Symptom problem domain classes, the mappings to the RDBMS tables were more complicated. Consequently, the number of dependencies from the data access and manipulation classes (Patient-DAM and Symptom-DAM) to the RDBMS tables (Patient table, Insurance Carrier table, Suffer table, and the Symptom table) has increased. Furthermore, because the Patient problem domain class is associated with the other three problem domain classes, the actual retrieval of all information necessary to create an instance of the Patient class could involve joining information from all six RDBMS tables. To accomplish this, the Patient-DAM class must first retrieve information from the Patient table, Insurance Carrier table, Suffer table, and the Appointment table. Because the primary key of the Patient table and the Participant table are identical, the Patient-DAM class can either directly retrieve the information from the Participant table, or the information can be joined using the participantNumber attributes of the two tables, which act as both primary and foreign keys. Finally, using the information contained in the Suffer table, the information in the Symptom table can also be retrieved. Obviously, the farther we get from the object-oriented problem domain class representation, the more work must be performed. However, as in the case of the ORDBMS example, notice that absolutely no modifications were made to the problem domain classes. Therefore, the data access and manipulation classes again have prevented data management functionality from creeping into the problem domain classes.

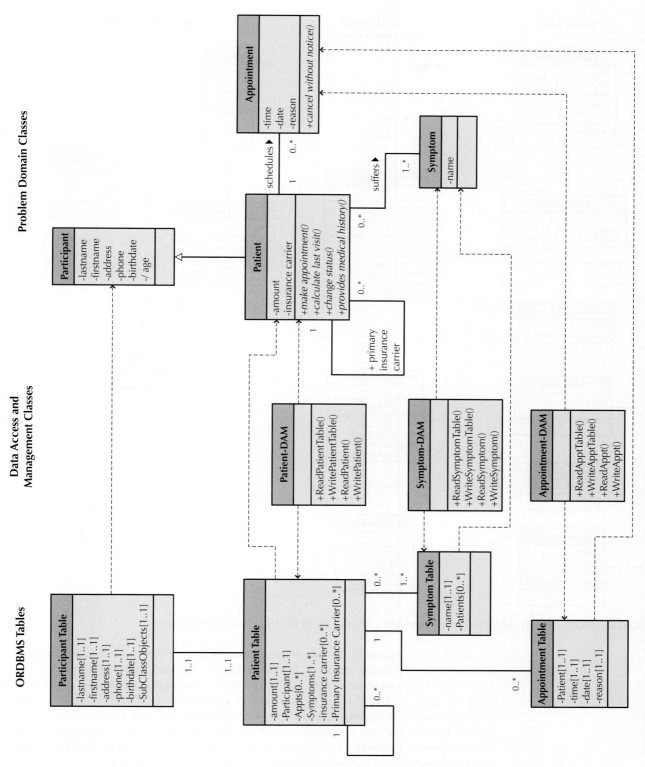

Problem Domain Classes

FIGURE 9-21   Managing Problem Domain Objects to ORDBMS using DAM Classes

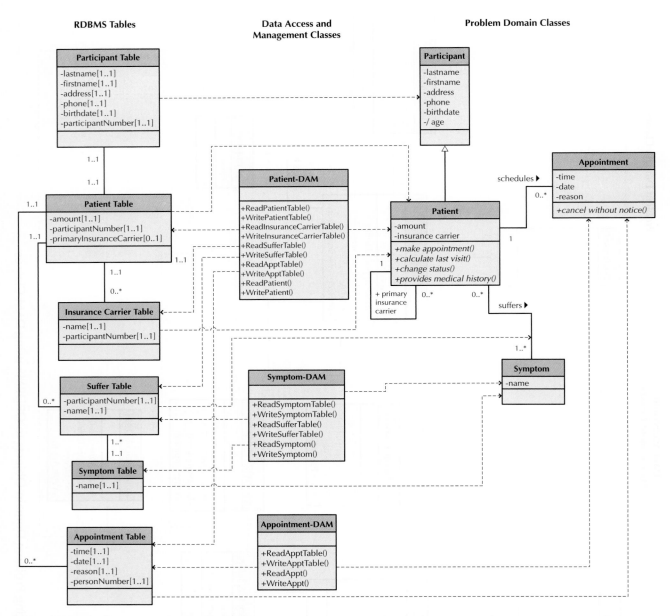

FIGURE 9-22   Mapping Problem Domain Objects to RDBMS using DAM Classes

One specific approach that has been suggested to support the implementation of data access and manipulation classes is to use an object-relational mapping library such as Hibernate.[11] Hibernate developed within the JBoss community, allows the mapping of objects written in Java that are to be stored in an RDBMS. Instead of using an object-oriented programming language to implement the data access and manipulation classes, with Hibernate, they are implemented in XML files that contain the mapping. As in the above approach, modeling the mapping in an XML file prevents the details on data access and manipulation from sneaking into the problem domain representation.

[11] For more information on Hibernate, see www.hibernate.org/.

NONFUNCTIONAL REQUIREMENTS AND DATA MANAGEMENT LAYER DESIGN[12]

Recall that nonfunctional requirements refer to behavioral properties that the system must have. These properties include issues related to performance, security, ease of use, operational environment, and reliability. In this text, we have grouped nonfunctional requirements into four categories: operational, performance, security, and cultural and political requirements. We describe each of these in relation to the data management layer.

The *operational requirements* for the data management layer include issues that deal with the technology being used to support object persistence. However, the choice of the *hardware and operating system* limits the choice of the technology and format of the object persistence available. This is especially true when you consider mobile computing. Given the limited memory and storage available on these devices, the choices to support object persistence are limited. One possible choice to support object persistence that works both on Google's Android and Apple's iOS-based platforms is SQLite. SQLite is a lightweight version of SQL that supports RDBMS. However, there are many different approaches to support object persistence that are more platform dependent; for example, Android supports storing objects with shared preferences (a key-value pair-based NoSQL approach), internal storage, on an SD card, in a local cache, or on a remote system. This, in turn, determines which set of the mapping rules described earlier will have to be used. Another operational requirement could be the ability to import and export data using XML. Again, this could limit the object stores under consideration.

The primary *performance requirements* that affect the data management layer are speed and capacity. As described before, depending on the anticipated—and, afterwards, actual—usage patterns of the objects being stored, different indexing and caching approaches may be necessary. When considering distributing objects over a network, speed considerations can cause objects to be replicated on different nodes in the network. Thus, multiple copies of the same object may be stored in different locations on the network. This raises the issue of update anomalies described before in conjunction with normalization. Depending on the application being built, NoSQL data stores that support an eventually consistent update model may be appropriate. Also, depending on the estimated size and growth of the system, different DBMSs may need to be considered. An additional requirement that can affect the design of the data management layer deals with the availability of the objects being stored. It might make sense to limit the availability to different objects based on the time of day. For example, one class of users may be allowed to access a set of objects only from 8 to 12 in the morning and a second set of users may be able to access them only from 1 to 5 in the afternoon. Through the DBMS, these types of restrictions could be set.

The *security requirements* deal primarily with access controls, encryption, and backup. Through a modern DBMS, different types of access can be set (e.g., Read, Update, or Delete) granting access only to users (or class of users) who have been authorized. Furthermore, *access control* can be set to guarantee that only users with "administrator" privileges are allowed to modify the object storage schema or access controls. Encryption requirements on this layer deal with whether the object should be stored in an encrypted format or not. Even though encrypted objects are more secure than unencrypted objects, the process of encrypting and decrypting the objects will slow down the system. Depending on the physical architecture being used, the cost of encryption may be negligible. For example, if we plan on encrypting the objects before transmitting them over a network, there may be no additional cost of storing them in the encrypted format. Backup requirements deal with ensuring that

---

[12] Because the vast majority of nonfunctional requirements affect the physical architecture layer, we provide additional details in Chapter 11.

the objects are routinely copied and stored in case the object store becomes corrupted or unusable. Having a backup copy made on a periodic basis and storing the updates that have occurred since the last backup copy was made ensures that the updates are not lost and that the object store can be reconstituted by running the copies of the updates against the backup copy to create a new, current copy.

The only *political and cultural requirements* that can affect the data management layer deal with how the detailed format of the data is to be stored. For example, in what format should a date be stored? Or how many characters should be allocated for a last name field that is part of an Employee object? There could be a corporate IT bias toward different hardware and software platforms. If so, this could limit the type of object store available.

APPLYING THE CONCEPTS AT CD SELECTIONS

In the previous installments of the CD Selections case, we saw how Alec, Margaret, and the development team had worked through developing models and designs of the problem domain classes. Now that the design of the problem domain layer is somewhat stable, the team has moved into developing the models and designs of the solution domain (data management, human–computer interaction, and physical architecture) classes. In this installment, we follow the team members that have been assigned to the development of the data management layer classes for the web-based system being developed for CD Selections.

SUMMARY

Object Persistence Formats

There are five basic types of object persistence formats: files (sequential and random access), object-oriented databases, object-relational databases, relational databases, and NoSQL data stores. Files are electronic lists of data that have been optimized to perform a particular transaction. There are two different access methods (sequential and random), and there are five different application types: master, look-up, transaction, audit, and history. Master files typically are kept for long periods of time because they store important business information, such as order information or customer mailing information. Look-up files contain static values that are used to validate fields in the master files, and transaction files temporarily hold information that will be used for a master file update. An audit file records before and after images of data as they are altered so that an audit can be performed if the integrity of the data is questioned. Finally, the history file stores past transactions (e.g., old customers, past orders) that are no longer needed by the system.

A database is a collection of information groupings related to one another in some way, and a DBMS (database management system) is software that creates and manipulates databases. There are three types of databases that are likely to be encountered during a project: relational, object-relational, and object-oriented. The relational database is the most popular kind of database for application development today. It is based on collections of tables that are related to each other through common fields, known as foreign keys. Object-relational databases are relational databases that have extensions that provide limited support for object orientation. The extensions typically include some support for the storage of objects in the relational table structure. Object-oriented databases come in two flavors: full-blown DBMS products and extensions to an object-oriented programming language. Both approaches typically fully support object orientation. Finally, NoSQL data stores represent a new class of approaches to support object persistence. These include key-value stores, document stores, column-oriented stores, and object databases.

The application's data should drive the storage format decision. Relational databases support simple data types very effectively, whereas object databases are best for complex data. The type of system also should be considered when choosing among data storage formats (e.g., relational databases have matured to support transactional systems). Although less critical to the format selection decision, the project team needs to consider what technology exists within the organization and the kind of technology likely to be used in the future.

Mapping Problem Domain Objects to Object Persistence Formats

There are many different approaches to support object persistence. Each of the different formats for object persistence has some conversion requirements. The complexity of the conversion requirements increases the farther the format is from an object-oriented format. An OODBMS has the fewest conversion requirements, whereas an RDBMS tends to have the most. No matter what format is chosen, all data management functionality should be kept out of the problem domain classes to minimize the maintenance requirements of the system and to maximize the portability and reusability of the problem domain classes.

Optimizing RDBMS-Based Object Storage

There are two primary dimensions by which to optimize a relational database: storage efficiency and speed of access. The most efficient relational database tables in terms of data storage are those that have no redundant data and very few null values. Normalization is the process whereby a series of rules are applied to the data management layer to determine how well it is formed. A model is in first normal form (1NF) if it does not lead to repeating fields, which are fields that repeat within a table to capture multiple values. Second normal form (2NF) requires that all tables be in 1NF and lead to fields whose values are dependent on the *whole* primary key. Third normal form (3NF) occurs when a model is in both 1NF and 2NF and none of the resulting fields in the tables depend on nonprimary key fields (i.e., transitive dependency). With each violation, additional tables should be created to remove the repeating fields or improper dependencies from the existing tables.

Once we have optimized the design of the object persistence for storage efficiency, the data may be spread out across a number of tables. To improve speed, the project team may decide to denormalize—add redundancy back into—the design. Denormalization reduces the number of joins that need to be performed in a query, thus speeding up data access. Denormalization is best in situations where data are accessed frequently and updated rarely. Three modeling situations are good candidates for denormalization: lookup tables, entities that share one-to-one relationships, and entities that share one-to-many relationships. In all three cases, attributes from one entity are moved or repeated in another entity to reduce the joins that must occur while accessing the database.

Clustering occurs when similar records are stored close together on the storage medium to speed up retrieval. In intrafile clustering, similar records in the table are stored together in some way, such as in sequence. Interfile clustering combines records from more than one table that, typically, are retrieved together. Indexes also can be created to improve the access speed of a system. An index is a minitable that contains values from one or more columns in a table and tells where the values can be found. Instead of performing a table scan, which is the most inefficient way to retrieve data from a table, an index points directly to the records that match the requirements of a query.

Finally, the speed of the system can be improved if the right hardware is purchased to support it. Analysts can use volumetrics to estimate the current and future size of the database and then share these numbers with the people who are responsible for buying and configuring the database hardware.

Designing Data Access and Manipulation Classes

Once the object persistence has been designed, a translation layer between the problem domain classes and the object persistence should be created. The translation layer is implemented through data access and manipulation classes. In this manner, any changes to the object persistence format chosen will require changes only to the data access and manipulation classes. The problem domain classes will be completely isolated from the changes. One popular approach to supporting data access and manipulation classes is through the use of a mapping library, such as Hibernate.

Nonfunctional Requirements and Data Management Layer Design

Nonfunctional requirements can affect the design of the data management layer. Operational requirements can limit the viability of different object persistence formats. Performance requirements can cause various indexing and caching approaches to be considered. Furthermore, performance requirements can create situations where denormalization must be considered. Performance considerations are especially important in the area of mobile computing. The security requirements can cause the design to include varying types of access to be controlled and the possibility of using different encryption algorithms to make the data more difficult to use by unauthorized users. Finally, political and cultural concerns can influence the design of certain attributes and objects.

KEY TERMS

QUESTIONS

1. Describe the four steps in object persistence design.
2. How are a file and a database different from each other?
3. What is the difference between an end-user database and an enterprise database? Provide an example of each one.
4. What are the differences between sequential and random access files?
5. Name five types of files and describe the primary purpose of each type.
6. What is the most popular kind of database today? Provide three examples of products that are based on this database technology.
7. What is referential integrity and how is it implemented in an RDBMS?
8. List some of the differences between an ORDBMS and an RDBMS.
9. What are the advantages of using an ORDBMS over an RDBMS?
10. List some of the differences between an ORDBMS and an OODBMS.
11. What are the advantages of using an ORDBMS over an OODBMS?
12. What are the advantages of using an OODBMS over an RDBMS?
13. What are the advantages of using an OODBMS over an ORDBMS?
14. What are the factors in determining the type of object persistence format that should be adopted for a system? Why are these factors so important?
15. Why should you consider the storage formats that already exist in an organization when deciding upon a storage format for a new system?
16. When implementing the object persistence in an ORDBMS, what types of issues must you address?

17. When implementing the object persistence in an RDBMS, what types of issues must you address?
18. Name three ways null values can be interpreted in a relational database. Why is this problematic?
19. What are the two dimensions in which to optimize a relational database?
20. What is the purpose of normalization?
21. How does a model meet the requirements of third normal form?
22. Describe three situations that can be good candidates for denormalization.
23. Describe several techniques that can improve performance of a database.
24. What is the difference between interfile and intrafile clustering? Why are they used?
25. What is an index and how can it improve the performance of a system?
26. Describe what should be considered when estimating the size of a database.
27. Why is it important to understand the initial and projected size of a database during design?
28. What are some of the nonfunctional requirements that can influence the design of the data management layer?
29. What are the key issues in deciding between using perfectly normalized databases and denormalized databases?
30. What is the primary purpose of the data access and manipulation classes?
31. Why should the data access and manipulation classes be dependent on the problem domain classes instead of the other way around?
32. Why should the object persistence classes be dependent on the problem domain classes instead of the other way around?

EXERCISES

A. Using the web or other resources, identify a product that can be classified as an end-user database and a product that can be classified as an enterprise database. How are the products described and marketed? What kinds of applications and users do they support? In what kinds of situations would an organization choose to implement an end-user database over an enterprise database?

B. Visit a commercial website (e.g., Amazon.com). If files were being used to store the data supporting the application, what types of files would be needed? What access type would be required? What data would they contain?

C. Using the web, review one of the following products. What are the main features and functions of the software? In what companies has the DBMS been implemented, and for what purposes? According to the information that you found, what are three strengths and weaknesses of the product?
  1. Relational DBMS
  2. Object-relational DBMS
  3. Object-oriented DBMS

D. You have been given a file that contains the following fields relating to CD information. Using the steps of normalization, create a model that represents this file in third normal form. The fields include:

Musical group name	CD title 2
Musicians in group	CD title 3
Date group was formed	CD 1 length
Group's agent	CD 2 length
CD title 1	CD 3 length

Assumptions:
  • Musicians in group contains a list of the members of the people in the musical group.
  • Musical groups can have more than one CD, so both group name and CD title are needed to uniquely identify a particular CD.

E. Jim Smith's dealership sells Fords, Hondas, and Toyotas. The dealership keeps information about each car manufacturer with whom they deal so that they can get in touch with them easily. The dealership also keeps information about the models of cars that they carry from each manufacturer. They keep information like list price, the price the dealership paid to obtain the model, and the model name and series (e.g., Honda Civic LX). They also keep information about all sales that they have made (for instance, they record a buyer's name, the car bought, and the amount paid for the car). To contact the buyers in the future, contact information is also kept (e.g., address, phone number).

Create a class diagram for this situation. Apply the rules of normalization to the class diagram to check the diagram for processing efficiency.

F. Describe how you would denormalize the model that you created in exercise E. Draw the new class diagram based on your suggested changes. How would performance be affected by your suggestions?

G. Examine the model that you created in exercise F. Develop a clustering and indexing strategy for this model. Describe how your strategy will improve the performance of the database.

H. Calculate the size of the database that you created in exercise F. Provide size estimates for the initial size of the database as well as for the database in one year's time. Assume that the dealership sells ten models of cars from each manufacturer to approximately 20,000 customers a year. The system will be set up initially with one year's worth of data.

L. For the A Real Estate Inc. problem in Chapter 4 (exercises I, J, and K), Chapter 5 (exercises P and Q), Chapter 6 (exercise D), Chapter 7 (exercise A), and Chapter 8 (exercise A):
  1. Apply the rules of normalization to the class diagram to check the diagram for processing efficiency.
  2. Develop a clustering and indexing strategy for this model. Describe how your strategy will improve the performance of the database.

M. For the A Video Store problem in Chapter 4 (exercises L, M, and N), Chapter 5 (exercises R and S), Chapter 6 (exercise E), Chapter 7 (exercise B), and Chapter 8 (exercise B):
  1. Apply the rules of normalization to the class diagram to check the diagram for processing efficiency.
  2. Develop a clustering and indexing strategy for this model. Describe how your strategy will improve the performance of the database.

N. For the gym membership problem in Chapter 4 (exercises O, P, and Q), Chapter 5 (exercises T and U), Chapter 6 (exercise F), Chapter 7 (exercise C), and Chapter 8 (exercise C):
  1. Apply the rules of normalization to the class diagram to check the diagram for processing efficiency.
  2. Develop a clustering and indexing strategy for this model. Describe how your strategy will improve the performance of the database.

O. For the Picnics R Us problem in Chapter 4 (exercises R, S, and T), Chapter 5 (exercises V and W), Chapter 6 (exercise G), Chapter 7 (exercise D), and Chapter 8 (exercise D):

1. Apply the rules of normalization to the class diagram to check the diagram for processing efficiency.
2. Develop a clustering and indexing strategy for this model. Describe how your strategy will improve the performance of the database.

N. For the Of-the-Month-Club problem in Chapter 4 (exercises U, V, and W), Chapter 5 (exercises X and Y), Chapter 6 (exercise H), Chapter 7 (exercise E), and Chapter 8 (exercise E):

1. Apply the rules of normalization to the class diagram to check the diagram for processing efficiency.
2. Develop a clustering and indexing strategy for this model. Describe how your strategy will improve the performance of the database.

MINICASES

1. The system development team at Hoffmann Baer, a small Swiss textile firm, is developing a new customer order entry system. In the process of designing the new system, the team has identified the following class and its attributes:

```
Inventory Order
Order Number (PK)
Order Date
Customer Name
Street Address
City
Canton-State
Postal Code
Country
Customer Type
Sales District Number
1 to 22 occurrences of:
    Item Name
    Quantity Ordered
    Item In
    Quantity Shipped
    Item Out
    Quantity Received
```

a. State the rule that is applied to place a class in first normal form (1NF). Revise the above class diagram so that it is in 1NF.

b. State the rule that is applied to place a class in second normal form (2NF). Create a class diagram using the class and attributes described (if necessary) to place it in 2NF.

c. State the rule that is applied to place a class in third normal form (3NF). Revise the class diagram to place it in 3NF.

d. When planning for the physical design of this database, can you identify any likely situations where the project team may choose to denormalize the class diagram? After going through the work of normalizing, why should this be considered?

2. Professor Takato decided to use a relational database management system (RDBMS) as a persistent storage for his CMS. He is now faced with the challenge of mapping the classes he defined in the previous stages of the project development to relational tables in the CMS database. For the part of the CMS that deals with the management of the paper submissions and reviews, Professor Takato ended up with a solution based on the following classes:

Person – representing a conference participant either as author or as reviewer or both.

Author – subclass of Person, representing the role of a participant as a paper author.

Reviewer – subclass of Person, representing the role of a participant as a paper reviewer.

Paper – representing a paper submitted to the conference.

Professor Takato also finalized the specific attributes/data to be represented within each class (like first/lastname and affiliation for the Person class, areas of competence for the Reviewer class, title, list of authors, list of keywords and description for the Paper class, and so on). Although important, the specific class attributes are of less concern for Professor Takato in the process of mapping the classes to relational database tables. Rather, he focused his attention on the relationships between the classes above and identified the following:

a. Classes Author and Reviewer are subclasses of Person representing is-a-kind-of inheritance.

b. An author can submit several papers.

c. A paper can have several authors.

d. A reviewer can review several papers.

e. A paper is reviewed by several reviewers.

Using the object-relational mapping rules defined in the text, help Professor Takato define his relational schema corresponding to the classes: Person, Author, Reviewer and Paper.

3. Refer to the Professional and Scientific Staff Management (PSSM) minicase in Chapters 4, 6, 7, and 8.

a. Apply the rules of normalization to the class diagram to check the diagram for processing efficiency.

b. Develop a clustering and indexing strategy for this model. Describe how your strategy will improve the performance of the database.

CHAPTER 10

USER INTERFACE DESIGN

A user interface is the part of the system with which the users interact. From the user's point of view, the user interface is the system. It includes the screen displays that provide navigation through the system, the screens and forms that capture data, and the reports that the system produces (whether on paper, on the screen, or via some other medium). This chapter introduces the basic principles and processes of interface design and discusses how to design the interface structure and standards, navigation design, input design, and output design. The chapter introduces the issues related to designing user interfaces for the mobile computing environment and social media. It also introduces the issues that need to be considered when designing user interfaces for a global audience. Finally, the chapter describes the affect of the nonfunctional requirements on designing the human–computer interaction layer.

OBJECTIVES

- Understand several fundamental user interface design principles.
- Understand the process of user interface design.
- Understand how to design the user interface structure.
- Understand how to design the user interface standards.
- Understand commonly used principles and techniques for navigation design.
- Understand commonly used principles and techniques for input design.
- Understand commonly used principles and techniques for output design.
- Be able to design a user interface.
- Understand the affect of nonfunctional requirements on the human-computer interaction layer.

CHAPTER OUTLINE

Introduction Principles for User Interface Design
  Layout
  Content Awareness
  Aesthetics
  User Experience
  Consistency
  Minimizing User Effort

User Interface Design Process
  Use Scenario Development
  Interface Structure Design
  Interface Standards Design
  Interface Design Prototyping
  Interface Evaluation
  Commonsense Approach to User
    Interface Design

INTRODUCTION

Interface design is the process of defining how a system will interact with external entities (e.g., customers, suppliers, other systems). In this chapter, we focus on the design of *user interfaces,* but it is also important to remember that there are sometimes *system interfaces,* which exchange information with other systems. System interfaces are typically designed as part of a systems integration effort. They are defined in general terms as part of the physical architecture and data management layers.

The human–computer interface layer defines the way the users will interact with the system and the nature of the inputs and outputs that the system accepts and produces. The user interface includes three fundamental parts. The first is the *navigation mechanism,* the way the user gives instructions to the system and tells it what to do (e.g., buttons, menus). The second is the *input mechanism,* the way the system captures information (e.g., forms for adding new customers). The third is the *output mechanism,* the way the system provides information to the user or to other systems (e.g., reports, web pages). Each of these is conceptually different, but they are closely intertwined. All computer displays contain navigation mechanisms, and most contain input and output mechanisms. Therefore, navigation design, input design, and output design are tightly coupled.

This chapter introduces several fundamental user interface design principles. It provides an overview of the design process for the human–computer interaction layer and an overview of the navigation, input, and output components that are used in interface design. We focus on the design of web-based interfaces and *graphical user interfaces (GUI)* that use windows, menus, icons, and a mouse (e.g., Windows, Macintosh).[1] Although text-based interfaces are still commonly used on mainframes and Unix systems, GUI interfaces are probably the most common type of interfaces that we use, with the possible exception of printed reports.[2] The chapter describes the issues related to

---

[1] Many people attribute the origin of GUI interfaces to Apple or Microsoft. Some people know that Microsoft copied from Apple, which, in turn, "borrowed" the whole idea from a system developed at the Xerox Palo Alto Research Center (PARC) in the 1970s. Very few know that the Xerox system was based on a system developed by Doug Englebart of Stanford that was first demonstrated at the Western Computer Conference in 1968. Around the same time, Doug also invented the mouse, desktop video conferencing, groupware, and a host of other things we now take for granted. Doug is a legend in the computer science community and has won too many awards to count, but is relatively unknown by the general public.

[2] A good book on GUI design is Susan Fowler, *GUI Design Handbook* (New York: McGraw-Hill, 1998).

designing user interfaces for mobile devices and social media. Finally, the chapter introduces the issues that must be considered when developing user interfaces for a global audience.

PRINCIPLES FOR USER INTERFACE DESIGN

In many ways, user interface design is an art. The goal is to make the interface pleasing to the eye and simple to use while minimizing the effort the users need to accomplish their work. The system is never an end in itself; it is merely a means to accomplish the business of the organization.

We have found that the greatest problem facing experienced designers is using space effectively. Simply put, often there is much more information that needs to be presented on a screen or report or form than will fit comfortably. Analysts must balance the need for simplicity and pleasant appearance against the need to present the information across multiple pages or screens, which decreases simplicity. In this section, we discuss some fundamental interface design principles, which are common for navigation design, input design, and output design[3] (see Figure 10-1).

Layout

The first element of design is the basic *layout* of the *screen, form,* or *report*. Most software designed for personal computers follows the standard Windows or Macintosh approach for screen design. The screen is divided into three boxes. The top box is the navigation area, through which the user issues commands to navigate through the system. The bottom box is the status area, which displays information about what the user

Principle	Description
Layout	The interface should be a series of areas on the screen that are used consistently for different purposes—for example, a top area for commands and navigation, a middle area for information to be input or output, and a bottom area for status information.
Content Awareness	Users should always be aware of where they are in the system and what information is being displayed.
Aesthetics	Interfaces should be functional and inviting to users through careful use of white space, colors, and fonts. There is often a trade-off between including enough white space to make the interface look pleasing without losing so much space that important information does not fit on the screen.
User Experience	Although ease of use and ease of learning often lead to similar design decisions, there is sometimes a trade-off between the two. Novice users or infrequent users of software prefer ease of learning, whereas frequent users prefer ease of use.
Consistency	Consistency in interface design enables users to predict what will happen before they perform a function. It is one of the most important elements in ease of learning, ease of use, and aesthetics.
Minimal User Effort	The interface should be simple to use. Most designers plan on having no more than three mouse clicks from the starting menu until users perform work.

FIGURE 10-1
Principles of User
Interface Design

[3] A good book on the design of interfaces is Susan Weinschenk, Pamela Jamar, and Sarah Yeo, *GUI Design Essentials* (New York: Wiley, 1997).

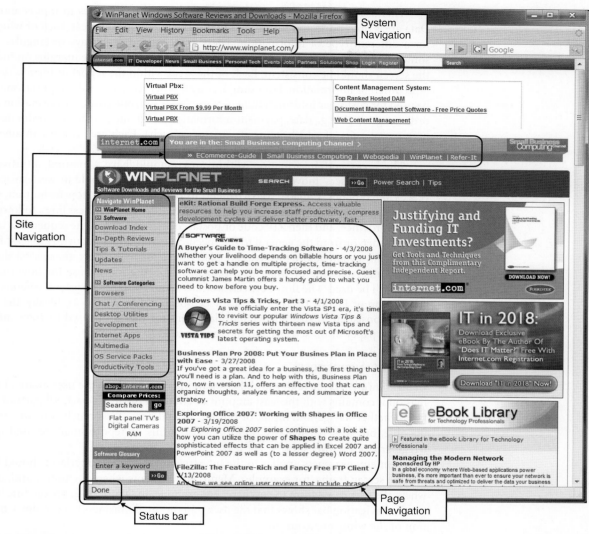

FIGURE 10-2 Layout with Multiple Navigation Areas

is doing. The middle—and largest—box is used to display reports and present forms for data entry.

In many cases (particularly on the web), multiple layout areas are used. Figure 10-2 shows a screen with five navigation areas, each of which is organized to provide different functions and navigation within different parts of the system. The top area provides the standard web browser navigation and command controls that change the contents of the entire system. The navigation area on the left edge maneuvers between sections and changes all content to its right. The other two section navigation areas at the top and middle of the page provide other ways to navigate between sections. The content in the middle of the page displays the results (i.e., software review articles) and provides additional navigation within the page about these reviews.

This use of multiple layout areas for navigation also applies to inputs and outputs. Data areas on reports and forms are often subdivided into subareas, each of which is used for a different type of information. These areas are almost always rectangular, although sometimes space constraints require odd shapes. Nonetheless, the margins on the edges of the screen should be consistent. Each of the areas within the report or form is designed to hold different information. For example, on an order form (or order report), one part may be used for customer information (e.g., name, address), one part for information about the order in general (e.g., date, payment information), and one part for the order details (e.g., how many units of which items at what price each). Each area is self-contained so that information in one area does not run into another.

The areas and information within areas should have a natural intuitive flow to minimize the users' movement from one area to the next. People in westernized nations (e.g., United States, Canada, Mexico) tend to read left-to-right, top-to-bottom, so related information should be placed so it is used in this order (e.g., address lines, followed by city, state or province, and then ZIP code or postal code). Sometimes, the sequence is in chronological order, or from the general to the specific, or from most frequently to least frequently used. In any event, before the areas are placed on a form or report, the analyst should have a clear understanding of what arrangement makes the most sense for how the form or report will be used. The flow between sections should also be consistent, whether horizontal (see top of Figure 10-3) or vertical (see bottom of Figure 10-3). Ideally, the areas will remain consistent in size, shape, and placement for the forms used to enter information (whether paper or on screen) and the reports used to present it.

Content Awareness

Content awareness refers to the ability of an interface to make the user aware of the information it contains with the least amount of effort on the user's part. All parts of the interface, whether navigation, input, or output, should provide as much content awareness as possible, but it is particularly important for forms or reports that are used quickly or irregularly (e.g., a website).

Content awareness applies to the interface in general. All interfaces should have titles (on the screen frame, for example). Menus should show where the user is and, if possible, where the user came from to get there. For example, in Figure 10-2, the top line in the center site navigation bar shows that the user is in the Small Business Computing Channel section of the winplanet.com site.

Content awareness also applies to the areas within forms and reports. All areas should be clear and well defined (with titles if space permits) so that it is difficult for the user to become confused about the information in any area. Then users can quickly locate the part of the form or report that is likely to contain the information they need. Sometimes the areas are marked by lines, colors, or headings (e.g., the site navigation links on the left side in Figure 10-2); in other cases, the areas are only implied (e.g., the page links in the center of Figure 10-2).

Content awareness also applies to the *fields* within each area. Fields are the individual elements of data that are input or output. The *field labels* that identify the fields on the interface should be short and specific—objectives that often conflict. There should be no uncertainty about the format of information within fields, whether for entry or display. For example, a date of 10/5/07 is different depending on whether you are in the United States (October 5, 2007) or in Canada (May 10, 2007). Any fields for which there is the possibility of uncertainty or multiple interpretations should provide explicit explanations.

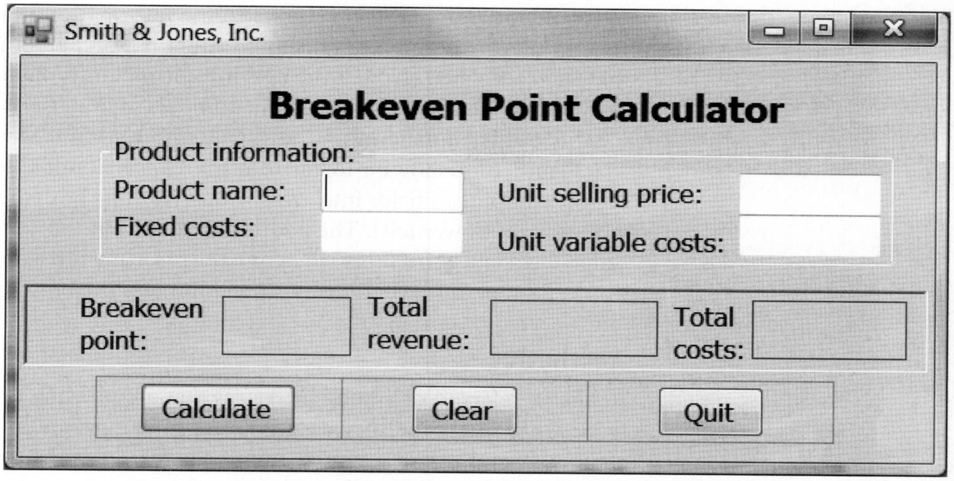

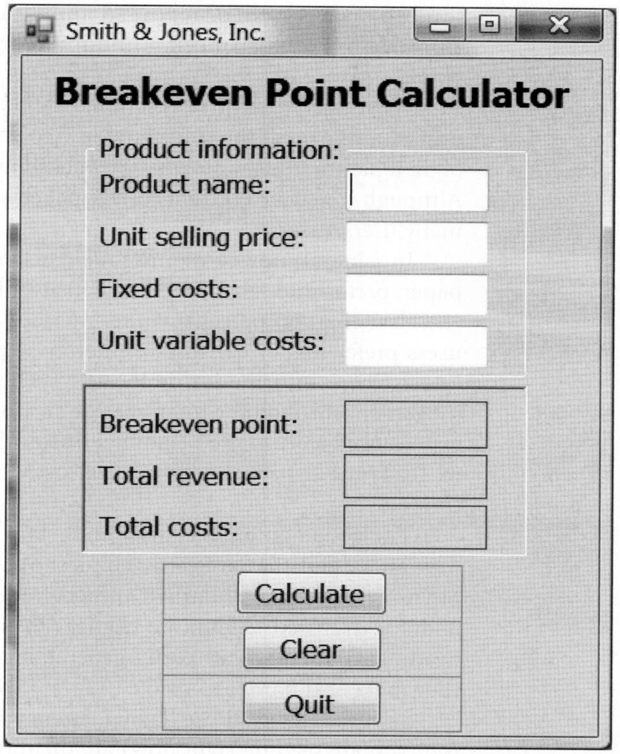

FIGURE 10-3   Flow Between Interface Sections

Content awareness also applies to the information that a form or report contains. In general, all forms and reports should contain a preparation date (i.e., the date printed or the date completed) so that the age of information is obvious. Likewise, all printed forms and software should provide version numbers so that users, analysts, and programmers can identify outdated materials.

Figure 10-4 shows a form from the University of Georgia. This form illustrates the logical grouping of fields into areas with an explicit box (top left), as well as an implied area with no box (lower left). The address fields within the address area follow a clear, natural order. Field labels are short where possible (see the top left) but long where more information is needed to prevent misinterpretation (see the bottom left).

Aesthetics

Aesthetics refers to designing interfaces that are pleasing to the eye. Interfaces do not have to be works of art, but they do need to be functional and inviting to use. In most cases, less is more, meaning that a simple, minimalist design is the best.

Space is usually at a premium on forms and reports, and often there is the temptation to squeeze as much information as possible onto a page or a screen. Unfortunately, this can make a form or report so unpleasant that users do not want to use it. In general, all forms and reports need a minimum amount of *white space* that is intentionally left blank.

What was your first reaction when you looked at Figure 10-4? This is the most unpleasant form at the University of Georgia, according to staff members. Its *density* is too high; it has too much information packed into a too-small space with too little white space. Although it may be efficient to save paper by using one page, not two, it is not effective for many users.

In general, novice or infrequent users of an interface, whether on a screen or on paper, prefer interfaces with low density, often one with a density of less than 50 percent (i.e., less than 50 percent of the interface occupied by information). More experienced users prefer higher densities, sometimes approaching 90 percent occupied, because they know where information is located and high densities reduce the amount of physical movement through the interface. We suspect the form in Figure 10-4 was designed for the experienced staff in the personnel office who use it daily, rather than for the clerical staff in academic departments with less personnel experience who use the form only a few times a year.

The design of text is equally important. As a general rule, all text should be in the same font and about the same size. Fonts should be no smaller than 8 points, but 10 points is often preferred, particularly if the interface will be used by older people. Changes in font and size are used to indicate changes in the type of information that is presented (e.g., headings, status indicators). In general, italics and underlining should be avoided because they make text harder to read.

Serif fonts (i.e., those having letters with serifs, or tails, such as Times Roman, or the font you are reading right now) are the most readable for printed reports, particularly for small letters. Sans serif fonts (i.e., those without serifs, such as Helvetica or Arial, like those used for the chapter titles in this book) are the most readable for computer screens and are often used for headings in printed reports. Never use all capital letters, except possibly, for titles.

Color and patterns should be used carefully and sparingly, and only when they serve a purpose. (About 10 percent of men are color blind, so the improper use of color can impair their ability to read color text.) A quick trip around the web will demonstrate the problems caused by indiscriminate use of colors and patterns. Remember, the goal is

EMPLOYEE PERSONNEL REPORT

UNIVERSITY OF GEORGIA

| DOCUMENT NO. | PAGE | DATE | FY | DEPARTMENT PHONE | COLLEGE OR DIVISION |

UGA EMPLOYMENT HISTORY
☐ (C) CURRENT ☐ (P) PREVIOUS
DATE

PAY TYPE

ACTION
MO DA YR

DEPARTMENT/PROJECT

| | PRI·DEPT | HIGH DEGREE | INSTITUTION | YEAR |

SOC.SEC.NUM. | LAST NAME | FIRST NAME/INITIAL | MIDDLE INITIAL/NAME | SUF

☐ (1) REGULAR ☐ (3) TEMPORARY
☐ (2) UGA STUDENT ☐ (4) NR-ALIEN
UGA % TIME
☐ (T) TIPPED

STREET OR ROUTE NO. (LINE 1) | NON-WORK PHONE | BIRTH DATE | SPOUSE'S NAME | CHAIR

☐ (E) EXEMPT ☐ (N) NON-EXEMPT ☐ (Y) FACULTY·RANK
☐ (M) MALE ☐ (S) SINGLE ☐ (N) NON-FACULTY
☐ (F) FEMALE ☐ (M) MARRIED

STREET OR ROUTE NO. (LINE 2) | UNIVERSITY PHONE | CITIZEN OF | I-9 | VISA | COUNTY

☐ (1) WHITE ☐ (3) ORIENTAL/ASIAN ☐ (5) HISPANIC
☐ (2) BLACK ☐ (4) AMERICAN INDIAN ☐ (6) MULTIRACIAL
☐ (9)

CITY | STATE | ZIP + 4 | UNIVERSITY BUILDING NAME | BLDG.NO/FLOOR/ROOM

COUNTY MONEY
(PER PAY PERIOD)

COOP. EXT. EMPLOYEES ONLY
UGA SALARY
COUNTY MONEY
TOTAL

PAYROLL PAYMENT DISTRIBUTION
☐ (1) SEND TO DEPT (DIST CODE)
☐ (2) DIRECT DEPOSIT(SEND PR105 TO PAYROLL)
☐ (3) PICK UP AT PAYROLL WINDOW

FOR PAYROLL DEPT USE ONLY
FED EXM | STATE EXM

OASDI
HI
RETIRE
EIC
GDCP

| TRX | HOME DEPT | SHORT TITLE | POSN NO. | APPT. BEGIN MO DA YR HR | APPT. END MO DA YR HR | JOBCLASS CODE | POSITION TITLE | POS % TIME | C N | FULL TIME ANNUAL SALARY | S C | SUPPLEMENT AMOUNT |
| MO DA YR HR | MO DA YR HR | MO DA YR HR | MO DA YR HR |

PAYROLL AUTHORIZATION

| TRX | HOME DEPT | SHORT TITLE | POSN NO | ACCOUNT | FISCAL YEAR EFT | BUDGET |

FROM
THRU
AMOUNT PER PAY PERIOD OR HOURLY RATE

☐ (C) PROMOTION
☐ (K) CHANGE TITLE FROM _____ TO _____
☐ (L) CHANGE NAME FROM _____ TO _____
☐ (M) CHANGE SSN FROM _____ TO _____
☐ (N) LEAVE W/O PAY FROM _____ TO _____
☐ (O) CHG COUNTY $ FROM _____ TO _____
☐ (P) TERMINATION-REASON
☐ (Q) OTHER (SPECIFY)

TOTALS

☐ (A) NEW UGA EMPLOYEE ☐ (B) LATERAL TRANSFER
☐ (D) REPLACEMENT POSN-NAME OF LAST INCUMBENT
☐ (E) APPOINTMENT TO NEW POSITION
☐ (F) CHANGE % TIME EMPLOYED FROM _____ TO _____
☐ (G) CONTINUATION WITHIN EXISTING BUDGET POSITION
☐ (H) REVISE DISTRIBUTION OF SALARY
☐ (I) TRANSFER FROM DEPT _____ TO _____
☐ (J) CHANGE PAY TYPE FROM _____ TO _____

REMARKS

| DEPARTMENT HEAD | DATE | VICE PRESIDENT | DATE | BUDGET REVIEW | DATE | BUDGET OFFICE | DATE |
| DEAN/DIRECTOR | DATE | FACULTY RECORDS | DATE | CONTRACTS & GRANTS | DATE | PERSONNEL | DATE |

FIGURE 10-4 Example of a Form

pleasant readability, not art; color and patterns should be used to strengthen the message, not overwhelm it. Color is best used to separate and categorize items, such as showing the difference between headings and regular text, or to highlight important information. Therefore, colors with high contrast should be used (e.g., black and white). In general, black text on a white background is the most readable, and blue on red is the least readable. (Most experts agree that background patterns on web pages should be avoided). Color has been shown to affect emotion, with red provoking intense emotion (e.g., anger) and blue provoking lowered emotions (e.g., drowsiness).

User Experience

User experience can essentially be broken down into two levels: those with experience and those without. Interfaces should be designed for both types of users. Novice users usually are most concerned with *ease of learning*—how quickly they can learn new systems. Expert users are usually most concerned with *ease of use*—how quickly they can use the system once they have learned how to use it. Often, these two are complementary and lead to similar design decisions, but sometimes there are trade-offs. Novices, for example, often prefer menus that show all available system functions, because these promote ease of learning. Experts, on the other hand, sometimes prefer fewer menus organized around the most commonly used functions.

Systems that will end up being used by many people on a daily basis are more likely to have a majority of expert users (e.g., order-entry systems). Although interfaces should try to balance ease of use and ease of learning, these types of systems should put more emphasis on ease of use rather than ease of learning. Users should be able to access the commonly used functions quickly, with few keystrokes or a small number of menu selections.

In many other systems (e.g., decision-support systems), most people remain occasional users for the lifetime of the system. In this case, greater emphasis may be placed on ease of learning rather than ease of use.

Ease of use and ease of learning often go hand-in-hand—but sometimes they don't. Research shows that expert and novice users have different requirements and behavior patterns in some cases. For example, novices virtually never look at the bottom area of a screen that presents status information, whereas experts refer to the status bar when they need information. Most systems should be designed to support frequent users, except for systems designed to be used infrequently or when many new users or occasional users are expected (e.g., the web). Likewise, systems that contain functionality that is used only occasionally must contain a highly intuitive interface or an interface that contains clear, explicit guidance regarding its use.

The balance of quick access to commonly used and well-known functions and guidance through new and less-well-known functions is challenging to the interface designer, and this balance often requires elegant solutions. Microsoft Office, for example, addresses this issue through the use of the "show-me" functions, which demonstrate the menus and buttons for specific functions. These features remain in the background until they are needed by novice users (or even experienced users when they use an unfamiliar part of the system).

Consistency

Consistency in design is probably the single most important factor in making a system simple to use because it enables users to predict what will happen. When interfaces are consistent, users can interact with one part of the system and then know how to interact with the rest, aside from elements unique to those parts. Consistency usually refers to the interface within one computer system, so that all parts of the same system work in the same way. Ideally, the system should also be consistent with other computer systems in the organization and with commercial software that is used (e.g., Windows). For example, many users are familiar with the web, so the use of web-like interfaces can reduce the amount of learning required by the user. In this way, the user can reuse web knowledge, thus significantly reducing the learning curve for a new system. Many software development tools support consistent system interfaces by providing standard interface objects (e.g., list boxes, pull-down menus, and radio buttons).

Consistency occurs at many different levels. Consistency in the *navigation controls* conveys how actions in the system should be performed. For example, using the same icon or command to change an item clearly communicates how changes are made throughout the system. Consistency in terminology is also important. This refers to using the same words for elements on forms and reports (e.g., not customer in one place and client in another). We also believe that consistency in report and form design is important, although a study suggests that being *too* consistent can cause problems.[4] When reports and forms are very similar except for very minor changes in titles, users sometimes mistakenly use the wrong form and either enter incorrect data or misinterpret its information. The implication for design is to make the reports and forms similar, but give them some distinctive elements (e.g., color, size of titles) that enable users to immediately detect differences.

Minimizing User Effort

Interfaces should be designed to minimize the amount of effort needed to accomplish tasks. This means using the fewest possible mouse clicks or keystrokes to move from one part of the system to another. Most interface designers follow the *three-clicks rule:* Users should be able to go from the start or main menu of a system to the information or action they want in no more than three mouse clicks or three keystrokes. However, with regard to this point, you need to be aware of Krug's principles (discussed later).

USER INTERFACE DESIGN PROCESS

User interface design is a five-step process that is iterative; analysts often move back and forth between steps rather than proceeding sequentially from step 1 to step 5 (see Figure 10-5). First, the analysts examine the *use cases* (see Chapter 4) and *sequence diagrams* (see Chapter 6) developed in analysis, and interview users to develop *use scenarios* that describe commonly employed patterns of actions the users will perform so the interface enables users to quickly and smoothly perform these scenarios. Second, the analysts develop the *windows navigation diagram* (*WND*) that defines the basic structure of the interface. These diagrams show all the interfaces (e.g., screens, forms, and reports) in the system and how they are connected. Third, the analysts design *interface standards,* which are the basic design elements on which interfaces in the system are based. Fourth, the analysts create an *interface design prototype* for each of the individual interfaces in the system, such as navigation controls (including the conversion of the *essential use cases* to *real use cases*), input screens, output screens, forms (including preprinted paper forms), and reports. Finally, the individual interfaces are subjected to *interface evaluation* to determine if they are satisfactory and how they can be improved.

---

[4] John Satzinger and Lorne Olfman, "User Interface Consistency Across End-User Application: The Effects of Mental Models," *Journal of Management Information Systems* (Spring 1998): 167–193.

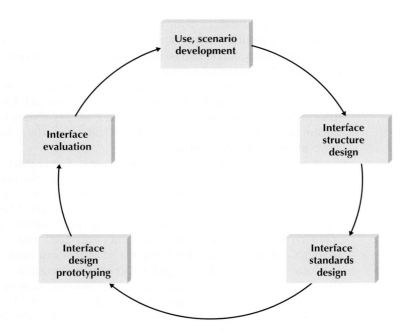

FIGURE 10-5
User Interface
Design Process

Interface evaluations almost always identify improvements, so the interface design process is repeated in a cyclical process until no new improvements are identified. In practice, most analysts interact closely with the users during the interface design process so that users have many chances to see the interface as it evolves, rather than waiting for one overall interface evaluation at the end of the interface design process. It is better for all concerned (both analysts and users) if changes are identified sooner rather than later. For example, if the interface structure or standards need improvements, it is better to identify changes before most of the screens that use the standards have been designed.[5]

Use Scenario Development

A *use scenario* is an outline of the steps that the users perform to accomplish some part of their work. A use scenario is one path through an essential use case. For example, Figure 10-6 shows the use-case diagram for the Appointment System. This figure shows that the Create New Patient use case is distinct from the Make Payment Arrangements use case. We model these two use cases separately because they represent separate processes that are used by the Make New Patient Appt use case.

The use-case diagram was designed to model all possible uses of the system—its complete functionality or all possible paths through the use case at a fairly high level of abstraction. In one use scenario, a patient makes a request with the receptionist regarding an appointment with the doctor. The receptionist looks up the patient and checks to see if the patient has any bills to be paid. The receptionist then asks the patient whether he or she wants to set up a new appointment, cancel an existing appointment, or change an

---

[5] A good source for more information on user interface evaluation is Deborah Hix and H. Rex Hartson, *Developing User Interfaces, Ensuring Usability Through Product & Process* (New York: Wiley, 1993).

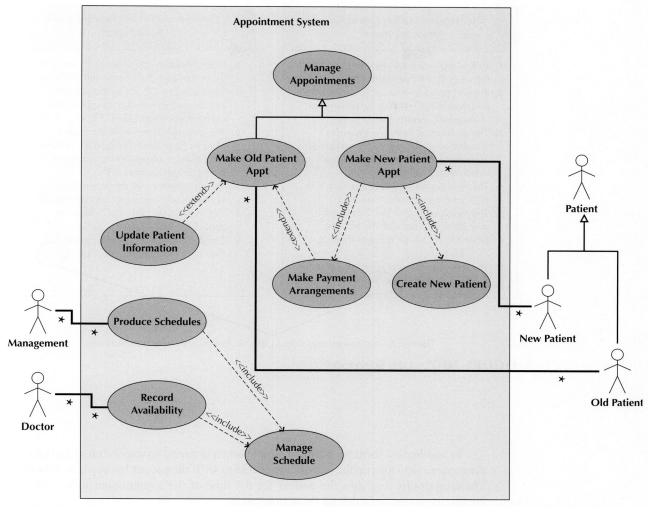

FIGURE 10-6 Appointment System Use-Case Diagram (see Figures 4-15 and 7-11)

existing appointment. If the patient wants to make a new appointment, the receptionist asks the patient for some suggested appointment times, which the receptionist matches against potential times available. The receptionist finally creates a new appointment (see Figures 6-1 and 6-9).

YOUR TURN

10-1 Web Page Critique

*V*isit the web home page for your university and navigate through several of its web pages. Evaluate the extent to which they meet the six design principles.

Use scenario: Existing Patient Makes New Appointment	Use scenario: Existing Patient Cancels Appointment
1. Patient requests appointment (1) and gives the receptionist their name and address (2). 2. The receptionist looks up the patient (3) and determines whether the patient has changed any information (3 & 4). 3. The receptionist then asks the patient whether he or she is going to set up a new appointment, change an appointment, or delete an appointment (5). 4. The receptionist asks the patient for a list of potential appointment times (S-1, 1). 5. The receptionist matches the potential appointment times with the available times and schedules the appointment (S-1, 2). 6. The receptionist informs the patient of their appointment time (6).	1. Patient requests appointment (1) and gives the receptionist their name and address (2). 2. The receptionist looks up the patient (3) and determines whether the patient has changed any information (3 & 4). 3. The receptionist then asks the patient whether he or she is going to set up a new appointment, change an appointment, or delete an appointment (5). 4. The receptionist asks the patient for the appointment time to be canceled (S-2, 1). 5. The receptionist finds and deletes the appointment (S-2, 2). 6. The receptionist informs the patient that their appointment time was canceled (6).

The numbers in parentheses refer to specific events in the essential use case.

FIGURE 10-7   Use Scenarios

In another use scenario, a patient simply wants to cancel an appointment. In this case, the receptionist looks up the patient and checks to see if the patient has any bills to be paid. The receptionist then asks the patient for the time of the appointment to be canceled. Finally, the receptionist deletes the appointment.

Use scenarios are presented in a simple narrative description that is tied to the essential use cases developed during analysis (see Chapter 4). Figure 10-7 shows the two use scenarios just described. The key point with using use cases for interface design is *not* to document all possible use scenarios within a use case. The goal is to document two or three of the most common use scenarios so the interface can be designed to enable the most common uses to be performed simply and easily.

YOUR TURN   10-2 Use Scenario Development for the Web

*V*isit the Web home page for your university and navi- scenarios for it.
gate through several of its Web pages. Develop two use

YOUR	10-3 Use Scenario Development for an ATM
TURN	

*S*uppose you have been charged with the task of   Develop two use scenarios for it.
redesigning the interface for the ATM at your local bank.

Interface Structure Design

The interface structure defines the basic components of the interface and how they work together to provide functionality to users. A *windows navigation diagram (WND)*[6] is used to show how all the screens, forms, and reports used by the system are related and how the user moves from one to another. Most systems have several WNDs, one for each major part of the system.

A WND is very similar to a behavioral state machine (see Chapter 6), in that they both model state changes. A behavioral state machine typically models the *state* changes of an object, whereas a WND models the state changes of the user interface. In a WND, each state where the user interface might be located is represented as a box. A box typically corresponds to a user interface component, such as a *window, form, button,* or *report.* For example, in Figure 10-8, there are five separate states: Client Menu, Find Client Form, Add Client Form, Client List, and Client Information Report.

Transitions are modeled as either a single-headed or double-headed arrow. A single-headed arrow indicates that a return to the calling state is not required, whereas a double-headed arrow represents a required return. For example, in Figure 10-8, the transition from the Client Menu state to the Find Client Form state does not require a return. The arrows are labeled with the action that causes the user interface to move from one state to another. For example, in Figure 10-8, to move from the Client Menu state to the Find Client Form state, the user must click the Find Client Button on the Client Menu.

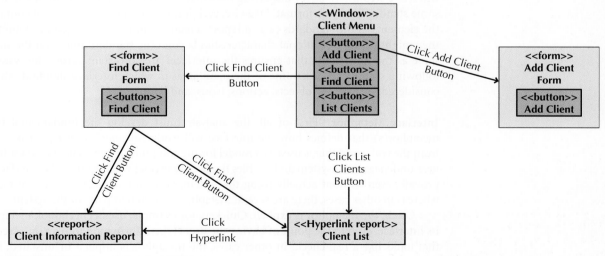

FIGURE 10-8   Sample WND

---

[6] A WND is actually an adaptation of the behavioral state machine and object diagrams [see Meilir Page-Jones, *Fundamentals of Object-Oriented Design in UML* (New York: Dorset House, 2000)].

The last item to be described in a WND is the *stereotype*. A stereotype is modeled as a text item enclosed within guillemets or angle brackets (<< >>). The stereotype represents the type of user interface component of a box on the diagram. For example, the Client Menu is a window, whereas Find Client Form is a form.

The basic structure of an interface follows the basic structure of the business process itself, as defined in the use cases and behavioral model. The analyst starts with the essential use cases and develops the fundamental flow of control of the system as it moves from object to object. The analyst then examines the use scenarios to see how well the WND supports them. Quite often, the use scenarios identify paths through the WND that are more complicated then they should be. The analyst then reworks the WND to simplify the ability of the interface to support the use scenarios, sometimes by making major changes to the menu structure, sometimes by adding shortcuts.

YOUR TURN

10-4 Interface Structure Design

Suppose you have been charged with the task of redesigning the interface for the ATM at your local bank. Design an interface structure design using a WND that shows how a user would navigate among the screens.

Interface Standards Design

Interface standards are the basic design elements that are common across the individual screens, forms, and reports within the system. Depending on the application, there may be several sets of interface standards for different parts of the system (e.g., one for web screens, one for paper reports, one for input forms). For example, the part of the system used by data entry operators might mirror other data entry applications in the company, whereas a web interface for displaying information from the same system might adhere to some standardized web format. Likewise, each individual interface might not contain all of the elements in the standards (e.g., a report screen might not have an edit capability), and they might contain additional characteristics beyond the standard ones, but the standards serve as the touchstone that ensures the interfaces are consistent across the system. The following sections discuss some of the main areas in which interface standards should be considered: metaphors, objects, actions, icons, and templates.

Interface Metaphor First of all, the analysts must develop the fundamental interface metaphor(s) that defines how the interface will work. An *interface metaphor* is a concept from the real world that is used as a model for the computer system. The metaphor helps the user understand the system and enables the user to predict what features the interface might provide, even without actually using the system. Sometimes, systems have one metaphor, whereas in other cases, there are several metaphors in different parts of the system.

Often, the metaphor is explicit. Quicken, for example, uses a checkbook metaphor for its interface, even to the point of having the users type information into an on-screen form that looks like a real check. In other cases, the metaphor is implicit or unstated, but it is there, nonetheless. Many Windows systems use the paper form or table as a metaphor.

In some cases, the metaphor is so obvious that it requires no thought. For example, most online stores use a shopping cart metaphor to temporarily store the items that the customer is considering purchasing. In other cases, a metaphor is hard to identify. In general, it is better not to force a metaphor that really doesn't fit a system, because an ill-fitting metaphor will confuse users by promoting incorrect assumptions.

Interface Objects   The template specifies the names that the interface will use for the major *interface objects,* the fundamental building blocks of the system, such as the classes. In many cases, the object names are straightforward, such as calling the shopping cart the "shopping cart." In other cases, it is not so simple. For example, Amazon.com sells much more than books. In some cases, the user might not know whether he or she is looking for a book, CD, DVD, or Kindle download. In those cases, the user can use a catchall search item: All Departments. In the case that the user knows the type of item that he or she wants to buy, the user can limit the search by selecting more-specific types of search items, such as Apps for Android, Books, Kindle Store, or Music. Obviously, the object names should be easily understood and help promote the interface metaphor.

In general, in cases of disagreements between the users and the analysts over names, whether for objects or actions (discussed later), the users should win. A more understandable name always beats a more precise or more accurate one.

Interface Actions   The template also specifies the navigation and command language style (e.g., menus) and grammar (e.g., object-action order; see the navigation design section later in this chapter). It gives names to the most commonly used *interface actions* in the navigation design (e.g., buy versus purchase or modify versus change).

Interface Icons   The interface objects and actions and their status (e.g., deleted or overdrawn) may be represented by *interface icons.* Icons are pictures that appear on command buttons as well as in reports and forms to highlight important information. Icon design is very challenging because it means developing a simple picture less than half the size of a postage stamp that needs to convey an often-complex meaning. The simplest and best approach is to simply adopt icons developed by others (e.g., a blank page to indicate create a new file, a diskette to indicate save). This has the advantage of quick icon development, and the icons might already be well understood by users because they have seen them in other software.

Commands are actions that are especially difficult to represent with icons because they are in motion, not static. Many icons have become well known from widespread use, but icons are not as well understood as first believed. Use of icons can sometimes cause more confusion than insight. [For example, did you know that a picture of a sweeping broom (paintbrush?) in Microsoft Word means format painter?] Icon meanings become clearer with use, but sometimes a picture is not worth even one word; when in doubt, use a word, not a picture.

Interface Templates   An *interface template* defines the general appearance of all screens in the information system and the paper-based forms and reports that are used. The template design, for example, specifies the basic layout of the screens [e.g., where the navigation area(s), status area, and form/report area(s) will be placed] and the color scheme(s) that will be applied. It defines whether windows will replace one another on the screen or will cascade over the top of each other. The template defines a standard placement and order for common interface actions (e.g., File Edit View rather than File View Edit). In short, the template draws together the other major interface design elements: metaphors, objects, actions, and icons.

Interface Design Prototyping

An interface design prototype is a mock-up or a simulation of a computer screen, form, or report. A prototype is prepared for each interface in the system to show the users and the programmers how the system will perform. In the "old days," an interface design prototype was usually specified on a paper form that showed what would be displayed on each part of the screen. Paper forms are still used today, but more and more interface design prototypes are being built using computer tools instead of paper. The four most common approaches to interface design prototyping are storyboards, windows layout diagrams, HTML prototypes, and language prototypes.

Storyboard   At its simplest, an interface design prototype is a paper-based *storyboard*. The storyboard shows hand-drawn pictures of what the screens will look like and how they flow from one screen to another, in the same way a storyboard for a cartoon shows how the action will flow from one scene to the next (see Figure 10-9). Storyboards are the simplest technique because all they require is paper (often a flip chart) and a pen—and someone with some artistic ability.

YOUR

TURN

10-5 Interface Standards Development

*S*uppose you have been charged with the task of redesigning the interface for the ATM at your local bank. Develop an interface standard that includes metaphors, objects, actions, icons, and a template.

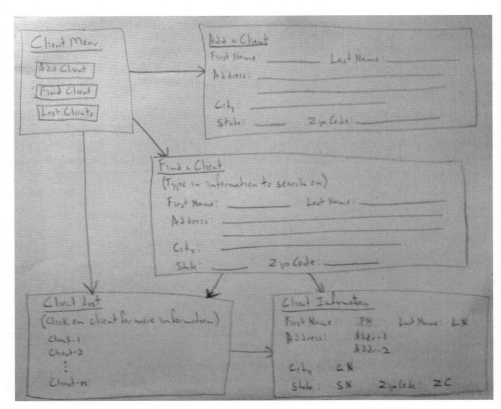

FIGURE 10-9   Sample Storyboard

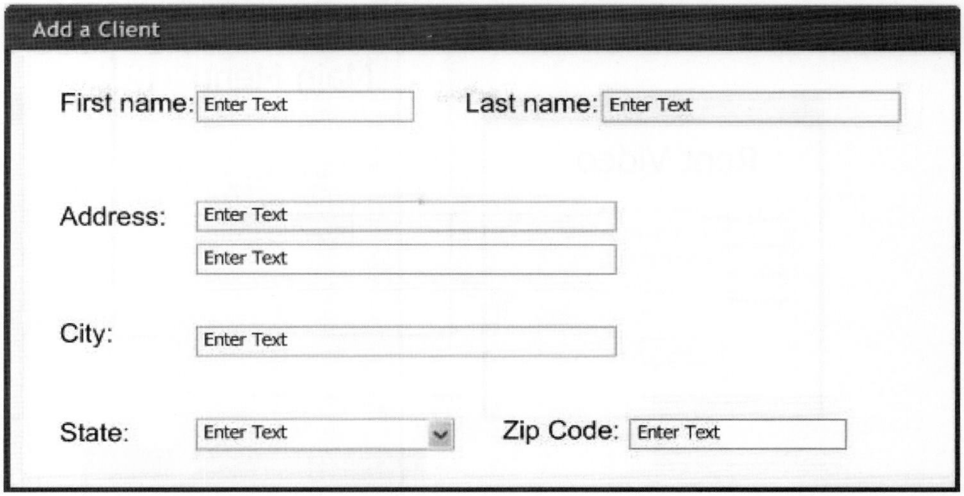

FIGURE 10-10
Sample Windows
Layout Diagram

Windows Layout Diagram   A slight step up from a storyboard is a *windows layout diagram*. From our perspective, a windows layout diagram is a storyboard that more closely resembles the actual user interface that the user will gradually receive. Typically, it is created using a tool such as Microsoft's Visio. Using this type of tool, the designer can quickly drag and drop the user interface components onto the canvas to lay out the design of the user interface. For example, in Figure 10-10, an equivalent diagram for the Add a Client window for the storyboard is portrayed. By combining the windows layout diagrams with the windows navigation diagram, the designer can work effectively with a set of users to design the look and feel of the evolving system without having to actually implement anything. In some cases, it makes sense to combine the ideas of the windows navigation diagram with the windows layout diagram to create a better storyboard (see Figure 10-11).

HTML Prototype   One of the most common types of interface design prototypes used today is the *HTML prototype* (see Figure 10-12). As the name suggests, an HTML prototype is built using web pages created in HTML (hypertext markup language). The designer uses HTML to create a series of web pages, which show the fundamental parts of the system. The users can interact with the pages by clicking on buttons and entering pretend data into forms (but because there is no system behind the pages, the data are never processed). The pages are linked together so that as the user clicks on buttons, the requested part of the system appears. HTML prototypes are superior to storyboards in that they enable users to interact with the system and gain a better sense of how to navigate among the different screens. However, HTML has limitations—the screens shown in HTML will never appear exactly like the real screens in the system (unless, of course, the real system will be a web system in HTML).

Language Prototype   A *language prototype* is an interface design prototype built using the actual language or tool that will be used to build the system. Language prototypes are designed in the same way as HTML prototypes (they enable the user to move from screen to screen but perform no real processing). For example, in Visual Basic, it is possible to create and view screens without actually attaching program code to the screens (see Figure 10-13). Language prototypes take longer to develop than storyboards or HTML prototypes, but have the distinct advantage of showing *exactly* what the screens will look like. The user does not have to guess about the shape or position of the elements on the screen.

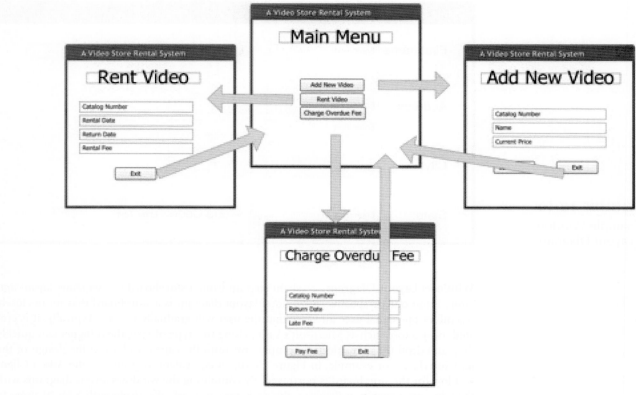

FIGURE 10-11   Sample Combined Windows Navigation and Layout Diagrams

Sign In

Username: [   ]
Password: [   ]
(Submit)

Welcome Employee!

Rent Video
List Videos
Find Video

Home

Video Rental

Title: The Blind Side
Serial#: 72348
Customer AVS Card#: 93156
Due Date: 3/21/10

(Rent) (Cancel)

Home

Find a Video

Title: [   ]
Genre: [   ]
Rating: [   ]
Actor: [   ]

Search

Home

FIGURE 10-12
Sample HTML
Prototype

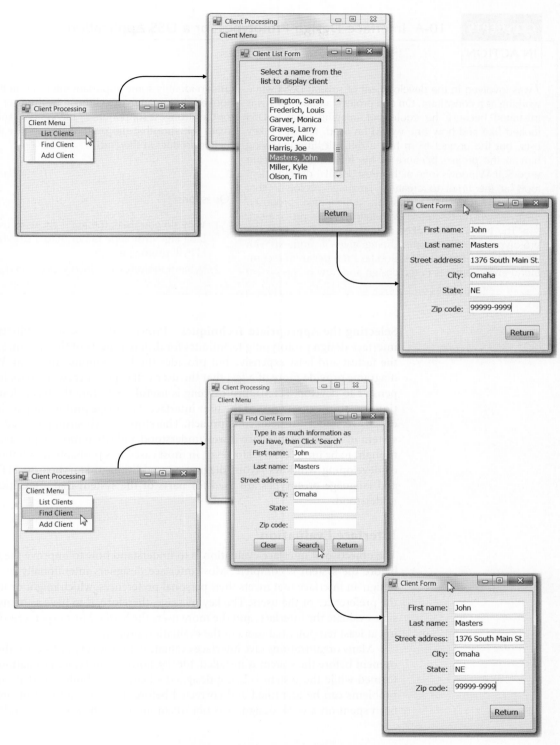

FIGURE 10-13   Sample Visual Basic Language Prototype

CONCEPTS

IN ACTION

10-A Interface Design Prototypes for a DSS Application

I was involved in the development of several DSSs while working as a consultant. On one project, a future user was frustrated because he could not imagine what a DSS looked like and how one would be used. He was a key user, but the project team had a difficult time involving him in the project because of his frustration. The team used SQL Windows (one of the most popular development tools at the time) to create a language prototype that demonstrated the future system's appearance, proposed menu system, and screens (with fields, but no processing).

The team was amazed at the user's response to the prototype. He appreciated being given a context with which to visualize the DSS, and he soon began to recommend improvements to the design and flow of the system

and to identify some important information that was overlooked during analysis. Ultimately, the user became one of the strongest supporters of the system, and the project team felt sure that the prototype would lead to a much better product in the end.

—*Barbara Wixom*

Questions

1. Why do you think the team chose to use a language prototype rather than a storyboard or HTML prototype?
2. What trade-offs were involved in the decision?

Selecting the Appropriate Techniques   Projects often use a combination of different interface design prototyping techniques for different parts of the system. Storyboarding is the fastest and least expensive but provides the least amount of detail. Windows layout diagrams provide more of a feel that the user will experience, while remaining fairly inexpensive to develop. HTML prototyping is useful for testing the basic design and navigation (see the next section) of the user interface. Language prototyping is the slowest, most expensive, and most detailed approach. Therefore, storyboarding is used for parts of the system in which the interface is well understood and when more expensive prototypes are thought to be unnecessary. However, in most cases, it is probably worth the additional cost of developing windows layout diagrams in addition to storyboards. HTML prototypes and language prototypes are used for parts of the system that are critical, yet not well understood.

Interface Evaluation[7]

The objective of interface evaluation is to understand how to improve the interface design before the system is complete. Most interface designers intentionally or unintentionally design an interface that meets their personal preferences, which might or might not match the preferences of the users. The key message, therefore, is to have as many people as possible evaluate the interface, and the more users the better. Most experts recommend involving at least ten potential users in the evaluation process.

Many organizations save interface evaluation for the very last step in the systems development before the system is installed. Ideally, however, interface evaluation should be performed while the system is being designed—before it is built—so that any major design problems can be identified and corrected before the time and cost of programming has been spent on a weak design. It is not uncommon for the system to undergo one or two

---

[7] Verifying and validation approaches, in general, were described in Chapters 4 through 7. Also, further approaches to testing the evolving system are described in Chapter 12. In this section, we describe approaches that have been customized to the human–computer interaction layer.

major changes after the users see the first interface design prototype because they identify problems that are overlooked by the project team.

As with interface design prototyping, interface evaluation can take many different forms, each with different costs and different amounts of detail. Four common approaches are heuristic evaluation, walkthrough evaluation, interactive evaluation, and formal usability testing. As with interface design prototyping, the different parts of a system can be evaluated using different techniques.

Heuristic Evaluation   A *heuristic evaluation* examines the interface by comparing it to a set of heuristics or principles for interface design. The project team develops a checklist of interface design principles—from the list at the start of this chapter, for example, as well as the list of principles in the navigation, input, and output design sections later in this chapter. At least three members of the project team then individually work through the interface design prototype, examining every interface to ensure it satisfies each design principle on a formal checklist. After each has gone through the prototype separately, they meet as a team to discuss their evaluations and identify specific improvements that are required.

YOUR TURN

10-6 Prototyping and Evaluation

*S*uppose you have been charged with the task of redesigning the interface for the ATM at your local bank. What type of prototyping and interface evaluation approach would you recommend? Why?

Walkthrough Evaluation   An interface design *walkthrough evaluation* is a meeting conducted with the users who ultimately have to operate the system. The project team presents the prototype to the users and walks them through the various parts of the interface. The project team shows the storyboard and windows layout diagrams or actually demonstrates the HTML or language prototype and explains how the interface will be used. The users identify improvements to each of the interfaces that are presented.

Interactive Evaluation   With an *interactive evaluation,* the users themselves actually work with the HTML or language prototype in a one-person session with member(s) of the project team (an interactive evaluation cannot be used with a storyboard or windows layout diagrams). As the user works with the prototype (often by going through the use scenarios, using the real use cases described later in this chapter, or just navigating at will through the system), he or she tells the project team member(s) what he or she likes and doesn't like and what additional information or functionality is needed. As the user interacts with the prototype, team member(s) record the cases when he or she appears to be unsure of what to do, makes mistakes, or misinterprets the meaning of an interface component. If the pattern of uncertainty, mistakes, or misinterpretations reoccurs across several of the users participating in the evaluation, it is a clear indication that those parts of the interface need improvement.

Formal Usability Testing   Formal *usability testing* is commonly done with commercial software products and products developed by large organizations that will be widely used through the organization. As the name suggests, it is a very formal—almost scientific—process that can be used only with language prototypes (and systems that have been completely built awaiting installation or shipping).[8] As with interactive evaluation, usability testing is done in one-person sessions in which a user works directly with the software. However, it is typically done in a special lab equipped with video cameras and special software that records every keystroke and mouse operation so they can be replayed to understand exactly what the user did.

The user is given a specific set of tasks to accomplish (usually the use scenarios), and after some initial instructions, the project team's members are not permitted to interact with the user to provide assistance. The user must work with the software without help, which can be hard on the users if they become confused with the system. It is critical that users understand that the goal is to test the interface, not their abilities, and if they are unable to complete the task, the interface—not the user—has failed the test.

Formal usability testing is very expensive, because each one-user session can take one to two days to analyze depending on the volume of detail collected in the computer logs and videos. Sessions typically last one to two hours. Most usability testing involves five to ten users, because if there are fewer than five users, the results depend too much on the specific individual users who participated, and more than ten users are often too expensive to justify (unless a large commercial software developer is involved).

Common Sense Approach to User Interface Design

When you consider all of the above material, to design an effective user interface design can be a daunting and very time-consuming task. An interesting book by Steve Krug,[9] however, provides us with a set of guiding principles for web usability. In this section, we adapt his principles to general user interface design.

First, the user should never have to think about how to navigate the user interface. As Krug puts it, "Don't make me think." Cognitively speaking, any time the user has to stop and figure out how to use the user interface, the creator of the user interface has failed. That might seem a little harsh, but it is true. From the user's perspective, the user interface is the system. If the developers have done their homework, the user interface should be intuitive to use. For example, even though it would be better to redesign the form shown in Figure 10-4, if the user's job is to input data into the system from the form, then the user interface should mimic that form. From a practical perspective, we should study how the user really uses the system. Based on Krug's observations of users, he found that users do not read web pages; instead, they tend to scan them. As a general user interface design guideline, we suggest that you make it easy for users to identify the different parts of the user interface so that they simply scan the screen to see the section of the interface that is applicable to the problem that they are solving. Given the user's tendency to simply scan the user interface, Krug suggests that we should consider studying billboards for inspiration. Billboards are designed to be "read" at 70 mph as you drive down the highway. Obviously, the most relevant information must catch your attention for the billboard advertisement to work. He suggests that we should use the set of conventions that we have grown up with. For example, when looking at a newspaper you know that it is organized into different sections. In the case of the *Wall Street Journal,* you know that the front page acts as an index into the rest of the paper. Consequently, we should look for conventions that we can employ to aid the user.

---

[8] A good source for usability testing is Jakob Nielsen and Robert Mack (eds.), *Usability Inspection Methods* (New York: Wiley, 1994). See also www.useit.com/papers.

[9] Steve Krug, *Don't Make Me Think: A Common Sense Approach to Web Usability, 2nd Ed.* (Berkeley, CA: New Riders, 2006).

CONCEPTS

IN ACTION

10-B ERP User Interfaces Drive

*I*t used to take workers at Hydro Agri's Canadian Fertilizer stores about 20 seconds to process an order. After installing SAP, it now takes 90 seconds. Entering an order requires users to navigate through six screens to find the data fields that were on one screen in the old system. The problem became so critical during the spring planting rush that the project team installing the SAP system was pressed into service to take telephone orders.

Many other customers have complained about similar problems in SAP and the other leading ERP systems. Ontario-based Algoma Steel uses PeopleSoft and now has to use a dozen screens to enter employee data that were contained in two screens in their old, custom-built personnel system. A-dec, a dental equipment maker based in Oregon, discovered the hard way that its Baan inventory system was still counting products that had been shipped in its on-hand inventories; the system required users to confirm the order shipments before the inventories were recorded as shipped, but it didn't automatically take them to the confirmation screen. So why have companies implemented ERP systems? The driving force behind most

implementations was not to simplify the users' jobs but, instead, to improve the quality of the data, simplify system maintenance, and/or beat the Y2K problem. Ease of use wasn't a consideration, and what makes ERP systems so hard to use is that in the attempt to make them one-size-fits-all, developers had to include many little-used data items and processes. Instead of having a small custom system collecting only the data needed by the company itself (which could be condensed to fit one or two screens), companies now find themselves using a system designed to collect all the data items that any company could possibly use—data items that now require six to twelve screens.

Source: "ERP user interfaces drive workers nuts," *ComputerWorld* (November 2, 1998).

Question

Suppose you were a systems analyst at one of the leading ERP vendors (e.g., SAP, PeopleSoft, Baan). How could you apply the interface design principles and techniques in this chapter to improve the ease of use of your system?

Second, he suggests that the number of clicks that a user must perform to complete the task is somewhat irrelevant. Instead, building on his first guiding principle, the important thing is to design the user interface such that the choices (clicks) to be made are unambiguous. Making a lot of obvious choices is a lot quicker and easier than a few vague and ambiguous ones. Consequently, don't worry about the number of screens that the user must work through. However, like any other rule, this can be taken to an extreme. Too many clicks is still too many clicks. The overall goal is to minimize the user's effort. Simply focus on making it easier for the user to complete the task.

Third, minimize the number of words on the screen. Given that users scan the screen to find what they are searching for, make it easier by not cluttering the screen with lots of noise. He suggests that in the case of web interfaces, that 50%–75% of the words can be eliminated without losing any information contained on the screen. Obviously, this may be somewhat extreme, but it does suggest that following the KISS[10] principle is critical when designing effective user interfaces.

NAVIGATION DESIGN

The navigation component of the interface enables the user to enter commands to navigate through the system and perform actions to enter and review information it contains. The navigation component also presents messages to the user about the success or failure of his

[10] Keep it simple, stupid!

or her actions. The goal of the navigation system is to make the system as simple as possible to use. A good navigation component is one the user never really notices. It simply functions the way the user expects, and thus the user gives it little thought. In other words, keep Krug's three guiding principles in mind as you work through the next three sections of the text.

Basic Principles

One of the hardest things about using a computer system is learning how to manipulate the navigation controls to make the system do what you want. Analysts usually need to assume that users have not read the manual, have not attended training, and do not have external help readily at hand. All controls should be clear and understandable and placed in an intuitive location on the screen. Ideally, the controls should anticipate what the user will do and simplify his or her efforts. For example, many setup programs are designed so that for a typical installation, the user can simply keep pressing the Next button.

Prevent Mistakes   The first principle of designing navigation controls is to prevent the user from making mistakes. A mistake costs time and causes frustration. Worse still, a series of mistakes can cause the user to discard the system. Mistakes can be reduced by labeling commands and actions appropriately and by limiting choices. Too many choices can confuse the user, particularly when they are similar and hard to describe in the short space available on the screen. When there are many similar choices on a menu, consider creating a second menu level or a series of options for basic commands.

Never display a command that cannot be used. For example, many Windows applications gray out commands that cannot be used; they are displayed on pull-down menus in a very light-colored font, but they cannot be selected. This shows that they are available but cannot be used in the current context. It also keeps all menu items in the same place.

When the user is about to perform a critical function that is difficult or impossible to undo (e.g., deleting a file), it is important to confirm the action with the user (and make sure the selection was not made by mistake). Having the user respond to a confirmation message, which explains what the user has requested and asks the user to confirm that this action is correct, usually does this.

Simplify Recovery from Mistakes   No matter what the system designer does, users will make mistakes. The system should make it as easy as possible to correct these errors. Ideally, the system has an Undo button that makes mistakes easy to override; however, writing the software for such buttons can be very complicated.

Use Consistent Grammar Order   One of the most fundamental decisions is the *grammar order*. Most commands require the user to specify an object (e.g., file, record, word), and the action to be performed on that object (e.g., copy, delete). The interface can require the user to first choose the object and then the action (an *object–action order*), or first choose the action and then the object (an *action–object order*). Most Windows applications use an object–action grammar order (e.g., think about copying a block of text in your word processor).

The grammar order should be consistent throughout the system, both at the data element level and at the overall menu level. Experts debate about the advantages of one approach over the other, but because most users are familiar with the object–action order, most systems today are designed using that approach.

Types of Navigation Controls

There are two traditional hardware devices that can be used to control the user interface: the keyboard, and a pointing device such as a mouse, trackball, or touch screen. Voice-recognition systems have made an appearance, but they are not yet common. There are three basic software approaches for defining user commands: languages, menus, and direct manipulation.

Languages   With a *command language,* the user enters commands using a special language developed for the computer system (e.g., UNIX and SQL both use command languages). Command languages sometimes provide greater flexibility than other approaches because the user can combine language elements in ways not predetermined by developers. However, they put a greater burden on users because users must learn syntax and type commands rather than select from a well-defined, limited number of choices. Systems today use command languages sparingly, except in cases where there is an extremely large number of command combinations that make it impractical to try to build all combinations into a menu (e.g., SQL queries for databases).

Natural language interfaces are designed to understand the user's own language (e.g., English, French, Spanish). These interfaces attempt to interpret what the user means, and often, they present back to the user a list of interpretations from which to choose. An example of the use of natural language is Microsoft's Help System, which enables users to ask free-form questions for help.

Menus   The most common type of navigation system today is the *menu.* A menu presents a user with a list of choices, each of which can be selected. Menus are easier to learn than languages because a limited number of available commands are presented to the user in an organized fashion. Clicking on an item with a pointing device or pressing a key that matches the menu choice (e.g., a function key) takes very little effort. Therefore, menus are usually preferred to languages.

Menus need to be designed with care because the submenus behind a main menu are hidden from users until they click on the menu item. It is better to make menus broad and shallow (i.e., each menu containing many items with only one or two layers of menus) rather than narrow and deep (i.e., each menu containing only a few items, but each leading to three or more layers of menus). A broad and shallow menu presents the user with the most information initially, so that he or she can see many options and requires only a few mouse clicks or keystrokes to perform an action. A narrow and deep menu makes users hunt for items hidden behind menu items and requires many more clicks or keystrokes to perform an action.

Research suggests that in an ideal world, any one menu should contain no more than eight items, and it should take no more than two mouse clicks or keystrokes from any menu to perform an action (or three from the main menu that starts a system).[11] However, analysts sometimes must break this guideline in the design of complex systems. In this case, menu items are often grouped together and separated by a horizontal line (see

YOUR TURN   10-7 Design a Navigation System

*D*esign a navigation system for a system into which users must enter information about customers, products, and orders. For all three, users will want to change, delete, find one specific record, and list all records.

[11] Kent L. Norman, *The Psychology of Menu Selection* (Norwood NJ.: Ablex Publishing Corp., 1991).

Figure 10-14). Often menu items have *hot keys* that enable experienced users to quickly invoke a command with keystrokes in lieu of a menu choice (e.g., Ctrl-F in Word invokes the Find command or Alt-F opens the file menu).

Menus should put together like items so that the user can intuitively guess what each menu contains. Most designers recommend grouping menu items by interface objects (e.g., customers, purchase orders, inventory) rather than by interface actions (e.g., new, update, format), so that all actions pertaining to one object are in one menu, all actions for another object are in a different menu, and so. However, this is highly dependent on the specific interface. As Figure 10-14 shows, Microsoft Visual Studio groups menu items by interface objects (e.g., File, Tools, Window) *and* by interface

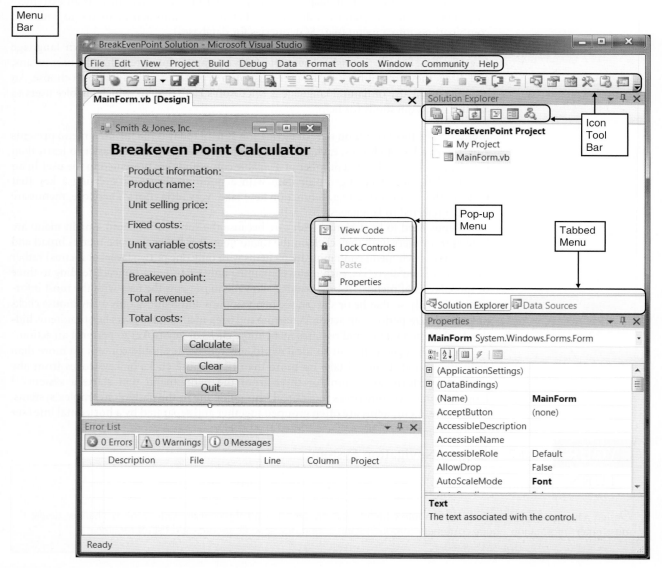

FIGURE 10-14   Common Types of Menus

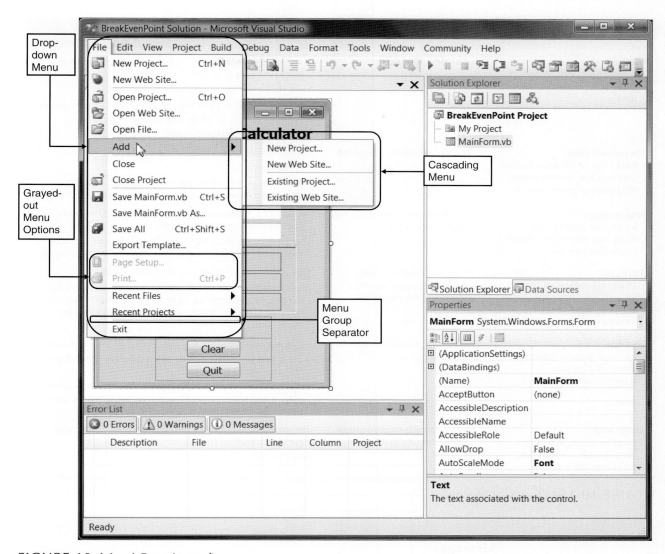

FIGURE 10-14   (*Continued*)

actions (e.g., Edit, Build, Format) on the same menu. Some of the more common types of menus include *menu bars, drop-down menus, pop-up menus, tab menus, toolbars,* and *image maps* (see Figure 10-15).

Direct Manipulation   With *direct manipulation,* the user enters commands by working directly with interface objects. For example, users can change the size of objects in Microsoft PowerPoint by clicking on them and moving their sides, or they can move files in Windows Explorer by dragging the filenames from one folder to another. Direct manipulation can be simple, but it suffers from two problems. First, users familiar with language- or menu-based interfaces don't always expect it. Second, not all commands are intuitive. [How do you copy (not move) files in Windows Explorer? On the Macintosh, why does moving a folder to the trash delete the file if it is on the hard disk, but eject the DVD if the file is on a DVD?]

Type of Menu	When to Use	Notes
Menu bar List of commands at the top of the screen; always on-screen	Main menu for system	Use the same organization as the operating system and other packages (e.g., File, Edit, View). Menu items are always one word, never two. Menu items lead to other menus rather than perform action. Never allow users to select actions they can't perform (instead, use grayed-out items).
Drop-down menu Menu that drops down immediately below another menu; disappears after one use	Second-level menu, often from menu bar	Menu items are often multiple words. Avoid abbreviations. Menu items perform action or lead to another cascading drop-down menu, pop-up menu, or tab menu.
Pop-up menu Menu that pops up and floats over the screen; disappears after one use	As a shortcut to commands for experienced users	Pop-up menus often (not always) are invoked by a right click in Windows-based systems. These menus are often overlooked by novice users, so usually they should duplicate functionality provided in other menus.
Tab menu Multipage menu with one tab for each page that pops up and floats over the screen; remains on-screen until closed	When user needs to change several settings or perform several related commands	Menu items should be short to fit on the tab label. Avoid more than one row of tabs, because clicking on a tab to open it can change the order of the tabs and in virtually no other case does selecting from a menu rearrange the menu itself.
Tool bar Menu of buttons (often with icons) that remains on screen until closed	As a shortcut to commands for experienced users	All buttons on the same tool bar should be the same size. If the labels vary dramatically in size, then use two different sizes (small and large). Buttons with icons should have a tool tip, an area that displays a text phrase explaining the button when the user pauses the mouse over it.
Image map Graphic image in which certain areas are linked to actions or other menus	Only when the graphic image adds meaning to the menu	The image should convey meaning to show which parts perform action when clicked. Tool tips can be helpful.

FIGURE 10-15 Types of Menus

Messages

Messages are the way the system responds to a user and informs him or her of the status of the interaction. There are many different types of messages, such as *error messages, confirmation messages, acknowledgment messages, delay messages,* and *help messages* (see Figure 10-16). In general, messages should be clear, concise, and complete, which are sometimes conflicting objectives. All messages should be grammatically correct and free of jargon and abbreviations (unless they are the users' jargon and abbreviations). Avoid negatives because they can be confusing (e.g., replace Are you sure you do not want to continue? with Do you want to quit?). Likewise, avoid humor, because it wears off quickly after the same message appears dozens of times.

Messages should require the user to acknowledge them (by clicking, for example), rather than being displayed for a few seconds and then disappearing. The exceptions are messages that inform the user of delays in processing, which should disappear once the delay has passed. In general, messages are text, but sometimes, standard icons are used. For example, Windows displays an hourglass when the system is busy. All messages should be carefully crafted, but error and help messages require particular care. Messages (and

Type of Messages	When to Use	Notes
Error message Informs the user that he or she has attempted to do something to which the system cannot respond	When the user does something that is not permitted or not possible	Always explain the reason and suggest corrective action. Traditionally, error messages have been accompanied by a beep, but many applications now omit it or permit users to remove it.
Confirmation message Asks users to confirm that they really want to perform the action they have selected	When user selects a potentially dangerous choice, such as deleting a file	Always explain the cause and suggest possible action. Often include several choices other than OK and cancel.
Acknowledgment message Informs the user that the system has accomplished what it was asked to do	Seldom or never. Users quickly become annoyed with all the unnecessary mouse clicks	Acknowledgment messages are typically included because novice users often like to be reassured that an action has taken place. The best approach is to provide acknowledgment information without a separate message on which the user must click. For example, if the user is viewing items in a list and adds one, then the updated list on the screen showing the added item is sufficient acknowledgment.
Delay message Informs the user that the computer system is working properly	When an activity takes more than seven seconds	Should permit the user to cancel the operation in case he or she does not want to wait for its completion. Should provide some indication of how long the delay will last.
Help message Provides additional information about the system and its components	In all systems	Help information is organized by table of contents and/or keyword search. Context-sensitive help provides information that depends on what the user was doing when help was requested. Help messages and online documentation are discussed in Chapter 12.

FIGURE 10-16   Types of Messages

especially error messages) should always explain the problem in polite, succinct terms (e.g., what the user did incorrectly) and explain corrective action as clearly and as explicitly as possible so the user knows exactly what needs to be done. In the case of complicated errors, the error message should display what the user entered, suggest probable causes for the error, and propose possible user responses. When in doubt, provide either more information than the user needs or the ability to get additional information. Error messages should provide a message number. Message numbers are not intended for users, but their presence makes it simpler for help desks and customer support lines to identify problems and help users because many messages use similar wording.

Navigation Design Documentation

The design of the navigation for a system is done through the use of WNDs and real use cases. Real use cases are derived from the essential use cases (see Chapter 4), use scenarios, and WNDs. Recall that an essential use case is one that describes only the minimum essential issues necessary to understand the required functionality. A real use case describes a specific set of steps that a user performs to use a specific part of a system. Real use cases are implementation dependent (i.e., they are detailed descriptions of how to use the system once it is implemented).

To evolve an essential use case into a real use case, two changes must be made. First, the use-case type must be changed from essential to real. Second, all events must be specified in terms of the actual user interface. Therefore, the normal flow of events, sub-flows, and alternative/exceptional flows must be modified. The normal flow of events, sub-flows, and alternative/exceptional flows for the real use case associated with the storyboard user interface prototype given in Figure 10-9 is shown in Figure 10-17. For example, step 2

Use-Case Name: Maintain Client List	ID: 12	Importance Level: High
Primary Actor: Sales Rep	Use-Case Type: Detail, Real	

Stakeholders and Interests:	Sales Rep - wants to add, find, or list clients

Brief Description: This use case describes how sales representatives can search and maintain the client list.

Trigger: Patient calls and asks for a new appointment or asks to cancel or change an existing appointment.

Type: External

Relationships:
 Association: Sales Rep
 Include:
 Extend:
 Generalization:

Normal Flow of Events:

1. The Sales Rep starts up the system.
2. The System provides the Sales Rep with the Main Menu for the System.
3. The System asks Sales Rep if he or she would like to Add a client. Find an existing Client, or to List all existing clients.
  If the Sales Rep wants to add a client, they click on the Add Client Link and execute S-1: New Client.
  If the Sales Rep wants to find a client, they click on the Find Client Link and execute S-2: Find Client.
  If the Sales Rep wants to list all clients, they click on the List Client Link and execute S-3: List Clients.
4. The System returns the Sales Rep to the Main Menu of the System.

Subflows:
 S-1: New Client
  1. The System asks the Sales Rep for relevant information.
  2. The Sales Rep types in the relevant information into the Form.
  3. The Sales Rep submits the information to the System.
 S-2: Find Client
  1. The System asks the Sales Rep for the search information.
  2. The Sales Rep types in the search information into the Form.
  3. The Sales Rep submits the information to the System.
  4. If the System finds a single Client that meets the search information,
    the System produces a Client Information report and returns the Sales Rep to the Main Menu of the System.
   Else If the System finds a list of Clients that meet the search information, the System executes S-3: List Clients.
 S-3: List Clients
  1. If this Subflow is executed from Step 3
    The System creates a List of All clients
   Else
    The System creates a List of clients that matched the S-2: Find Client search criteria.
  2. The Sales Rep selects a client.
  3. The System produces a Client Information report.

Alternate/Exceptional Flows:
 S-2 4a. The System produces an Error Message.

FIGURE 10-17

Real Use Case Example

of the normal flow of events states that "The System provides the Sales Rep with the Main Menu for the System," which allows the Sales Rep to interact with the Maintain Client List aspect of the system.

INPUT DESIGN

Inputs facilitate the entry of data into the computer system, whether highly structured data, such as order information (e.g., item numbers, quantities, costs), or unstructured information (e.g., comments). Input design means designing the screens used to enter the information as well as any forms on which users write or type information (e.g., timecards, expense claims).

Basic Principles

The goal of the input mechanism is to simply and easily capture accurate information for the system. The fundamental principles for input design reflect the nature of the inputs (whether batch or online) and ways to simplify their collection.

Online versus Batch Processing There are two general formats for entering inputs into a computer system: online processing and batch processing. With *online processing* (sometimes

---

CONCEPTS IN ACTION

10-C Public Safety Depends on a Good User Interface

*P*olice officers in San Jose, California, experienced a number of problems with a new mobile dispatch system that included a Windows-based touch-screen computer in every patrol car. Routine tasks were difficult to perform, and the essential call for assistance was considered needlessly complicated.

The new system, costing $4.7 million, was an off-the-shelf system purchased from Intergraph Corp. It replaced a 14-year-old text-based system that was custom developed. Initially, the system was unstable, periodically crashing a day or two after installation and down for the next several days.

At the request of the San Jose police union, a user-interface design consulting firm was brought in to evaluate the new system. A number of errors were discovered in the system, including inaccurate map information, screens cluttered with unnecessary information, difficult-to-read on-screen type, and difficult-to-perform basic tasks, such as license plate checks. In addition, the police officers themselves were not consulted about the design of the interface. Many users felt that the Windows desktop GUI with its complex hierarchical menu structure was not suitable for in-vehicle use. While driving, officers

found that the repeated taps on the screen required to complete tasks were very distracting, and one officer crashed his vehicle into a parked car because of the distraction of working with the system.

Further complicating the transition to the new system was the bare-bones training program. Just three hours of training were given on a desktop system, using track pads on the keyboards, not the 12-inch touch screen that would be found in the patrol cars.

After the rocky start, the software vendor worked closely with the city of San Jose to fix bugs and smooth out workflows. It seems clear, however, that the rollout could have been much easier if the officers and dispatchers had been involved in planning the system in the first place.

Source: Katie Hafner, "Wanted by the Police: A Good Interface," *New York Times,* November 11, 2004.

Question

If you were involved in the acquisition of a new system for the police force in your community, what steps could you take to ensure the success of the project?

called *transaction processing*), each input item (e.g., a customer order, a purchase order) is entered into the system individually, usually at the same time as the event or transaction prompting the input. For example, when you check a book out from the library, buy an item at the store, or make an airline reservation, the computer system that supports that process uses online processing to immediately record the transaction in the appropriate database(s). Online processing is most commonly used when it is important to have *real-time information* about the business process. For example, when you reserve an airline seat, the seat is no longer available for someone else to use.

With *batch processing,* all the inputs collected over some time period are gathered together and entered into the system at one time in a batch. Some business processes naturally generate information in batches. For example, most hourly payrolls are done using batch processing because time cards are gathered together in batches and processed at once. Batch processing is also used for transaction processing systems that do not require real-time information. For example, most stores send sales information to district offices so that new replacement inventory can be ordered. This information can be sent in real time as it is captured in the store so that the district offices are aware within a second or two that a product is sold. If stores do not need this up-to-the-second real-time data, they will collect sales data throughout the day and transmit it every evening in a batch to the district office. This batching simplifies the data communications process and often saves in communications costs, but it does mean that inventories are not accurate in real time, but rather, are accurate only at the end of the day after the batch has been processed.

Capture Data at the Source   Perhaps the most important principle of input design is to capture the data in an electronic format at its original source or as close to the original source as possible. In the early days of computing, computer systems replaced traditional manual systems that operated on paper forms. As these business processes were automated, many of the original paper forms remained, either because no one thought to replace them or because it was too expensive to do so. Instead, the business process continued to contain manual forms that were taken to the computer center in batches to be typed into the computer system by a *data entry operator.*

Many business processes still operate this way today. For example, most organizations have expense claim forms that are completed by hand and submitted to an accounting department, which approves them and enters them into the system in batches. There are three problems with this approach. First, it is expensive because it duplicates work (the form is filled out twice, once by hand, once by keyboard).[12] Second, it increases processing time because the paper forms must be physically moved through the process. Third, it increases the cost and probability of error, because it separates the entry from the processing of information; someone might misread the handwriting on the input form, data may be entered incorrectly, or the original input could contain an error that invalidates the information.

Most transaction-processing systems today are designed to capture data at its source. *Source data automation* refers to using special hardware devices to automatically capture data without requiring anyone to type it. Stores commonly use *bar-code readers* that automatically scan products and enter data directly into the computer system. No intermediate formats such as paper forms are used. Similar technologies include *optical character*

---

[12] Or in the case of the University of Georgia, three times: first by hand on an expense form, a second time when it is typed onto a new form for the "official" submission because the accounting department refuses handwritten forms, and, finally, when it is typed into the accounting computer system.

recognition, which can read printed numbers and text (e.g., on checks), *magnetic stripe readers,* which can read information encoded on magnetic strip (e.g., credit cards), and *smart cards,* which contain microprocessors, memory chips, and batteries (much like credit card–sized calculators). As well as reducing the time and cost of data entry, these systems reduce errors because they are far less likely to capture data incorrectly. Today, portable computers and scanners allow data to be captured at the source even in mobile settings (e.g., air courier deliveries, use of rental cars).

These automatic systems are not capable of collecting a lot of information, so the next-best option is to capture data immediately from the source using a trained entry operator. Many airline and hotel reservations, loan applications, and catalog orders are recorded directly into a computer system while the customer provides the operator with answers to questions. Some systems eliminate the operator altogether and allow users to enter their own data. For example, many universities no longer accept paper-based applications for admissions; all applications are typed by students into electronic forms.

The forms for capturing information (on a screen, on paper, etc.) should support the data source. That is, the order of the information on the form should match the natural flow of information from the data source, and data-entry forms should match paper forms used to initially capture the data.

Minimize Keystrokes   Another important principle is to minimize keystrokes. Key-strokes cost time and money, whether they are performed by a customer, user, or trained data-entry operator. The system should never ask for information that can be obtained in another way (e.g., by retrieving it from a database or by performing a calculation). Like-wise, a system should not require a user to type information that can be selected from a list; selecting reduces errors and speeds entry.

In many cases, some fields have values that often recur. These frequent values should be used as the *default value* for the field so that the user can simply accept the value and not have to retype it time and time again. Examples of default values are the current date, the area code held by the majority of a company's customers, and a billing address, which is based on the customer's residence. Most systems permit changes to default values to handle data-entry exceptions as they occur.

YOUR TURN | 10-8 Career Services

*S*uppose you are designing the new interface for a career services system at your university that accepts student résumés and presents them in a standard format to recruiters. Describe how you could incorporate the basic principles of input design into your interface design. Remember to include the use of online versus batch data input, the capture of information, and plans to minimize keystrokes.

Types of Inputs

Each data item that has to be input is linked to a field on the form into which its value is typed. Each field also has a field label, which is the text beside, above, or below the field that tells the user what type of information belongs in the field. Often, the field label is similar

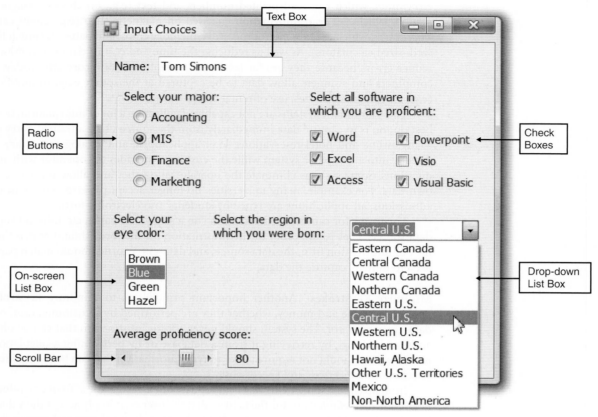

FIGURE 10-18 Use Input Options

to the name of the data element, but they do not have to have identical words. In some cases, a field displays a template over the entry box to show the user exactly how data should be typed. There are many different types of inputs, in the same way that there are many different types of fields (see Figure 10-18).

Text As the name suggests, a *text box* is used to enter text. Text boxes can be defined to have a fixed length, or can be scrollable and can accept a virtually unlimited amount of text. In either case, boxes can contain single or multiple lines of textual information. We never use a text box if we can use a selection box.

Text boxes should have field labels placed to the *left* of the entry area, their size clearly delimited by a box (or a set of underlines in a non-GUI interface). If there are multiple text boxes, their field labels and the left edges of their entry boxes should be aligned. Text boxes should permit standard GUI functions, such as cut, copy, and paste.

Numbers A *number box* is used to enter numbers. Some software can automatically format numbers as they are entered, so that 3452478 becomes $34,524.78. Dates are a special form of numbers that sometimes have their own type of number box. Never use a number box if you can use a selection box.

Type of Box	When to Use	Notes
Check box Presents a complete list of choices, each with a square box in front	When several items can be selected from a list of items	Check boxes are not mutually exclusive. Do not use negatives for box labels. Check box labels should be placed in some logical order, such as that defined by the business process, or failing that, alphabetically or most commonly used first. Use no more than ten check boxes for any particular set of options. If you need more boxes, group them into subcategories.
Radio button Presents a complete list of mutually exclusive choices, each with a circle in front	When only one item can be selected from a set of mutually exclusive items	Use no more than six radio buttons in any one list; if you need more, use a drop-down list box. If there are only two options, one check box is usually preferred to two radio buttons, unless the options are not clear. Avoid placing radio buttons close to check boxes to prevent confusion between different selection lists.
On-screen list box Presents a list of choices in a box	Seldom or never—only if there is insufficient room for check boxes or radio buttons	This type of box can permit only one item to be selected (in which case, it is an ugly version of radio buttons). This type of box can also permit many items to be selected (in which case, it is an ugly version of check boxes), but users often fail to realize they can choose multiple items. This type of box permits the list of items to be scrolled, thus reducing the amount of screen space needed.
Drop-down list box Displays selected item in one-line box that opens to reveal list of choices	When there is insufficient room to display all choices	This type of box acts like radio buttons but is more compact. This type of box hides choices from users until it is opened, which can decrease ease of use; conversely, because it shelters novice users from seldom-used choices, it can improve ease of use. This type of box simplifies design if the number of choices is unclear, because it takes only one line when closed.
Combo box A special type of drop-down list box that permits users to type as well as scroll the list	Shortcut for experienced users	This type of box acts like drop-down list but is faster for experienced users when the list of items is long.
Slider Graphic scale with a sliding pointer to select a number	Entering an approximate numeric value from a large continuous scale	The slider makes it difficult for the user to select a precise number. Some sliders also include a number box to enable the user to enter a specific number.

FIGURE 10-19 Types of Selection Boxes

YOUR TURN

10-9 Career Services

Consider a web form that a student would use to input student information and résumé information into a career services application at your university. First, sketch out how this form would look and identify the fields that the form would include. What types of validity checks would you use to make sure that the correct information is entered into the system?

Selection Box   A *selection box* enables the user to select a value from a predefined list. The items in the list should be arranged in some meaningful order, such as alphabetical for long lists or in order of most frequently used. The default selection value should be chosen with care. A selection box can be initialized as unselected. However, it is better to start with the most commonly used item already selected.

Input Validation

All data entered into the system need to be validated to ensure their accuracy. Input *validation* (also called *edit checks*) can take many forms. Ideally, computer systems should not accept data that fail any important validation check to prevent invalid information from entering the system. However, this can be very difficult, and invalid data often slip past data-entry operators and the users providing the information. It is up to the system to identify invalid data and either make changes or notify someone who can resolve the information problem.

There are six different types of validation checks: *completeness check, format check, range check, check digit check, consistency check,* and *database check* (see Figure 10-20). Every system should use at least one validation check on all entered data and, ideally, performs all appropriate checks where possible.

OUTPUT DESIGN

Outputs are the reports that the system produces, whether on the screen, on paper, or in other media, such as the web. Outputs are perhaps the most visible part of any system because a primary reason for using an information system is to access the information that it produces.

Basic Principles

The goal of the output mechanism is to present information to users so they can accurately understand it with the least effort. The fundamental principles for output design reflect how the outputs are used and ways to make it simpler for users to understand them.

Understand Report Usage   The first principle in designing reports is to understand how they are used. Reports can be used for many different purposes. In some cases—but not very often—reports are read cover to cover because all information is needed. In most cases, reports are used to identify specific items or used as references to find information, so the order in which items are sorted on the report or grouped within categories is critical. This is particularly important for the design of electronic or web-based reports. Web reports that are intended to be read from start to finish should be presented in one long scrollable page, whereas reports that are used primarily to find specific information should be broken into multiple pages, each with a separate link. Page numbers and the date on which the report was prepared are also important for reference reports.

The frequency of the report can also play an important role in its design and distribution. *Real-time reports* provide data that are accurate to the second or minute at which they were produced (e.g., stock market quotes). *Batch reports* are those that report historical information that may be months, days, or hours old, and they often provide

Type of Validation	When to Use	Notes
Completeness check Ensures all required data have been entered	When several fields must be entered before the form can be processed	If required information is missing, the form is returned to the user unprocessed.
Format check Ensures data are of the right type (e.g., numeric) and in the right format (e.g., month, day, year)	When fields are numeric or contain coded data	Ideally, numeric fields should not permit users to type text data, but if this is not possible, the entered data must be checked to ensure it is numeric. Some fields use special codes or formats (e.g., license plates with three letters and three numbers) that must be checked.
Range check Ensures numeric data are within correct minimum and maximum values	With all numeric data, if possible	A range check permits only numbers between correct values. Such a system can also be used to screen data for "reasonableness"—e.g., rejecting birth dates prior to 1880 because people do not live to be a great deal over 100 years old (most likely, 1980 was intended).
Check digit check Check digits are added to numeric codes	When numeric codes are used	Check digits are numbers added to a code as a way of enabling the system to quickly validate correctness. For example, U.S. Social Security numbers and Canadian Social Insurance numbers assign only eight of the nine digits in the number. The ninth number—the check digit—is calculated using a mathematical formula from the first eight numbers. When the identification number is typed into a computer system, the system uses the formula and compares the result with the check digit. If the numbers don't match, then an error has occurred.
Consistency checks Ensure combinations of data are valid	When data are related	Data fields are often related. For example, someone's birth year should precede the year in which he or she was married. Although it is impossible for the system to know which data are incorrect, it can report the error to the user for correction.
Database checks Compare data against a database (or file) to ensure they are correct	When data are available to be checked	Data are compared against information in a database (or file) to ensure they are correct. For example, before an identification number is accepted, the database is queried to ensure that the number is valid. Because database checks are more expensive than the other types of checks (they require the system to do more work), most systems perform the other checks first and perform database checks only after the data have passed the previous checks.

FIGURE 10-20   Types of Input Validation

additional information beyond the reported information (e.g., totals, summaries, historical averages).

There are no inherent advantages to real-time reports over batch reports. The only advantages lie in the time value of the information. If the information in a report is time critical (e.g., stock prices, air-traffic control information), then real-time reports have value. This is particularly important because real-time reports are often expensive to produce; unless they offer some clear business value, they might not be worth the extra cost.

Manage Information Load   Most managers get too much information, not too little (i.e., the *information load* that the manager must deal with is too great). The goal of a well-designed report is to provide all the information needed to support the task for which it was designed. This does not mean that the report needs to provide all the information available on the subject—just what the users decide they need in order to perform their jobs. In some cases, this can result in the production of several different reports on the same topics for the same users because they are used in different ways. This is not a bad design.

For users in westernized countries, the most important information should always be presented first in the top-left corner of the screen or paper report. Information should be provided in a format that is usable without modification. The user should not need to re-sort the report's information, highlight critical information to find it more easily amid a mass of data, or perform additional mathematical calculations.

Minimize Bias   No analyst sets out to design a biased report. The problem with bias is that it can be very subtle; analysts can introduce it unintentionally. *Bias* can be introduced by the way lists of data are sorted because entries that appear first in a list can receive more attention than those later in the list. Data are often sorted in alphabetical order, making those entries starting with the letter *A* more prominent. Data can be sorted in chronological order (or reverse chronological order), placing more emphasis on older (or most recent) entries. Data may be sorted by numeric value, placing more emphasis on higher or lower values. For example, consider a monthly sales report by state. Should the report be listed in alphabetical order by state name, in descending order by the amount sold, or in some other order (e.g., geographic region)? There are no easy answers to this, except to say that the order of presentation should match the way the information is used.

Graphical displays and reports can present particularly challenging design issues.[13] The scale on the axes in graphs is particularly subject to bias. For most types of graphs, the scale should always begin at zero; otherwise, comparisons among values can be misleading. For example, have sales increased by very much since year 1 (see Figure 10-21a and 10-21b)? The numbers in both charts are the same, but the visual images the two present are quite different. A glance at Figure 10-21a would suggest only minor changes, whereas a glance at Figure 10-21b might suggest that there have been some significant increases. In fact, sales have increased by a total of 15 percent over five years, or 3 percent per year. Figure 10-21a presents the most accurate picture; Figure 10-21b is biased because the scale starts very close to the lowest value in the graph and misleads the eye into inferring that there have been major changes. Figure 10-21b is the default graph produced by Microsoft Excel. You should also beware of the so-called 3D effects in Microsoft Excel. For example, the pie charts in Figures 11-21c and 11-21d represent the same data; in fact, the data itself are constant. However, owing to the "3D" pie chart, the slices nearer the front look bigger.

YOUR TURN	10-10 Finding Bias

*R*ead through recent copies of a newspaper or popular press magazine, such as *Time, Newsweek,* or *BusinessWeek,* and find four graphs. Are any of these biased? if so, why?

---

[13] Two of the best books on the design of charts and graphical displays are by Edward R. Tufte, *The Visual Display of Quantitative Information, Envisioning Information* (Cheshire, CT: Graphics Press, 2001) and *Visual Explanations*: Images and Quantities, Evidence and Narrative (Cheshire, CT: Graphics Press, 1997). Another good book is by William Cleveland, *Visualizing Data* (Summit, NJ: Hobart Press, 1993).

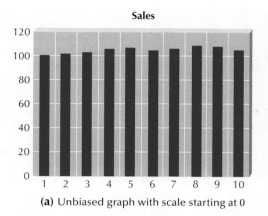

(a) Unbiased graph with scale starting at 0

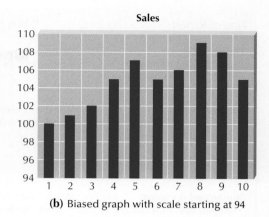

(b) Biased graph with scale starting at 94

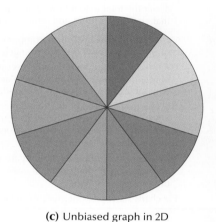

(c) Unbiased graph in 2D

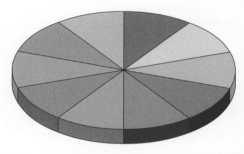

(d) Biased graph in 3D

FIGURE 10-21   Bias in Graphs

Types of Outputs

There are many different types of reports, such as *detail reports, summary reports, exception reports, turnaround documents,* and *graphs* (see Figure 10-22). Classifying reports is challenging because many reports have characteristics of several different types. For example, some detail reports also produce summary totals, making them summary reports.

Media

Many different types of media are used to produce reports. The two dominant media in use today are paper and electronic. Paper is the more traditional medium. For almost as long as there have been human organizations, there have been reports on paper or similar media (e.g., papyrus, stone). Paper is permanent, easy to use, and accessible in most situations. It also is highly portable, at least for short reports.

Type of Report	When to Use	Notes
Detail report Lists detailed information about all the items requested	When user needs full information about the items	This report is usually produced only in response to a query about items matching some criteria. This report is usually read cover to cover to aid understanding of one or more items in depth.
Summary report Lists summary information about all items	When user needs brief information on many items	This report is usually produced only in response to a query about items matching some criteria, but it can be a complete database. This report is usually read for the purpose of comparing several items to each other. The order in which items are sorted is important.
Turnaround document Outputs that "turn around" and become inputs	When a user (often a customer) needs to return an output to be processed	Turnaround documents are a special type of report that are both outputs and inputs. For example, most bills sent to consumers (e.g., credit-card bills) provide information about the total amount owed and also contain a form that consumers fill in and return with payment.
Graphs Charts used in addition to and instead of tables of numbers	When users need to compare data among several items	Well-done graphs help users compare two or more items or understand how one has changed over time. Graphs are poor at helping users recognize precise numeric values and should be replaced by or combined with tables when precision is important. Bar charts tend to be better than tables of numbers or other types of charts when it comes to comparing values between items (but avoid three-dimensional charts that make comparisons difficult). Line charts make it easier to compare values over time, whereas scatter charts make it easier to find clusters or unusual data. Pie charts show proportions or the relative shares of a whole.

FIGURE 10-22   Types of Reports

Paper also has several rather significant drawbacks. It is inflexible. Once the report is printed, it cannot be sorted or reformatted to present a different view of the information. Likewise, if the information on the report changes, the entire report must be reprinted. Paper reports are expensive, are hard to duplicate, and require considerable supplies (paper, ink) and storage space. Paper reports also are hard to move long distances quickly (e.g., from a head office in Toronto to a regional office in Bermuda).

Many organizations are, therefore, moving to electronic production of reports, whereby reports are "printed" but stored in electronic format on file servers or web servers so that users can easily access them. Often, the reports are available in more predesigned formats than their paper-based counterparts because the cost of producing and storing different formats is minimal. Electronic reports also can be produced on demand as needed, and they enable the user to more easily search for certain words. Furthermore, electronic reports can provide a means of supporting ad hoc reports, where users customize the contents of the report at the time the report is generated. Some users still print the electronic report on their own printers, but the reduced cost of electronic delivery over distance and the ease of enabling more users to access the reports than when they were only in paper form usually offsets the cost of local printing.

CONCEPTS

IN ACTION

10-D Selecting the Wrong Students

I helped a university department develop a small DSS to analyze and rank students who applied to a specialized program. Some of the information was numeric and could easily be processed directly by the system (e.g., grade-point average, standardized test scores). Other information required the faculty to make subjective judgments among the students (e.g., extracurricular activities, work experience). The users entered their evaluations of the subjective information via several data analysis screens in which the students were listed in alphabetical order.

To make the system easier to use, it was designed so that the reports listing the results of the analysis were also presented in alphabetical order by student name, rather than in order from the highest-ranked student to the lowest-ranked student. In a series of tests prior to installation, the users selected the wrong students to admit in 20 percent of

the cases. They assumed, wrongly, that the students listed first were the highest-ranked students and simply selected the first students on the list for admission. Neither the title on the report nor the fact that all the students' names were in alphabetical order made users realize that they had read the report incorrectly.

—*Alan Dennis*

Question

This system was biased because users assumed that the list of students implied ranking. Suppose that you are an analyst charged with minimizing bias in this application. Where else may you find bias in the application? How would you eliminate it?

MOBILE COMPUTING AND USER INTERFACE DESIGN[14]

From a user interface design perspective, going mobile is both exciting and challenging. Obviously, with today's smartphones, such as the Droid™ or iPhone™, there are many possibilities. However, just because these phones have the ability to surf the web doesn't

CONCEPTS

IN ACTION

10-E Cutting Paper to Save Money

*O*ne of the Fortune 500 firms with which I have worked had an eighteen-story office building for its world headquarters. It devoted two full floors of this building to nothing more than storing "current" paper reports (a separate warehouse was maintained outside the city for archived reports such as tax documents). Imagine the annual cost of office space in the headquarters building tied up in these paper reports. Now imagine how a staff member would gain access to the reports, and you can quickly understand the driving force behind electronic reports, even if most users end up printing them. Within one year of switching to electronic reports

(for as many reports as practical), the paper report storage area was reduced to one small storage room.

—*Alan Dennis*

Questions

1. What types of reports are most suited to electronic format?

2. What types of reports are less suited to electronic reports?

---

[14] Obviously, in a short section, we cannot cover all of the issues related to developing mobile applications. For anyone who is seriously considering developing mobile applications, we recommend that you begin by looking at books that deal with the specific devices on which you will be deploying your application. For example, Donn Felker, *Android™ Application Development For Dummies™* (Hoboken, NJ: Wiley, 2011); Neal Goldstein and Tony Bove, *iPhone™ Application Development All-In-One For Dummies™* (Hoboken, NJ: Wiley, 2010); Neal Goldstein and Tony Bove, *iPad™ Application Development For Dummies™* (Hoboken, NJ: Wiley, 2010); Chris Stevens, *Designing for the iPad™: Building Applications that Sell* (Chichester, UK: Wiley, 2011).

FIGURE 10-23 iPad Tablet Screens (Three Apple iPad 3G Tablet Computers)
(*Source:* Oleksiy Maksymenko/ACP International/Glow Images)

mean that a simple web interface is the answer. These devices have limited screen space and capabilities, such as *touch screens* and *haptic feedback*, which regular computers do not. Consequently, you really need to focus on designing the interface for the device and not simply porting the web interface over to it. Furthermore, you need to realize that a *tablet*, such as the iPad™, is not a big *smartphone*; it is in its own category with its own challenges and capabilities (see Figure 10-23). Consequently, you really need to design the interface for *mobile devices* from the ground up. In this section, we discuss some challenges and provide some guidelines to develop effective mobile interfaces. However, before we begin, you should realize that all of the material described previously is still applicable. It's just that when you are dealing with these devices, additional issues must be considered.

Tidwell[15] identifies six challenges that a mobile user interface designer must face. The screen of a phone is tiny. There simply is not a lot of "real estate" available to use (see Figure 10-24). Not only are the screens tiny, but they come in different sizes. What works on one screen might not work on another screen. Some screens have haptic abilities: they respond to touch and orientation, and in some cases, they vibrate. Obviously, these abilities are not available on all mobile devices. However, they do provide interesting possibilities for user interface design. Virtual and actual physical keypads are tiny. Consequently, too much typing can be challenging for the user to input the right information. People use their mobile devices, especially their phones, in all kinds of environments. They use them in dark places (like a poorly lit classroom). They use them in bright sunlight. They use them in quiet places (like the library or movie theater) and they use them in noisy places (such as at a football game). These devices are simply used everywhere today. Because these devices are used everywhere, the users can be easily distracted from the device. For example, have you ever texted someone when you aren't supposed to be using your phone, like during class? Or what about out on a date? In other words, users are typically multitasking while using their phone. They do not want to spend a lot of energy on trying to navigate your mobile site or app. Consequently, Krug's three design principles described earlier are very important, especially his first one: Don't make me think!

---

[15] Jenifer Tidwell, *Designing Interfaces: Patterns for Effective Design*, 2nd ed. (Sebastopol, CA: O'Reilly, 2010).

Based on these challenges, Tidwell provides a set of suggestions that you should follow in designing a user interface for these devices. First, given the mobile context, you really need to focus on what the user needs and not what the user might want. In other words, you really should go back to business process and functional modeling (Chapter 4). In this case, only focus on the tasks that users need to perform when they are in the mobile context. This is a good example of a nonfunctional requirement (mobile computing) affecting the possible functional requirements.

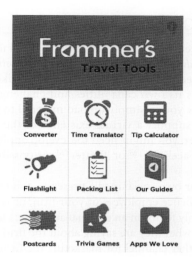

FIGURE 10-24
iPhone™ Smartphone
Example

Second, if you are porting an application or website to a mobile device, remove all "fluff" from the site: Strip the site down to its bare essentials. If the user needs access to the full site, be sure to provide a link to it in an obvious location. Alternatively, you could provide a complete mobile version of the application or website to the user. Obviously, the design of the user interface will be different, but the functionality should be the same.

Third, whenever possible, take advantage of the unique capabilities built into these devices. Some of the devices have *GPS* built in. Depending on your application, knowing where the user is could change the results. In other cases, devices such as the iPad™ have an accelerometer that allows the app to "know" the orientation of the device. Many of these devices have speech recognition capabilities, cameras that can be used for scanning, touch screens that allow sophisticated gestures to be used, and haptic feedback, such as bumps and vibrations. All of these capabilities could prove useful in developing different mobile applications.

Fourth, when considering a phone, you tend to have a limited width from which to work. Consequently, you should try to linearize the content of the application (see Figure 10-25). By that we mean, take advantage of vertical scrolling and try to minimize, if not eliminate, horizontal scrolling. It is simply more natural for users to scroll up and down instead of left to right on these devices.

Fifth, optimize your mobile application for the user. This includes minimizing the number of times the device must interact with a server to download or upload information with a server. Not everyone has access to 3G, alone true 4G, networks. In many cases, uploading and downloading are still very slow. Optimization also includes the user's interaction with the device. Instead of using a lot of typing, scrolling, and taps on a touch screen, consider using the speech recognition capability. It's a lot easier to speak slowly to a smartphone than it is to have to type a lot into a virtual or physical keyboard.

FIGURE 10-25
Linearization Example

Finally, Tidwell provides a set of reusable patterns that have been customized for mobile devices. These include things such as a vertical stack, filmstrip, and bottom navigation.[16]

---

[16] Tidwell also suggests that the Design for Mobile (patterns.design4mobile.com) pattern library provides many good patterns to use when developing mobile applications.

SOCIAL MEDIA AND USER INTERFACE DESIGN[17]

Given the impact that Facebook™ and Twitter™ have had in today's world, developing applications for *social media* has obviously come to the forefront. In many ways, mobile computing and social media have grown up together. Like mobile computing, each social media platform has its own capabilities and challenges. Social media platforms range from sites that allow you to simply upload material to them, such as Flickr™ and YouTube™, to sites that support a virtual existence in the metaverse, such as Second Life™. During your career, you might need to develop applications for a specific social media platform, such as Facebook™ or Twitter™, or possibly develop your own social media site.

When developing your own social media presence, you must understand who is your target audience. Is the audience employees of your firm, or is the audience outside of the firm? In this section, we only focus on an external audience. Once you know who the audience is, you need to know what they are saying about the firm. In many ways, social media is nothing more than another channel for marketing the firm's products and capabilities. Before you can deploy a social media presence, you really need to understand what the users' needs (desires) are. In other words, back to requirements determination. In this case, the problem is that the users are "out there" somewhere, so typical approaches to gathering requirements, such as interviews and observation, don't work. Instead, you need to hunt through the web to root out your requirements. Some of the more useful places to look are blogs or other social media outlets that address issues that would be of interest to your firm. When all else fails, you can always use a search engine such as Google™. Regardless, you obviously have to understand the functional requirements before you can design your social media presence.

Once you understand your functional requirements, you need to determine what type of social media presence is necessary to address the requirements effectively. Each social media platform has its own niche. Consequently, you might need to deploy many different applications across different platforms to effectively meet the firm's social media presence requirements. Also, you must look at your social media site as a means for your firm to build and maintain a positive image or brand. Therefore, the social media site must contain material that your potential customers want to consume. You must remember that the underlying purpose of marketing is to "manufacture" wants and then "convert" the wants into needs. Given that your social media site is effectively another marketing channel, your site must be able to draw in new customers and get current customers to regularly return. In this section, we provide some general guidelines for developing your own social media site so that both new customers visit and current customers return.[18]

First, you really need to post to your site regularly. If the content of the site becomes stagnant, no one will want to visit. The content of the site should contain a mixture of media: videos, podcasts, sound clips, and so on. The site's material should include a mixture of firm-driven material, material from customers, and links to

---

[17] Much of the material in this section has been adapted from Jenifer Tidwell, *Designing Interfaces: Patterns for Effective Design*, 2nd ed. (Sebastopol, CA: O'Reilly, 2010).

[18] Two good books devoted to developing applications for social media in general include Erin Malone, *Designing Social Interfaces* (Sebastopol, CA: O'Reilly, 2009) and Gavin Bell, *Building Social Web Applications: Establishing Community at the Heart of Your Site* (Sebastopol, CA: O'Reilly, 2009). A couple of books devoted to two specific social media platforms are Jesse Stay, *Facebook™ Application Development for Dummies™* (Hoboken, NJ: Wiley, 2011) and Dusty Reagan, *Twitter™ Application Development for Dummies™* (Hoboken, NJ: Wiley, 2010).

FIGURE 10-26   The Home Pages of Facebook™ and Twitter™.
(*Source:* Hermes Images/Glow Images)

relevant content that is located on other sites. Also, be sure to include ways for visitors to join in a "conversation" with the firm, such as Facebook™ comments or Twitter™ Tweets (see Figure 10-26).

Second, make sure you understand the difference between *push* and *pull* approaches. If the user must come to you to find out something, then you are using a pull-based approach. On the other hand, if you put the information out to the user, then you are using a push-based approach. When it comes to social media, you really need to use a combination of the approaches. For example, in Facebook™, if someone posts on your wall or sends you a request, Facebook™ will send you an e-mail message to try an entice you back to the Facebook™ site. The act of posting to your site was a pull-based action, and the e-mail message sent to you is a push-based action. In a nutshell, you want to focus on more of a push-based approach. You want your content to get to your customers in as an effective manner as possible. You don't want them to have to come looking for you. Encourage them to opt in for update notifications to come to them in a form that they prefer. Some might prefer e-mail notifications, and others might prefer you post to their Facebook™ or Twitter™ accounts. Also, be sure to include links to your social media sites on your home page. But be sure not to overwhelm the customer. Not every customer wants to know every tidbit regarding the firm. Only give the customer what the customer wants. Remember, Krug's first principle: Don't make me think! A corollary to this principle for social media would be: Don't make me work! Make it easy for the customer to find only what they want (or maybe what we want them to want).

Third, be sure that your home page and your social media sites are all synced together so that when one is updated, the other sites "know" about the update. This makes your job of maintaining the different sites much easier and it allows your customers to have a consistent experience across all sites. However, don't overdo this. It is obvious that different sites have different media and, potentially, different audiences. You aren't going to use Facebook™ in the same way you would use Twitter™, YouTube™, or a blog. Be sure to

FIGURE 10-27
John Wiley & Sons
Website

include crosslinks among the different sites. This enables your customer to easily navigate through your different sites. For example in Figure 10-27, we see that Wiley has supported the ability to join e-mail lists and the ability to link to Apple's iTunes™ store to download apps produced by Wiley on their home page. However, they do not provide links to either their Facebook™ or Twitter™ sites.

Fourth, enable the customers to share the great content that you have created. You can include buttons that allow them to e-mail the content to their closest "friends" or other followers in their own social network. You also should provide a means to gather feedback from your customers regarding your content. One way is to include the ability for customers to make and share comments regarding your content. Another way is to provide a voting or "like" mechanism to encourage the customer to become engaged with your site.

Fifth, be sure to design your sites so that not only your customers can easily find the material for which they are searching, but also search engines can find the material. Search engines are at least as likely to bring new customers to your sites as other customers. Design the site so that once the customer lands on your site, he or she stays there for a while. One way that you can accomplish this is by providing the customer with links to "related" material. If you decided to include a voting or "like" mechanism, be sure to enable the customer to see the "best" or, at least, the most popular material first. Another possibility is to create a leaderboard that displays the most shared material. You need to leverage the information gained by implementing the fourth guideline.

Sixth, one of the more difficult things to accomplish is to have your sites become a place that your customers feel that they belong. You want your customers to feel that they are members of something; you want to try to build a feeling of community. The more they feel that they belong, the more likely they will recommend your site to their friends. One way to accomplish this is to encourage employees, at least the "right" employees, to author their own "independent" sites that discuss topics of interest to your customers. This will give a more personal feel to the firm and possibly entice customers to stick around on the site longer.

Finally, in most cases, your customers visit your sites using a variety of hardware platforms. The platforms range from the desktop to the notebook to the tablet to the smartphone. Consequently, all of the material related to general user interface design and to mobile computing is applicable. Because you have a global audience, you need to be sure to take into account international and cultural issues in your design. We turn to this topic in the next section.

INTERNATIONAL AND CULTURAL ISSUES AND USER INTERFACE DESIGN[19]

With the World Wide Web, virtually any firm can have a global presence. With this capability, a firm must be cognizant of a set of international and cultural issues. These issues include multilingual requirements, color, and cultural differences.

Multilingual Requirements

The first and most obvious difference between applications used in one region and those designed for global use is language. Global applications often have *multilingual requirements,* which means that they have to support users who speak different languages and write using non-English letters (e.g., those with accents, Cyrillic, Japanese). One of the most challenging aspects in designing global systems is getting a good translation of the original language messages into a new language. Words often have similar meanings but can convey subtly different meanings when they are translated, so it is important to use translators skilled in translating technical words. A few rules that you should follow are to:

- Keep the writing short and simple. It is much easier to avoid mistranslations.[20]
- Avoid humor, jargon, slang, clichés, puns, analogies, and metaphors. These tend to be too culturally specific. Consequently, the underlying point being made will most likely be lost in translation.
- Use good grammar. Be sure to punctuate everything correctly. Even though you might be tempted to ignore grammar and punctuation rules to try to make a point, it makes translating more difficult, especially for automated translation systems. Don't depend on automated spelling and grammar checkers to enforce this. At this time, they simply aren't good enough.

Another challenge is often screen space. In general, English-language messages usually take 20 percent to 30 percent fewer letters than their French or Spanish counterparts. Designing global systems requires allocating more screen space to messages than might be used in the English-language version.

---

[19] A set of books that provide a good introduction to building information systems for a multicultural audience includes Elisa M. del Galdo and Jakob Nielsen, *International User Interfaces* (New York, NY: Wiley, 1996), Nitish Singh and Arun Pereira, *The Culturally Customized Web Site: Customizing Web Sites for the Global Marketplace* (Oxford, UK: Elsevier Butterworth Heinemann, 2005), and John Yunker, *Beyond Borders: Web Globalization Strategies* (Berkley, CA: New Riders, 2003).

[20] However, even this does not guarantee good translations if you use an automatic translation facility. For example, type the text "I would like my steak cooked rare" into babel fish (http://babelfish.yahoo.com/) and translate it to Russian and back to English. You will get back "I wanted would be my rare welded [steykom] done"— not exactly the most useful translation.

Some systems are designed to handle multiple languages on the fly so that users in different countries can use different languages concurrently; that is, the same system supports several different languages simultaneously (a concurrent multilingual system). Other systems contain separate parts that are written in each language and must be reinstalled before a specific language can be used; that is, each language is provided by a different version of the system so that any one installation will use only one language (i.e., a discrete multilingual system). Either approach can be effective, but this functionality must be designed into the system well in advance of implementation.

Finally, one other consideration that must be considered is reading direction. In most Western societies, readers read from left to right and top to bottom. This is not true for many other cultures. For example, in Arabic countries, readers typically read right to left and top to bottom.

CONCEPTS

IN ACTION

10-F Developing Multilingual Systems

I've had the opportunity to develop two multilingual systems. The first was a special-purpose decision support system to help schedule orders in paper mills called BCW-Trim. The system was installed by several dozen paper mills in Canada and the United States, and it was designed to work in either English or French. All messages were stored in separate files (one set English, one set French), with the program written to use variables initialized either to the English or French text. The appropriate language files were included when the system was compiled to produce either the French or English version.

The second program was a groupware system called GroupSystems, for which I designed several modules. The system has been translated into dozens of different languages, including French, Spanish, Portuguese, German, Finnish, and Croatian. This system enables the user to switch between languages at will by storing messages in simple text files. This design is far more flexible because each individual installation can revise the messages at will. Without this approach, it is unlikely that there would have been sufficient demand to warrant the development of versions to support less commonly used languages (e.g., Croatian).

—*Alan Dennis*

Questions

1. How would you decide how much to support users who speak languages other than English?
2. Would you create multilingual capabilities for any application that would be available to non-English-speaking people? Think about websites that exist today.

Color

To begin with, color is not black and white. The meaning associated with a color is totally culturally dependent. In fact, black and white isn't necessarily black and white; they could be white and black. In most Western cultures, black is associated with death, mourning, and grief or with respect and formality. For example, in the United States, we typically wear black to a funeral, or you would expect to see religious leaders in black (think about the robes typically worn by a Catholic priest). In many Eastern cultures, on the other hand, white is associated with death or the color of robes worn by religious leaders. In an example reported by Singh and Pereira, when senior citizens in the United States and India were asked to "visualize the following statement: A lady dressed in white, in a place of worship," the results that came back were as near to the opposite as one could get. In India, the lady would be a widow, but in the United States, she would be expected to be a bride.

Other colors that have meanings that are culturally driven include green, blue, red, yellow, and purple. In the United States, red implies excitement, spice, passion, sex, and even anger; in Mexico, it indicates religion; in the United Kingdom, it indicates authority, power, and government; in Scandinavian countries, it indicates strength; and in China, it means communism, joy, and good luck. Blue is associated with holiness in Israel, cleanliness in Scandinavia, love and truth in India, loyalty in Germany, and trust, justice, and "official" business in the United States. In Ireland, green signifies nationalism and Catholicism, and in the United States, it denotes health, environmentalism, safety, greed, and envy. Green is a very confusing color for Americans. In the Arab Middle East, green is a sign of holiness; in France, green represents criminality; and in Malaysia, it signifies danger and disease. Yellow also has many culturally dependent meanings. In the United States, it is associated with caution and cowardice; in Scandinavia, warmth; in Germany, envy; and in India, commerce. Purple signifies death, nobility, or the Church in Latin America, the United States, and Italy, respectively. Obviously, when building a website for a global audience, colors must be chosen carefully; otherwise, unintentional messages will be sent.

Cultural Differences

The *New York Times* columnist Tom Friedman talks about the need for a firm to use its own local capabilities as a basis for competitive advantage in a global market. He refers to this process as *glocalization*. In some ways, when developing a website for an international audience, you need to consider the opposite of glocalization. You need to think about what message needs to be sent to a local culture from your global organization to achieve the business goals of the firm. Consequently, you need to be able to understand the different local cultures. Cultural issues have been studied at both organizational and national levels. Different researchers have emphasized different dimensions on which to focus our attention. In this section, we limit our discussion to cultural issues that affect designing effective user interfaces. In particular, we only address the research of Edward Hall and Geert Hofstede.[21]

Hall identified three dimensions that are directly relevant to user interface design: speed of messages, context, and time. The *speed of messages* dimension deals with how fast a member of a culture is expected to understand a message and how "deep" the content of a typical message will be in a culture. The deeper the message content, the longer it will take for a member of a culture to understand the message. For example, two different approaches to describe a historical event would be a news headline (fast and shallow) and a documentary (slow and deep). According to Hall, different cultures have different expectations of the content of and response to a message. This particular dimension has implications for the content of the message contained in the user interface. Krug's third design principle turns out to be culturally driven. For a Western audience, minimizing the number of words contained in a user interface makes sense. Westerners prefer to get to the point as fast as possible. However, this is not true for Eastern cultures.[22] Consequently, for a firm like Amazon.com, providing detailed reviews and short excerpts from a book provides support for a slow and deep culture, while providing bullet point types of comments supports the fast and shallow culture. By providing both, Amazon.com addresses both needs.

The second dimension, *context*, deals with the level of implicit information that is used in the culture versus the information needing to be made explicit. In *high-context* cultures, most information is known intrinsically and does not have to be made explicit. Therefore, the actual content of the message is fairly limited. However, in *low-context* cultures, everything must be spelled out explicitly to avoid any ambiguity, and therefore the message needs to be very detailed. You will find this dimension causing problems when attempting

---

[21] See Geert Hofstede, *Culture's Consequences: Comparing Values, Behaviors, Institutions and Organizations Across Nations,* 2nd ed. (Thousand Oaks, CA: Sage, 2001), Geert Hofstede, Gert Jan Hofstede, and Michael Minkov, *Cultures and Organizations: Software of the Mind,* 3rd ed. (New York: McGraw-Hill, 2010), Edward T. Hall, *Beyond Culture* (New York: Anchor Books, 1981).

[22] See Richarde E. Nisbett, *The Geography of Thought: How Asians and Westerners Think Differently … And Why* (New York: Free Press, 2003).

to close a business deal. In most Western societies, the lawyers want everything spelled out. In contrast, in most Eastern societies, it may, in fact, be considered insulting to have to spell everything out. From a website design perspective, Singh and Pereira point out that in a high-context culture, focusing the design on aesthetics, politeness, and humility produces an effective website, but in a low-context culture, things such as the terms and conditions of a purchase, the "rank" of the product and firm, and the use of superlatives in describing the product and firm are critical attributes of a successful website.

Hall's third dimension, *time*, addresses how a culture deals with many different things going on simultaneously. In a *polychronic time* culture, members of the culture tend to do many things at the same time but are easily distracted, and view time commitments as very flexible. With *monochronic time* cultures, members of the culture solve many things by focusing on one thing at a time, are single-minded, and consider time commitments as something that is set in stone. When designing for a polychronic culture, the liberal use of "pop-up" messages might be fun and engaging, while in a monochronic culture, pop-up messages simply annoy the user. In the past, Northern Hemisphere cultures have been monochronic and Southern Hemisphere cultures have been polychronic. However, with the use of e-mail interruptions and text messaging, this could change over time.

Hofstede has identified cultural dimensions that are relevant to the user interface. These include power distance, uncertainty avoidance, individualism versus collectivism, and masculinity versus femininity. The first dimension, *power distance*, addresses how the distribution of social power is dealt with in the culture. In cultures with a high power distance, members of the culture believe in the authority of the social hierarchy. In cultures with low power distance, members of the culture believe that power should be more equally distributed. Consequently, in cultures with a high power distance, emphasis on the "greatness" of the leaders of the firm, the use of "proper titles" for members of the firm, and the posting of testimonials on behalf of the firm by "prominent" members of society is important. International awards won by the firm, its members, or its products should also be posted prominently on the website.

The second dimension, *uncertainty avoidance*, addresses to what degree a culture is comfortable with uncertainty. In a culture with a high uncertainty avoidance, members avoid taking risks, value tradition, and are much more comfortable in a rule-driven society. In cultures that score high on uncertainty avoidance, more customer service needs to be provided, more important "local" contacts need to be available, the firm's and product's history and tradition need to be provided on the website, and, in the case of software, the use of free trials and downloads is critical. In other words, you need to build trust and reduce perceived risk between the customer and the firm. This can be supported through product seals of approval or the use of WebTrust™ and SysTrust™ certifications for the website.[23] Merely translating a website from a low uncertainty avoidance culture to a high uncertainty avoidance culture is not sufficient. You also need to point out relationships between the local culture and the firm's products.

The third dimension, *individualism* versus *collectivism*, is based on the level of emphasis the culture places on the individual or the collective, or group. In North America and Europe, individualism is rewarded. However, in East Asia, it is believed that by focusing on optimizing the group, the individual will be most successful. In other words, it is the group that is the most important. In a collective society, presenting information on how the firm "gives back" to the community, supports "member" clubs, "loyalty" programs, and "chat" facilities, and provides links to "local" sites of interest are very important characteristics for a website. In contrast, in an individualistic society, providing support for personalization

---

[23] See www.webtrust.org

of the user's experience with the website, emphasizing the uniqueness of the products that the user is viewing, and emphasizing the privacy policy of the site are critical.

Hofstede's fourth dimension, *masculinity* versus *femininity*, does not mean how men and women are treated by the culture. But, instead, this dimension addresses how well masculine and feminine characteristics are valued by the culture. For example, in a masculine culture, characteristics such as being assertive, ambitious, aggressive, and competitive are valued, whereas in a feminine culture, characteristics such as being encouraging, compassionate, thoughtful, gentle, and cooperative are valued. In masculine cultures, a focus on the effectiveness of the firm's products is essential. Also, clearly separating male- and female-oriented topics and placing them on different sections of a website can be critical. According to Singh and Pereira, feminine cultures value a focus on aesthetics and using more of a soft-sell approach, where the focus on more affective, intangible aspects of the firm, its members, and its products is more appropriate.

Obviously, operationalizing Hall's and Hofstede's dimensions for effective user interface design is not easy. However, in a global market, ignoring cultural issues in user interface design, whether it is for an internal system used only by employees of the firm or an external system that is used by customers, will most certainly cause a system to fail. This is especially true when you consider mobile and social media sites.

NONFUNCTIONAL REQUIREMENTS AND HUMAN–COMPUTER INTERACTION LAYER DESIGN

The human–computer interaction layer is heavily influenced by nonfunctional requirements. In this chapter, we dealt with issues such as layout of the user interface, awareness of content, aesthetics, user experience, and consistency. We also have provided information on how to design the navigation, inputs, and outputs of the user interface. Finally, we have considered mobile computing, social media, and international and cultural issues in user-interface design. None of these have anything to do with the functional requirements of the system. However, if they are ignored, the system can be unusable. As with the data management layer, there are four primary types of nonfunctional requirements that can be important in designing the human–computer interaction layer: operational, performance, security, and cultural and political requirements.

Operational requirements, such as choice of hardware and software platforms, influence the design of the human–computer interaction layer. For example, something as simple as the number of buttons on a mouse (one, two, three, or more) changes the interaction that the user will experience. Other operational nonfunctional requirements that can influence the design of the human–computer interaction layer include system integration and portability. In these cases, a web-based solution may be required, which can affect the design; not all features of a user interface can be implemented efficiently and effectively on the web. This can require additional user-interface design. Obviously, the entire area of mobile computing can affect the success or failure of the system.

Performance requirements, over time, have become less of an issue for this layer. However, speed requirements are still paramount; especially with mobile computing. Most users do not care for hitting return or clicking the mouse and having to take a coffee break while they are waiting for the system to respond, so efficiency issues must be still addressed. Depending on the user-interface toolkit used, different user-interface components may be required. Furthermore, the interaction of the human–computer interaction layer with the other layers must be considered. For example, if the system response is slow, incorporating more-efficient data structures with the problem domain layer, including indexes in the tables with the data management layer, and/or replicating objects across the physical architecture layer could be required.

Security requirements affecting the human–computer interaction layer deal primarily with the access controls implemented to protect the objects from unauthorized access. Most of these controls are enforced through the DBMS on the data management layer and the operating system on the physical architecture layer. However, the human–computer interaction layer design must include appropriate log-on controls and the possibility of encryption.

In addition to the international and cultural issues described previously, unstated norms affect the cultural and political requirements that can affect the design of the human–computer interaction layer. Unstated norm requirements include having the date displayed in the appropriate format (MM/DD/YYYY versus DD/MM/YYYY). For a system to be truly useful in a global environment, the user interface must be customizable to address local cultural requirements.

APPLYING THE CONCEPTS AT CD SELECTIONS

Previously, Alec had the development team focusing on developing the analysis models of the problem domain. In the previous chapter's installment, Alec had split part of the team and had assigned them to work on the data management layer and to develop its design. In this installment, we follow the development team members that have been assigned to the human–computer interaction layer. Based on what Margaret has learned about mobile computing, social media, and globalization, she really wants to be able to deploy across multiple platforms in such a way that CD Selections will be able to reach a global market. However, Alec isn't quite sure that trying to deploy over multiple incompatible platforms is a good idea.

SUMMARY

Principles for User Interface Design

The first element of the user interface design is the layout of the screen, form, or report, which is usually depicted using rectangular shapes with a top area for navigation, a central area for inputs and outputs, and a status line at the bottom. The design should help the user be aware of content and context, both between different parts of the system as they navigate through it and within any one form or report. All interfaces should be aesthetically pleasing (not works of art) and need to include significant white space, use colors carefully, and be consistent with fonts. Most interfaces should be designed to support both novice or first-time users and experienced users. Consistency in design (both within the system and across other systems used by the users) is important for the navigation controls, terminology, and the layout of forms and reports. All interfaces should attempt to minimize user effort—for example, by requiring no more than three clicks from the main menu to perform an action.

User Interface Design Process

First, analysts develop use scenarios that describe commonly used patterns of actions that the users will perform. Second, they design the interface structure via a WND based on the essential use cases. The WND is then tested with the use scenarios to ensure that it enables users to quickly and smoothly perform these scenarios. Third, analysts define the interface standards in terms of interface metaphor(s), objects, actions, and icons. These elements are drawn together by the design of a basic interface template for each major section of the system. Fourth, the designs of the individual interfaces are prototyped, either through a simple storyboard, an HTML prototype, or a prototype using the development language of the system itself (e.g., Visual Basic). Finally, interface evaluation is conducted using heuristic evaluation, walkthrough evaluation, interactive evaluation, or formal usability testing. This evaluation almost always identifies improvements, so the interfaces are redesigned and evaluated further.

Navigation Design

The fundamental goal of the navigation design is to make the system as simple to use as possible, by preventing the user from making mistakes, simplifying the recovery from mistakes, and using a consistent grammar order (usually object-action order). Command languages, natural languages, and direct manipulation are used in navigation, but the most common approach is menus. Error messages, confirmation messages, acknowledgment messages, delay messages, and help messages are common types of messages. Once the navigation design is agreed upon, it is documented in the form of WNDs and real use cases.

Input Design

The goal of the input mechanism is to simply and easily capture accurate information for the system, typically by using online or batch processing, capturing data at the source, and minimizing keystrokes. Input design includes both the design of input screens and all preprinted forms that are used to collect data before they are entered into the information system. There are many types of inputs, such as text fields, number fields, check boxes, radio buttons, on-screen list boxes, drop-down list boxes, and sliders. Most inputs are validated by using some combination of completeness checks, format checks, range checks, check digits, consistency checks, and database checks.

Output Design

The goal of the output mechanism is to present information to users so they can accurately understand it with the least effort, usually by understanding how reports will be used and designing them to minimize information overload and bias. Output design means designing both screens and reports in other media, such as paper and the web. There are many types of reports, such as detail reports, summary reports, exception reports, turnaround documents, and graphs.

Mobile Computing and User Interface Design

Mobile computing provides many exciting capabilities for user interface design. However, mobile computing also brings a set of challenges that need to be addressed including tiny screen sizes, touch screens, haptic feedback capabilities, and tiny virtual and physical keypads. Additionally, they are used everywhere and user distraction can be a problem. From a mobile design perspective, you should only support what the user needs, you should take advantage of each devices' unique capabilities, and you should minimize network traffic.

Social Media and User Interface Design

In today's world, firms must take advantage of social media to be competitive. Potential social media outlets include Facebook™, Flickr™, Google™, LinkedIn™, Second Life™, Twitter™, and YouTube™. Only your imagination will be your limitation. However, each of these outlets come with their own capabilities and challenges. To develop a successful social media experience, you must understand who your target audience is. You also must keep everything on the sites fresh and up-to-date and synced with each other, and you must know when it is appropriate to "push" versus "pull" material to and from your community. Notice the word we just used; community. In many ways, social media is about building a trusting community where users can share with one another.

International and Cultural Issues and User Interface Design

Today, by definition, we live in a multicultural, global world where our firm's customers are "out there" somewhere. Given globalization, many different issues that were never important in the past are now critical. Supporting multilingual systems is now the norm. When supporting multilingual systems, a set of issues arise including language translation problems, data field size, and reading direction. The proper use of color can be a minefield. Something as simple as using the "wrong" color can change the meaning of the message being portrayed in the user interface. Finally, cultural differences are real. Hall identified three relevant dimensions of culture that can affect the effectiveness of a user interface: speed of messages, context, and time. Hofstede identified four dimensions that can affect the usability of a user interface: power distance, uncertainty avoidance, individualism versus collectivism, and masculinity versus femininity.

Nonfunctional Requirements and the Human–Computer Interaction Layer

Nonfunctional requirements can affect the usefulness of the human–computer interaction layer. Because the user or client sees the system as the human–computer interaction layer, not paying sufficient attention to the nonfunctional requirements in the design of this layer can cause the entire system development effort to fail. These requirements include operational, performance, security, and cultural and political issues, and they are intertwined with the design of the data management and physical architecture layers.

KEY TERMS

Acknowledgment message, 440
Action-object order, 436
Aesthetics, 418
Bar-code reader, 444
Batch processing, 444
Batch report, 448
Bias, 450
Button, 425
Check box, 447
Check digit check, 448
Collectivism, 462
Color, 460
Combo box, 447

Command language, 437
Completeness check, 448
Confirmation message, 440
Consistency, 420
Consistency check, 448
Content awareness, 416
Context, 461
Cultural differences, 461
Database check, 448
Data-entry operator, 445
Default value, 445
Delay message, 440
Density, 418

Detail report, 451
Direct manipulation, 439
Drop-down list box, 447
Drop-down menu, 439
Ease of learning, 420
Ease of use, 420
Edit check, 448
Error message, 440
Essential use case, 421
Exception report, 451
Femininity, 463
Field, 416
Field label, 416

QUESTIONS

1. Explain three important user interface design principles.
2. What are three fundamental parts of most user interfaces?
3. Why is content awareness important?
4. What is white space, and why is it important?
5. Under what circumstances should densities be low? High?
6. How can a system be designed to be used by both experienced and first-time users?
7. Why is consistency in design important? Why can too much consistency cause problems?
8. How can different parts of the interface be consistent?
9. Describe the basic process of user interface design.
10. What are use cases, and why are they important?
11. What is a WND, and why is it used?
12. Why are interface standards important?
13. Explain the purpose and contents of interface metaphors, interface objects, interface actions, interface icons, and interface templates.
14. Why do we prototype the user interface design?
15. Compare and contrast the three types of interface design prototypes.
16. Why is it important to perform an interface evaluation before the system is built?
17. Compare and contrast the four types of interface evaluation.
18. Under what conditions is heuristic evaluation justified?

19. What type of interface evaluation did you perform in the Your Turn 10-1?
20. What are Krug's three design principles?
21. Describe three basic principles of navigation design.
22. How can you prevent mistakes?
23. Explain the differences between object-action order and action-object order.
24. Describe four types of navigation controls
25. Why are menus the most commonly used navigation control?
26. Compare and contrast four types of menus.
27. Under what circumstances would you use a drop-down menu versus a tab menu?
28. Under what circumstances would you use an image map versus a simple list menu?
29. Describe five types of messages.
30. What are the key factors in designing an error message?
31. What is context-sensitive help? Does your word processor have context-sensitive help?
32. How do an essential use case and a real use case differ?
33. What is the relationship between essential use cases and use scenarios?
34. What is the relationship between real use cases and use scenarios?
35. Explain three principles in the design of inputs.
36. Compare and contrast batch processing and online processing. Describe one application that would use batch processing and one that would use online processing.
37. Why is capturing data at the source important?
38. Describe four devices that can be used for source data automation.
39. Describe five types of inputs.
40. Compare and contrast check boxes and radio buttons. When would you use one versus the other?
41. Compare and contrast on-screen list boxes and drop-down list boxes. When would you use one versus the other?
42. Why is input validation important?
43. Describe five types of input validation methods.
44. Explain three principles in the design of outputs.
45. Describe five types of outputs.
46. When would you use electronic reports rather than paper reports, and vice versa?
47. What do you think are three common mistakes that novice analysts make in navigation design?
48. What do you think are three common mistakes that novice analysts make in input design?
49. What do you think are three common mistakes that novice analysts make in output design?
50. How would you improve the form in Figure 10-4?
51. What are the six challenges you face when developing mobile applications?
52. What are the six suggestions to address the mobile computing challenges?
53. With regard to social media, what is the difference between "push" and "pull" approaches to interacting with customers?
54. Why is it important to keep your social media sites synced?
55. How can you keep your customers engaged with your social media sites?
56. What are some of the multilingual issues that you may face when developing for a global audience?
57. How important is the proper use of color when developing websites for a global audience? Give some examples of potential pitfalls that you could run into.
58. Name the three cultural dimensions that are relevant to user interface design identified by Hall. Why are they relevant?
59. Name the four cultural dimensions that are relevant to user interface design identified by Hofstede. Why are they relevant?
60. What are some of the nonfunctional requirements that can influence the design of the human–computer interaction layer?

EXERCISES

A. Develop two use scenarios for a website that sells some retail products (e.g., books, music, clothes).
B. Create a storyboard for a website that sells some retail products (e.g., books, music, clothes).
C. Draw a WND for a website that sells some retail products (e.g., books, music, clothes).
D. Create a Windows layout diagram for the home page of a website that sells some retail products (e.g., books, music, clothes).
E. Describe the primary components of the interface standards for a website that sells some retail products (metaphors, objects, actions, icons and template).

F. Ask Jeeves (http://www.ask.com) is an Internet search engine that uses natural language. Experiment with it and compare it to search engines that use key words.

G. Draw a WND for Your Turn 10-7 using the opposite grammar order from your original design (if you didn't do it, draw two WNDs, one in each grammar order). Which is better? Why?

H. In Your Turn 10-7, you probably used menus. Design the navigation system again using a command language.

I. For the A Real Estate Inc. problem in Chapter 4 (exercises I, J, and K), Chapter 5 (exercises P and Q), Chapter 6 (exercise D), Chapter 7 (exercise A), Chapter 8 (exercise A), and Chapter 9 (exercise L):
1. Develop two use scenarios.
2. Draw a WND.
3. Design a storyboard.

J. Based on your solution to exercise I:
1. Create Windows layout diagrams for the interface design.
2. Develop an HTML prototype.
3. Develop a real use case.

K. For the A Video Store problem in Chapter 4 (exercises L, M, and N), Chapter 5 (exercises R and S), Chapter 6 (exercise E), Chapter 7 (exercise B), Chapter 8 (exercise B), and Chapter 9 (exercise M):
1. Develop two use scenarios.
2. Draw a WND.
3. Design a storyboard.

L. Based on your solution to exercise K:
1. Create Windows layout diagrams for the interface design.
2. Develop an HTML prototype.
3. Develop a real use case.

M. For the gym membership problem in Chapter 4 (exercises O, P, and Q), Chapter 5 (exercises T and U), Chapter 6 (exercise F), Chapter 7 (exercise C), Chapter 8 (exercise C), and Chapter 9 (exercise N):
1. Develop two use scenarios.
2. Draw a WND.
3. Design a storyboard.

N. Based on your solution to exercise M:
1. Create Windows layout diagrams for the interface design.
2. Develop an HTML prototype.
3. Develop a real use case.

O. For the Picnics R Us problem in Chapter 4 (exercises R, S, and T), Chapter 5 (exercises V and W), Chapter 6 (exercise G), Chapter 7 (exercise D), Chapter 8 (exercise D), and Chapter 9 (exercise O):
1. Develop two use scenarios.
2. Draw a WND.
3. Design a storyboard.

P. Based on your solution to exercise O:
1. Create Windows layout diagrams for the interface design.
2. Develop an HTML prototype.
3. Develop a real use case.

Q. For the Of-the-Month-Club problem in Chapter 4 (exercises U, V, and W), Chapter 5 (exercises X and Y), Chapter 6 (exercise H), Chapter 7 (exercise E), Chapter 8 (exercise E), and Chapter 9 (exercise N):
1. Develop two use scenarios.
2. Draw a WND.
3. Design a storyboard.

R. Based on your solution to exercise Q:
1. Create Windows layout diagrams for the interface design.
2. Develop an HTML prototype.
3. Develop a real use case.

S. Create a user interface design for a mobile solution for the:
1. A Real Estate Inc. problem.
2. A Video Store problem.
3. Gym membership problem.
4. Picnics R Us problem.
5. Of-the-Month-Club problem.

T. How would your answers change to exercises I through S if you were developing for a global marketplace?

MINICASES

1. Tots to Teens is a catalog retailer specializing in children's clothing. A project has been under way to develop a new order entry system for the company's catalog clerks. The old system had a character-based user interface that corresponded to the system's COBOL underpinnings. The new system will feature a graphical user interface more in keeping with up-to-date PC products in use today. The company hopes that this new user interface will help reduce the turnover they have experienced with their order entry clerks. Many newly hired order entry staff found the old system very difficult to learn and were overwhelmed by the numerous mysterious codes that had to be used to communicate with the system.

A user interface walkthrough evaluation was scheduled for today to give the user a first look at the new system's interface. The project team was careful to invite several key users from the order entry department. In particular, Norma was included because of her years of experience with the order entry system. Norma was known to be an informal leader in the department; her opinion influenced many of her associates. Norma had let it be known that she was less than thrilled with the ideas she had heard for the new system. Owing to her experience and good memory, Norma worked very effectively with the character-based system and was able to breeze through even the most convoluted transactions with ease. Norma had trouble suppressing a sneer when she heard talk of such things as "icons" and "buttons" in the new user interface.

Cindy was also invited to the walkthrough because of her influence in the order entry department. Cindy has been with the department for just one year, but she quickly became known because of her successful organization of a sick child daycare service for the children of the department workers. Sick children are the number-one cause of absenteeism in the department, and many of the workers could not afford to miss workdays. Never one to keep quiet when a situation needed improvement, Cindy has been a vocal supporter of the new system.

a. Drawing upon the design principles presented in the text, describe the features of the user interface that will be most important to experienced users like Norma.

b. Drawing upon the design principles presented in the text, describe the features of the user interface that will be most important to novice users like Cindy.

2. The members of a systems development project team have gone out for lunch together and, as often happens, the conversation turns to work. The team has been working on the development of the user interface design, and so far, work has been progressing smoothly. The team should be completing work on the interface prototypes early next week. A combination of storyboards and language prototypes has been used in this project. The storyboards depict the overall structure and flow of the system, but the team went on to develop language prototypes of the actual screens because they felt that the users would gain from looking at the actual screens.

Robin (the youngest member of the project team): I read an article last night about a really cool way to evaluate a user interface design. It's called usability testing, and it's done by all the major software vendors. I think we should use it to evaluate our interface design.

Dayita (systems analyst): I've heard of that too, but isn't it really expensive?

Pankaj (project manager): I'm afraid it is, and I'm not sure if we can justify the expense for this project.

Robin: But we really need to know if the interface works. I thought this usability testing technique would help us prove we have a good design.

Man Yee (systems analyst): It would, Robin, but there are other ways too. I suggest we do a thorough walk-through with our users and present the interface to them at a meeting. We can project each interface screen so that the users can see it and give us their reactions. This is probably the most efficient way to get their feedback to our work.

Dayita: That's true, but I'd sure like to see the users sit down and work with the system. I've always learned a lot by watching what they do, noting where they are stuck, and listening to their comments and feedback.

Manuel (systems analyst): We've put so much work into this interface design that we really need to review it ourselves. Let's just make a list of the key design principles and ensure that we've followed them consistently. We want to get on with the implementation, you know.

Pankaj: These are all good ideas. It seems like we've all got different views of evaluating the interface design. Let's brainstorm and decide on the technique that is best for our project.

Develop a set of guidelines that can help the project team to select the most appropriate interface evaluation technique for their project.

3. In order to kick-off the work on the Web UI of his CMS, Professor Takato (see cases for chapters 5, 6, 7, 8, and 9) called a meeting with his closest collaborators, students at the City University of Osaka. Aiko and Katsumi are junior students in the Human Computer Interaction (HCI) program, and Mark is a senior student visiting from the US with a bachelor degree in Computer Information Systems, working on his Masters at the City University of Osaka in Human Computer Interaction. Keenly aware of the limited resources available for this project and of the rather severe time constraints for finalization, Professor Takato is trying to make the best out of the given circumstances while getting his team to develop the best possible Web UI for his CMS. For that purpose, and in order to give a head start to his team, Professor Takato did some preliminary work of his own by identifying the main screens and forms of the interface and the navigation patterns across them. He identified the following:

a. General Conference Presentation screen
b. Conference Topics screen
c. Conference venue, traveling and accommodation info screen
d. Program committees screen
e. Important dates and other info screen
f. Paper Registration Submission form – input title, description and list of keywords
g. Participant (author or reviewer) Registration form – first name/last name, affiliation, e-mail, areas of competence (if reviewer)
h. Review System – input rating of reviewed paper, comments and suggestions to the authors
i. Post-conference Survey System – list of multiple-choice survey questions.

What he is asking of his students is to prototype the specified screens and forms in the most effective way possible. How should Aiko, Katsumi, and Mark distribute the work among themselves? What prototyping approaches should they use for the various screens and forms? Should they use different prototyping approaches at different stages of prototype development?

4. Mark, Professor Takato's visiting student from the US (see minicase 3, Chapter 10) has been assigned the important task of implementing the conference participant registration form for the Web UI of the Conference Management System (CMS). This form collects the input data for the Person class, defined as the base class for the Author and Reviewer classes (see minicase 2, Chapter 9). The following pieces of information are expected to be input in this form:

– Title;
– First name;
– Last name;
– Middle initial;
– Address;
– City;
– Zip (Postal Code);
– Country;
– State;
– E-mail;
– Phone;
– Reviewer(Yes/No);
– Areas of competence (if reviewer, up to 5);
– Author(Yes/No);
– Papers submitted (if author, up to 3).

Drawing upon the discussion in the text, help Mark to figure out what kind of controls to use and what types of validations to apply for correctly collecting data through this form. What approach should be used to ensure that a person is not registered more than once in the CMS and that no author will be allowed to review a paper on which he/she is listed as an author or coauthor?

5. Refer to the Professional and Scientific Staff Management (PSSM) Minicase in Chapters 4, 6, 7, 8, and 9.
   a. Develop two use scenarios, draw a WND, and design a storyboard.
   b. Based on your answers to part a, create windows layout diagrams for the user interface, develop an HTML prototype of the user interface, and develop a set of real use cases for the user interface.

c. How would your user interface design have to be modified if you were to deploy it on a tablet? What about a smartphone?

d. What, if any, social media sites should PSSM consider?

e. How would your answers change if you were developing the system for a global audience?

CHAPTER 11

ARCHITECTURE

An important component of the design of an information system is the design of the physical architecture layer, which describes the system's hardware, software, and network environment. The physical architecture layer design flows primarily from the nonfunctional requirements, such as operational, performance, security, cultural, and political requirements. The deliverable from the physical architecture layer design includes the architecture and the hardware and software specification.

OBJECTIVES

- Understand the different physical architecture components
- Understand server-based, client-based, and client–server physical architectures
- Be familiar with cloud computing and Green IT
- Be able to create a network model using a deployment diagram
- Be familiar with how to create a hardware and software specification
- Understand how operational, performance, security, cultural, and political requirements affect the design of the physical architecture layer

CHAPTER OUTLINE

INTRODUCTION

In today's environment, most information systems are spread across two or more computers. A web-based system, for example, runs in the browser on a desktop computer but interacts with the web server (and possibly other computers) over the Internet. A system that operates completely inside a company's network may have a Visual Basic program installed on one computer but interact with a database server elsewhere on the network.

473

Therefore, an important step of design is the creation of the physical architecture layer design, the plan for how the system will be distributed across the computers, and what hardware and software will be used for each computer (e.g., Windows, Linux).

Most systems are built to use the existing hardware and software in the organization, so often the current architecture and hardware and software infrastructure restricts the choice. Other factors, such as corporate standards, existing site-licensing agreements, and product–vendor relationships also can mandate what architecture, hardware, and software the project team must design. However, many organizations now have a variety of infrastructures available or are openly looking for pilot projects to test new architectures, hardware, and software, which enable a project team to select an architecture on the basis of other important factors.

Designing a physical architecture layer can be quite difficult; therefore, many organizations hire expert consultants or assign very experienced analysts to the task.[1] In this chapter, we examine the key factors in physical architecture layer design, but it is important to remember that it takes lots of experience to do it well. The nonfunctional requirements developed during analysis (see Chapter 3) play a key role in physical architecture layer design. These requirements are reexamined and refined into more detailed requirements that influence the system's architecture. In this chapter, we explain how the designers think about application architectures, and we describe the three primary architectures: server-based, client-based, and client–server. Next, we consider the impact of cloud computing and Green IT on the physical architecture layer. We consider how the requirements and architecture can be used to develop the hardware and software specifications that define exactly what hardware and systems software (e.g., database systems) are needed to support the information system being developed. We look at using UML's deployment diagram as a way to model the physical architecture layer. Finally, we examine how the very general nonfunctional requirements from analysis are refined into more specific requirements and the implications that they have for physical architecture layer design.

ELEMENTS OF THE PHYSICAL ARCHITECTURE LAYER

The objective of designing the physical architecture layer is to determine what parts of the application software will be assigned to what hardware. In this section, we first discuss the major architectural elements to understand how the software can be divided into different parts. Then we briefly discuss the major types of hardware onto which the software can be placed. Although there are numerous ways the software components can be placed on the hardware components, there are three principal application architectures in use today: *server-based architectures, client-based architectures,* and *client–server architectures.* The most common architecture is the client–server architecture, so we focus on that.

Architectural Components

The major *architectural components* of any system are the software and the hardware. The major software components of the system being developed have to be identified and then allocated to the various hardware components on which the system will operate. Each of these components can be combined in a variety of different ways.

All software systems can be divided into four basic functions. The first is *data storage* (associated with the object persistence located on the data management layer—see Chapter 9).

---

[1] For more information on the physical architecture layer, see Stephen D. Burd, *Systems Architecture,* 6th ed. (Boston: Course Technology, 2011); Irv Englander, *The Architecture of Computer Hardware and Systems Software: An Information Technology Approach,* 4th ed. (Hoboken, NJ: Wiley, 2009); and Kalani Kirk Hausman and Susan L. Cook, *IT Architecture for Dummies*[TM] (Hoboken, NJ: Wiley, 2011).

Most application programs require data to be stored and retrieved, whether the information is a small file such as a memo produced by a word processor, or a large database that stores an organization's accounting records. These are the data documented in the structural model (CRC cards and class diagrams). The second function is *data access logic* (associated with the data access and manipulation classes located on the data management layer—see Chapter 9), the processing required to access data, which often means database queries in *SQL (structured query language)*. The third function is the *application logic* (located on the problem domain layer—see Chapters 4 through 8), which can be simple or complex, depending on the application. This is the logic documented in the functional (activity diagrams and use cases) and behavioral models (sequence, communication, and behavioral state machines). The fourth function is the *presentation logic* (located on the human–computer interaction layer—see Chapter 10), the presentation of information to the user, and the acceptance of the user's commands (the user interface). These four functions (data storage, data access logic, application logic, and presentation logic) are the basic building blocks of any application.

The three primary hardware components of a system are *client computers, servers,* and the *network* that connects them. Client computers are the input/output devices employed by the user and are usually desktop or laptop computers, but they can also be handheld devices, cell phones, special-purpose terminals, and so on. Servers are typically larger computers that are used to store software and hardware that can be accessed by anyone who has permission. Servers can come in several types: *mainframes* (very large, powerful computers usually costing millions of dollars), *minicomputers* (large computers costing hundreds of thousands of dollars) and *microcomputers* (small desktop computers like the ones we all use to those costing $50,000 or more). The network that connects the computers can vary in speed from a slow cell phone or modem connection that must be dialed, to medium-speed always-on frame relay networks, to fast always-on broadband connections such as cable modem, DSL, or T1 circuits, to high-speed always-on ethernet, T3, or ATM circuits.[2]

Server-Based Architectures

The very first computing architectures were server-based architectures, with the server (usually a central mainframe computer) performing all four functions. The clients (usually terminals) enabled users to send and receive messages to and from the server computer. The clients merely captured keystrokes and sent them to the server for processing and accepted instructions from the server on what to display (see Figure 11-1).

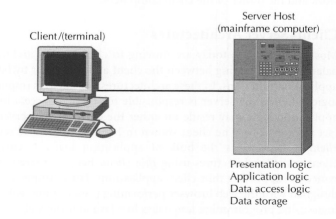

Client /(terminal)

Server Host
(mainframe computer)

Presentation logic
Application logic
Data access logic
Data storage

FIGURE 11-1
Server-Based
Architecture

---

[2] For more information on networks, see Alan Dennis, *Networking in the Internet Age* (New York: Wiley, 2002).

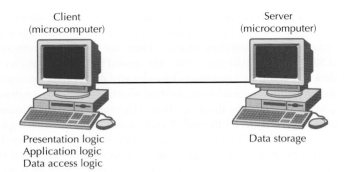

Client
(microcomputer)

Server
(microcomputer)

FIGURE 11-2
Client-Based
Architectures

Presentation logic
Application logic
Data access logic

Data storage

This very simple architecture often works very well. Application software is developed and stored on one computer, and all data are on the same computer. There is one point of control, because all messages flow through the one central server. The fundamental problem with server-based networks is that the server must process all messages. As the demands for more and more applications grow, many server computers become overloaded and unable to quickly process all the users' demands. Response time becomes slower, and network managers are required to spend increasingly more money to upgrade the server computer. Unfortunately, upgrades come in large increments and are expensive; it is difficult to upgrade "a little."

Client-Based Architectures

With client-based architectures, the clients are personal computers on a local area network (LAN), and the server computer is a server on the same network. The application software on the client computers is responsible for the presentation logic, the application logic, and the data access logic; the server simply stores the data (see Figure 11-2).

This simple architecture also often works well. However, as the demand for more and more network applications grow, the network circuits can become overloaded. The fundamental problem in client-based networks is that all data on the server must travel to the client for processing. For example, suppose the user wishes to display a list of all employees with company life insurance. All the data in the database must travel from the server where the database is stored over the network to the client, which then examines each record to see if it matches the data requested by the user. This can overload both the network and the power of the client computers.

Client–Server Architectures

Most organizations today are moving to client–server architectures, which attempt to balance the processing between the client and the server by having both do some of the application functions. In these architectures, the client is responsible for the presentation logic, whereas the server is responsible for the data access logic and data storage. The application logic may reside on either the client or the server or be split between both (see Figure 11-3). The client shown in Figure 11-3 can be referred to as a *thick,* or *fat, client* if it contains the bulk of application logic. A current practice is to create client–server architectures using *thin clients* because there is less overhead and maintenance in supporting thin-client applications. For example, many web-based systems are designed with the web browser performing presentation, with only minimal application logic using programming languages like Java and the web server having the application logic, data access logic, and data storage.

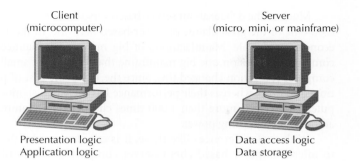

Client
(microcomputer)

Server
(micro, mini, or mainframe)

FIGURE 11-3
Client–Server
Architecture

Presentation logic
Application logic

Data access logic
Data storage

Client–server architectures have four important benefits. First, they are *scalable*. That means it is easy to increase or decrease the storage and processing capabilities of the servers. If one server becomes overloaded, you simply add another server so that many servers are used to perform the application logic, data access logic, or data storage. The cost to upgrade is much more gradual, and you can upgrade in smaller steps rather than spending hundreds of thousands to upgrade a mainframe server.

Client–server architectures can support many different types of clients and servers. It is possible to connect computers that use different operating systems so that users can choose which type of computer they prefer (e.g., combining both Windows computers and Apple Macintoshes on the same network). We are not locked into one vendor, as is often the case with server-based networks. *Middleware* is a type of system software designed to translate between different vendors' software. Middleware is installed on both the client computer and the server computer. The client software communicates with the middleware, which can reformat the message into a standard language that can be understood by the middleware assisting the server software.

For thin-client server architectures that use Internet standards, it is simple to clearly separate the presentation logic, the application logic, and the data access logic and design so each is somewhat independent. For example, the presentation logic can be designed in HTML or XML to specify how the page will appear on the screen (e.g., the colors, fonts, order of items, specific words used, command buttons, the type of selection lists; see Chapter 10). Simple program statements are used to link parts of the interface to specific application logic modules that perform various functions. These HTML or XML files defining the interface can be changed without affecting the application logic. Likewise, it is possible to change the application logic without changing the presentation logic or the data, which are stored in databases and accessed using SQL commands.

Finally, because no single server computer supports all the applications, the network is generally more reliable. There is no central point of failure that will halt the entire network if it fails, as there is in server-based computing. If any one server fails in a client–server environment, the network can continue to function using all the other servers (but, of course, any applications that require the failed server will not work).

Client–server architectures also have some critical limitations, the most important of which is its complexity. All applications in client–server computing have two parts, the software on the client and the software on the server. Writing this software is more complicated than writing the traditional all-in-one software used in server-based architectures. Updating the network with a new version of the software is more complicated, too. In server-based architectures, there is one place where application software is stored; to update the software, we simply replace it there. With client–server architectures, we must update all clients and all servers.

Much of the debate about server-based versus client–server architectures has centered on cost. One of the great claims of server-based networks in the 1980s was that they provided economies of scale. Manufacturers of big mainframes claimed it was cheaper to provide computer services on one big mainframe than on a set of smaller computers. The personal computer revolution changed this. Since the 1980s, the cost of personal computers has continued to drop, whereas their performance has increased significantly. Today, personal computer hardware is more than 1,000 times cheaper than mainframe hardware for the same amount of computing power.

With cost differences like these, it is easy to see why there has been a sudden rush to microcomputer-based client–server computing. The problem with these cost comparisons is that they ignore the *total cost of ownership,* which includes factors other than obvious hardware and software costs. For example, many cost comparisons overlook the increased complexity associated with developing application software for client–server networks. Most experts believe that it costs four to five times more to develop and maintain application software for client–server computing than it does for server-based computing.

Client–Server Tiers

There are many ways the application logic can be partitioned between the client and the server. The example in Figure 11-3 is one of the most common. In this case, the server is responsible for the data, and the client is responsible for the application and presentation. This is called a *two-tiered architecture* because it uses only two sets of computers, clients, and servers.

A *three-tiered architecture* uses three sets of computers (see Figure 11-4). In this case, the software on the client computer is responsible for presentation logic, an application server (or servers) is responsible for the application logic, and a separate database server (or servers) is responsible for the data access logic and data storage.

An *n-tiered architecture* uses more than three sets of computers. In this case, the client is responsible for presentation, database servers are responsible for the data access logic and data storage, and the application logic is spread across two or more different sets of servers. This type of architecture is common in today's e-commerce systems (see Figure 11-5). The first is the web browser on the client computer employed by a user to access the system and enter commands (presentation logic). The second component is a web server that responds to the user's requests, either by providing (HTML) pages and graphics (application logic) or by sending the request to the third component on another application server that performs various functions (application logic). The fourth component is a database

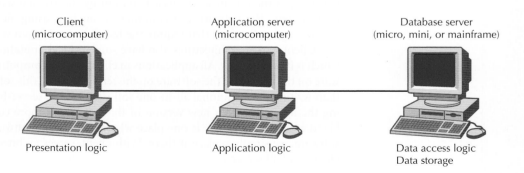

Client (microcomputer)	Application server (microcomputer)	Database server (micro, mini, or mainframe)
Presentation logic	Application logic	Data access logic Data storage

FIGURE 11-4

Three-Tiered Client–Server Architecture

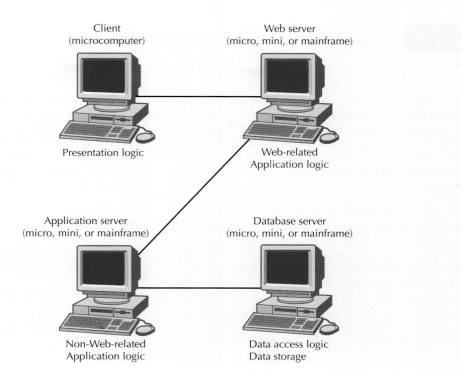

FIGURE 11-5
Four-Tiered
Client–Server
Architecture

server that stores all the data (data access logic and data storage). Each of these four components is separate, making it easy to spread the different components on different servers and to partition the application logic on two different servers.

The primary advantage of an *n*-tiered client–server architecture compared with a two-tiered architecture (or a three-tiered with a two-tiered) is that it separates the processing that occurs to better balance the load on the different servers; it is more scalable. In Figure 11-5, we have three separate servers, a configuration that provides more power than if we had used a two-tiered architecture with only one server. If we discover that the application server is too heavily loaded, we can simply replace it with a more powerful server or just put in several more application servers. Conversely, if we discover the database server is underused, we could store data from another application on it.

There are two primary disadvantages to an *n*-tiered architecture compared with a two-tiered architecture (or a three-tiered with a two-tiered). First, the configuration puts a greater load on the network. If you compare Figures 11-3, 11-4, and 11-5, you will see that the *n*-tiered model requires more communication among the servers; it generates more network traffic, so you need a higher-capacity network. It is also much more difficult to program and test software in *n*-tiered architectures than in two-tiered architectures because more devices have to communicate to complete a user's transaction.

Selecting a Physical Architecture

Most systems are built to use the existing infrastructure in the organization, so often the current infrastructure restricts the choice of architecture. For example, if the new system will be built for a mainframe-centric organization, a server-based architecture may be the best option. Other factors such as corporate standards, existing licensing agreements, and product/vendor relationships can also mandate what architecture the project team needs to

---

CONCEPTS

IN ACTION

11-A A Monster Client–Server Architecture

*E*very spring, Monster.com, one of the largest job sites in the United States with an average of more than 3 million visitors per month, experiences a large increase in traffic. Aaron Braham, vice president of operations, attributes the spike to college students who increase their job-search activities as they approach graduation.

Monster.com uses a three-tier client–server architecture that has 150 web servers and 30 databaase servers in its main site in Indianapolis. It plans to move that to 400 in 2001 by gradually growing the main site and adding a new site with servers in Maynard, Massachusetts, just in time for the spring rush. The main website has a set of load-balancing devices that forward web requests to the different servers, depending on how busy they are.

Braham says the major challenge is that 90 percent of the traffic is not simple requests for web pages but, rather, search requests (e.g., network jobs available in New Mexico), which require more processing and access

to the database servers. Monster.com has more than 350,000 job postings and more than 3 million résumés on file, spread across its database servers. Several copies of each posting and résumé are kept on several database servers to improve access speed and provide redundancy in case a server crashes, so just keeping the database servers in sync so that they contain correct data is a challenge.

Questions

1. What are the two or three primary nonfunctional requirements that have influenced Monster.com's application architecture?

2. What alternatives do you think Monster.com considered?

Source: "Resume Influx Tests Mettle of Job Sites' Scalability," *Internet Week* (May 29, 2000).

---

design. However, many organizations now have a variety of infrastructures available or are openly looking for pilot projects to test new architectures and infrastructures, enabling a project team to select an architecture based on other important factors.

Each of the computing architectures just discussed has its strengths and weaknesses, and no architecture is inherently better than the others. Thus, it is important to understand the strengths and weaknesses of each computing architecture and when to use each. Figure 11-6 presents a summary of the important characteristics of each.

Cost of Infrastructure   One of the strongest driving forces to client–server architectures is cost of infrastructure (the hardware, software, and networks that will support the application system). Simply put, personal computers are more than 1,000 times cheaper than mainframes for the same amount of computing power. The personal computers on our desks today have more processing power, memory, and hard disk space than the typical mainframe of the past, and the cost of the personal computers is a fraction of the cost of the mainframe.

FIGURE 11-6

Characteristics
of Computing
Architectures

Characteristic	Server-based	Client-based	Client–Server
Cost of infrastructure	Very high	Medium	Low
Cost of development	Medium	Low	High
Ease of development	Low	High	Low to medium
Interface capabilities	Low	High	High
Control and security	High	Low	Medium
Scalability	Low	Medium	High

Therefore, the cost of client–server architectures is low compared to server-based architectures that rely on mainframes. Client–server architectures also tend to be cheaper than client-based architectures because they place less of a load on networks and thus require less network capacity.

YOUR TURN

11-1 Course Registration System

*T*hink about the course registration system in your university. What physical architecture does it use? If you had to create a new system today, would you use the architecture or change to a different one? describe the criteria that you would consider when making your decision.

Cost of Development   The cost of developing systems is an important factor when considering the financial benefits of client–server architectures. Developing application software for client–server computing is extremely complex, and most experts believe that it costs four to five times more to develop and maintain application software for client–server computing than it does for server-based computing. Developing application software for client-based architectures is usually cheaper still, because there are many GUI development tools for simple stand-alone computers that communicate with database servers (e.g., Visual Basic, Access).

The cost differential might change as more companies gain experience with client–server applications, new client–server products are developed and refined, and client–server standards mature. However, given the inherent complexity of client–server software and the need to coordinate the interactions of software on different computers, there is likely to remain a cost difference.

Ease of Development   In most organizations today, there is a huge backlog of mainframe applications, systems that have been approved but that lack the staff to implement them. This backlog signals the difficulty in developing server-based systems. The tools for mainframe-based systems often are not user friendly and require highly specialized skills (e.g., COBOL/CICS)—skills that new graduates often don't have and aren't interested in acquiring. In contrast, client-based and client–server architectures can rely on *graphical user interface (GUI)* development tools that can be intuitive and easy to use. The development of applications for these architectures can be fast and painless. Unfortunately, the applications for client–server systems can be very complex because they must be built for several layers of hardware (e.g., database servers, web servers, client workstations) that need to communicate effectively with one another. Project teams often underestimate the effort involved in creating secure, efficient client–server applications.

Interface Capabilities   Typically, server-based applications contain plain, character-based interfaces. For example, think about airline reservation systems such as SABRE, which can be quite difficult to use unless the operator is well trained on the commands and hundreds of codes that are used to navigate through the system. Today, most users of systems expect a GUI or a web-based interface that they can operate using a mouse and graphical objects (e.g., pushbuttons, drop-down lists, icons, and so on). GUI and web development tools typically are created to support client-based or client–server applications; rarely can server-based environments support these types of applications.

Control and Security   The server-based architecture was originally developed to control and secure data, and it is much easier to administer because all the data are stored in a single location. In contrast, client–server computing requires a high degree of coordination among many components, and the chance for security holes or control problems is much more likely. Also, the hardware and software used in client–server architecture is still maturing in terms of security. When an organization has a system that absolutely must be secure (e.g., an application used by the Department of Defense), then the project team may be more comfortable with the server-based alternative on highly secure and control-oriented mainframe computers.

Scalability   *Scalability* refers to the ability to increase or decrease the capacity of the computing infrastructure in response to changing capacity needs. The most scalable architecture is client–server computing because servers can be added to (or removed from) the architecture when processing needs change. Also, the types of hardware that are used in client–server situations (e.g., minicomputers) typically can be upgraded at a pace that most closely matches the growth of the application. In contrast, server-based architectures rely primarily on mainframe hardware that needs to be scaled up in large, expensive increments, and client-based architectures have ceilings above which the application cannot grow because increases in use and data can result in increased network traffic to the extent that performance is unacceptable.

CLOUD COMPUTING[3]

Cloud computing is the idea of treating IT as a utility or commodity. Essentially, cloud computing is the latest approach to support distributed computing in a client-server type of architecture (see previous section) where the server is "in the cloud" and the client is on the desktop. The cloud can be the firm's corporate data center, an external data center, or some combination of the two; however, more and more it generally is seen as an external, rather than an internal, service. Consequently, the idea of *multitenancy*, where the cloud vendor has multiple customers using the same resource at the same time, becomes a real issue for both the cloud vendor and the cloud customer. Cloud computing may become the greatest enabler for IT *outsourcing* (see Chapter 7).

There are three different classifications of clouds: private, public, and hybrid. *Private clouds* are available only to employees of the firm, *public clouds* are available to the general public, and *hybrid clouds* combine the private and public cloud ideas to form a single cloud. In some senses, all e-commerce sites could run in a hybrid cloud environment where the customer sales transaction portion of the system would need to be public while all other aspects would be private.

Fundamentally, cloud computing is an umbrella technology that encompasses the ideas of virtualization, service-oriented architectures, and grid computing. The idea of virtualization is not new. *Virtualization* is the idea of treating any computing resource, regardless of where it is located, as if it is "in" the client machine. This idea evolved from *virtual memory*. Virtual memory was developed originally in the 1960s. Virtual memory allowed the user/programmer to act as if the amount of main memory in the computer was unlimited. This was done by swapping pages of main memory out to disk when the content of the pages was not being used and swapping a page from disk back to main memory when it was needed. Before virtual memory was created, the programmer had to write code to

---

[3] Judith Hurwitz, Marcia Kaufman, Fern Halper, and Robin Bloor, *Cloud Computing for Dummies*™ (Hoboken, NJ: Wiley 2010).

perform the paging function for each application. Virtualization is simply the scaling up of this idea to all computing resources, not simply main memory. This includes treating a mainframe computer as if it is a set of virtual servers, each of which can be running different operating and/or application systems.

Web services basically support connections between different *services* to form *service-oriented architectures.*[4] Basically, a service is a piece of software that supports some aspect of a business process. A service can be an implementation of part of a business process, it can be an implementation of an entire business process (for example, salesforce.com), or it can be object persistence support for the data management layer (see Chapter 9). These services can be either internal or external to the firm. Services can be combined to support *business processes.* We suggest modeling business process with use cases, use-case diagrams, and activity diagrams. A service-oriented architecture allows business processes to be supported by "plugging and playing" services together in a static and/or dynamic manner.[5] Some of the pluggable and playable services can be purchased outright or they can be billed to the firm based on their use, a sort of pay-as-you-go model.

Grid computing[6] tends to be the underlying hardware technology that supports the cloud. A grid is a very large set of networked computers that tend to be geographically dispersed. For example, the grid that supports SalesForce.com's CRM application contains about 1,000 computers. The computers do not have to be of the same type. For example, they can be a mixture of Linux servers and mainframes. With grid computing, firms have the ability to add and remove computers to support a business process based on the current level of activity taking place in that particular business process. This provides an enormous amount of flexibility in configuring the underlying physical architecture that supports business processes.

Combining virtualization, service-oriented architectures, and grid computing is what all the hoopla is about with regard to cloud computing. Cloud computing is highly elastic and scalable, it supports a demand-driven approach to provisioning and deprovisioning of resources, and it supports a billing model that only charges for the resources being used. From a business perspective, cloud computing supports the idea of IT being a commodity.

The cloud can contain the firm's IT infrastructure, IT platform, and software. *Infrastructure as a Service (IaaS)* refers to the cloud providing the computing hardware to the firm as a remote service. The hardware typically includes the computing hardware that supports application servers, networking, and data storage. Amazon's EC2 (aws.amazon.com/ec2/) service is a good example of this. With *Platform as a Service (PaaS)*, the cloud vendor not only provides hardware support to a customer but also provides the customer with either package-based solutions, different services that can be combined to create a solution, or the development tools necessary to create custom solutions in the PaaS vendor's cloud. SalesForce.com is a good example of the vendor providing a package-based solution, Amazon's SimpleDB and Simple Query Service is a good example of different services being supported, and Google's App Engine is a good example of a cloud vendor providing good development tools. Like most things in IT, *Software as a Service (SaaS)* is not a new idea. SaaS has been around for more than 30 years. In the 1970s, there were many "service bureaus" that supported *timesharing* of hardware and software to many different customers; that is, they supported multitenancy. For example, ADP has supported payroll

---

[4] Douglas K. Barry, *Web Services and Service-Oriented Architectures* (San Francisco: Morgan Kaufman, 2003).

[5] P. Ghandforoush, T.K. Sen, and D. Tegarden, R. Ramaswamy, "Designing Systems Using Business Components: A Case Study in Call Center Automation." *International Journal of Electronic Customer Relationship Management,* 4(2) (2010), pp. 161–179.

[6] Pawel Plaszczak and Richard Welner, Jr., *Grid Computing* (San Francisco: Morgan Kaufman, 2006).

functions for many firms for a very long time. Today, SalesForce.com's CRM system is a good example of a SaaS cloud-based solution.

However, cloud computing must overcome certain obstacles before it becomes the primary approach to provision the physical architecture layer.[7] The first obstacle is the mixed level of cloud performance. One issue is whether the vendor has the resources to provide the firm with enough "power" during a peak load. The issue here is that a typical cloud vendor is supporting many different firms. If the vendor does not have enough computing resources to handle all of the firms' peak loads at the same time, then there will have to be some degradation of some or all of the firms' support. This is primarily a result of the unpredictability of the overall performance requirements with disk I/O and network traffic. Given the multitenancy typical of a cloud vendor's hardware, bottlenecks with disks will occur. However, given the dependency on networks, data transfer rates are critical. In an enlightening example, Armbrust and colleagues show that when dealing with large volumes of data, it is faster to transfer data using overnight shipping. In their example, they showed that if you were to transfer 10 terabytes of data with an average transfer rate of 20 Mbits/sec that it would take more than 45 days to complete the transfer. If you shipped the data overnight instead, you would effectively be using a transfer rate of 1,500 Mbits/sec.

A second obstacle deals with the level of dependency that a customer's firm has on a cloud vendor. Firms are dependent on cloud vendors based on the type of service that they are using (IaaS, PaaS, and SaaS), the actual level of service availability, and the potential of data lock-in. Currently, most cloud vendors' API to storage are proprietary. Consequently, the customer's data becomes "locked in" to the specific cloud vendor's storage. This is also true for much of the actual service APIs. Consequently, customers find themselves hoping that the cloud vendor will be the equivalent of a benevolent dictator that will act in the interest of the customer; otherwise, actual level of service being provided could suffer. Given the potential for data and/or service lock-in, a customer must pay close attention to the viability of the cloud vendor. If the vendor goes out of business, the customer could be following suit very quickly. If the cloud vendor also has outsourced to other cloud vendors, such as to a disk farm company, then they could find themselves in the same situation. This could lead to a cascading effect of business failures. Consequently, when a firm is considering outsourcing their IT area into the cloud, the firm had better understand the total risk involved.

A third major obstacle to cloud adoption is the perceived level of security available in the cloud. Not only does a firm have to worry about security from the outside, but when you consider multitenancy, the firm must seriously consider potential attacks from within their cloud from other cloud users. From a service availability perspective, a denial-of-service attack against another tenant within the cloud can cause performance degradation of the firm's systems. Finally, a firm must consider protecting itself from the cloud vendor. The cloud vendor is responsible only for physical security and firewalls. All application-level security tends to be the responsibility of the cloud customer. Obviously, security in the cloud is a very complex endeavor. Given the confidentiality and auditability requirements of Sarbanes-Oxley (SOX) and the Health and Human Services Health Insurance Portability and Accountability Act (HIPAA), security in the cloud becomes a major concern for a firm to move any of its confidential data, including e-mail, to the cloud. In many ways, when using a cloud a firm is simply taking a leap of faith that the cloud is secure.

[7] Michael Armbrust, Armando Fox, Rean Griffith, Anthony D. Joseph, Randy Katz, Andy Konwinski, Gunho Lee, David Patterson, Areil Rabkin, Ion Stoica, and Matei Zahara, "A View of Cloud Computing," *Communications of the ACM*, 53(4) (2010), pp. 50–58.

GREEN IT[8]

Given all of the computing power being deployed to solve today's business problems, Green IT has become important. *Green IT* is a broad term that encompasses virtually anything that helps reduce the environmental impact of IT. Some of the topics included are e-waste, greening data centers, and the dream of the paperless office. In this section, we describe how Green IT can affect the physical architecture layer.

First, when it comes to disposing old electronic devices, care must be taken. Old computers contain very toxic material including lead, PCBs, mercury, and cadmium. One of the major Green IT issues is how to dispose of this *e-waste*. One of the most disturbing trends in dealing with e-waste is the shipping of the e-waste from the developed world to the developing world where environmental standards are virtually nonexistent. Owing to "backyard recycling" techniques used in these locations, the toxic material contained in the e-waste shows up in the soil, water, and air. Alternatives to simply dumping old computers into the trash include extending the replacement cycles of the machines by converting the machines from Windows-based machines to Linux-based machines. Linux takes less "horsepower" to run than Windows. Therefore, for certain applications, a Linux-based desktop is more than sufficient to implement parts of the physical architecture layer.

Second, large data centers use as much electricity in a day as a small city. Consequently, given this level of power consumption, creating *green data centers* in the future will be crucial. There are a whole set of ways to create a green data center. One way is to pay very close attention to where the data center is to be located. Placing the data center in the shade of a mountain or tall building will reduce the cost of energy required. For example, HP placed one of its new data centers in northeast England so that it could be cooled by the cold winds that blow onto shore from the North Sea.[9] Looking into alternative energy possibilities is another way to deal with energy consumption. For example, Google has been in the business of buying wind farms to generate the power for its data centers and HP has shown how a cow manure–based methane power plant could be created to generate the power to run a data center in dairy country.[10]

A third way to consider making your IT infrastructure greener is to consider the cloud (see previous section). With the cloud's virtualization capabilities, the number of high-powered servers and desktops can be reduced. However, you will need to perform some tradeoffs between the obstacles of moving to the cloud and the move toward a greener IT. A fourth way to address the power demands for a modern IT infrastructure is by only purchasing Energy Star compliant electronics. A fifth way is to encourage employees to have their machines go to "sleep" to save energy when the machines have been idle for some period of time.

The *paperless office* idea has been around for a very long time. However, up until now, the idea has been more fantasy than reality. Today, with the advent of multiuse tablets, such as Apple's iPad™, the paperless office is becoming a reality. When considering the cloud and the apps available on the iPad™, it is possible not only to create a paperless office but also to have the paperless office effectively be a portable office.

[8] Caril Baroudi, Jeffrey Hill, Arnold Reinhold, and Jhana Senxian, *Green IT for Dummies*™ (Hoboken, NJ: Wiley, 2009).

[9] www.smartplanet.com/blog/smart-takes/hp-opens-first-wind-cooled-green-data-center-most-efficient-to-date/4191.

[10] See http://www.google.com/corporate/datacenter/renewable-energy.html and greentechnolog.com/2010/05/post_102.html.

CONCEPTS

IN ACTION

11-B Greener IT @ Microsoft

*B*y 2012, Microsoft plans on reducing its carbon emissions by 30% in comparison to its 2007 levels. By deploying a power management strategy to 165,000 desktop and laptop computers, they have been able to realize a 27% drop in power usage by their managed desktop and laptop computers and a 12.33 kilowatt hours per computer savings, which provides them with a $12 to $14 savings per year per computer.

Microsoft has been designing the Zi Zhu campus in Shanghai, China, with sustainability in mind from the beginning. Some of the environmentally friendly features include a passive optical lighting system, demand-driven ventilation, innovative heating and cooling systems, power management systems, virtualization in a cloud-based infrastructure, and unified communication technology. In other efforts to become greener, Microsoft is reducing travel through the use of Office Communicator, Live Meeting, and RoundTable software, creating policies to move toward a paperless office, and taking a proactive approach to managing their e-waste.

Source: www.microsoft.com/environment/our-commitment/greener-it.aspx.

INFRASTRUCTURE DESIGN

In most cases, a system is built for an organization that has a hardware, software, and communications infrastructure already in place. Thus, project teams are usually more concerned with how an existing infrastructure needs to be changed or improved to support the requirements that were identified during analysis, as opposed to how to design and build an infrastructure from scratch. Coordination of infrastructure components is very complex, and it requires highly skilled technical professionals. As a project team, it is best to allow the infrastructure analysts to make changes to the computing infrastructure. In this section, we summarize key elements of infrastructure design to create a basic understanding of what it includes. We describe UML's deployment diagram and the network model.

Deployment Diagram

Deployment diagrams are used to represent the relationships between the hardware components used in the physical infrastructure of an information system. For example, when designing a distributed information system that will use a wide area network, a deployment diagram can be used to show the communication relationships among the different nodes in the network. They also can be used to represent the software components and how they are deployed over the physical architecture or infrastructure of an information system. In this case, a deployment diagram represents the environment for the execution of the software.

The elements of a deployment diagram include nodes, artifacts, and communication paths (see Figure 11-7). Other elements can also be included in this diagram. In our case, we include only the three primary elements and the element that portrays an artifact being deployed onto a node.

A *node* represents any piece of hardware that needs to be included in the model of the physical architecture layer design. For example, nodes typically include client computers, servers, separate networks, or individual network devices. Typically, a node is labeled with its name and, possibly, with a stereotype. The stereotype is modeled as a text item surrounded by "<< >>" symbols. The stereotype represents the type of node being represented on the diagram. For example, typical stereotypes include device, mobile device, database server, web server, and application server. There are times that the notation of a node should be extended to better communicate the design of the physical architecture layer. Figure 11-8 includes a set of typical network node symbols that can be used instead of the standard notation.

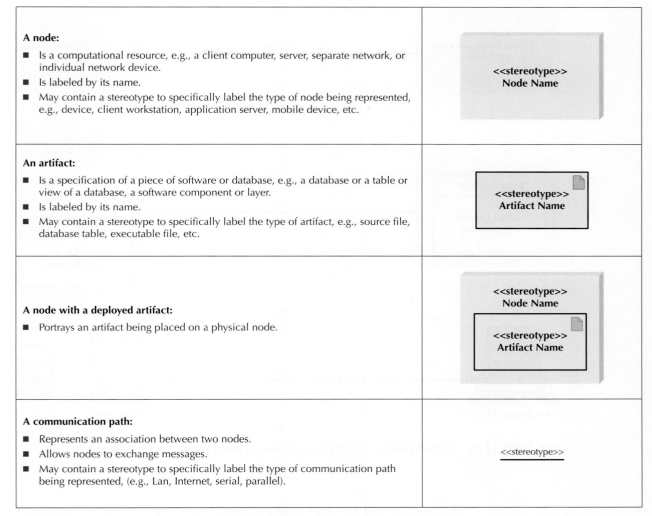

A node: ■ Is a computational resource, e.g., a client computer, server, separate network, or individual network device. ■ Is labeled by its name. ■ May contain a stereotype to specifically label the type of node being represented, e.g., device, client workstation, application server, mobile device, etc.	<<stereotype>> **Node Name**
An artifact: ■ Is a specification of a piece of software or database, e.g., a database or a table or view of a database, a software component or layer. ■ Is labeled by its name. ■ May contain a stereotype to specifically label the type of artifact, e.g., source file, database table, executable file, etc.	<<stereotype>> **Artifact Name**
A node with a deployed artifact: ■ Portrays an artifact being placed on a physical node.	<<stereotype>> **Node Name** <<stereotype>> **Artifact Name**
A communication path: ■ Represents an association between two nodes. ■ Allows nodes to exchange messages. ■ May contain a stereotype to specifically label the type of communication path being represented, (e.g., Lan, Internet, serial, parallel).	<<stereotype>>

FIGURE 11-7   Development Diagram Syntax

An *artifact* represents a piece of the information system that is to be deployed onto the physical architecture (see Figure 11-7). Typically, an artifact represents a software component, a subsystem, a database table, an entire database, or a layer (data management, human–computer interaction, or problem domain). Artifacts, like nodes, can be labeled with both a name and a stereotype. Stereotypes for artifacts include source file, database table, and executable file.

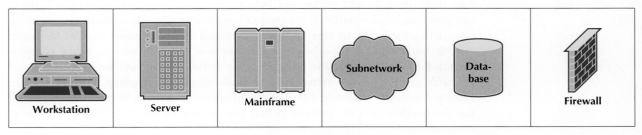

FIGURE 11-8   Extended Node Syntax for Development Diagram

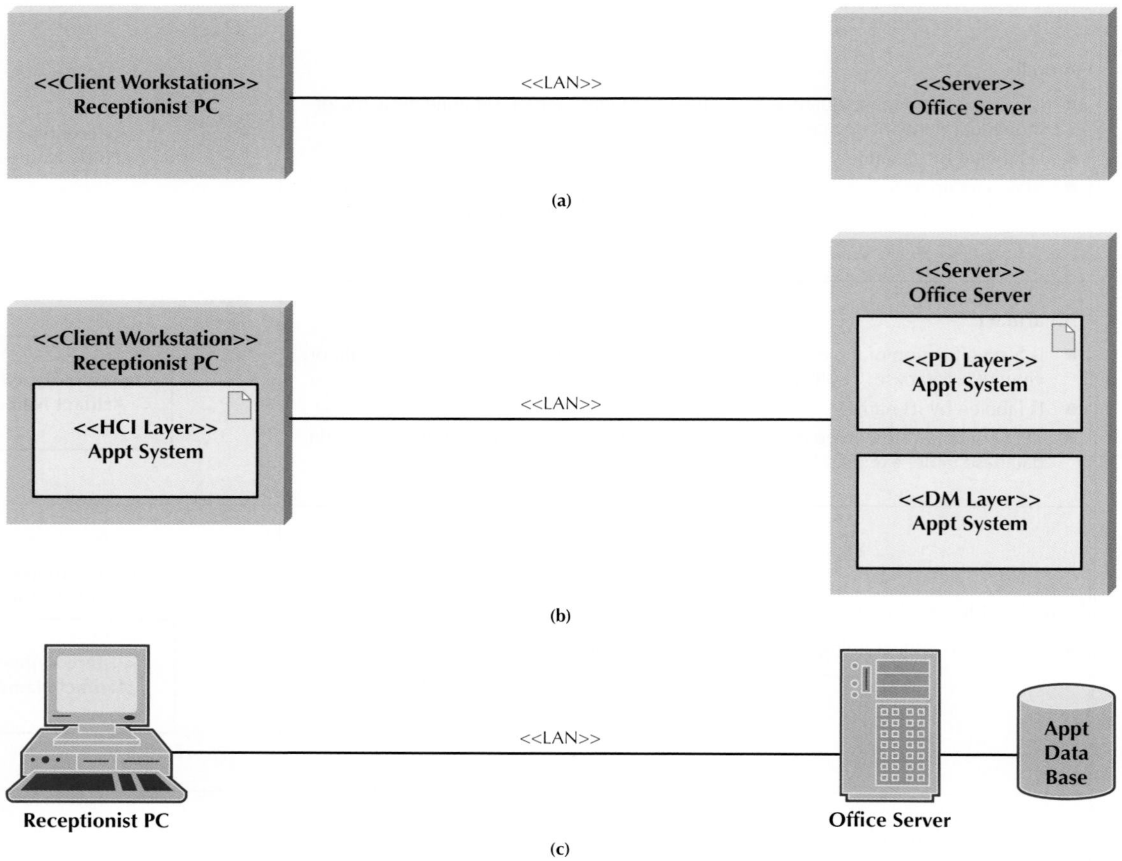

FIGURE 11-9 Three Versions of Appointment System Deployment Diagram

A *communication path* represents a communication link between the nodes of the physical architecture (see Figure 11-7). Communication paths are stereotyped based on the type of communication link they represent (e.g., LAN, Internet, serial, parallel, or USB) or the protocol that is being supported by the link (e.g., TCP/IP).

Figure 11-9 portrays three different versions of a deployment diagram. Version a uses only the basic standard notation. Version b introduces the idea of deploying an artifact onto a node (see Figure 11-7). In this case, the artifacts represent the different layers of the appointment system described in earlier chapters. Version c uses the extended notation to represent the same architecture. As you can see, all three versions have their strengths and weaknesses. When comparing version a and version b, the user can glean more information from version b with little additional effort. However, when comparing version a to version c, the extended node notation enables the user to quickly understand the hardware requirements of the architecture. When comparing version b to version c, version b supports the software distribution explicitly but forces the user to rely on the stereotypes to understand the required hardware, whereas version c omits the software distribution information entirely. We recommend that you use the combination of symbols to best portray the physical architecture to the user community.

Network Model

The *network model* is a diagram that shows the major components of the information system (e.g., servers, communication lines, networks) and their geographic locations throughout the organization. There is no one way to depict a network model, and in our experience, analysts create their own standards and symbols, using presentation applications (e.g., Power-Point) or diagramming tools (e.g., Visio). In this text, we use UML's deployment diagram.

The purpose of the network model is twofold: to convey the complexity of the system and to show how the system's software components will fit together. The diagram also helps the project team develop the hardware and software specification that is described later in this chapter.

The components of the network model are the various clients (e.g., personal computers, kiosks), servers (e.g., database, network, communications, printer), network equipment (e.g., wires, dial-up connections, satellite links), and external systems or networks (e.g., Internet service providers) that support the application. *Locations* are the geographic sites related to these components. For example, if a company created an application for users at four of its plants in Canada and eight plants in the United States, and it used one external system to provide Internet service, the network model to depict this would contain twelve locations $(4 + 8 = 12)$.

Creating the network model is a top-down exercise whereby we first graphically depict all the locations where the application will reside. Placing symbols that represent the locations for the components on a diagram and then connecting them with lines that are labeled with the approximate amount of data or types of network circuits between the separated components accomplish this.

Companies seldom build networks to connect distant locations by buying land and laying cable (or sending up their own satellites). Instead, they usually lease services provided by large telecommunications firms such as AT&T, Sprint, and Verizon. Figure 11-10 shows a typical network. The clouds in the diagram represent the networks at different locations (e.g., Toronto, Atlanta). The lines represent network connections between specific points (e.g., Toronto to Brampton). In other cases, a company might lease connections from many points to many others, and rather than trying to show all the connections, a separate cloud may be drawn to represent this many-to-many type of connection (e.g., the cloud in the center of Figure 11-10 represents a network of many-to-many connections provided by a telecom firm like Verizon).

This high-level diagram has several purposes. First, it shows the locations of the components needed to support the application; therefore, the project team can get a good understanding of the geographic scope of the new system and how complex and costly the communications infrastructure will be to support. (For example, an application that supports one site will probably have less communications costs as compared to a more-complex application that will be shared all over the world.) The diagram also indicates the external components of the system (e.g., customer systems, supplier systems), which may impact security or global needs (discussed later in this chapter).

The second step of the network model is to create low-level network diagrams for each of the locations shown on the top-level diagram. First, hardware is drawn on the model in a way that depicts how the hardware for the new system will be placed throughout the location. It usually helps to use symbols that resemble the hardware that will be used. The amount of detail to include on the network model depends on the needs of the project. Some low-level network models contain text descriptions below each of the hardware components that describe in detail the proposed hardware configurations and processing needs; others include only the number of users that are associated with the different parts of the diagram.

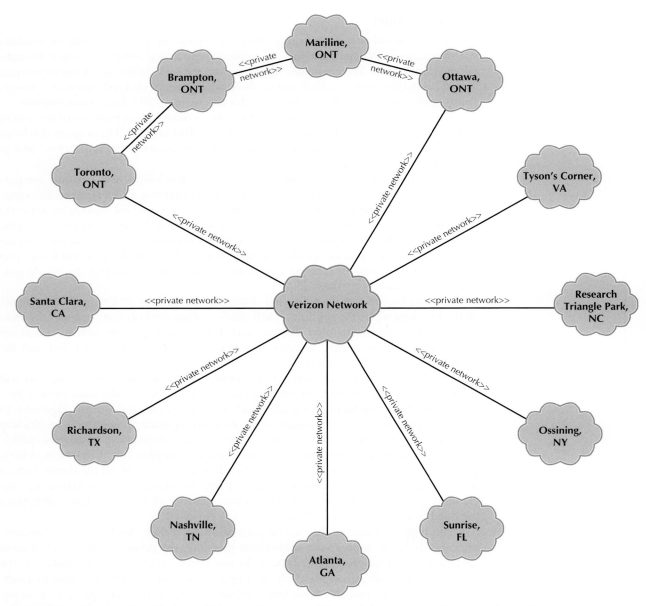

FIGURE 11-10   Deployment Diagram Representation of a Top-Level Network Model

Next, lines are drawn connecting the components that will be physically attached to each other. In terms of software, some network models list the required software for each network model component right on the diagram, whereas other times, the software is described in a memo attached to the network model and stored in the project binder. Figure 11-11 shows a deployment diagram that portrays two levels of detail of a low-level network model. Notice, we use both the standard and extended node notation in this figure. In this case, we have included a package (see Chapter 7) to represent a set of connections to the router in the MFA building. By including a package, we show only the detail necessary. The extended notation, in many cases, aids the user in understanding the

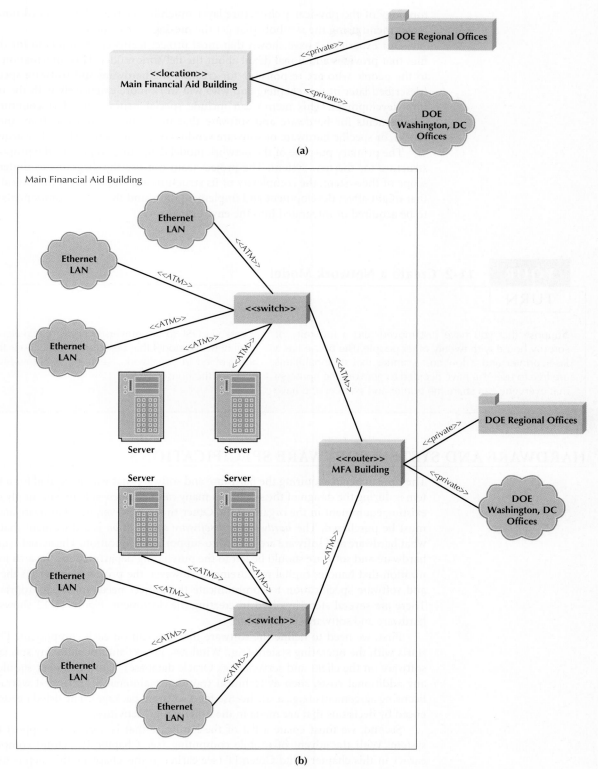

FIGURE 11-11   Deployment Diagram Representation of a Low-Level Network Model

topology of the physical architecture layer much better than the standard notation. We recommend using the symbols that get the message across best.

Our experiences have shown that most project teams create a memo for the project files that provides additional detail about the network model. This information is helpful to the people who are responsible for creating the hardware and software specifications (described later in this chapter) and who will work more extensively with the infrastructure development. This memo can include special issues that affect communications, requirements for hardware and software that might not be obvious from the network model, or specific hardware or software vendors or products that should be acquired.

The primary purpose of the network model diagram is to present the proposed infrastructure for the new system. The project team can use the diagrams to understand the scope of the system, the complexity of its structure, any important communication issues that might affect development and implementation, and the actual components that need to be acquired or integrated into the environment.

YOUR

TURN

11-2 Create a Network Model

Suppose that you have just moved into a fraternity or sorority house with twenty other people. The house has a laser printer and a low-cost scanner that its inhabitants are free to use. You have decided to network the house so that everyone can share the printer and scanner and have access via DSL to the university network. Create a high-level network model that describes the locations that will be involved in your work. Next, create a low-level diagram for the house itself.

HARDWARE AND SYSTEM SOFTWARE SPECIFICATIONS

The time to begin acquiring the hardware and software that will be needed for a future system is during the design of the system. In many cases, the new system will simply run on the existing equipment in the organization. Other times, however, new hardware and software must be purchased. The *hardware and software specification* is a document that describes what hardware and software are needed to support an application. The actual acquisition of hardware and software should be left to the purchasing department or the area in the organization that handles capital procurement. However, the project team writes the hardware and software specification to communicate the project needs to the appropriate people. There are several steps involved in creating the document. Figure 11-12 shows a sample hardware and software specification.

First, we need to define the software that will run on each component. This usually starts with the operating system (e.g., Windows, Linux) and includes any special-purpose software on the client and servers (e.g., Oracle database). This document should consider any additional costs, such as technical training, maintenance, extended warranties, and licensing agreements (e.g., a site license for a software package). The listed needs are influenced by decisions that are made in the other design activities.

Second, we must create a list of the hardware that is needed to support the future system. With the advent of mobile computing (see Chapter 10), cloud computing (see earlier in this chapter), and Green IT (see earlier in this chapter), this step is much more involved than it used to be. However, the low-level network model provides a good starting point for recording the project's hardware needs because each component on the diagram

Specification	Standard Client	Standard Web Server	Standard Application Server	Standard Database Server
Operating System	• Windows • Internet Explorer	• Linux	• Linus	• Linux
Special Software	• Acrobat Reader • Adobe Flash • QuickTime	• Apache	• Java	• Oracle
Hardware	• 8 GB Memory • 500 GB disk drive	• 16 GB Memory • 1TB disk drive	• 24 GB Memory • 2–600 GB disk drives	• 32 GB Memory • 4–600 GB Hotplug disk drives

	• Intel Core i5
	• 18.5-inch
	  HD monitor
		• Intel Xenon X3470
		• 18.5-inch
		  HD monitor
			• Intel Xenon X5620
			• 18.5-inch
			  HD monitor
				• Intel Xenon X5650
				  18.5-inch
				  HD monitor

Network	• 100 Mbps Ethernet High-speed Wireless	• 100 Mbps Ethernet	• 100 Mbps Ethernet	• 100 Mbps Ethernet

FIGURE 11-12   Sample Hardware and Software Specification

corresponds to an item on this list. In general, the list can include things like database servers, network servers, peripheral devices (e.g., printers, scanners), backup devices, storage components, and any other hardware component that is needed to support an application. At this time, you also should note the quantity of each item that will be needed.

Third, we must describe, in as much detail as possible, the minimum requirements for each piece of hardware. Typically, the project team must convey requirements like the amount of processing capacity, the amount of storage space, and any special features that should be included. Many organizations have standard lists of approved hardware and software that must be used, so in many cases, this step simply involves selecting items from the lists. Other times, however, the team is operating in new territory and is not constrained by the need to select from an approved list. This step becomes easier with experience; however, there are some hints that can help you describe hardware needs (see Figure 11-13). For example, consider the hardware standards within the organization or those recommended by vendors. Talk with experienced system developers or other companies with similar systems. Finally, think about the factors that affect hardware performance, such as the response-time expectations of the users, data volumes, software memory requirements, the number of users accessing the system, the number of external connections, and growth projections.

FIGURE 11-13
Factors in Hardware and Software Selection

Functions and Features   What specific functions and features are needed (e.g., size of monitor, software features)?

Performance   How fast does the hardware and software operate (e.g., processor, number of database writes per second)?

Legacy Databases and Systems   How well does the hardware and software interact with legacy systems (e.g., can it write to this database)?

Hardware and OS Strategy   What are the future migration plans (e.g., the goal is to have all of one vendor's equipment)?

Cost of Ownership   What are the costs beyond purchase (e.g., incremental license costs, annual maintenance, training costs, salary costs)?

Political Preferences   People are creatures of habit and are resistant to change, so changes should be minimized.

Vendor Performance   Some vendors have reputations or future prospects that are different from those of a specific hardware or software system they currently sell.

The last step to consider is to evaluate vendor proposals (see Chapter 7). The easiest way to do this is to create an *alternative matrix* (see Chapters 2 and 7). In this case, the evaluation criteria in the alternative matrix should include all architectural requirements, both optional and mandatory, and each criterion should be weighted. Some general criteria include CPU speed, bus speed, disk size, disk access time, cache size, cache speed, RAM size, RAM speed, data transfer rate, video RAM size and speed, and printer and display resolution. Of course, in today's connected world, the networking hardware and software would also need to be specified including routers, print servers, hubs, and switches. Mobile devices such as smartphones and tablets may be part of the physical architecture solution. Depending on the problem domain requirements, additional hardware and system software could be required, such as speech recognition and generation software and hardware, digitizing tablets, and possibly head-mounted displays, shutter glasses, force feedback pointing devices, and 3D printers. Each of these types of specialized devices have their own specialized evaluation criteria. In a nutshell, when creating a hardware and system software specification, most systems analysts find that they need help from IT and CS personnel.

Depending on the overall cost and size of the project, one thing that should be seriously considered is the use of a benchmark. A *benchmark* is essentially a sample of programs that would be expected to run on the new physical architecture. Even though benchmarks can be expensive to create, they tend to provide a more realistic picture of how the proposed physical architecture layer will perform.

When evaluating hardware, there is a set of errors that you should avoid.[11]

- Not only should you provide sample programs for the benchmarks, but you also need to provide actual data. Otherwise, the benchmark results could be misleading.
- You need to carefully review the mix of system software and hardware. For example, in many cases, Linux performs better on the same hardware when compared against Windows, but some applications might not be available under Linux. Consequently, there may be some tradeoffs that should be considered.
- When considering adding additional hardware, be sure to evaluate the additional hardware based on marginal utility, not actual utility.
- Do not specify the physical architecture before you understand the problem domain requirements. This might seem obvious, but when you consider the time it takes for a mainframe computer, a large number of servers, or a large number of client machines to be specified, ordered, and delivered, it can be tempting to specify the hardware and system software prematurely. This could lead to either under- or over-specification.
- Recognize the reality of *Parkinson's Law*. From an IT perspective, Parkinson's Law implies that regardless of the users' real needs, their imagined needs will always fill up whatever capacity the system has. Consequently, it is imperative that the physical architecture layer design be based on the current and expected future architecture of the problem domain layer.
- Do not limit choices to a single vendor. This is especially true when you consider commodity hardware, such as displays, desktops, and department-size servers.
- Given the rate of technological change that is taking place in IT, consider leading-edge ideas. For example, even though tablet computers have been around for a

[11] Alton R. Kindred, *Data Systems and Management: An Introduction to Systems Analysis and Design,* 2nd ed. (Englewood Cliffs, NJ: Prentice-Hall, 1980).

while, the iPad™ was not on most people's radar. Today, it is considered to be a game changer when considering client-based hardware. Consequently, you really must stay up to date when it comes to the design of the physical architecture layer.

YOUR
TURN

11-3 University Course Registration System

Develop a hardware and software specification for the university course registration system described in Your Turn 11-1.

YOUR
TURN

11-4 Create a Hardware and Software Specification

You have decided to purchase a computer, printer, and low-cost scanner to support your academic work. Create a hardware and software specification for these components that describes your hardware and software needs.

NONFUNCTIONAL REQUIREMENTS AND PHYSICAL ARCHITECTURE LAYER DESIGN

The design of the physical architecture layer specifies the overall architecture and the placement of software and hardware that will be used. Each of the architectures discussed before has its strengths and weaknesses. Most organizations are trying to move to client–server architectures for cost reasons, so in the event that there is no compelling reason to choose one architecture over another, cost usually suggests client–server.

Creating a physical architecture layer design begins with the nonfunctional requirements. The first step is to refine the nonfunctional requirements into more detailed requirements that are then used to help select the architecture to be used (server-based, client-based, or client–server) and what software components will be placed on each device. In a client–server architecture, one also has to decide whether to use a two-tier, three-tier, or n-tier architecture. Then the nonfunctional requirements and the architecture design are used to develop the hardware and software specification.

Four primary types of nonfunctional requirements can be important in designing the architecture: operational requirements, performance requirements, security requirements, and cultural/political requirements. We describe each in turn, and then explain how they can affect the physical architecture layer design.

Operational Requirements

Operational requirements specify the operating environment(s) in which the system must perform and how those might change over time. This usually refers to operating systems, system software, and information systems with which the system must interact, but on occasion, it also includes the physical environment if the environment is important to the application (e.g., it's located on a noisy factory floor, so no audible alerts can be heard). Figure 11-14 summarizes four key operational requirement areas and provides some examples of each.

Type of Requirement	Definition	Examples
Technical Environment Requirements	Special hardware, software, and network requirements imposed by business requirements	• The system will work over the web environment with Internet Explorer. • All office locations will have an always-on network connection to enable real-time database updates. • A version of the system will be provided for customers connecting over the Internet via a tablet or smartphone.
System Integration Requirements	The extent to which the system will operate with other systems	• The system must be able to import and export Excel spreadsheets. • The system will read and write to the main inventory database in the inventory system.
Portability Requirements	The extent to which the system will need to operate in other environments	• The system must be able to work with different operating systems (e.g., Linux, Mac OS, and Windows). • The system might need to operate with handheld devices such as Android and Apple iOS devices.
Maintainability Requirements	Expected business changes to which the system should be able to adapt	• The system will be able to support more than one manufacturing plant with six months' advance notice. • New versions of the system will be released every six months.

FIGURE 11-14 Operational Requirements

Technical Environment Requirements *Technical environment requirements* specify the type of hardware and software system on which the system will work. These requirements usually focus on the operating system software (e.g., Windows, Linux, Mac OS), database system software (e.g., Oracle), and other system software (e.g., Firefox). In today's distributed world, issues related to mobile computing (see Chapter 10), cloud computing (see the earlier section in this chapter), and Green IT (see the earlier section in this chapter) are very relevant. Consequently, it also includes all of the different types of hardware from mainframe computers all the way to smartphones. Depending on the

CONCEPTS IN ACTION

11-E Power Outage Costs a Million Dollars

*L*ithonia Lighting, located just outside of Atlanta, is the world's largest manufacturer of light fixtures, with more than $1 billion in annual sales. One afternoon, the power transformer at its corporate headquarters exploded, leaving the entire office complex, including the corporate data center, without power. The data center's backup power system immediately took over and kept critical parts of the data center operational. However, it was insufficient to power all systems, so the system supporting sales for all of Lithonia Lighting's North American agents, dealers, and distributors had to be turned off.

The transformer was quickly replaced and power was restored. However, the three-hour shutdown of the sales system cost $1 million in potential sales lost. Unfortunately, it is not uncommon for the cost of a disruption to be hundreds or thousands of times the cost of the failed components.

Question

What would you recommend to avoid similar losses in the future?

applications being deployed over the physical architecture, specialized hardware could be required, such as 3D displays, 3D printing, 3D sound systems, and tablets with accelerometers. With today's technology, the possible combinations of hardware that can be brought to bear to solve a problem are nearly endless. Consequently, this is one area where additional expertise might be required.

CONCEPTS

IN ACTION

11-F Citywide Broadband

*S*eattle officials might finally be ready to start crunching numbers and nailing down specifics of a long-mulled, but slow moving, proposal to build a citywide broadband network. But they might yet need to build a case before the city council to justify such a network. The idea has been studied by the city council, a task force, and the mayor's information technology office since 2004. Having estimated last year that it would cost $500 million to build and connect all of Seattle to a fiber broadband network, the city is now looking to invite private companies to do that work—perhaps with taxpayer help.

The mayor is asking the city council to free up $180,000 he says his information technology office will use to invite companies to bid on the sweeping and still-undefined job.

Questions

1. How much money (if any) will taxpayers contribute in subsidies or access to public facilities?
2. Should the city start with a small-scale pilot project, as it did with Wi-Fi, before deciding about a citywide wireless network?
3. Is such a network even a feasible, cost-effective idea?

CONCEPTS

IN ACTION

11-G Complex Electrical Systems

*S*ystems integration across platforms and companies grows more complex with time. In a case study from Florida in 2008, an electrical company's real-time system detected a minor problem in the power grid and shut down the entire system—plunging more than 2 million people into the dark. The system experts place the blame on a substation software system that detected the minor fluctuation but had the ability to immediately shut down the entire system. Although there may be times where such a rapid response is vital (such as the nuclear disasters in Chernobyl, Ukraine, and on Three Mile Island), this was a case where such response was not warranted.

Questions

1. Because software controls substation operations, how might a systems analyst approach this as a systems project?
2. Are there special considerations that a systems analyst needs to think about when dealing with real-time systems?

System Integration Requirements   *System integration requirements* are those that require the system to operate with other information systems, either inside or outside the company. These typically specify interfaces through which data will be exchanged with other systems.

Portability Requirements   Information systems never remain constant. Business needs change and operating technologies change, so the information systems that support them and run on them must change, too. *Portability requirements* define how the technical operating environments might change over time and how the system must respond (e.g., the system currently runs on Windows Vista, whereas in the future the system might have to be deployed on Linux). Portability requirements also refer to potential changes in business requirements that drive technical environment changes. For example, users might want to access a website from their cell phones.

Maintainability Requirements *Maintainability requirements* specify the business requirement changes that can be anticipated. Not all changes are predictable, but some are. For example, suppose a small company has only one manufacturing plant but is anticipating the construction of a second plant in the next five years. All information systems must be written to make it easy to track each plant separately, whether for personnel, budgeting, or inventory systems. The maintainability requirements attempt to anticipate future requirements so that the systems designed today will be easy to maintain if and when those future requirements appear. Maintainability requirements can also define the update cycle for the system, such as the frequency with which new versions will be released.

Performance Requirements

Performance requirements focus on performance issues, such as response time, capacity, and reliability. Figure 11-15 summarizes three key performance requirement areas and provides some examples.

Speed Requirements *Speed requirements* are exactly what they say: How fast should the system operate? First is the *response time* of the system: how long it takes the system to respond to a user request. Although everyone would prefer low response times, with the system responding immediately to each user request, this is not practical. We could design such a system, but it would be expensive. Most users understand that certain parts of a system will respond quickly, whereas others are slower. Actions that are performed locally on the user's computer must be almost immediate (e.g., typing, dragging and dropping), whereas others that require communicating across a network can have higher response times (e.g., a web request). In general, response times less than seven seconds are considered acceptable when they require communication over a network.

Type of Requirement	Definition	Examples
Speed Requirements	The time within which the system must perform its functions	• Response time must be less than 7 seconds for any transaction over the network. • The inventory database must be updated in real time. • Orders will be transmitted to the factory floor every 30 minutes.
Capacity Requirements	The total and peak number of users and the volume of data expected	• There will be a maximum of 100–200 simultaneous users at peak use times. • A typical transaction will require the transmission of 10K of data. • The system will store data on approximately 5,000 customers for a total of about 2 MB of data.
Availability and Reliability Requirements	The extent to which the system will be available to the users and the permissible failure rate due to errors	• Scheduled maintenance shall not exceed one 6-hour period each month. • The system shall have 99% uptime performance.

FIGURE 11-15 Performance Requirements

The second aspect of speed requirements is how long it takes transactions in one part of the system to be reflected in other parts. For example, how soon after an order is placed will the items it contained be shown as no longer available for sale to someone else? If the inventory is not updated immediately, then someone else could place an order for the same item, only to find out later it is out of stock. Or how soon after an order is placed is it sent to the warehouse to be picked from inventory and shipped? In this case, some time delay might have little impact.

Capacity Requirements *Capacity requirements* attempt to predict how many users the system will have to support, both in total and simultaneously. Capacity requirements are important in understanding the size of the databases, the processing power needed, and so on. The most important requirement is usually the peak number of simultaneous users, because this has a direct impact on the processing power of the computer(s) needed to support the system.

It is often easier to predict the number of users for internal systems designed to support an organization's own employees than it is to predict the number of users for customer-facing systems, especially those on the web. How *does* Weather.com estimate the peak number of users who will simultaneously seek weather information? This is as much an art as a science, so often the team provides a range of estimates, with wider ranges used to signal a less accurate estimate.

CONCEPTS IN ACTION 11-H The Importance of Capacity Planning

At the end of 1997, Oxford Health Plans posted a $120 million loss to its books. The company's unexpected growth was its undoing because the system, which was originally planned to support the company's 217,000 members, had to meet the needs of a membership that exceeded 1.5 million.

System users found that processing a new member sign-up took 15 minutes instead of the proposed 6 seconds. Also, the computer problems left Oxford unable to send out bills to many of its customer accounts and rendered it unable to track payments to hundreds of doctors and hospitals. In less than a year, uncollected payments from customers tripled to more than $400 million, and the

payments owed to caregivers amounted to more than $650 million. Mistakes in infrastructure planning can cost far more than the cost of hardware, software, and network equipment alone.

Source: Ron Winslow and George Anders, "Management: How New Technology was Oxford's Nemesis," *Wall Street Journal* (December 11, 1997, page A.1).

Question

If you had been in charge of the Oxford project, what things would you have considered when planning the system capacity?

Availability and Reliability Requirements *Availability and reliability requirements* focus on the extent to which users can assume that the system will be available for them to use. Although some systems are intended to be used only during the forty-hour workweek, some systems are designed to be used by people around the world. For such systems, project team members need to consider how the application can be operated, supported, and maintained 24/7 (i.e., 24 hours a day, 7 days a week). This 24/7 requirement means that users might need help or have questions at any time, and a support desk that is available eight hours a day will not be sufficient support. It is also important to consider what reliability is needed in the system. A system that requires high reliability (e.g., a medical device or telephone switch) needs far greater planning and testing than one that does not have such high-reliability needs (e.g., personnel system, web catalog).

It is more difficult to predict the peaks and valleys in use of the system when the system has a global audience. Typically, applications are backed up on weekends or late evenings when users are no longer accessing the system. Such maintenance activities need to be rethought with global initiatives. The development of web interfaces, in particular, has escalated the need for 24/7 support; by default, the web can be accessed by anyone at any time. For example, the developers of a web application for U.S. outdoor gear and clothing retailer, Orvis, were surprised when the first order after going live came from Japan.

Security Requirements[12]

Security is the ability to protect the information system from disruption and data loss, whether caused by an intentional act (e.g., a hacker, a terrorist attack) or a random event (e.g., disk failure, tornado). Security is primarily the responsibility of the operations group—the staff responsible for installing and operating security controls, such as firewalls, intrusion-detection systems, and routine backup and recovery operations. Nonetheless, developers of new systems must ensure that the system's *security requirements* produce reasonable precautions to prevent problems; system developers are responsible for ensuring security within the information systems themselves.

Security is an ever-increasing problem in today's Internet-enabled world. Historically, the greatest security threat has come from inside the organization itself. Ever since the early 1980s when the FBI first began keeping computer crime statistics and security firms began conducting surveys of computer crime, organizational employees have perpetrated the vast majority of computer crimes. For years, 80 percent of unauthorized break-ins, thefts, and sabotage have been committed by insiders, leaving only 20 percent to hackers external to the organizations.

In 2001, that changed. Depending on what survey you read, the percentage of incidents attributed to external hackers in 2001 increased to 50 to 70 percent of all incidents, meaning that the greatest risk facing organizations is now from the outside. Although some of this shift may be due to better internal security and better communications with employees to prevent security problems, much of it is simply due to an increase in activity by external hackers. With cloud computing, security has become even more important.

Developing security requirements usually starts with some assessment of the value of the system and its data. This helps pinpoint extremely important systems so that the operations staff is aware of the risks. Security within systems usually focuses on specifying who can access what data, identifying the need for encryption and authentication, and ensuring the application prevents the spread of viruses (see Figure 11-16).

System Value   The most important computer asset in any organization is not the equipment; it is the organization's data. For example, suppose someone destroyed a mainframe computer worth $10 million. The mainframe could be replaced, simply by buying a new one. It would be expensive, but the problem would be solved in a few weeks. Now suppose someone destroyed all the student records at your university so that no one knew what courses anyone had taken, or their grades. The cost would far exceed the cost of replacing a $10 million computer. The lawsuits alone would easily exceed $10 million, and the cost of

---

[12] For more information, see Brett C. Tjaden, *Fundamentals of Secure Computer Systems* (Wilsonville, OR: Franklin, Beedle, and Associates, 2004); for security controls associated with the Sarbanes–Oxley act, see Dennis C. Brewer, *Security Controls for Sarbanes–Oxley Section 404 IT Compliance: Authorization, Authentication, and Access* (Indianapolis: Wiley, 2006).

Type of Requirement	Definition	Examples
System Value Estimates	Estimated business value of the system and its data	• The system is not mission critical but a system outage is estimated to cost $50,000 per hour in lost revenue. • A complete loss of all system data is estimated to cost $20 million.
Access Control Requirements	Limitations on who can access what data	• Only department managers will be able to change inventory items within their own department. • Telephone operators will be able to read and create items in the customer file but cannot change or delete items.
Encryption and Authentication Requirements	Defines what data will be encrypted Where and whether authentication will be needed for user access	• Data will be encrypted from the user's computer to the website to provide secure ordering. • Users logging in from outside the office will be required to authenticate.
Virus Control Requirements	Requirements to control the spread of viruses	• All uploaded files will be checked for viruses before being saved in the system.

FIGURE 11-16   Security Requirements

staff to find paper records and reenter the data from them would be enormous and certainly would take more than a few weeks.

In some cases, the information system itself has value that far exceeds the cost of the equipment as well. For example, for an Internet bank that has no brick-and-mortar branches, the website is a *mission-critical system*. If the website crashes, the bank cannot conduct business with its customers. A mission-critical application is an information system that is literally critical to the survival of the organization. It is an application that cannot be permitted to fail, and if it does fail, the network staff drops everything else to fix it. Mission-critical applications are usually clearly identified so their importance is not overlooked.

Even temporary disruptions in service can have significant costs. The cost of disruptions to a company's primary website or the LANs and backbones that support telephone sales operations are often measured in the millions of dollars. Amazon.com, for example, has revenues of more than $10 million per hour, so if their website were unavailable for an hour or even part of an hour, they would lose millions of dollars in revenue. Companies that do less e-business or do telephone sales have lower costs, but recent surveys suggest losses of $100,000 to $200,000 per hour are not uncommon for major customer-facing information systems.

Access Control Requirements   Some of the data stored in the system need to be kept confidential; some data need special controls on who is allowed to change or delete it. Personnel records, for example, should be able to be read only by the personnel department and the employee's supervisor; changes should be permitted to be made only by the personnel department. *Access control requirements* state who can access what data and what type of access is permitted: whether the individual can create, read, update, and/or delete the data. The requirements reduce the chance that an authorized user of the system can perform unauthorized actions. One approach to address these requirements is through the use of *access control lists,* which can be implemented via the operating system or database management system.

Encryption and Authentication Requirements   One of the best ways to prevent unauthorized access to data is *encryption,* which is a means of disguising information by the use of mathematical algorithms (or formulas). Encryption can be used to protect data stored in databases or data that are in transit over a network from a database to a computer. There are two fundamentally different types of encryption: symmetric and asymmetric. A *symmetric encryption algorithm* [such as Data Encryption Standard (DES) or Advanced Encryption Standard (AES)] is one in which the key used to encrypt a message is the *same* as the one used to decrypt it, which means that it is essential to protect the key and that a separate key must be used for each person or organization with whom the system shares information (or else everyone can read all the data).

An *asymmetric encryption algorithm* (such as *public key encryption*) is one in which the key used to encrypt data (called the *public key*) is different from the one used to decrypt it (called the *private key*). Even if everyone knows the public key, once the data are encrypted, they cannot be decrypted without the private key. Public key encryption greatly reduces the key-management problem. Each user has its public key that is used to encrypt messages sent to it. These public keys are widely publicized (e.g., listed in a telephone book style directory)—that's why they're called public keys. The private key, in contrast, is kept secret (which is why it's called private).

Public key encryption also permits *authentication* (or digital signatures). When one user sends a message to another, it is difficult to legally prove who actually sent the message. Legal proof is important in many communications, such as bank transfers and buy/sell orders in currency and stock trading, which normally require legal signatures. Public key encryption algorithms are *invertible*, meaning that text encrypted with either key can be decrypted by the other. Normally, we encrypt with the public key and decrypt with the private key. However, it is possible to do the reverse: encrypt with the private key and decrypt with the public key. Because the private key is secret, only the real user can use it to encrypt a message. Thus, a digital signature or authentication sequence is used as a legal signature on many financial transactions. This signature is usually the name of the signing party plus other unique information from the message (e.g., date, time, or dollar amount). This signature and the other information are encrypted by the sender using the private key. The receiver uses the sender's public key to decrypt the signature block and compares the result to the name and other key contents in the rest of the message to ensure a match.

The only problem with this approach lies in ensuring that the person or organization that sent the document with the correct private key is the actual person or organization. Anyone can post a public key on the Internet, so there is no way of knowing for sure who actually used it. For example, it would be possible for someone other than Organization A in this example to claim to be Organization A when, in fact, he or she is an imposter.

This is where the Internet's public key infrastructure (PKI) becomes important.[13] The PKI is a set of hardware, software, organizations, and policies designed to make public key encryption work on the Internet. PKI begins with a *certificate authority (CA)*, which is a trusted organization that can vouch for the authenticity of the person or organization using authentication (e.g., VeriSign). A person wanting to use a CA registers with the CA and must provide some proof of identify. There are several levels of certification, ranging from a simple confirmation from a valid e-mail address to a complete police-style background check with an in-person interview. The CA issues a digital certificate that is the requestor's public key, encrypted using the CA's private key as proof of identify. This certificate is then attached to the user's e-mail or web transactions in addition to the authentication information. The receiver then verifies the certificate by decrypting it with the CA's public key and must also contact the CA to ensure that the user's certificate has not been revoked by the CA.

---

[13] For more on the PKI, see www.ietf.org/html.charters/pkix-charter.html.

The encryption and authentication requirements state what encryption and authentication requirements are needed for what data. For example, will sensitive data such as customer credit-card numbers be stored in the database in encrypted form, or will encryption be used to take orders over the Internet from the company's website? Will users be required to use a digital certificate in addition to a standard password?

CONCEPTS

IN ACTION

11-I Securing the Environment

Quinnipiac University is a four-year university in Hamden, Connecticut, with about 7,400 students. Because the university has residence halls, the IT staff has to support academic functions and also has to be an Internet service provider (ISP) for students. The IT staff can shape much of the academic usage of the Internet, but students living in residence halls can cause havoc. Students (and faculty) inadvertently open the campus to all kind of attacks, such as viruses, malware, worms, spybots, and other intrusions through accessing various websites. A particularly trying time is after the semester break in late January, when students return to campus and plug in laptops that have been corrupted with other viruses from home networks. These viruses try to infect the entire campus. Quinnipiac University installed an intrusion prevention system (IPS) by Tipping Point Technology in August 2006. On a daily basis, this IPS detects and drops thousands of destructive messages and packets. But the real proof was

in 2007 when students returned to campus from semester break. In the previous year, the viruses and spyware virtually took the campus network down for three days. But in January 2007, there were no outages, and the network remained strong and functioning at full speed. Brian Kelly, information security officer at Quinnipiac University said, "Without the IPS solution, this campus would have struggled under this barrage of malicious packets and may have shut down. With the IPS system, we were able to function at full speed without any problems."

Questions

1. What might be some of the tangible and intangible costs of having the Internet down for three days on a busy college campus?
2. What benefits would an IPS mean to a campus and to end users (e.g., faculty, staff, and students)?

Virus Control Requirements *Virus control requirements* address the single most common security problem: *viruses*. Studies have shown that almost 90 percent of organizations suffer a virus infection each year. Viruses cause unwanted events—some harmless (such as nuisance messages), some serious (such as the destruction of data). Any time a system permits data to be imported or uploaded from a user's computer, there is the potential for a virus infection. Many systems require that all information systems that permit the import or upload of user files to check those files for viruses before they are stored in the system.

Cultural and Political Requirements

Cultural and political requirements are those specific to the countries in which the system will be used. In today's global business environment, organizations are expanding their systems to reach users around the world. Although this can make great business sense, its impact on application development should not be underestimated. Yet another important part of the design of the system's physical architecture is understanding the global cultural and political requirements for the system (see Chapter 10 and Figure 11-17).

Customization Requirements For global applications, the project team needs to give some thought to *customization requirements:* How much of the application will be controlled by a central group, and how much of the application will be managed locally? For example, some companies allow subsidiaries in some countries to customize the application by omitting or adding certain features. This decision has trade-offs between flexibility

Type of Requirement	Definition	Examples
Customization Requirements	Specification of what aspects of the system can be changed by local users	• Country managers will be able to define new fields in the product database to capture country-specific information. • Country managers will be able to change the format of the telephone number field in the customer database.
Legal Requirements	The laws and regulations that impose requirements on the system	• Personal information about customers cannot be transferred out of European Union countries into the United States. • It is against U.S. federal law to divulge information on who rented what videotape, so access to a customer's rental history is permitted only to regional managers.

FIGURE 11-17 Cultural and Political Requirements

and control because customization often makes it more difficult for the project team to create and maintain the application. It also means that training can differ among different parts of the organization, and customization can create problems when staff moves from one location to another.

Owing to the use of different languages, in some cases, specialized hardware that has been customized to the local culture is required. For example, the use of Kanji keyboards makes a lot of sense for Japanese users. Basically, having specialized keyboards makes sense for any language that does not use the typical Roman alphabet, e.g., Arabic, Hebrew, Greek, or Russian. There are also emulators available for many different languages. Depending on the users being served, assistive devices could be required, such as Braille devices, eye-tracking devices, head pointers, head/mouth stick keyboards, or adaptive ability switches. Depending on the cultural and political requirements, many different hardware platforms might need to be considered.

Legal Requirements   *Legal requirements* are requirements imposed by laws and government regulations. System developers sometimes forget to think about legal regulations; unfortunately, forgetting comes at some risk because ignorance of the law is no defense. For example, in 1997 a French court convicted the Georgia Institute of Technology of violating French language law. Georgia Tech operated a small campus in France that offered summer programs for American students. The information on the campus web server was primarily in English because classes are conducted in English, which violated the law requiring French to be the predominant language on all Internet servers in France. By formally considering legal regulations, you are less likely to overlook them.

Synopsis

In many cases, the technical environment requirements as driven by the business requirements can simply define the physical architecture layer. In this case, the choice is simple: Business requirements dominate other considerations. For example, the business requirements might specify that the system needs to work over the web using the customer's web browser. In this case, the architecture probably should be a thin client–server. Such business requirements are most likely in systems designed to support external customers. Internal systems can also impose business requirements, but usually, they are not as restrictive.

Requirements	Server-Based	Client-Based	Thin Client–Server	Thick Client–Server
Operational Requirements				
System Integration Requirements	✓		✓	✓
Portability Requirements			✓	
Maintainability Requirements	✓		✓	
Performance Requirements				
Speed Requirements			✓	✓
Capacity Requirements			✓	✓
Availability/Reliability Requirements	✓		✓	✓
Security Requirements				
High System Value	✓		✓	
Access Control Requirements	✓			
Encryption/Authentication Requirements			✓	✓
Virus Control Requirements	✓			
Cultural/Political Requirements				
Customization Requirements			✓	
Legal Requirements	✓		✓	✓

FIGURE 11-18
Nonfunctional Requirements and Their Implications for Architecture Change

In the event that the technical environment requirements do not stipulate a specific architecture, then the other nonfunctional requirements become important. Even in cases when the business requirements drive the architecture, it is still important to work through and refine the remaining nonfunctional requirements because they are important in later stages of design and implementation. Figure 11-18 summarizes the relationship between requirements and recommended architectures.

Operational Requirements System integration requirements can lead to one architecture being chosen over another, depending on the architecture and design of the system(s) with which the system needs to integrate. For example, if the system must integrate with a desktop system (e.g., Excel), this might suggest a thin or thick client–server architecture, whereas if it must integrate with a server-based system, a server-based architecture may be indicated. Systems that have extensive portability requirements tend to be best suited for a thin client–server architecture because it is simpler to write for web-based standards (e.g., HTML, XML) that extend the reach of the system to other platforms, rather than trying to write and rewrite extensive presentation logic for different platforms in the server-based, client-based, or thick client–server architectures. Systems with extensive maintainability requirements might not be well suited to client-based or thick client–server architectures because of the need to reinstall software on the desktops.

Performance Requirements   Generally speaking, information systems that have high performance requirements are best suited to client–server architectures. Client–server architectures are more scalable, which mean they respond better to changing capacity needs and thus enable the organization to better tune the hardware to the speed requirements of the system. Client–server architectures that have multiple servers in each tier should be more reliable and have greater availability, because if any one server crashes, requests are simply passed to other servers, and users might not even notice (although response time could be worse). In practice, however, reliability and availability depend greatly on the hardware and operating system, and Windows-based computers tend to have lower reliability and availability than Linux or mainframe computers.

Security Requirements   Generally speaking, because all software is in one location and because mainframe operating systems are more secure than microcomputer operating systems, server-based architectures tend to be more secure. For this reason, high-value systems are more likely to be found on mainframe computers, even if the mainframe is used as a server in client–server architectures. In today's Internet-dominated world, authentication and encryption tools for Internet-based client–server architectures are more advanced than those for mainframe server–based architectures. Viruses are potential problems in all architectures because they easily spread on desktop computers. If a server-based system can reduce the functions needed on desktop systems, then they may be more secure.

Cultural and Political Requirements   As cultural and political requirements become more important, the ability to separate the presentation logic from the application logic and the data becomes important. Such separation makes it easier to develop the presentation logic in different languages while keeping the application logic and data the same. It also makes it easier to customize the presentation logic for different users and to change it to better meet cultural norms. To the extent that the presentation logic provides access to the application and data, it also makes it easier to implement different versions that enable or disable different features required by laws and regulations in different countries. This separation is the easiest in thin client–server architectures, so systems with many cultural and political requirements often use thin client–server architectures. As with system integration requirements, the impact of legal requirements depends on the specific nature of the requirements, but in general, client-based systems tend to be less flexible.

YOUR TURN   **11-5 Global e-Learning System**

Many multinational organizations provide global web-based e-learning courses for their employees. First, develop a set of nonfunctional requirements for such a system. Consider the operational requirements, performance requirements, security requirements, and cultural and political requirements. Then create an architecture design to satisfy these requirements.

YOUR	11-6 University Course Registration System
TURN	

*T*hink about the course registration system in your university. First, develop a set of nonfunctional requirements if the system were to be developed today. Consider the operational requirements, performance requirements, security requirements, and cultural and political requirements. Then create an architecture design to satisfy these requirements.

YOUR	11-7 Global e-Learning System
TURN	

*D*evelop a hardware and software specification for the global e-learning system described in Your Turn 11-5.

APPLYING THE CONCEPTS AT CD SELECTIONS

As with the previous two chapters, in this installment of the CD Selections case, we see that Alec has spun off part of his team to focus on designing the physical architecture layer. However, given the dependence among the human–computer interaction, data management, and physical architecture layers, this group finds that they must be in relatively constant contact with the other groups. Otherwise, deploying the problem domain, human–computer interaction, and data management layers over the architecture could prove to be difficult. Consequently, Alec has decided to focus his coordination efforts among the different layer groups by heading up the physical architecture layer's group. He saw this as a way to better understand the implications of deploying the system over multiple, and possibly incompatible, platforms.

SUMMARY

An important component of design is the design of the physical architecture layer, which includes the hardware, software, and communications infrastructure for the new system, and the way the information system will be distributed over the architecture. The physical architecture layer design is described in a deliverable that contains the network model and the hardware and software specification.

Elements of the Physical Architecture Layer

All software systems can be divided into four basic functions: data storage, data access logic, application logic, and presentation logic. There are three fundamental computing architectures that place these functions on different computers. In server-based architectures, the server performs all the functions. In client-based architectures, the client computers are responsible for presentation logic, application logic, and data access logic, with data stored on a file server. In client–server architectures, the client is responsible for the presentation logic and the server is responsible for the data access logic and data storage. In thin client–server architectures, the server performs the application logic, whereas in thick client–server architectures, the application logic is shared between the servers and clients. In a two-tiered client–server architecture, there are two groups of computers: one client and a set of servers. In a three-tiered client–server architecture, there are three groups of computers: a client, a set of application servers, and a set of database servers.

Each of the computing architectures has its strengths and weaknesses, and no architecture is inherently better than the others. The choice that the project team makes should be based on several criteria, including cost of development, ease of development, need for GUI applications, network capacity, central control and security, and scalability. The project team should also take into consideration the existing architecture and special software requirements of the project.

Cloud Computing

Cloud computing is the newest approach to support distributed processing. In many ways, it also could become the greatest enabler of IT outsourcing. Cloud computing is based on three established technologies: virtualization, service-oriented architectures, and grid computing. There are three different classifications of clouds: private, public, and hybrid. Currently, the cloud is being touted as a host for the firm's IT infrastructure, IT platform, and software. However, there are still a number of obstacles for cloud vendors to overcome before the cloud can become a mainstream solution. These obstacles include issues related to the actual level of performance in the cloud, the level of dependence that a customer firm will have on a cloud vendor, and the real levels of security available in the cloud.

Green IT

Owing to the overall impact that IT has had on the environment, there has been a renewed interest in so-called Green IT. The motivation behind the Green IT movement is to minimize the harm to the environment that IT causes. There are three general questions that Green IT tries to address. First, how should e-waste be discarded? Second, given the power consumption of IT, how can we reduce the power consumption and the carbon footprint of IT? Third, how can IT enable the paperless office? IT firms, such as Google, HP, and Microsoft, are providing leadership in this area.

Infrastructure Design

Deployment diagrams are used to portray the design of the physical architecture layer. The diagrams are composed of nodes, artifacts, and communication links. Also, the diagram can be extended with graphical icons that represent different types of nodes.

The network model is a diagram that shows the technical components of the information system (e.g., servers, personal computers, networks) and their geographic locations throughout the organization. The components of the network model are the various clients (e.g., personal computers, kiosks), servers (e.g., database, network, communications, printer), network equipment (e.g., wire, dial-up connections, satellite links), and external systems or networks (e.g., Internet service provider) that support the application. Creating a network model is a top-down process whereby a high-level diagram is first created to show the geographic sites, or locations, that house the various components of the future system. Next, a low-level diagram is created to describe each location in detail, and it shows the system's hardware components and how they are attached to one another.

Hardware and Software Specification

The hardware and software specification is a document that describes what hardware and software are needed to support the application. To create a specification document, the hardware that is needed to support the future system is listed and then described in as much detail as possible. Next, the software to run on each hardware component is written down, along with any additional associated costs, such as technical training, maintenance, extended warranties, and licensing agreements. Although the project team may suggest specific products or vendors, ultimately, the hardware and software specification is turned over to the people who are in charge of procurement.

Nonfunctional Requirements and Physical Architecture Layer Design

Creating an architecture design begins with the nonfunctional requirements. Operational requirements specify the operating environment(s) in which the system must perform and how those might change over time (i.e., technical environment, system integration, portability, and maintainability). Performance requirements focus on performance issues such as system speed, capacity, and availability and reliability. Security requirements attempt to protect the information system from disruption and data loss (e.g., system value, access control, encryption and authentication, and virus control). Cultural and political requirements are those that are specific to the specific countries in which the system will be used (e.g., customization and legal).

KEY TERMS

QUESTIONS

1. What are the four basic functions of any information system?
2. What are the three primary hardware components of any physical architecture?
3. Name two examples of a server.
4. Compare and contrast server-based architectures, client-based architectures, and client–server-based architectures.
5. What is the biggest problem with server-based computing?
6. What is the biggest problem with client-based computing?
7. Describe the major benefits and limitations of thin client–server architectures.
8. Describe the major benefits and limitations of thick client–server architectures.
9. Describe the differences among two-tiered, three-tiered, and *n*-tiered architectures.
10. Define *scalable*. Why is this term important to system developers?
11. What six criteria are helpful to use when comparing the appropriateness of computing alternatives?
12. Why should the project team consider the existing physical architecture in the organization when designing the physical architecture layer of the new system?
13. Name the three different types of clouds. How do they differ from one another?
14. What is meant by a service-oriented architecture?
15. Define virtualization. How does it relate to the cloud?
16. What are the differences among IaaS, PaaS, and SaaS?
17. What are the obstacles for provisioning the physical architecture layer with cloud technologies?
18. What, if any, are the issues related to security in the cloud?
19. What are SOX and HIPAA and how could they affect a firm's decision to adopt cloud technology?
20. How do tablets, such as the iPad™, enable the paperless office?
21. What additional hardware- and software-associated costs might need to be included on the hardware and software specification?
22. Who is ultimately in charge of acquiring hardware and software for a project?
23. What is a benchmark and why is it important?
24. Why is Parkinson's Law relevant to the design of the physical architecture layer?
25. What do you think are three common mistakes that novice analysts make in architecture design, and hardware and software specification?
26. Describe the major nonfunctional requirements and how they influence physical architecture layer design.
27. Why is it useful to define the nonfunctional requirements in more detail, even if the technical environment requirements dictate a specific architecture?
28. What does the network model communicate to the project team?
29. What are the differences between the top-level network model and the low-level network model?
30. Are some nonfunctional requirements more important than others in influencing the architecture design and hardware and software specification?
31. What do you think are the most important security issues for a system?

EXERCISES

A. Using the web (or past issues of computer industry magazines, such as *Computerworld*), locate a system that runs in a server-based environment. Based on your reading, why do you think the company chose that computing environment?

B. Using the web (or past issues of computer industry magazines, such as *Computerworld*), locate a system that runs in a client–server environment. Based on your reading, why do you think the company chose that computing environment?

C. Using the web, locate examples of a mainframe component, a minicomputer component, and a microcomputer component. Compare the components in terms of price, speed, available memory, and disk storage. Did you find large differences in prices when the performances of the components are considered?

D. You have been selected to find the best client–server architecture for a web-based order entry system that is being developed for L.L. Bean. Write a short memo that describes to the project manager your reason for selecting an *n*-tiered architecture over a two-tiered architecture. In the memo, give some idea as to what different components of the architecture you would include.

E. Think about the system that your university currently uses for career services, and suppose that you are in charge of replacing the system with a new one. Describe how you would decide on the computing architecture for the new system using the criteria presented in this chapter. What information will you need to find out before you can make an educated comparison of the alternatives?

F. Using the web, find information on the effects that e-waste and backyard recycling has on developing countries. Based on what you find, what Green IT policies would you suggest a firm put in place to minimize the negative effects of e-waste?

G. Using the web, find examples of company's pursuing a Green IT strategy. Describe what they are doing.

H. Energy Star is a joint program between the U.S. Department of Energy and the Environmental Protection Agency. What are the requirements for various IT devices to be certified as being Energy Star compliant?

I. Using the web, find examples of firms using the cloud as a basis for the physical architecture layer. Describe exactly what they are doing.

J. Locate a consumer products company on the web and read its company description (so that you get a good understanding of the geographic locations of the company). Pretend that the company is about to create a new application to support retail sales over the web. Create a high-level network model that depicts the locations that would include components that support this application.

K. Create a low-level network diagram for the building that houses the computer labs at your university. Choose an application (e.g., course registration, student admissions) and include only the components that are relevant to that application.

L. An energy company with headquarters in Dallas, Texas, is thinking about developing a system to track the efficiency of its oil refineries in North America.

Each week, the ten refineries—as far as Valdez, Alaska, and as close as San Antonio, Texas—will upload performance data via satellite to the corporate mainframe in Dallas. Production managers at each site will use a personal computer to connect to an Internet service provider and access reports via the web. Create a high-level network model that depicts the locations that have components supporting this system.

M. Suppose that your mother is a real estate agent, and she has decided to automate her daily tasks using a laptop computer. Consider her potential hardware and software needs, and create a hardware and software specification that describes them. The specification should be developed to help your mother buy her hardware and software on her own.

N. Suppose that the admissions office in your university has a web-based application so that students can apply for admission online. Recently, there has been a push to admit more international students into the university. What do you recommend that the application include to ensure that it supports this global requirement?

O. Based on the A Real Estate Inc. problem in Chapter 4 (exercises I, J, and K), Chapter 5 (exercises P and Q), Chapter 6 (exercise D), Chapter 7 (exercise A), Chapter 8 (exercise A), Chapter 9 (exercise L), and Chapter 10 (exercises I and J), suggest a physical architecture design and portray it with a deployment diagram.

P. Based on the A Video Store problem in Chapter 4 (exercises L, M, and N), Chapter 5 (exercises R and S), Chapter 6 (exercise E), Chapter 7 (exercise B), Chapter 8 (exercise B), Chapter 9 (exercise M), and Chapter 10 (exercises K and L), suggest a physical architecture design and portray it with a deployment diagram.

Q. Based on the gym membership problem in Chapter 4 (exercises O, P, and Q), Chapter 5 (exercises T and U), Chapter 6 (exercise F), Chapter 7 (exercise C), Chapter 8 (exercise C), Chapter 9 (exercise N), and Chapter 10 (exercises M and N), suggest a physical architecture design and portray it with a deployment diagram.

R. Based on the Picnics R Us exercises in Chapter 4 (exercises R, S, and T), Chapter 5 (exercises V and W), Chapter 6 (exercise G), Chapter 7 (exercise D), Chapter 8 (exercise D), Chapter 9 (exercise O), and Chapter 10 (exercises O and P), suggest a physical architecture design and portray it with a deployment diagram.

S. Based on the Of-the-Month-Club problem in Chapter 4 (exercises U, V, and W), Chapter 5 (exercises X and Y), Chapter 6 (exercise H), Chapter 7 (exercise E), Chapter 8 (exercise E), Chapter 9 (exercise N), and Chapter 10 (exercises Q and R), suggest a physical architecture design and portray it with a deployment diagram.

MINICASES

1. The New York City municipality decided to adopt an ambitious project to automate the monitoring, management, and user services of the city's Public Transportation System (PTS). The project, scheduled to be implemented by the year 2030, is called the New York City Global Public Transportation System 2030 (NYC-GPTS2030) and is intended to be comprehensive by including all forms of public transportation and by automating all activities related to the functioning of the Public Transportation System in the city of New York.

   The NYC-GPTS2030 aims to implement a variety of innovative features in many activity areas related to public transportation. Among other things, in terms of user services, it is expected that people will be able to check arrivals/departures/delays of all PTS vehicles (buses, subway trains, etc.) at any public transportation station in the city or use a mobile device to check out transportation details for any point in the city at any time. In addition, mobile devices will be used for purchasing electronic tickets on the spot at any time, thus eliminating the need for cashiers.

   In terms of traffic management, the system will be able to detect any incidents or traffic delay and their locations live, and make recommendations for countermeasures, such as supplementing subway trains in the affected sections.

   The NYC-GPTS2030 system will also benefit the PTS employees as well by the introduction of an automated time-tracking system, which allows flexible working schedules for most categories of PTS employees.

   The designers of the NYC-GPTS2030 system realize that the potential for innovation and the ramifications of how this system will change public transportation in the city of New York are hard to fully predict at this point. Accordingly, they are looking for a complete set of specifications of the system requirements for NYC-GPTS2030. Drawing upon the criteria and categories from the text, highlight the main issues and the most important concerns the design team should focus on for each type of system requirement.

2. Clara Silva is a relatively new member of a project team that is developing a retail store management system for a chain of grocery stores. The company's headquarters is in Rio de Janeiro, and the chain has 33 locations throughout Brazil, including multiple stores in several cities.

   The new system will be a networked, client-server architecture. Stores will be linked to one of four regional servers, which will in turn be linked to the corporate headquarters. The regional servers are also interlinked. Each retail store will be outfitted with similar configurations of 6–20 point-of-sale terminals and 10–20 personal computers (based on the size of the store) networked to a local file server. Ms. Silva has been given the task of developing a network model that will document the geographic structure of this system. She has not faced a system of such scope before and is unsure of where to begin.
   a. Prepare a set of instructions for Ms. Silva to follow in developing this network model.
   b. Using a deployment diagram, draw a network model for this organization.
   c. Prepare a set of instructions for Ms. Silva to follow in developing hardware and software specifications.

3. Refer to the Professional and Scientific Staff Management (PSSM) minicase in Chapters 4, 6, 7, 8, 9, and 10. Based on the solutions developed for those problems, suggest a physical architecture design and portray it with a deployment diagram.

4. Given his limited budget for the development of the FRDI's CMS, Professor Takato plans to rely, at least in the first phase of the project, on hardware and infrastructure resources that are currently available at the City University of Osaka. Clearly, Professor Takato and his CMS will share those resources with other projects the university has. Consider the functionality and resources needed by the main modules Professor Takato identified for his system (listed in minicase 4, Chapter 7):
   a. General Presentation and Marketing – presents the conference, conference topics, venue, traveling and accommodation information, program committees, important dates, and other.
   b. Communication Automation – supports generation and submission of possibly personalized e-mails or other types of messages to various categories of recipients: potential participants drawn from a database of previous conference participants to advertise the upcoming conference, submitting authors (e.g., for confirmation of paper acceptance), reviewers (e.g., for sending list of papers to review), and other.
   c. Submission System – supports submission of papers.
   d. Review System – supports management of the paper review process.
   e. Registration System – supports conference registration and payment of conference fees.
   f. Post-conference Survey System – supports the collection and evaluation of post-conference online feedback from participants.

   What are the minimal resources Professor Takato should request for? What are the resources more likely to be impacted by each of the modules above? Which modules may raise scalability issues and require additional resources? Specify the type of those resources.

PART THREE

IMPLEMENTATION

During construction, the actual system is built. Building a successful information system requires a set of activities: programming, testing, and documenting the system. In today's global economy, cultural issues also play an important role in managing these activities. Installing an information system requires switching from the current system to the new system. This conversion process can be quite involved; for example, cultural differences among the users, the development team, and the two groups can be quite challenging. Furthermore, not only does conversion involve shutting the old system down and turning the new one on, it also can involve a significant training effort. Finally, operating the system may uncover additional requirements that may have to be addressed by the development team.

CHAPTER 12
DEVELOPMENT

Programs
Test Results
Documentation

CHAPTER 13
INSTALLATION

Conversion Plan
Change Management Plan
Support Plan
Project Assessment

Programs

Test Plan

Documentation

Conversion Plan

Change Management Plan

Support Plan

Project Assessment

CHAPTER 12

DEVELOPMENT

This chapter discusses the activities needed to successfully build an information system: programming, testing, and documenting the system. Programming is time consuming and costly, but except in unusual circumstances, it is the simplest for the systems analyst because it is well understood. For this reason, the systems analyst focuses on testing (proving that the system works as designed) and developing documentation.

OBJECTIVES

- Be familiar with the system construction process
- Understand different types of tests and when to use them
- Understand how to develop documentation

CHAPTER OUTLINE

INTRODUCTION

When people first learn about developing information systems, they usually think immediately about writing programs. Programming can be the largest single component of any systems development project in terms of time and cost. However, it also can be the best understood component and therefore—except in rare circumstances—offers the fewest problems of all aspects of system development. When projects fail, it is usually not because the programmers were unable to write the programs, but because the analysis, design, installation, and/or project management were done poorly. In this chapter, we focus on the construction and testing of the software and the documentation.

Construction is the development of all parts of the system, including the software itself, documentation, and new operating procedures. Looking back at Figure 1-18, we see that the Construction phase of the Enhanced Unified Process deals predominantly with the Implementation, Testing, and Configuration and Change Management workflows. Implementation obviously deals with programming. Programming is often seen as the focal point of systems development. After all, systems development *is* writing programs. It is the reason we do all the analysis and design. And it's fun. Many beginning programmers see testing and documentation as bothersome afterthoughts. Testing and documentation aren't fun, so they often receive less attention than the creative activity of writing programs.

However, programming and testing are very similar to writing and editing. No professional writer (or student writing an important term paper) would stop after writing the first draft. Rereading, editing, and revising the initial draft into a good paper are the hallmarks of good writing. Likewise, thorough testing is the hallmark of professional software developers. Most professional organizations devote more time and money to testing (and the subsequent revision and retesting) than to writing the programs in the first place.

The reasons are simple economics: Downtime and failures caused by software bugs[1] are extremely expensive. Many large organizations estimate the costs of downtime of critical applications at $50,000 to $200,000 *per hour.*[2] One serious bug that causes an hour of downtime can cost more than one year's salary of a programmer—and how often are bugs found and fixed in one hour? Testing is, therefore, a form of insurance. Organizations are willing to spend a lot of time and money to prevent the possibility of major failures after the system is installed.

Therefore, a program is usually not considered finished until the test for that program is passed. For this reason, programming and testing are tightly coupled, and because programming is the primary job of the programmer (not the analyst), testing (not programming) often becomes the focus of the construction stage for the systems analysis team.

The Configuration and Change Management workflow keeps track of the state of the evolving system. The evolving information system comprises a set of artifacts that includes, for example, diagrams, source code, and executables. During the development process, these artifacts are modified. The amount of work, and hence dollars, that goes into the development of the artifacts is substantial. Therefore, the artifacts themselves should be handled as any expensive asset would be handled: Access controls must be put into place to safeguard the artifacts from being stolen or destroyed. Because the artifacts are modified on a regular, if not continuous, basis, good version control mechanisms should be established. The *traceability* of the artifacts back through the various artifacts developed, such as data management layer designs, class diagrams, package diagrams, and use-case diagrams, to the specific requirements is also very important. Without this traceability, we will not know which aspects of a system to modify when—not if—the requirements change.

---

[1] When I (Alan Dennis) was an undergraduate, I had the opportunity to hear Admiral Grace Hopper tell how the term *bug* was introduced. She was working on one of the early Navy computers when suddenly it failed. The computer would not restart properly so she began to search for failed vacuum tubes. She found a moth inside one tube and recorded in the log book that a bug had caused the computer to crash. From then on, every computer crash was jokingly blamed on a bug (as opposed to programmer error), and eventually, the term *bug* entered the general language of computing.

[2] See Billie Shea, "Quality Patrol: Eye on the Enterprise," *Application Development Trends* (November 5, 1998): 31–38.

In this chapter, we discuss three aspects of construction: managing programming, testing, and writing the documentation. Because programming is primarily the job of programmers, not systems analysts, and because this is not a programming book, we devote less time to programming than to testing and documentation. Furthermore, we do not delve into the details of configuration and change management in this chapter (see Chapter 13).[3]

MANAGING PROGRAMMING

In general, systems analysts do not write programs; programmers write programs. Therefore, the primary task of the systems analysts during programming is . . . waiting. However, the project manager is usually very busy managing the programming effort by assigning the programmers, coordinating the activities, and managing the programming schedule.[4]

Assigning Programmers

The first step in programming is assigning modules to the programmers. As discussed in Chapter 8, each module (class, object, or method) should be as separate and distinct as possible from the other modules (i.e., cohesion should be maximized and coupling should be minimized). The project manager first groups together classes that are related so that each programmer is working on related classes. These groups of classes are then assigned to programmers. A good place to start is to look at the package diagrams.

CONCEPTS

IN ACTION

12-A The Cost of a Bug

My first programming job in 1977 was to convert a set of application systems from one version of COBOL to another version of COBOL for the government of Prince Edward Island. The testing approach was to first run a set of test data through the old system and then run it through the new system to ensure that the results from the two matched. If they matched, then the last three months of production data were run through both to ensure they, too, matched.

Things went well until I began to convert the gas tax system that kept records on everyone authorized to purchase gasoline without paying tax. The test data ran fine, but the results using the production data were peculiar. The old and new systems matched, but rather than listing several thousand records, the report listed only fifty. I checked the production data file and found it listed only fifty records, not the thousands that were supposed to be there.

The system worked by copying the existing gas tax records file into a new file and making changes in the new file. The old file was then copied to tape backup. There was a bug in the program such that if there were no changes to the file, a new file was created, but no records were copied into it.

I checked the tape backups and found one with the full set of data that were scheduled to be overwritten three days after I discovered the problem. The government was only three days away from losing all gas tax records.

— *Alan Dennis*

Question

What would have been the cost of this bug if it hadn't been caught?

---

[3] A good reference for information on configuration and change management is Jessica Keyes, *Software Configuration Management* (Boca Raton, FL: Auerbach, 2004).

[4] One of the best books on managing programming (even though it was first written more than 30 years ago) is that by Frederick P. Brooks, Jr. *The Mythical Man-Month*, 20th Anniversary Edition (Reading, MA: Addison-Wesley, 1995).

One of the rules of systems development is that the more programmers who are involved in a project, the longer the system will take to build. This is because as the size of the programming team increases, the need for coordination increases exponentially, and the more coordination required, the less time programmers can spend actually writing systems. The best size is the smallest possible programming team. When projects are so complex that they require a large team, the best strategy is to try to break the project into a series of smaller parts that can function as independently as possible.

CONCEPTS **12-B Identifying the Best Talent**

IN ACTION

Quantitative analysis provides the data for making decisions. Some major league baseball teams have used Sabermetrics (and similar data analysis measures) to quantify the value of a player to his salary. For example, is a .250 hitter with a strong base-stealing background getting paid $5 million per year worth more in terms of salary dollars as compared to a .330 hitter with slower speed who is getting paid $15 million per year? Is Alex Rodriguez of the Yankees worth more than Joe Maurer of the Minnesota Twins? Is Ryan Dempster of the Chicago Cubs more valuable than Tim Hudson of the Atlanta Braves? Statistics are gathered and analyzed on all aspects of the game: hitting, fielding, injuries, leadership, coachability, age (and expected lifetime contributions), and much more.

Now take that concept to the business marketplace. Can quantitative measures be placed on IT workers? Is a systems analyst who manages six major projects in a year worth more than an analyst who manages twelve much smaller projects? Is a project leader who is known for

effectively managing teams worth more than one who frequently has disagreements with his or her team? Is a developer who can write 1,000 lines of Java code in a week worth more, or less, than a developer who writes 500 lines of Visual Basic code? When we discussed tangible and intangible benefits, the authors suggested putting a dollar amount on to intangible benefits, if possible.

Questions

1. How could you put quantitative measures on the qualitative efforts of a systems analyst?
2. What might be some of the costs in implementing a statistical analysis of employees?
3. What data might be needed to do such a statistical analysis on employees?
4. Could this system replace the conventional human resources department roles for recruiting and evaluating corporate talent?

Coordinating Activities

Coordination can be done through both high-tech and low-tech means. The simplest approach is to have a weekly project meeting to discuss any changes to the system that have arisen during the past week—or just any issues that have come up. Regular meetings, even if they are brief, encourage the widespread communication and discussion of issues before they become problems.

Another important way to improve coordination is to create and follow standards that can range from formal rules for naming files, to forms that must be completed when goals are reached, to programming guidelines (see Chapter 2). When a team forms standards and then follows them, the project can be completed faster because task coordination is less complex.

The analysts also must put mechanisms in place to keep the programming effort well organized. Many project teams set up three areas in which programmers can work: a development area, a testing area, and a production area. These areas can be different directories on a server hard disk, different servers, or different physical locations, but the point is that files, data, and programs are separated based on their status of completion. At first, programmers access and build files within the development area and then copy them to the testing area when the programmers are finished. If a program does not pass a test, it is sent back to development. Once all programs are tested and ready to support the new system, they are copied into the production area—the location where the final system will reside.

Keeping files and programs in different places based on completion status helps manage *change control*, the action of coordinating a system as it changes through construction. Another change control technique is keeping track of which programmer changes which classes and packages by using a *program log*. The log is merely a form on which programmers sign out classes and packages to write and sign in when they are completed. Both the programming areas and program log help the analysts understand exactly who has worked on what and the system's current status. Without these techniques, files can be put into production without the proper testing (e.g., two programmers can start working on the same class or package at the same time).

If a CASE tool is used during the construction step, it can be very helpful for change control because many CASE tools are set up to track the status of programs and help manage programmers as they work. In most cases, maintaining coordination is not conceptually complex. It just requires a lot of discipline and attention to tracking small details.

Managing the Schedule

The time estimates that were produced during project identification and refined during analysis and design almost always need to be refined as the project progresses during construction, because it is virtually impossible to develop an exact assessment of the project's schedule. As we discussed in Chapter 2, a well-done set of time estimates usually has a 10 percent margin of error by the time we reach the construction step. It is critical that the time estimates be revised as construction proceeds. If a program module takes longer to develop than expected, then the prudent response is to move the expected completion date later by the same amount.

CONCEPTS **12-C Finishing the Process**

IN ACTION

As a great analyst, you've planned, analyzed, and designed a good solution. Now you need to implement. As part of implementation, do you think that training is just a wasted expense?

Stress is common in a help-desk call center. Users of computing services call to get access to locked accounts or get help when technology isn't working as planned, and they often are very upset. Employees of the help-desk call center can get stressed out, and this can result in more sick days, less productivity, and higher turnover. Max Productivity Incorporated is a training company that works with people in high-stress jobs. Their training program helps train employees how to relax, how to shake

off tough users, and how to create win–win scenarios. They claim to be able to reduce employee turnover by 50 percent, increase productivity by 20 percent, and reduce stress, anger, and depression by 75 percent.

Questions

1. How would you challenge Max Productivity to verify its claims for reducing turnover, increasing productivity, and decreasing stress and anger?
2. How would you conduct a cost–benefit analysis on hiring Max Productivity Incorporated to do ongoing training for your help-desk call-center employees?

PRACTICAL	12-1 Avoiding Classic Implementation Mistakes
TIP	

In previous chapters, we discussed classic mistakes and how to avoid them. Here, we summarize four classic mistakes in implementation:[5]

- *Research-oriented development:* Using state-of-the-art technology requires research-oriented development that explores the new technology because "bleeding edge" tools and techniques are not well understood, are not well documented, and do not function exactly as promised.

  Solution: If you use state-of-the-art technology, you need to significantly increase the project's time and cost estimates, even if (some experts would say *especially if*) such technologies claim to reduce time and effort.

- *Using low-cost personnel:* You get what you pay for. The lowest-cost consultant or staff member is significantly less productive than the best staff. Several studies have shown that the best programmers produce software six to eight times faster than the least productive (yet cost only 50 to 100 percent more).

  Solution: If cost is a critical issue, assign the best, most expensive personnel; never assign entry-level personnel in an attempt to save costs.

- *Lack of code control:* On large projects, programmers need to coordinate changes to the program source code (so that two programmers don't try to change the same program at the same time and overwrite each other's changes). Although manual procedures appear to work (e.g., sending e-mail notes to others when you work on a program to tell them not to), mistakes are inevitable.

  Solution: Use a source code library that requires programmers to "check out" programs and prohibits others from working on them at the same time.

- *Inadequate testing:* The number one reason for project failure during implementation is ad hoc testing—where programmers and analysts test the system without formal test plans.

  Solution: Always allocate sufficient time in the project plan for formal testing.

One of the most common causes for schedule problems is scope creep. Scope creep occurs when new requirements are added to the project after the system design was finalized. Scope creep can be very expensive because changes made late in system development can require much of the completed system design (and even programs already written) to be redone. Any proposed change during construction must require the approval of the project manager and should only be done after a quick cost–benefit analysis has been done.

Another common cause is the unnoticed day-by-day slippages in the schedule. One package is a day late here; another one is a day late there. Pretty soon, these minor delays add up and the project is noticeably behind schedule. Once again, the key to managing the programming effort is to watch these minor slippages carefully and update the schedule accordingly.

Typically, a project manager creates a risk assessment that tracks potential risks along with an evaluation of their likelihood and potential impact. As the construction step moves to a close, the list of risks changes as some items are removed and others surface. The best project managers, however, work hard to keep risks from having an impact on the schedule and costs associated with the project.

Cultural Issues

One of the major issues facing information systems development organizations is the offshoring of the implementation aspects of information systems development. Conflicts caused by different national and organizational cultures are now becoming a real area of concern. With the potential of cloud computing (see Chapter 11) potentially enabling even more outsourcing, the potential of cultural conflict is even greater.

[5] Adapted from Steve McConnell, *Rapid Development*, (Redmond, WA: Microsoft Press, 1996).

CONCEPTS

IN ACTION

12-D Managing a Late Project: When to Say When?

*S*ystems projects are notorious for being late and over budget. When should management stop a project that is late or costing more than the intended budget? Consider this case.

Valley Enterprises opted to implement Voice over Internet Protocol (VoIP) service in its Phoenix, Arizona, service area. The company has fifteen locations in the Phoenix area—all with LANs and all with secure Wi-Fi connections. Their current phone system was designed and implemented in the 1950s, when they operated in three locations. As they expanded to additional locations, they generally implemented standard telecommunications solutions with little thought for compatibility. Over the years, they added phone services as they added new buildings and facilities. The CEO, Doug Wilson, heard of VoIP at a trade show and contacted TMR Telecommunications Consultants for a bid. TMR spent a week with the CIO of Valley Enterprises gathering data, and submitted a bid for $50,000 in late 2007. The project was to be started by March 2008 and completed by January 2009. The bid was accepted.

TMR started the project in March 2008. In late July 2008, TMR was bought out by Advanced Communications

of Scottsdale, Arizona. This merger delayed the project initially by over a month. In early September 2008, some of the same personnel from TMR, as well as a new project manager from Advanced Communications, went back to the project.

By March 2009, the project had already cost $150,000 and only eight of the locations had VoIP implemented. Advanced Communications insisted that the LANs were out of date and were unable to carry the expanded load without major upgrades to the bandwidth, the routers, and other telecommunications equipment.

Questions

1. Is it time to end this project? Why or why not?
2. What negotiations should have occurred between TMR and Valley Enterprises prior to December 2008?
3. What should a project manager or project coordinator from Valley Enterprises do when the project first starts to slip?

A simple example that can demonstrate cultural differences with regard to student learning is the idea of plagiarism. What exactly does plagiarism really imply? Different cultures have very different views. In some cultures, one of the highest forms of respect is simply to quote an expert. However, in these same cultures, there is no need to reference the expert. The act of quoting the expert itself is the act of respect. In some cases, actually referencing the expert through the use of quotation marks and a footnote may be viewed as an insult to the expert and the reader, because it is obvious to the reader that the writer did not expect the reader to recognize the expert's quote. This expectation was either caused by the reader's own ignorance or the expert's lack of reputation. Either way, the writer would be insulting someone through the use of quotation marks and footnotes. These cultures tend to be collectivist in nature (see Chapters 10 and 13). Consequently, since the collective owns all ideas, there is no concept of theft of ideas. However, in the United States, the opposite is true. If a writer does not use quotation marks and footnotes to appropriately give credit to the source of the quote (or paraphrase), then the writer is guilty of theft.[6] Obviously, in today's global world, plagiarism is not a simple issue.

Another simple example of cultural differences, with regard to student learning, is the idea of students working together to complete homework assignments. Even though we all know that research has shown that students learn better in groups, in the United States, we view students who turn in the same assignment as cheaters.[7] In other cultures, individual performance is not as important as the performance of the group. Again, these cultures are

---

[6] A wonderful little book on plagiarism is Richard A. Posner, *The Little Book of Plagiarism* (New York: Pantheon Books, 2007).

[7] In this case, the recent work of Roger Schank is very enlightening. For example, see Roger C. Schank, *Making Minds Less Well Educated than Our Own* (Mahwah, NJ: Lawrence Erlbaum Associates, 2004).

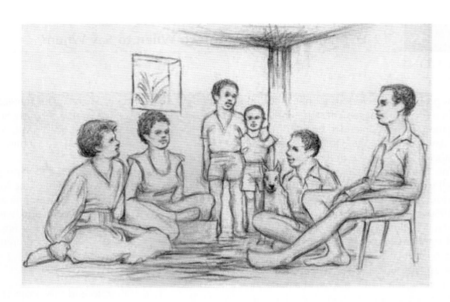

FIGURE 12-1
Cultural Optical
Illusion[8]

collectivist in nature. Consequently, helping a fellow student to understand the assignment and to perform better in the class would be the expectation. Furthermore, this attitude extends to test taking. If a fellow student is struggling on a test and if you were from a collectivist culture, it would be your duty to allow your fellow student to copy your answer. Obviously, this is another example of a substantive cultural difference. From a business perspective, these different views of plagiarism and cheating could have serious implications for the protection of intellectual property.

Cultural differences also extend to the way we literally see things. A classic image that demonstrates the way our visual perception is affected by our culture and environment is demonstrated in the sketch shown in Figure 12-1. Depending on where you are from, you will focus on and see different things. For example, where does the scene portrayed in the sketch take place? What is the item above the woman's head? Most Westerners say that the scene takes place in a room and the item above the woman's head is a window through which you can see a plant of some sort. However, if you are from East Africa, the scene is believed to take place outside under a tree and the item above the woman's head is actually a box that she is balancing on her head. In East Africa, there are very few homes with corners: Most homes are round. Therefore, the location of the scene must be outside under a tree. Consequently, the item above the women's head cannot be a window. Instead, it must be a box.

In a set of studies comparing East Asians and Americans, Richard Nisbett, of the University of Michigan, and his colleagues has demonstrated that there are cultural differences that not only affect the way individuals see things, but also what they focus on.[9] For example, in Figure 12-2a, we see an aquarium-like setting. In a recall test, a set of Japanese and American students watched an animation of the aquarium. After watching the aquarium animation, they were asked to recall what they had seen. Both groups were able to recall about the same number of "focal" fish, those that were moving faster and/or were larger. However, the Japanese students were able to recall a lot of the "background" material, such as the bubbles rising in the middle of the tank, the small frog and snail near the bottom of

[8] www.optical-illusionist.com/illusions/cultural-optical-illusion.
[9] Richard E. Nisbett, *The Geography of Thought: How Asians and Westerners Think Differently … and Why* (New York, NY: Free Press, 2003).

(a)

(b)

(c)

FIGURE 12-2

Asian and American
Differences

the tank, the plants, and the rocks, whereas the Americans generally could not. In a separate test, where the students were asked whether they had seen the fish before, both the Japanese and American students easily recognized the fish when it was displayed with its original background (see Figure 12-2b). However, the Japanese had a much more difficult time in recognizing the fish when it was portrayed in a different context with a novel background (see Figure 12-2c). According to Nisbett and his colleagues, this result, along with others, provided evidence that Asians tended to bind the focal objects to the environment within which they were first seen, whereas Americans simply focus on the individual objects themselves.

As we stated previously, with offshore outsourcing, information systems development teams can be geographically dispersed and multicultural in their membership. Given the above issues and when we consider the cultural differences Hall and Hofstede identified (see Chapter 10), cultural issues add a new wrinkle in the management of developing a successful information system.[10] Hall's context dimension partially explains Nisbett's results. From an information systems development perspective, *context* could influence the ability of a team member to see (or not see) potential creative solutions that are out of the box or affect a team member's ability (or inability) to understand the entire problem under consideration. Furthermore, given this dimension, the level of detail in direction could be varied between cultures. Hofstede's *individualism* and *collectivism* dimension partially explains the results regarding plagiarism and cheating described above. Given the importance that intellectual property plays in IT, this potentially could be a real problem when

---

[10] See Geert Hofstede, *Culture's Consequences: Comparing Values, Behaviors, Institutions and Organizations Across Nations,* 2nd ed. (Thousand Oaks, CA: Sage, 2001); Geert Hofstede, Gert Jan Hofstede, and Michael Minkov, *Cultures and Organizations: Software of the Mind,* 3rd ed. (New York, NY: McGraw-Hill, 2010); Edward T. Hall, *Beyond Culture* (New York: Anchor Books, 1981).

offshoring development to a collectivist culture. Furthermore, Hall's *speed of messages* and context dimensions could also affect the way this could be addressed. Depending on the culture, too much detail could be insulting, but attempting to put this issue into a contextual frame that is culturally sensitive is difficult.

When managing programmers in a multicultural setting, Hall's *time* dimension must also be considered. In *monochronic time* cultures, deadlines are critical. This is probably why *timeboxing* has been relatively successful as a method to control projects (see Chapter 2). However, in a *polychronic time* culture, a *deadline* is nothing more than a suggestion. Obviously, when managing programmers, understanding how the culture considers time is very important to have both a successful product delivery and a successful development process.

Hofstede's other previously mentioned dimensions are *power distance*, *uncertainty avoidance*, and *masculinity versus femininity*. Managing programmers in a culture with a high power distance value is different than with a culture with a low power distance. For example, in the United States, programmers see themselves as equals to their managers. In fact, in some firms, the president of the firm can be found "coding" solutions alongside of a brand new hire. This somewhat explains the growing popularity of agile methods (see Chapter 1). In comparison, in a high power distance culture, the president of the firm would never stoop to performing the same tasks as a new hire. It would be insulting to the president and embarrassing to the new hire.

With regard to uncertainty avoidance, the choice of systems development approach could be affected. In a culture that prefers everything to be neat and ordered, a systems development methodology that is very rule-driven would be beneficial. Also, development team member professional certification and team and firm ISO or CMMI certifications would lend credibility to the team, whereas in a culture that willingly takes on risk, certifications might not increase the perceived standing of the development team.

When managing programmers in a masculine culture, it is critical to provide recognition to the top-performing members of the development team and also to recognize the top-performing teams. On the other hand, when considering a feminine culture, it is more important to ensure that the workplace is a supportive, noncompetitive, and nurturing environment.

Hofstede has identified a fifth dimension, *long-* versus *short-term orientation*, which deals with how the culture views the past and the future. In a long-term focused culture, team development and a deep relationship with a client is very important, while in a culture that emphasizes the short term, delivering a high-quality product on time is all that really matters.

For years, project managers in the United States have had to bring together individuals from very different backgrounds. There was always a common spoken and written language, English, and the melting pot idea that guaranteed some level of commonality among the team members.[11] However, in today's "flat world," there is no longer any common culture or common spoken and written language. From an information systems development perspective, the common language today tends to be UML, Java, SQL, C++, Objective-C, and Visual Basic, not English. However, at this time, there is no common culture. Consequently, understanding cultural issues will be extremely important for the near future to successfully manage international and multicultural development teams.

---

[11] People who grew up in different areas of the United States (e.g., New York City, Nashville, Minneapolis, Denver, and Los Angeles) are, in a very real sense, culturally different. For an interesting take on this, see Joel Garreau, *The Nine Nations of North America* (New York NY: Avon Books, 1981). However, the prevalence of the Internet and cable TV has created much more of a shared culture in the United States than in many other parts of the world. Obviously, the Internet and cable TV also could affect the world in the long run.

CONCEPTS	12-E Managing Global Projects
IN ACTION	

Shamrock Foods is a major food distributor centered in Tralee, Ireland. Originally a dairy cooperative, they branched into various food components [dried milk, cheese solids, and flavorings (or flavourings, as they would spell it)]. They have had substantial growth, most coming by way of acquisition of existing companies or facilities. For example, Iowa Soybean in the United States in now a subsidiary of Shamrock Foods, as is a large dairy cooperative in Wisconsin.

They have processing facilities in more than twelve countries and distribution sales in more than thirty countries. With the rapid growth by acquisition, they have generally adopted a hands-off policy, keeping the systems separated and not integrated into unified ERP system. Thus, each acquired company is still largely autonomous, although it reports to Shamrock Foods and is managed by Shamrock Foods.

This separation concept has been a problem for the CFO of Shamrock Foods, Conor Lynch. The board of directors would like some aggregated data for direction and analysis of acquired businesses. Conor has the reports from the various subsidiaries but has to have his staff convert them to a consistent basic currency (generally either to euros or U.S. dollars).

Questions

1. When should a multinational or multisite business consolidate data systems?

2. There are costs associated with consolidating data systems—the various acquired companies had their own functioning accounting systems—with a variety of hardware and software systems. What justification should Conor use to push for a consolidated, unified ERP system?

3. Conor at times has to deal with incomplete and incompatible data. Inventory systems might be first in, first out (FIFO) for some of the subsidiaries and last in, first out (LIFO) or another accounting method for other subsidiaries. How might a multinational CFO deal with incomplete and incompatible data?

DESIGNING TESTS

In object-oriented systems, the temptation is to minimize testing. After all, through the use of patterns, frameworks, class libraries, and components, much of the system has been tested previously. Therefore, we should not have to test as much. Right? Wrong! Testing is more critical to object-oriented systems than systems developed in the past. Based on encapsulation (and information hiding), polymorphism (and dynamic binding), inheritance, reuse, and the actual object-oriented products, thorough testing is much more difficult and critical. Given the complexity of the development processes used and the global nature of information systems development, testing becomes even more crucial. Thus, object-oriented testing must be done systematically and the results must be documented so that the project team knows what has and has not been tested. Testing object-oriented systems is, therefore, very complex. Consequently, a complete coverage of the topic is beyond the scope of this book. However, given the importance of testing in developing quality software, in this section we provide a basic overview of object-oriented systems testing.[12]

The purpose of testing is not to demonstrate that the system is free of errors. It is not possible to prove that a system is error free, which is especially true with object-oriented systems. This is similar to theory testing. You cannot prove a theory. If a test fails to find problems with

[12] For a good introduction to testing object-oriented software, see John D. McGregor and David A. Sykes, *A Practical Guide to Testing Object-Oriented Software* (Boston: Addison-Wesley, 2001). For a thorough coverage of testing object-oriented software, see Robert V. Binder, *Testing Object-Oriented Systems: Models, Patterns, and Tools* (Reading, MA: Addison-Wesley, 1999). This book provides more than 1,000 pages of information with regard to how to test the different artifacts and processes included in object-oriented systems development.

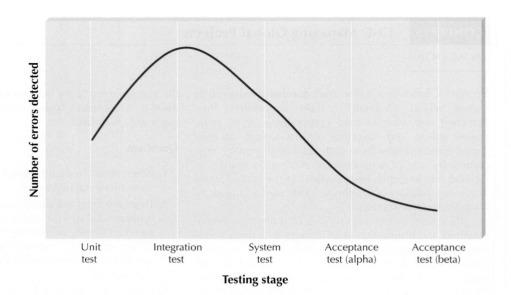

FIGURE 12-3

Error-Discovery Rates
for Different Stages
of Tests

a theory, your confidence in the theory is increased. However, if a test succeeds in finding a problem, then the theory has been falsified. Software testing is similar in that it can only show the existence of errors. The purpose of testing is to uncover as many errors as feasible.[13]

There are four general stages of tests: unit tests, integration tests, system tests, and acceptance tests. Although each application system is different, most errors are found during integration and system testing (see Figure 12-3).

In each of the following sections, we describe the four stages. However, before doing this, we describe the effect that the object-oriented characteristics have on testing and the necessary planning and management activities that must take place to have a successful testing program.

Testing and Object Orientation

Most testing techniques have been developed to support non–object-oriented development. Therefore, most of the testing approaches have had to be adapted to object-oriented systems. The characteristics of object-oriented systems that affect testing the most are encapsulation (and information hiding), polymorphism (and dynamic binding), inheritance, and the use of patterns, class libraries, frameworks, and components. Also, the sheer volume of products that come out of a typical object-oriented development process has increased the importance of testing in object-oriented systems development.

Encapsulation and Information Hiding Encapsulation and information hiding allow processes and data to be combined to create holistic entities (i.e., objects). They support hiding everything behind a visible interface. Although this allows the system to be modified and maintained in an effective and efficient manner, it makes testing the system problematic. What do you need to test to build confidence in the system's ability to meet the user's need? You need to test the business process that is represented in the use cases. However, the business process is distributed over a set of collaborating classes and contained in the methods of those classes. The

---

[13] It is not cost-effective to try to get every error out of the software. Except in simple examples, it is, in fact, impossible. There are simply too many combinations to check.

only way to know the effect that a business process has on a system is to look at the state changes that take place in the system. But in object-oriented systems, the instances of the classes hide the data behind a class boundary. How is it possible then to see the impact of a business process?

A second issue raised by encapsulation and information hiding is the definition of a "unit" for unit testing. What is the unit to be tested? Is it the package, class, or method? In traditional approaches, the answer would be the process that is contained in a function. However, the process in object-oriented systems is distributed over a set of classes. Therefore, testing individual methods makes no sense. The answer is the class. This dramatically changes the way unit testing is done.

A third issue raised is the impact on integration testing. In this case, objects can be aggregated to form aggregate objects; for instance, a car has many parts, or they can be grouped together to form collaborations. Furthermore, they can be used in class libraries, frameworks, and components. Based on all of these different ways classes can be grouped together, how does one effectively do integration testing?

Polymorphism and Dynamic Binding   Polymorphism and dynamic binding dramatically affect both unit and integration testing. Because an individual business process is implemented through a set of methods distributed over a set of objects, as shown before, the unit test makes no sense at the method level. However, with polymorphism and dynamic binding, the same method (a small part of the overall business process) can be implemented in many different objects. Therefore, testing individual implementations of methods makes no sense. Again, the unit that makes sense to test is the class. Except for trivial cases, dynamic binding makes it impossible to know which implementation is going to be executed until the system does it. Therefore, integration testing becomes very challenging.

Inheritance   When taking into consideration the issues raised about inheritance (see Chapter 8), it should not be a surprise that inheritance affects the testing of object-oriented systems. Through the use of inheritance, bugs can be propagated instantaneously from a superclass to all its direct and indirect subclasses. However, the tests that are applicable to a superclass are also applicable to all its subclasses. As usual, inheritance is a double-edged sword. Finally, even though we have stated this many times before, inheritance should support only a generalization and specialization type of semantics. Remember, when using inheritance, the principle of substitutability is critical (see Chapter 5). All these issues affect unit and integration testing.

Reuse   On the surface, reuse should decrease the amount of testing required. However, each time a class is used in a different context, the class must be tested again. Therefore, any time a class library, framework, or component is used, unit testing and integration testing are important. In the case of a component, the unit to test is the component itself. Remember that a component has a well-defined API (application program interface) that hides the details of its implementation.

Object-Oriented Development Process and Products   In virtually all textbooks, including this one, testing is covered near the end of system development. This seems to imply that testing is something that takes place only after the programming has ended. However, every product[14] that comes out of the object-oriented development process must

---

[14] For example, activity diagrams, use-case descriptions, use-case diagrams, CRC cards, class diagrams, object diagrams, sequence diagrams, communication diagrams, behavioral state machines, package diagrams, contracts, method specifications, use scenarios, window navigation diagrams, storyboards, windows layout diagrams, real use cases, and source code.

be tested. For example, it is a lot easier to ensure that the requirements are captured and modeled correctly through testing the use cases, and it is a lot cheaper to catch this type of error back in analysis than it is in implementation. Obviously, this is also true for testing collaborations. By the time we have implemented a collaboration as a set of layers and partitions, we could have expended a great deal of time—and time is money—on implementing the wrong thing. So testing collaborations by role-playing the CRC cards in analysis actually saves the team lots of time and money.

Testing is something that must take place throughout system development, not simply at the end. However, the type of testing that can take place on nonexecutable representations, such as use cases and CRC cards, is different from those on code written in an object-oriented programming language. The primary approach to testing nonexecutable representations is some form of an inspection or walkthrough of the representation.[15] Role-playing the CRC cards based on use cases is an example of one type of walkthrough.

Test Planning

Testing starts with the development of a *test plan*, which defines a series of tests that will be conducted. Because testing takes place throughout the development of an object-oriented system, a test plan should be developed at the very beginning of system development and continuously updated as the system evolves. For example, the representation of a class evolves from a simplistic CRC card to a set of classes that are implemented in a programming language. In Figure 12-4, we see a CRC card representation of an Order class that contains invariants. Each of these invariants must be tested and enforced for the Order class to be considered to be of sufficient quality. One simple invariant test would be to attempt to assign a value to the Cust ID attribute that was not associated with the Customer object that is contained in the Customer attribute. Another invariant test would be to try and assign more than one date to the Date attribute. Finally, a trickier invariant test would be to try to assign an integer value to the Shipping attribute. This one is more difficult because most programming languages allow an integer to be "cast" to a double. If the value contained in the Shipping attribute really is supposed to be a double, then casting the integer value to a double would be an error. These tests should be done using a walkthrough approach when the class is specified, as we did in Chapters 4, 5, 6, and 7, and a more rigorous approach once the class has been fully implemented. This is an example of unit testing a class, which is described in the next section. To ensure the quality of a class, it should be tested each time its representation is changed.

The test plan should address all products that are created during the development of the system. For example, tests should be created that can be used to test completeness of a CRC card. Each individual test has a specific objective and describes a set of very specific *test cases* to examine. In the case of invariant-based tests, a description of the invariant is given, and the original values of the attribute, the event that will cause the attribute value to change, the actual results observed, the expected results, and whether it passed or failed are shown. *Test specifications* are created for each type of constraint that must be met by the class. Also, similar types of specifications are done for integration, system, and acceptance tests.

Not all classes are likely to be finished at the same time, so the programmer usually writes *stubs* for the unfinished classes to enable the classes around them to be tested. A stub is a placeholder for a class that usually displays a simple test message on the screen or

---

[15] See Michael Fagan, "Design and Code Inspections to Reduce Errors in Program Development," *IBM Systems Journal*, 15, no. 3 (1976); and Daniel P. Freedman and Gerald M. Weinberg, *Handbook of Walkthrough, Inspections, and Technical Reviews: Evaluating Programs, Projects, and Products*, 3rd ed. (New York: Dorset House Publishing, 1990). Also, Chapters 4, 5, 6, and 7 describe the walkthrough process in detail in relation to the verification and validation of the analysis models.

Front:

Class Name: Order		ID: 2	Type: Concrete, Domain
Description: An Individual that needs to receive or has received medical attention			Associated Use Cases: 3

Responsibilities	Collaborators
Calculate subtotal	
Calculate tax	
Calculate shipping	
Calculate total	

(a)

Back:

Attributes:

Order Number	(1..1)	(unsigned long)	
Date	(1..1)	(Date)	
Sub Total	(0..1)	(double)	{Sub Total = ProductOrder.sum(GetExtension())}
Tax	(0..1)	(double)	(Tax = State.GetTaxRate() * Sub Total)
Shipping	(0..1)	(double)	
Total	(0..1)	(double)	
Customer	(1..1)	(Customer)	
Cust ID	(1..1)	(unsigned long)	{Cust ID = Customer. GetCustID()}
State	(1..1)	(State)	
StateName	(1..1)	(String)	{State Name = State. GetState()}

Relationships:

Generalization (a-kind-of): _____

Aggregation (has-parts): _____

Other Associations: Customer {1..1} State {1..1} Product {1..*}

(b)

FIGURE 12-4
Order CRC Card (see
Figure 8-19)

returns some *hardcoded* value[16] when it is selected. For example, consider an application system that provides creating, changing, deleting, finding, and printing functions for some object such as CDs, patients, or employees. Depending on the final design, these different functions could end up in different objects on different layers. Therefore, to test the functionality associated with the classes on the problem domain layer, a stub would be written for each of the classes on the other layers that interact with the problem domain classes. These stubs would be the minimal interface necessary to be able to test the problem domain classes. For example, they would have methods that could receive the messages being sent by the problem domain layer objects and methods that could send messages to the problem domain layer objects. Typically, the methods would display a message on the screen notifying the tester that the method was successfully reached (e.g., Delete item from Database method reached). In this way, the problem domain classes could pass class testing before the classes on the other layers were completed.

Unit Tests

Unit tests focus on a single unit—the class. There are two approaches to unit testing: black-box testing and white-box testing (see Figure 12-5). *Black-box testing* is the most commonly used because each class represents an encapsulated object. Black-box testing is driven by the CRC cards, behavior state machines, and contracts associated with a class, not by the programmers' interpretation. In this case, the test plan is developed directly from the specification of the class: each item in the specification becomes a test, and several test cases are developed for it. *White-box testing* is based on the method specifications associated with each class. However, white-box testing has had limited impact in object-oriented development. This is due to the rather small size of the individual methods in a class. Most approaches to testing classes use black-box testing to ensure their correctness.

Class tests should be based on the invariants on the CRC cards, the behavioral state machines associated with each class, and the pre- and post-conditions contained on each method's contract. Assuming all the constraints have been captured on the CRC cards and contracts, individual test cases can be developed fairly easily. For example, suppose the CRC card for an order class gave an invariant that the order quantity must be between ten and one hundred cases. The tester would develop a series of test cases to ensure that the quantity is validated before the system accepts it. It is impossible to test every possible combination of input and situation; there are simply too many possible combinations. In this example, the test requires a minimum of three test cases: one with a valid value (e.g., 15), one with a low invalid value (e.g., 7), and one with a high invalid value (e.g., 110). Most tests would also include a test case with a nonnumeric value to ensure the data types were checked (e.g., ABCD). A really good test would include a test case with nonsensical but potentially valid data (e.g. 21.4).

Using a behavioral state machine is a useful way to identify tests for a class. Any class that has a behavioral state machine associated with it has a potentially complex life cycle. It is possible to create a series of tests to guarantee that each state can be reached. For example, Figure 12-6 portrays the behavioral state machine for the Order class just discussed. In this case, there are many transitions between the different states of an instance of the Order class. Tests should be created to guarantee that the only transitions allowed from an instance of the Order class are the ones specifically defined. In this case, it should

---

[16] *Hardcoded* means written into the program. For example, suppose you were writing a unit to calculate the net present value of a loan. The stub might be written to always display (or return to the calling module) a value of 100 regardless of the input values.

Stage	Types of Tests	Test Plan Source	When to Use	Notes
Unit Testing	**Black-Box Testing** Treats class as a black box	CRC Cards Class Diagrams Contracts	For normal unit testing	• Tester focuses on whether the class meets the requirements stated in the specifications.
	White-Box Testing Looks inside the class to test its major elements	Method Specifications	When complexity is high	• By looking inside the class to review the code itself, the tester may discover errors or assumptions not immediately obvious to someone treating the class as a black box.
Integration Testing	**User Interface Testing** The tester tests each interface function	Interface Design	For normal integration testing	• Testing is done by moving through each and every menu item in the interface either in a top-down or bottom-up manner.
	Use-Case Testing The tester tests each use case	Use Cases	When the user interface is important	• Testing is done by moving through each use case to ensure they work correctly. • Usually combined with user interface testing because it does not test all interfaces.
	Interaction Testing Tests each process in step-by-step fashion	Class Diagrams Sequence Diagrams Communication Diagrams	When the system performs data processing	• The entire system begins as a set of stubs. Each class is added in turn and the results of the class compared to the correct result from the test data; when a class passes, the next class is added and the test rerun. This is done for each package. Once each package has passed all tests, then the process repeats integrating the packages.
	System Interface Testing Tests the exchange of data with other systems	Use-Case Diagram	When the system exchanges data	• Because data transfers between systems are often automated and not monitored directly by the users, it is critical to design tests to ensure they are being done correctly.
System Testing	**Requirements Testing** Tests to whether original business requirements are met	System Design, Unit Tests, and Integration Tests	For normal system testing	• Ensures that changes made as a result of integration testing did not create new errors. • Testers often pretend to be uninformed users and perform improper actions to ensure the system is immune to invalid actions (e.g., adding blank records).
	Usability Testing Tests how convenient the system is to use	Interface Design and Use Cases	When user interface is important	• Often done by analyst with experience in how users think and in good interface design. • Sometimes uses formal usability testing procedures discussed in Chapter 10.
	Security Testing Tests disaster recovery and unauthorized access	Infrastructure Design	When the system is important	• Security testing is a complex task, usually done by an infrastructure analyst assigned to the project. • In extreme cases, a professional firm may be hired.
	Performance Testing Examines the ability to perform under high loads	System Proposal Infrastructure Design	When the system is important	• High volumes of transactions are generated and given to the system. • Often done by using special purpose testing software.
	Documentation Testing Tests the accuracy of the documentation	Help System, Procedures, Tutorials	For normal system testing	• Analysts spot check or check every item on every page in all documentation to ensure the documentation items and examples work properly.
Acceptance Testing	**Alpha Testing** Conducted by users to ensure they accept the system	System Tests	For normal acceptance testing	• Often repeats previous tests but are conducted by users themselves to ensure they accept the system.
	Beta Testing Uses real data, not test data	System Requirements	When the system is important	• Users closely monitor system for errors or useful improvements.

FIGURE 12-5 Types of Tests

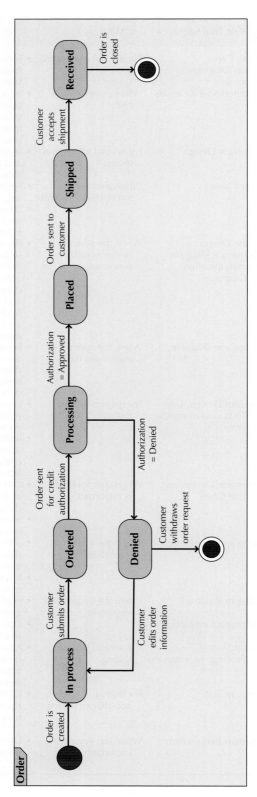

FIGURE 12-6 Order Behavioral State Machine (see Figure 6-17)

Method Name:	addOrder	Class Name:	Customer		ID:	36

Clients (Consumers):

Associated Use Cases:
    addCustomerOrder

Description of Responsibilities:
    Implement the necessary behavior to add a new order to an existing customer keeping
    the orders in sorted order by the order's order number.

Arguments Received:
    anOrder:Order

Type of Value Returned: void

Pre-Conditions:
    not Orders.includes(anOrder)

Post-Conditions:
    Orders = Orders@pre.including(anOrder)

FIGURE 12-7
addOrder Contract
(see Figure 8-25)

be impossible for an Order object to go from the In process state to the Placed state without traversing the Ordered and Processing states via the Customer submits order, Order sent for credit authorization, and Authorization = Approved transitions. This state-based testing can be done throughout the development of the class via walkthroughs and role-playing early in the evolution of the class and more rigorous testing once it has been implemented in a programming language.

Tests also can be developed for each contract associated with the class. In the case of a contract, a set of tests for each pre- and post-condition is required. For example, the contract of the addOrder method of the Customer class shown in Figure 12-7 has both a pre- and post-condition that essentially requires the new order to have not existed with the instance of the Customer class before the method executes and that it is associated with the Customer object after the method executes. Tests must be created to enforce these constraints. If the class is a subclass of another class, then all the tests associated with the superclass must be executed again. The interactions among the constraints, invariants, and the pre- and post-conditions in the subclass and the superclass(es) must be also addressed.

Finally, owing to good object-oriented design, to fully test a class, special testing methods might have to be added to the class being tested. For example, how can invariants be tested? The only way to really test them is to have methods that are visible to the outside of the class that can be used to manipulate the values of the class's attributes. However, adding these types of methods to a class does two things. First, they add to the testing requirements because they themselves have to be tested. Second, if they are not removed from the deployed version of the system, the system will be less efficient, and the advantage of information hiding effectively is lost. As is readily apparent, testing classes is complex. Therefore, great care must be taken when designing tests for classes.

Integration Tests

Integration tests assess whether a set of classes that must work together do so without error. They ensure that the interfaces and linkages between different parts of the system work properly. At this point, the classes have passed their individual unit tests, so the focus now is on the flow of control among the classes and on the data exchanged among them. Integration testing follows the same general procedures as unit testing: the tester develops a test plan that has a series of tests, which, in turn, have a test. Integration testing is often done by a set of programmers and/or systems analysts.

From an object-oriented systems perspective, integration testing can be difficult. A single class can be in many different aggregations, because of the way objects can be combined to form new objects, class libraries, frameworks, components, and packages. Where is the best place to start the integration? Typically, the answer is to begin with the set of classes, a collaboration, that are used to support the highest-priority use case (see Chapter 4). Also, dynamic binding makes it crucial to design the integration tests carefully to ensure the combinations of methods are tested.

There are four approaches to integration testing: *user interface testing,*[17] *use-case testing*, *interaction testing*, and *system interface testing* (see Figure 12-5). Most projects use all four approaches. However, like unit testing, integration testing must be carefully planned. In the case of use-case testing, only the aspects of the class and class invariants related to the specific use case are included in these use-case context-dependent class tests. In fact, typically, use-case testing is performed one scenario at a time. In many ways, use-case testing can be viewed as a more rigorous role-playing exercise (see Chapter 5). Like unit testing, integration testing should be performed throughout the evolution of the system. In the early stages of the system's development, you should be working with the CRC cards and role-playing them. Later on, you will have the contracts and method specifications completed. Gradually, you will have implemented the problem domain classes, the user interface classes, and the data management layer classes in a programming language. As in unit testing, each time a new representation (diagram, text, program) is created, a new integration test needs to be performed. Therefore, as the system evolves to more completely support the use case, we can more rigorously test whether the use case is fully supported or not.

One of the major problems with integration testing and object-oriented systems is the difficulty caused by the interaction of inheritance and dynamic binding. This specific problem has become known as the *yo-yo problem*. The yo-yo problem occurs when the analyst or designer must bounce up and down through the inheritance graph to understand the control flow through the methods being executed. In most cases, this is caused by a rather deep inheritance graph; that is, the subclass has many superclasses above it in the inheritance graph. The yo-yo problem becomes even more of a nightmare in testing object-oriented systems when inheritance conflicts exist and when multiple inheritance is used (See Chapter 8). About the only realistic approach to testing through the yo-yo problem is through an interactive debugger that is typically part of a systems development environment, such as Eclipse, Netbeans, or Visual Studio.

System Tests

System tests are usually conducted by the systems analysts to ensure that all classes work together without error. System testing is similar to integration testing but is much broader in scope. Whereas integration testing focuses on whether the classes work

[17] We describe some of the different types of user interface testing in Chapter 10.

together without error, system tests examine how well the system meets business requirements and its usability, security, and performance under heavy load (see Figure 12-5). It also tests the system's documentation.

Acceptance Tests

Acceptance testing is done primarily by the users with support from the project team. The goal is to confirm that the system is complete, meets the business needs that prompted the system to be developed, and is acceptable to the users. Acceptance testing is done in two stages: *alpha testing*, in which users test the system using made-up data, and *beta testing*, in which users begin to use the system with real data but are carefully monitored for errors (see Figure 12-5).

| YOUR | 12-1 Test Planning for an ATM |
| TURN | |

Suppose you are a project manager for a bank developing software for ATMs. Develop a unit test plan for the user interface component of the ATM.

DEVELOPING DOCUMENTATION

Like testing, developing documentation of the system must be done throughout system development. There are two fundamentally different types of documentation: system documentation and user documentation. *System documentation* is intended to help programmers and systems analysts understand the application software and enable them to build it or maintain it after the system is installed. System documentation is largely a by-product of the systems analysis and design process and is created as the project unfolds. Each step and phase produces documents that are essential in understanding how the system is or is to be built, and these documents are stored in the project binder(s). In many object-oriented development environments, it is possible to somewhat automate the creation of detailed documentation for classes and methods. For example, in Java, if the programmers use javadoc-style comments, it is possible to create HTML pages that document a class and its methods automatically by using the javadoc utility.[18] Because most programmers look on documentation with much distaste, anything that can make documentation easier to create is useful.

User documentation (such as user's manuals, training manuals, and online help systems) is designed to help the user operate the system. Although most project teams expect users to have received training and to have read the user's manuals before operating the system, unfortunately, this is not always the case. It is more common today—especially in the case of commercial software packages for microcomputers—for users to begin using the software without training or reading the user's manuals. In this section, we focus on user documentation.[19]

---

[18] For those who have used Java, javadoc is how the JDK documentation from Sun is created.

[19] For more information on developing documentation, see Thomas T. Barker, *Writing Software Documentation* (Boston: Allyn and Bacon, 1998).

User documentation is often left until the end of the project, which is a dangerous strategy. Developing good documentation takes longer than many people expect because it requires much more than simply writing a few pages. Producing documentation requires designing the documents (whether on paper or online), writing the text, editing the documents, and testing them. For good-quality documentation, this process usually takes about three hours per page (single-spaced) for paper-based documentation or two hours per screen for online documentation. Thus, a "simple" documentation, such as a ten-page user's manual and a set of twenty help screens, takes seventy hours. Of course, lower-quality documentation can be produced faster.

The time required to develop and test user documentation should be built into the project plan. Most organizations plan for documentation development to start once the interface design and program specifications are complete. The initial draft of documentation is usually scheduled for completion immediately after the unit tests are complete. This reduces (but doesn't eliminate) the chance that the documentation will need to be changed owing to software changes and still leaves enough time for the documentation to be tested and revised before the acceptance tests are started.

Although paper-based manuals are still important, online documentation is becoming more important. Paper-based documentation is simpler to use because it is more familiar to users, especially novices who have less computer experience; online documentation requires the users to learn one more set of commands. Paper-based documentation is also easier to flip through and gain a general understanding of its organization and topics and can be used far away from the computer itself.

There are four key strengths of online documentation that all but guarantee it will be the dominant form for the 21st century. Searching for information is often simpler (provided the help search index is well designed) because the user can type in a variety of keywords to view information almost instantaneously, rather than having to search through the index or table of contents in a paper document. The same information can be presented several times in many different formats, so that the user can find and read the information in the most informative way (such redundancy is possible in paper documentation, but the cost and intimidating size of the resulting manual make it impractical). Online documentation provides many new ways for the user to interact with the documentation that is not possible in static paper documentation. For example, it is possible to use links, or "tool tips" (i.e., pop-up text; see Chapter 10) to explain unfamiliar terms, and one can write "show-me" routines that demonstrate on the screen exactly what buttons to click and text to type. Finally, online documentation is significantly less expensive to distribute than paper documentation.

Types of Documentation

There are three fundamentally different types of user documentation: reference documents, procedures manuals, and tutorials. *Reference documents* (also called the help system) are designed to be used when the user needs to learn how to perform a specific function (e.g., updating a field, adding a new record). Often, people read reference information when they have tried and failed to perform the function; writing reference documents requires special care because the user is often impatient or frustrated when he or she begins to read them.

Procedures manuals describe how to perform business tasks (e.g., printing a monthly report, taking a customer order). Each item in the procedures manual typically guides the user through a task that requires several functions or steps in the system. Therefore, each entry is typically much longer than an entry in a reference document.

Tutorials—obviously—teach people how to use major components of a system (e.g., an introduction to the basic operations of the system). Each entry in the tutorial is typically longer still than the entries in procedures manuals, and the entries are usually designed to be read in sequence (whereas entries in reference documents and procedures manuals are designed to be read individually).

Regardless of the type of user documentation, the overall process for developing it is similar to the process of developing interfaces (see Chapter 10). The developer first designs the general structure for the documentation and then develops the individual components within it.

Designing Documentation Structure

In this section, we focus on the development of online documentation, because we believe it will become the most common form of user documentation. The general structure used in most online documentation, whether reference documents, procedures manuals, or tutorials, is to develop a set of *documentation navigation controls* that lead the user to *documentation topics*. The documentation topics are the material that user wants to read, whereas the navigation controls are the way the user locates and accesses a specific topic.

Designing the structure of the documentation begins by identifying the different types of topics and navigation controls that need to be included. Figure 12-8 shows a commonly used structure for online reference documents (i.e., the help system). The documentation topics generally come from three sources. The first and most obvious source of topics is the set of commands and menus in the user interface. This set of topics is very useful if the user wants to understand how a particular command or menu is used.

However, the users often don't know what commands to look for or where they are in the system's menu structure. Instead, users have tasks they want to perform, and rather than thinking in terms of commands, they think in terms of their tasks. Therefore, the second and, often, more useful set of topics focuses on how to perform certain tasks, usually those in the use scenarios, WND, and the real use cases from the user interface design (see Chapter 10). These topics walk the user through the set of steps (often involving several keystrokes or mouse clicks) needed to perform some task.

The third topic is definitions of important terms. These terms are usually the use cases and classes in the system, but sometimes, they also include commands.

There are five general types of navigation controls for topics, but not all systems use all five types (see Figure 12-8). The first is the table of contents that organizes the information in a logical form, as though the users were to read the reference documentation from start to finish. The index provides access into the topics based on important keywords, in the same way that the index at the back of a book helps us find topics. Text search provides the ability to search through the topics either for any text the user types or for words that match a developer-specified set of words that is much larger than the set of words in the index. Unlike the index, text search typically provides no organization to the words (other than alphabetical). Some systems provide the ability to use an intelligent agent to help in the search. The fifth and final navigation controls to topics are the hyperlinks between topics that enable the user to click and move among topics.

Procedures manuals and tutorials are similar but, often, simpler in structure. Topics for procedures manuals usually come from the use scenarios, WNDs, and the real use cases developed during interface design and from other basic tasks the users must perform. Topics for tutorials are usually organized around major sections of the system and the level of experience of the user. Most tutorials start with the basic, most commonly used commands and then move into more complex and less commonly used commands.

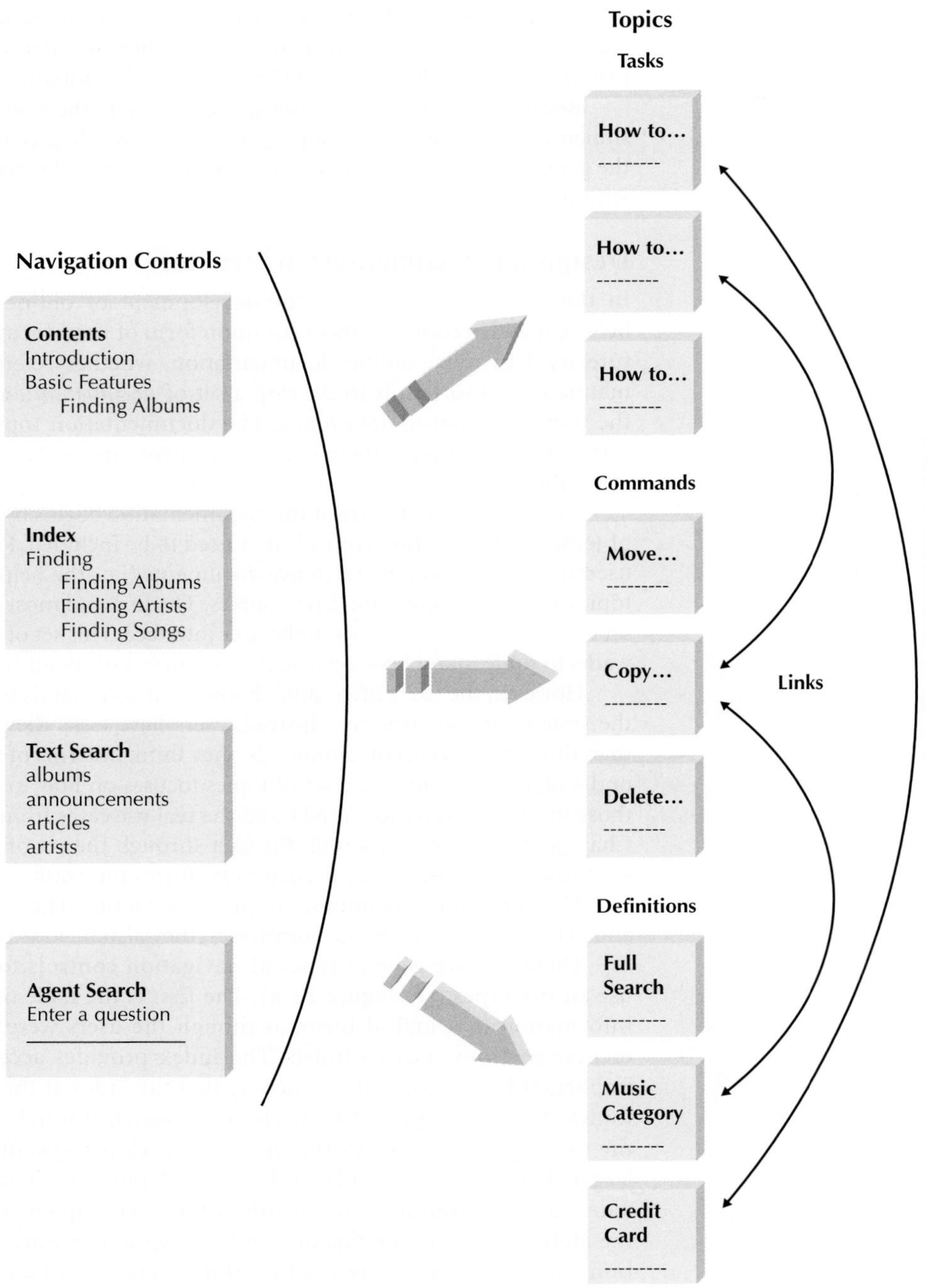

FIGURE 12-8
Organizing Online
Reference Documents

Writing Documentation Topics

The general format for topics is fairly similar across application systems and operating systems (see Figure 12-9). Topics typically start with very clear titles, followed by some introductory text that defines the topic and then by detailed, step-by-step instructions on how to perform what is being described. Many topics include screen images to help the user

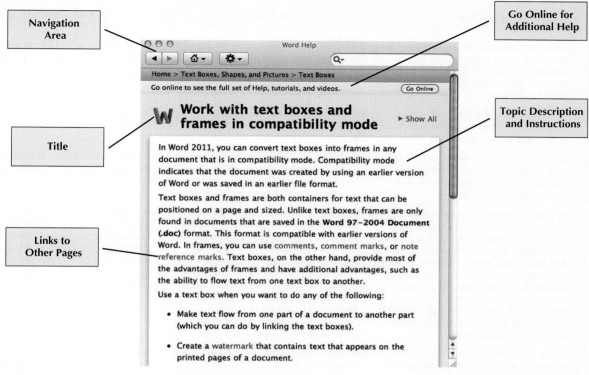

FIGURE 12-9   A Help Topic in Microsoft Word 2011

find items on the screen; some also have tutorials and videos available online that demonstrate the functions of interest to the user. Most also include navigation controls to enable the movement among topics, usually at the top of the window, plus links to other topics. Some also include links to related topics that include options or other commands and tasks the user might want to perform in concert with the topic being read.

Writing the topic content can be challenging. It requires a good understanding of the user (or more accurately, the range of users) and a knowledge of what skills the users currently have and can be expected to import from other systems and tools they are using or have used (including the system that the new system is replacing). Topics should always be written from the viewpoint of the user and describe what the user wants to accomplish, not what the system can do. Figure 12-10 provides some general guidelines to improve the quality of documentation text.[20]

Identifying Navigation Terms

As we write the documentation topics, we also begin to identify the terms that will be used to help users find topics. The table of contents is usually the most straightforward, because it is developed from the logical structure of the documentation topics, whether reference topics, procedure topics, or tutorial topics. The items for the index and search engine require more care because they are developed from the major parts of the system and the users'

---

[20] One of the best books to explain the art of writing is William Strunk and E. B. White, *Elements of Style*, 4th ed. (Needham Heights, MA: Allyn & Bacon, 2000).

Guideline	Before the Guideline	After the Guideline
Use the active voice: The active voice creates more active and readable text by putting the subject at the start of the sentence, the verb in the middle, and the object at the end.	Finding albums is done using the album title, the artist's name, or a song title.	You can find an album by using the album title, the artist's name, or a song title.
Use e-prime style: E-prime style creates more active writing by omitting all forms of the verb to be.	The text you want to copy must be selected before you click on the copy button.	Select the text you want to copy before you click on the copy button.
Use consistent terms: Always use the same term to refer to the same items, rather than switching among synonyms (e.g., change, modify, update).	Select the text you want to copy. Pressing the copy button will copy the marked text to the new location.	Select the text you want to copy. Pressing the copy button will copy the selected text to the new location.
Use simple language: Always use the simplest language possible to accurately convey the meaning. This does not mean you should "dumb down" the text but that you should avoid artificially inflating its complexity. Avoid separating subjects and verbs and try to use the fewest words possible. (When you encounter a complex piece of text, try eliminating words; you may be surprised at how few words are really needed to convey meaning.)	The Georgia Statewide Academic and Medical System (GSAMS) is a cooperative and collaborative distance learning network in the state of Georgia. The organization in Atlanta that administers and manages the technical and overall operations of the currently more than 300 interactive audio and video teleconferencing classrooms throughout Georgia system is the Department of Administrative Service (DOAS). (56 words)	The Department of Administrative Service (DOAS) in Atlanta manages the Georgia Statewide Academic and Medical System (GSAMS), a distance learning network with more than 300 teleconferencing classrooms throughout Georgia. (29 words)
Use friendly language: Too often, documentation is cold and sterile because it is written in a very formal manner. Remember, you are writing for a person, not a computer.	Blank disks have been provided to you by Operations. It is suggested that you ensure your data are not lost by making backup copies of all essential data.	You should make a backup copy of all data that are important to you. If you need more diskettes, contact Operations.
Use parallel grammatical structures: Parallel grammatical structures indicate the similarity among items in list and help the reader understand content.	Opening files Saving a document How to delete files	Opening a file Saving a file Deleting a file
Use steps correctly: Novices often intersperse action and the results of action when describing a step-by-step process. Steps are always actions.	1. Press the customer button. 2. The customer dialogue box will appear. 3. Type the customer ID and press the submit button and the customer record will appear.	1. Press the customer button. 2. Type the customer ID in the customer dialogue box when it appears. 3. Press the submit button to view the customer record for this customer.
Use short paragraphs: Readers of documentation usually quickly scan text to find the information they need, so the text in the middle of long paragraphs is often overlooked. Use separate paragraphs to help readers find information more quickly.		

Source: Adapted from T. T. Barker, *Writing Software Documentation* (Boston: Allyn & Bacon, 1998).

FIGURE 12-10 Guidelines for Crafting Documentation Topics

YOUR TURN   **12-2 Documentation for an ATM**

Suppose you are a project manager for a bank developing software for ATMs. Develop an online help system.

business functions. Every time we write a topic, we must also list the terms that will be used to find the topic. Terms for the index and search engine can come from four distinct sources.

The first source for index terms is the set of the commands in the user interface, such as open file, modify customer, and print open orders. All commands contain two parts (action and object). It is important to develop the index for both parts because users could search for information using either part. A user looking for more information about saving files, for example, might search by using the term *save* or the term *files.*

The second source is the set of major concepts in the system, which are often use cases and classes. In the case of the Appointment system, for example, this might include appointment, symptoms, or patient.

A third source is the set of business tasks the user performs, such as ordering replacement units or making an appointment. Often, these are contained in the command set, but sometimes they require several commands and use terms that do not always appear in the system. Good sources for these terms are the use scenarios and real use cases developed during interface design (see Chapter 10).

A fourth, often controversial, source is the set of synonyms for the three sets of preceding items. Users sometimes don't think in terms of the nicely defined terms used by the system. They might try to find information on how to stop or quit rather than exit, or erase rather than delete. Including synonyms in the index increases the complexity and size of the documentation system but can greatly improve the usefulness of the system to the users.

APPLYING THE CONCEPTS AT CD SELECTIONS

Because the material in this chapter actually takes place throughout the systems development process, this installment of the CD Selections case simply revisits some of the earlier installments and shows where this material has either already been described or where the development team should have performed these tasks.

SUMMARY

Managing Programming

Programming is done by programmers, so systems analysts have few responsibilities during this stage. The project manager, however, is usually very busy. The first task is to assign the programmers to complete the project, ideally the fewest possible because coordination problems increase as the size of the programming team increases. Coordination can be improved by having regular meetings, ensuring that standards are followed, implementing change control, and using CASE tools effectively. One of the key functions of the project manager is to manage the schedule and adjust it for delays. Two common causes of delays are scope creep and minor slippages that go unnoticed. Given today's global development of information systems, different cultural issues need to be taken into consideration.

Designing Tests

Tests must be carefully planned because the cost of fixing one major bug after the system is installed can easily exceed the annual salary of a programmer. A test plan contains several tests that examine different aspects of the system. A test, in turn, specifies several test cases that will be examined by the testers. A unit test examines a class within the system; test cases come from the class specifications or the class code itself. An integration test examines how well several classes work together; test cases come from the interface design, use cases, and the use-case, sequence, and collaboration diagrams. A system test examines the system as a whole and is broader than the unit and integration tests; test cases come from the system design, the infrastructure design, the unit, and integration tests. Acceptance testing is done by the users to determine whether the system is acceptable to them; it draws on the system test plans (alpha testing) and the real work the users perform (beta testing).

Developing Documentation

Documentation, both user documentation and system documentation, is moving away from paper-based documents to online documentation. There are three types of user documentation: Reference documents are designed to be used when the user needs to learn how to perform a specific function (e.g., an online help system), procedures manuals describe how to perform business tasks, and tutorials teach people how to use the system. Documentation navigation controls (e.g., a table of contents, index, find, intelligent agents, or links between pages) enable users to find documentation topics (e.g., how to perform a function, how to use an interface command, an explanation of a term).

KEY TERMS

Acceptance test, 535
Alpha test, 535
Beta test, 535
Black-box testing, 530
Change control, 519
Collectivism, 523
Construction, 516
Context, 523
Documentation navigation
  control, 537
Documentation topic, 537
Femininity, 524
Hardcoded, 530
Individualism, 523
Integration test, 534
Interaction testing, 534

Long-term orientation, 524
Masculinity, 524
Monochronic time, 524
Polychronic time, 524
Power distance, 524
Procedures manual, 536
Program log, 519
Reference document, 536
Requirements testing, 531
Security testing, 531
Short-term orientation, 524
Speed of messages, 524
Stub, 528
System documentation, 535
System interface testing, 534
System test, 534

Test case, 528
Test plan, 528
Test specification, 528
Time, 524
Timeboxing, 524
Traceability, 516
Tutorial, 536
Uncertainty avoidance, 524
Unit test, 530
Usability testing, 531
Use-case testing, 534
User documentation, 535
User interface testing, 534
White-box testing, 530
Yo-yo problem, 534

QUESTIONS

1. Why is testing important?
2. How can different national or organizational cultures affect the management of an information systems development project?
3. What is the primary role of systems analysts during the programming stage?

4. In *The Mythical Man-Month*, Frederick Brooks argues that adding more programmers to a late project makes it later. Why?
5. When offshoring development, how could differences in Hall's context dimension of culture affect the contribution of a team member to the successful

development of an information system? What about Hall's time or speed of messages dimensions?

6. What are Hofstede's five dimensions of cultural differences? How could differences in them influence the effectiveness of an information systems development team?

7. What are the common language or languages used today in information systems development?

8. What is the purpose of testing?

9. Describe how object orientation affects testing.

10. Compare and contrast the terms test, test plan, and test case.

11. What is a stub and why is it used in testing?

12. What is the primary goal of unit testing?

13. How are the test cases developed for unit tests?

14. Compare and contrast black-box testing and white-box testing.

15. What are the different types of class tests?

14. What is the primary goal of integration testing?

17. How are the test cases developed for integration tests?

18. Describe the yo-yo problem. Why does it make integration testing difficult?

19. What is the primary goal of system testing?

20. How are the test cases developed for system tests?

21. What is the primary goal of acceptance testing?

22. How are the test cases developed for acceptance tests?

23. Compare and contrast alpha testing and beta testing.

24. Compare and contrast user documentation and system documentation.

25. Why is online documentation becoming more important?

26. What are the primary disadvantages of online documentation?

27. Compare and contrast reference documents, procedures manuals, and tutorials.

28. What are five types of documentation navigation controls?

29. What are the commonly used sources of documentation topics? Which is the most important? Why?

30. What are the commonly used sources of documentation navigation controls? Which is the most important? Why?

EXERCISES

A. Different views of plagiarism and collaborative learning were described as examples of differences among different cultures today. Using the web, identify other differences that could affect the success of an information systems development team.

B. Besides Hall and Hofstede, both David Victor and Fons Trompenaars have identified a set of cultural dimensions that could be useful in information systems development. Using the web, identify their dimensions.

C. If the registration system at your university does not have a good online help system, develop one for one screen of the user interface.

D. Examine and prepare a report on the online help system for the calculator program in Windows (or a similar one on the Mac or Unix). (You will probably be surprised at the amount of help for such a simple program.)

E. Compare and contrast the online help at two different websites that enable you to perform some function (e.g., make travel reservations, order books).

F. Create an invariant test specification for the class you chose for the A Real Estate Inc. problem in exercise A in Chapter 8.

G. Create a use-case test plan, including the specific class plans and invariant tests, for a use case from the A Real Estate Inc. exercises in the previous chapters.

H. Create an invariant test specification for the class you chose for the A Video Store problem in exercise B in Chapter 8.

I. Create a use-case test plan, including the specific class plans and invariant tests, for a use case from the A Video Store exercises in the previous chapters.

J. Create an invariant test specification for the class you chose for the gym problem in exercise C in Chapter 8.

K. Create a use-case test plan, including the specific class plans and invariant tests for a use case from the health club exercises in previous chapters.

L. Create an invariant test specification for the class you chose for Picnics R Us in exercise D in Chapter 8.

M. Create a use-case test plan, including the specific class plans and invariant tests, for a use case from the Picnics R Us exercises in the previous chapters.

N. Create an invariant test specification for the class you chose for the Of-the-Month Club (OTMC) in exercise E in Chapter 8.

O. Create a use-case test plan, including the specific class plans and invariant tests, for a use case from the Of-the-Month Club (OTMC) exercises in the previous chapters.

MINICASES

1. Rajan Srikant is a project manager on a new systems development project for a major Indian outsourcing firm. This project is Rajan's first, and he has led his team successfully to the programming phase. The project has not always gone smoothly; Rajan himself has made a few mistakes, but he is generally pleased with the progress of his team and the quality of the system being developed. Now that programming has begun, Rajan is hoping for a break from the hectic pace at work.

   Prior to the programming phase, Rajan recognized that the time estimates made earlier in the project were too optimistic. However, he was firmly committed to meeting the project deadline because he wanted his first project to be a success. Anticipating time constraints, Rajan arranged with the human resources department to bring in two new college graduates and two college interns to increase the programming staff. He would have liked to recruit more experienced workers, but he was committed to keeping the already tight project budget under control.

   Rajan lined up the programming assignments, and work on the programs began about two weeks ago. Lately, he has received unfavorable feedback from the programming team leaders. Some bugs have been found when modules from different programmers were integrated. Several programmers have discovered to their dismay that someone had amended their programs without their knowledge.

   a. What problems can you identify in this situation? What advice do you have for the project manager?

   b. Will Rajan be able to achieve his desired goals of being on time and within budget?

2. Professor Takato, president of FRDI, has asked Mark, his visiting student from the U.S., to develop a test plan, part of the overall test plan for his CMS, that verifies the correct implementation of the state transitions of a Paper object. Professor Takato wants to make sure that the rules (listed in minicase 2, Chapter 6) by which a paper gets published in the conference proceedings are accurately reflected in the implementation of the CMS. He asked Mark to use as a reference the Behavioral State Machine developed during the analysis phase of the system (the solution to minicase 2, Chapter 6) and develop as many test scenarios as possible such that the more significant cases and possible failure scenarios are covered. In particular, Professor Takato asked Mark to come up with an exhaustive test plan for the part of the Behavioral State Machine for the Paper object that describes the rule "Paper must be reviewed by 3 independent reviewers and receive at least 2 favorable reviews for acceptance," by which the Review_Paper() method of the paper object, described in structured English-based algorithm specification in the solution for minicase 3 in Chapter 8, would essentially be tested as well.

   Help Mark develop a test plan for the state transitions of the Paper object and make it as complete as possible. Develop an exhaustive test plan for the Review_Paper() method implementing the "2-out-of-3" rule.

CHAPTER 13

INSTALLATION

This chapter examines the activities needed to install an information system and successfully convert an organization to using it. It also discusses post-implementation activities, such as system support, system maintenance, and project assessment. Installing the system and making it available for use from a technical perspective is relatively straightforward. However, the training and organizational issues surrounding the installation are more complex and challenging because they focus on people, not computers.

OBJECTIVES

- Be familiar with the system installation process
- Understand different types of conversion strategies and when to use them
- Understand several techniques for managing change
- Be familiar with post-installation processes

CHAPTER OUTLINE

INTRODUCTION

It must be remembered that there is nothing more difficult to plan, more doubtful of success, nor more dangerous to manage than the creation of a new system. For the initiator has the animosity of all who would profit by the preservation of the old institution and merely lukewarm defenders in those who would gain by the new.

—Niccolò Machiavelli, *The Prince*, 1513

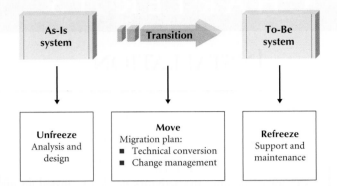

FIGURE 13-1
Implementing Change

Although written almost 500 years ago, Machiavelli's comments are still true today. Managing the change to a new system—whether it is computerized or not—is one of the most difficult tasks in any organization. Because of the challenges involved, most organizations begin developing their conversion and change management plans while the programmers are still developing the software. Leaving conversion and change management planning to the last minute is a recipe for failure.

In many ways, using a computer system or set of work processes is much like driving on a dirt road. Over time, with repeated use, the road begins to develop ruts in the most commonly used parts of the road. Although these ruts show where to drive, they make change difficult. As people use a computer system or set of work processes, those systems or work processes begin to become habits or norms; people learn them and become comfortable with them. These systems or work processes then begin to limit people's activities and make it difficult for them to change because they begin to see their jobs in terms of these processes rather than of the final business goal of serving customers.

One of the earliest models for managing organizational change was developed by Kurt Lewin.[1] Lewin argued that change is a three-step process: unfreeze, move, refreeze (Figure 13-1). First, the project team must *unfreeze* the existing habits and norms (the as-is system) so that change is possible. Most of system development to this point has laid the groundwork for unfreezing. Users are aware of the new system being developed, some have participated in an analysis of the current system (and so are aware of its problems), and some have helped design the new system (and so have some sense of the potential benefits of the new system). These activities have helped to unfreeze the current habits and norms.

The second step is to help the organization move to the new system via a *migration plan*. The migration plan has two major elements. One is technical, which includes how the new system will be installed and how data in the as-is system will be moved into the to-be system; this is discussed in the conversion section of this chapter. The second component is organizational, which includes helping users understand the change and motivating them to adopt it; this is discussed in the change management section of this chapter.

The third step is to *refreeze* the new system as the habitual way of performing the work processes—ensuring that the new system successfully becomes the standard way of performing the business function it supports. This refreezing process is a key goal of the *post-implementation* activities discussed in the final section of this chapter. By providing ongoing support for the new system and immediately beginning to identify improvements

---

[1] Kurt Lewin, "Frontiers in Group Dynamics," *Human Relations*, 1, no. 5 (1947): 5–41; and Kurt Lewin, "Group Decision and Social Change," in E. E. Maccoby, T. M. Newcomb, and E. L. Hartley (eds.), *Readings in Social Psychology* (New York: Holt, Rinehart & Winston, 1958), pp. 197–211.

for the next version of the system, the organization helps solidify the new system as the new habitual way of doing business. Post-implementation activities include system support, which means providing help desk and telephone support for users with problems; system maintenance, which means fixing bugs and improving the system after it has been installed; and project assessment, evaluating the project to identify what went well and what could be improved for the next system development project.

Change management is the most challenging of the three components because it focuses on people, not technology, and because it is the one aspect of the project that is the least controllable by the project team. Change management means winning the hearts and minds of potential users and convincing them that the new system actually provides value.

Maintenance is the most costly aspect of the installation process, because the cost of maintaining systems usually greatly exceeds the initial development costs. It is not unusual for organizations to spend 60 to 80 percent of their total IS development budget on maintenance. Although this might sound surprising initially, think about the software you use. How many software packages do you use that are the very first version? Most commercial software packages become truly useful and enter widespread use only in their second or third version. Maintenance and continual improvement of software is ongoing, whether it is a commercially available package or software developed in-house. Would you buy software if you knew that no new versions were going to be produced? Of course, commercial software is somewhat different from custom in-house software used by only one company, but the fundamental issues remain.

Project assessment is probably the least commonly performed part of system development, but is perhaps the one that has the most long-term value to the IS department. Project assessment enables project team members to step back and consider what they did right and what they could have done better. It is an important component in the individual growth and development of each member of the team, because it encourages team members to learn from their successes and failures. It also enables new ideas or new approaches to system development to be recognized, examined, and shared with other project teams to improve their performance.

In this chapter, we describe deploying the new system through the process of transitioning from the old to the new system (i.e., conversion). Next, we describe issues related to managing the changes necessary to adapt to the new business processes. Finally, we describe issues related to placing the system into production (i.e., post-implementation activities). However, before we address these issues, we describe how the cultural issues affect the deployment of a new system.

CULTURAL ISSUES AND INFORMATION TECHNOLOGY ADOPTION[2]

Cultural issues are one of the things that are typically identified as at least partially to blame when there is a failure in an organization. Cultural issues have been studied at both organizational and national levels. In previous chapters, we discussed the effect that cultural issues can have on designing the human–computer interaction and physical architecture layers (see Chapters 10 and 11) and the management of programmers (Chapter 12). The cultural dimensions identified by Hall and Hofstede included speed of messages, context, time, power distance, uncertainty avoidance, individualism versus collectivism, masculinity

[2] A good summary of cultural issues and information systems is Dorothy E. Leidner and Timothy Kayworth, "A Review of Culture in Information Systems Research: Toward a Theory of Information Technology Culture Conflict," *MIS Quarterly* 30, no. 2 (2006): 357–399.

versus femininity, and long- versus short-term orientation.[3] In this chapter, we describe how these dimensions can affect the successful deployment of an information system that supports a global information supply chain.

Hall's first dimension, *speed of messages*, has implications for the development of documentation (see Chapter 12) and training approaches (see later in this chapter). In a culture that values "deep" content, so that members of the culture can take their time to thoroughly understand the new system, simply providing an online help system is not going to be sufficient to ensure the successful adoption of the new information system. However, in a culture that prefers "fast" messages, an online help system could be sufficient.

Hall's second dimension, *context*, also affects the adoption and deployment of a new system. In high-context cultures, it is expected that the new information system will be placed into the entire context of the enterprise-wide system. Members of this type of society expect to be able to understand exactly where the system fits into the firm's overall picture. Again, like the speed of messages dimension, this affects the training approach used and the documentation developed.

Hall's third dimension, *time*, can also effect the adoption and deployment of a new system. In a *polychronic time* culture, the training could need to be spread out over a longer period of time, when compared to a *monochronic* time culture. In a monochromic time culture, interruptions would be considered rude. Consequently, training could be accomplished in a small set of intense sessions. However, with a polychronic time culture, because interruptions may occur frequently, maximum flexibility in setting up the training sessions may be necessary.

Hofstede's first dimension, *power distance*, addresses how power issues are dealt with in the culture. For example, if a superior in an organization has an incorrect belief about an important issue, can a subordinate point out this error? In some cultures, the answer is a resounding no. Consequently, this dimension could have major ramifications for the successful deployment of an information system. For example, in a culture with a high power distance, the deployment of a new information system is dependent on the impression of the most important stakeholder (see Chapter 2). Therefore, much care must be taken to ensure that this stakeholder is pleased with the system. Otherwise, it might never be used.

Hofstede's second dimension, *uncertainty avoidance*, is based on the degree to which the culture depends on rules for direction, how well individuals in the culture handle stress, and the importance of employment stability. For example, in a high-uncertainty-avoidance culture, the use of detailed procedures manuals (see Chapter 12) and good training (see later in this chapter) can reduce the uncertainty in adopting the new system.

Hofstede's third dimension, *individualism* versus *collectivism*, is based on the level of emphasis the culture places on the individual or the collective. The relationship between the individual and the group is important for the success of an information system. Depending on the culture's orientation, the success of an information system being transitioned into production can depend on whether the focus of the information system will benefit the individual or the group.

Hofstede's fourth dimension, *masculinity* versus *femininity*, addresses how well masculine and feminine characteristics are valued by the culture. Some of the differences that could affect the adoption of an information system include employee motivational issues. In a masculine culture, motivation would be based on advancement, earnings, and training, whereas in a feminine culture, motivations would include friendly atmosphere, physical

---

[3] See Geert Hofstede, *Culture's Consequences: Comparing Values, Behaviors, Institutions and Organizations Across Nations,* 2nd ed. (Thousand Oaks, CA: Sage, 2001); Geert Hofstede, Gert Jan Hofstede, and Michael Minkov, *Cultures and Organizations: Software of the Mind,* 3rd ed. (New York: McGraw-Hill, 2010); and Edward T. Hall, *Beyond Culture* (New York: Anchor Books, 1981).

conditions, and cooperation. Depending on how the culture views this dimension, different motivations might need to be used to increase the likelihood of the information system being successfully deployed.

The fifth dimension, *long-* versus *short-term orientation*, deals with how the culture views the past and the future. In East Asia, long-term thinking is highly respected, whereas in North America and Europe, short-term profits and the current stock price seem to be the only things that matter. Based on this dimension, all the political concerns raised previously in this text become very important. For example, if the local culture views success only in a short-term manner, then any new information system that is deployed to support one department of an organization may give that department a competitive advantage over other departments in the short run. If only short-run measures are used to judge the success of a department, then it would be in the interest of the other departments to fight the successful deployment of the information system. However, if a longer-run perspective is the norm, then the other departments could be convinced to support the new information system because they could have new supportive information systems in the future.

Obviously, when reviewing these dimensions, we can see they interact with each other. The most important thing to remember from an IT perspective is that we must be careful not to view the local user community through our eyes; in a global economy, we must take into consideration the local cultural concerns for the information system to be deployed in a successful manner.

CONVERSION[4]

Conversion is the technical process by which a new system replaces an old system. Users are moved from using the as-is business processes and computer programs to the to-be business processes and programs. The migration plan specifies what activities will be performed when and by whom, and includes both technical aspects (such as installing hardware and software and converting data from the as-is system to the to-be system) and organizational aspects (such as training and motivating the users to embrace the new system). Conversion refers to the technical aspects of the migration plan.

There are three major steps to the conversion plan before commencement of operations: Install hardware, install software, and convert data (Figure 13-2). Although it may be

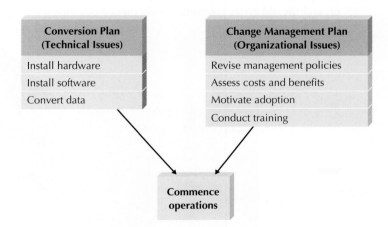

FIGURE 13-2

Elements of a Migration Plan

---

[4] The material in this section is related to the Enhanced Unified Process's Transition phase and the Deployment workflow (see Figure 1-18).

possible to do some of these steps in parallel, usually they must be done sequentially at any one location.

The first step in the conversion plan is to buy and install any needed hardware. In many cases, no new hardware is needed, but sometimes the project requires new hardware such as servers, client computers, printers, and networking equipment. It is critical to work closely with vendors who are supplying needed hardware and software to ensure that the deliveries are coordinated with the conversion schedule so that the equipment is available when it is needed. Nothing can stop a conversion plan in its tracks as easily as the failure of a vendor to deliver needed equipment.

Once the hardware is installed, tested, and certified as being operational, the second step is to install the software. This includes the to-be system under development and, sometimes, additional software that must be installed to make the system operational. At this point, the system is usually tested again to ensure that it operates as planned.

The third step is to convert the data from the as-is system to the to-be system. Data conversion is usually the most technically complicated step in the migration plan. Often, separate programs must be written to convert the data from the as-is system to the new formats required in the to-be system and store it in the to-be system files and databases. This process is often complicated by the fact that the files and databases in the to-be system do not exactly match the files and databases in the as-is system (e.g., the to-be system may use several tables in a database to store customer data that were contained in one file in the as-is system). Formal test plans are always required for data conversion efforts (see Chapter 12).

Conversion can be thought of along three dimensions: the style in which the conversion is done (*conversion style*), what location or work groups are converted at what time (*conversion location*), and what modules of the system are converted at what time (*conversion modules*). Figure 13-3 shows the potential relationships among these three dimensions.

Conversion Style

The conversion style is the way users are switched between the old and new systems. There are two fundamentally different approaches to the style of conversion: direct conversion and parallel conversion.

Direct Conversion   With *direct conversion* (sometimes called cold turkey, big bang, or abrupt cutover), the new system instantly replaces the old system. The new system is turned on and the old system is immediately turned off. This is the approach that we are likely to

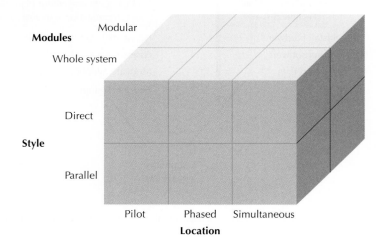

FIGURE 13-3
Conversion Strategies

use when we upgrade commercial software (e.g., Microsoft Word) from one version to another; we simply begin using the new version and stop using the old version.

Direct conversion is the simplest and most straightforward. However, it is the most risky because any problems with the new system that have escaped detection during testing can seriously disrupt the organization.

Parallel Conversion   With *parallel conversion*, the new system is operated side by side with the old system; both systems are used simultaneously. For example, if a new accounting system is installed, the organization enters data into both the old system and the new system and then carefully compares the output from both systems to ensure that the new system is performing correctly. After some time period (often one to two months) of parallel operation and intense comparison between the two systems, the old system is turned off and the organization continues using the new system.

This approach is more likely to catch any major bugs in the new system and prevent the organization from suffering major problems. If problems are discovered in the new system, the system is simply turned off and fixed and then the conversion process starts again. The problem with this approach is the added expense of operating two systems that perform the same function.

Conversion Location

Conversion location refers to the parts of the organization that are converted when the conversion occurs. Often, parts of the organization are physically located in different offices (e.g., Toronto, Atlanta, Los Angeles). In other cases, location refers to different organizational units located in different parts of the same office complex (e.g., order entry, shipping, purchasing). There are at least three fundamentally different approaches to selecting the way different organizational locations are converted: pilot conversion, phased conversion, and simultaneous conversion.

Pilot Conversion   With a *pilot conversion*, one or more locations or units or work groups within a location are selected to be converted first as part of a pilot test. The locations participating in the pilot test are converted (using either direct or parallel conversion). If the system passes the pilot test, then the system is installed at the remaining locations (again using either direct or parallel conversion).

Pilot conversion has the advantage of providing an additional level of testing before the system is widely deployed throughout the organization, so that any problems with the system affect only the pilot locations. However, this type of conversion obviously requires more time before the system is installed at all organizational locations. Also, it means that different organizational units are using different versions of the system and business processes, which can make it difficult for them to exchange data.

Phased Conversion   With *phased conversion*, the system is installed sequentially at different locations. A first set of locations is converted, then a second set, then a third set, and so on, until all locations are converted. Sometimes, there is a deliberate delay between the different sets (at least between the first and the second), so that any problems with the system are detected before too much of the organization is affected. In other cases, the sets are converted back to back so that as soon as those converting one location have finished, the project team moves to the next and continues the conversion.

Phased conversion has the same advantages and disadvantages of pilot conversion. In addition, it means that fewer people are required to perform the actual conversion (and any associated user training) than if all locations were converted at once.

CONCEPTS

IN ACTION

13-A Too Much Paper (Part 1)

The South Dakota Department of Worker's Compensation was sinking under a load of paper files. As a state agency that oversees that employees are treated fairly when they are injured on the job, the agency had a plethora of paper files and filing cabinets. If a person (or company) called to see the status of an injury claim, the clerk who received the call would have to take a message, get the paper file, review the status, and call the person back. Files were stored in huge filing cabinets and were entered by year and case number (e.g., the 415th person injured in 2008 would be in a file numbered 08-415). But most people did not remember their file number and would give a name, address, and date of injury. The clerk would look in a spiral notebook for the last name around the date that was given and then find the file number to retrieve the folder. Some folders were small—possibly a minor cut or minor injury that was taken care of quickly and the employee was back to work. Other folders could be very large, with medical reports from several doctors verifying the extent of the injury (such as an arm amputation). A digital solution

was suggested; reports could be submitted online using a secure website. Medical reports could be submitted electronically, either as a .pdf file or as a faxed digital file. This solution would also mean that the clerk taking the phone call could query the database by the person's name and access the information in a matter of seconds.

Questions

1. The digital solution was going to change the work process of filing injury claims, interacting with people who filed the claims (or companies) wanting to see the status of the claim, and with the process of claims. What might this mean from a work-flow analysis?

2. In many ways, this was a business process reengineering solution. The proposal was to throw out the old process for a completely electronic version. What might a systems analyst do in the data-gathering stage?

Simultaneous Conversion   *Simultaneous conversion*, as the name suggests, means that all locations are converted at the same time. The new system is installed and made ready at all locations; at a preset time, all users begin using the new system. Simultaneous conversion is often used with direct conversion, but it can also be used with parallel conversion.

Simultaneous conversion eliminates problems with having different organizational units using different systems and processes. However, it also means that the organization must have sufficient staff to perform the conversion and train the users at all locations simultaneously.

Conversion Modules

Although it is natural to assume that systems are usually installed in their entirety, this is not always the case.

Whole-System Conversion   A *whole-system conversion*, in which the entire system is installed at one time, is the most common. It is simple and the easiest to understand. However, if the system is large and/or extremely complex (e.g., an enterprise resource-planning system such as SAP or PeopleSoft), the whole system can prove too difficult for users to learn in one conversion step.

Modular Conversion   When the *modules*[5] within a system are separate and distinct, organizations sometimes choose to convert to the new system one module at a time—that is, using modular conversion. Modular conversion requires special care in developing the

[5] In this case, a module is typically a component or a package, i.e., a set of collaborating classes.

system (and usually adds extra cost). Each module either must be written to work with both the old and new systems or object wrappers (see Chapter 7) must be used to encapsulate the old system from the new. When modules are tightly integrated, this is very challenging and therefore is seldom done. However, when there is only a loose association between modules, module conversion is easier. For example, consider a conversion from an old version of Microsoft Office to a new version. It is relatively simple to convert from the old version of Word to the new version without simultaneously having to change from the old to the new version of Microsoft Excel.

Modular conversion reduces the amount of training required to begin using the new system. Users need training only in the new module being implemented. However, modular conversion does take longer and has more steps than does the whole-system process.

Selecting the Appropriate Conversion Strategy

Each of the three dimensions in Figure 13-3 is independent, so that a conversion strategy can be developed to fit in any one of the boxes in this figure. Different boxes can also be mixed and matched into one *conversion strategy*. For example, one commonly used approach is to begin with a pilot conversion of the whole system using parallel conversion in a handful of test locations. Once the system has passed the pilot test at these locations, it is then installed in the remaining locations using phased conversion with direct cutover. There are three important factors to consider in selecting a conversion strategy: *risk, cost,* and the *time* required (Figure 13-4).

Risk   After the system has passed a rigorous battery of unit, system, integration, and acceptance testing, it should be bug free . . . maybe. Because humans make mistakes, nothing built by people is ever perfect. Even after all these tests, there might still be a few undiscovered bugs. The conversion process provides one last step in which to catch these bugs before the system goes live and the bugs have the chance to cause problems.

Parallel conversion is less risky than is direct conversion because it has a greater chance of detecting bugs that have gone undiscovered in testing. Likewise, pilot conversion is less risky than is phased conversion or simultaneous conversion because if bugs do occur, they occur in pilot test locations whose staff are aware that they might encounter bugs. Because potential bugs affect fewer users, there is less risk. Likewise, converting a few modules at a time lowers the probability of a bug because there is more likely to be a bug in the whole system than in any given module.

How important the risk is depends on the system being implemented—the combination of the probability that bugs remain undetected in the system and the potential cost of those undetected bugs. If the system has indeed been subjected to extensive methodical testing, including alpha and beta testing, then the probability of undetected bugs is lower than if the testing was less rigorous. However, there still might have been mistakes made in

Characteristic	Conversion Style		Conversion Location			Conversion Modules	
	Direct Conversion	Parallel Conversion	Pilot Conversion	Phased Conversion	Simultaneous Conversion	Whole-System Conversion	Modular Conversion
Risk	High	Low	Low	Medium	High	High	Medium
Cost	Low	High	Medium	Medium	High	Medium	High
Time	Short	Long	Medium	Long	Short	Short	Long

FIGURE 13-4   Characteristics of Conversion Strategies

the analysis process, so that although there might be no software bugs, the software might fail to properly address the business needs.

Assessing the cost of a bug is challenging, but most analysts and senior managers can make a reasonable guess at the relative cost of a bug. For example, the cost of a bug in an automated stock market trading program or a heart–lung machine keeping someone alive is likely to be much greater than a bug in a computer game or word processing program. Therefore, risk is likely to be a very important factor in the conversion process if the system has not been as thoroughly tested as it might have been or if the cost of bugs is high. If the system has been thoroughly tested or the cost of bugs is not that high, then risk becomes less important to the conversion decision.

Cost As might be expected, different conversion strategies have different costs. These costs can include things such as salaries for people who work with the system (e.g., users, trainers, system administrators, external consultants), travel expenses, operation expenses, communication costs, and hardware leases. Parallel conversion is more expensive than direct cutover because it requires that two systems (the old and the new) be operated at the same time. Employees must then perform twice the usual work because they have to enter the same data into both the old and the new systems. Parallel conversion also requires the results of the two systems to be completely cross-checked to make sure there are no differences between the two, which entails additional time and cost.

Pilot conversion and phased conversion have somewhat similar costs. Simultaneous conversion has higher costs because more staff are required to support all the locations as they simultaneously switch from the old to the new system. Modular conversion is more expensive than whole-system conversion because it requires more programming. The old system must be updated to work with selected modules in the new system, and modules in the new system must be programmed to work with selected modules in both the old and new systems.

Time The final factor is the amount of time required to convert between the old and the new system. Direct conversion is the fastest because it is immediate. Parallel conversion takes longer because the full advantages of the new system do not become available until the old system is turned off. Simultaneous conversion is fastest because all locations are converted at the same time. Phased conversion usually takes longer than pilot conversion because once the pilot test is complete, all remaining locations are usually (but not always) converted simultaneously. Phased conversion proceeds in waves, often requiring several months before all locations are converted. Likewise, modular conversion takes longer than whole-system conversion because the models are introduced one after another.

YOUR TURN

13-1 Developing a Conversion Plan

Suppose you are leading the conversion from one word processor to another at your university. Develop a conversion plan (i.e., technical issues only). You have also been asked to develop a conversion plan for the university's new web-based course registration system. How would the second conversion plan be similar to and different from the one you developed for the word processor?

CONCEPTS IN ACTION

13-B U.S. Army Installation Support

Throughout the 1960s, 1970s, and 1980s, the U.S. Army automated its installations (army bases, in civilian terms). Automation was usually a local effort at each of the more than 100 bases. Although some bases had developed software together (or borrowed software developed at other bases), each base often had software that performed different functions or performed the same function in different ways. In 1989, the army decided to standardize the software so that the same software would be used everywhere. This would greatly reduce software maintenance and also reduce training when soldiers were transferred between bases.

The software took four years to develop. The system was quite complex, and the project manager was concerned that there was a high risk that not all requirements of all installations had been properly captured. Cost and time were less important because the project had already run four years and cost $100 million.

Therefore, the project manager chose a modular pilot conversion using parallel conversion. The manager selected seven installations, each representing a different type of army installation (e.g., training base, arsenal, depot) and began the conversion. All went well, but several new features were identified that had been overlooked during the analysis, design, and construction. These were added and the pilot testing resumed. Finally, the system was installed in the rest of the army installations using a phased direct conversion of the whole system.

—*Alan Dennis*

Questions

1. Do you think the conversion strategy was appropriate?
2. Regardless of whether you agree, what other conversion strategy could have been used?

CHANGE MANAGEMENT[6]

In the context of a systems development project, change management is the process of helping people to adopt and adapt to the to-be system and its accompanying work processes without undue stress. There are three key roles in any major organizational change. The first is the *sponsor* of the change—the person who wants the change. This person is the business sponsor who first initiated the request for the new system (see Chapter 2). Usually, the sponsor is a senior manager of the part of the organization that must adopt and use the new system. It is critical that the sponsor be active in the change management process because a change that is clearly being driven by the sponsor, not by the project team or the IS organization, has greater legitimacy. The sponsor has direct management authority over those who adopt the system.

The second role is that of the *change agent*—the person(s) leading the change effort. The change agent, charged with actually planning and implementing the change, is usually someone outside of the business unit adopting the system and therefore has no direct management authority over the potential adopters. Because the change agent is an outsider, he or she has less credibility than do the sponsor and other members of the business unit. After all, once the system has been installed, the change agent usually leaves and thus has no ongoing impact.

---

[6] The material in this section is related to the Enhanced Unified Process's Transition and Production phases and the Configuration and Change Management workflow (see Figure 1-18). Many books have been written on change management. Some of our favorites are the following: Patrick Connor and Linda Lake, *Managing Organizational Change*, 2nd ed. (Westport, CT: Praeger, 1994); Douglas Smith, *Taking Charge of Change* (Reading, MA: Addison-Wesley, 1996); Daryl Conner, *Managing at the Speed of Change* (New York: Villard Books, 1992); and Mary Lynn Manns and Linda Rising, *Fearless Change: Patterns for Introducing New Ideas* (Boston: Addison-Wesley, 2005).

The third role is that of *potential adopters,* or targets of the change—the people who actually must change. These are the people for whom the new system is designed and who will ultimately choose to use or not use the system.

In the early days of computing, many project teams simply assumed that their job ended when the old system was converted to the new system at a technical level. The philosophy was "build it and they will come." Unfortunately, that happens only in the movies. Resistance to change is common in most organizations. Therefore, the change management plan is an important part of the overall installation plan that glues together the key steps in the change management process. Successful change requires that people want to adopt the change and are able to adopt the change. The change management plan has four basic steps: revising management policies, assessing the cost and benefit models of potential adopters, motivating adoption, and enabling people to adopt through training (see Figure 13-2). However, before we can discuss the change management plan, we must first understand why people resist change.

Understanding Resistance to Change[7]

People resist change—even change for the better—for very rational reasons. What is good for the organization is not necessarily good for the people who work there. For example, consider an order-processing clerk who used to receive orders to be shipped on paper shipping documents but now uses a computer to receive the same information. Rather than typing shipping labels with a typewriter, the clerk now clicks on the print button on the computer and the label is produced automatically. The clerk can now ship many more orders each day, which is a clear benefit to the organization. The clerk, however, probably doesn't really care how many packages are shipped. His or her pay doesn't change; it's just a question of which the clerk prefers to use, a computer or typewriter. Learning to use the new system and work processes—even if the change is minor—requires more effort than continuing to use the existing, well-understood system and work processes.

So why do people accept change? Simply put, every change has a set of costs and benefits associated with it. If the benefits of accepting the change outweigh the costs of the change, then people change. And sometimes, the benefit of change is avoidance of the pain that might be experienced if the change were not adopted (e.g., if you don't change, you are fired, so one of the benefits of adopting the change is that you still have a job).

In general, when people are presented with an opportunity for change, they perform a cost–benefit analysis (sometime consciously, sometimes subconsciously) and decide the extent to which they will embrace and adopt the change. They identify the costs of and benefits from the system and decide whether the change is worthwhile. However, it is not that simple, because most costs and benefits are not certain. There is some uncertainty as to whether a certain benefit or cost will actually occur; so both the costs of and benefits from the new system need to be weighted by the degree of certainty associated with them (Figure 13-5). Unfortunately, most humans tend to overestimate the probability of costs and underestimate the probability of benefits.

There are also costs and, sometimes, benefits associated with the actual *transition process* itself. For example, suppose we found a nicer house or apartment than our current one. Even if we liked it better, we might decide not to move simply because the cost of moving outweighed the benefits from the new house or apartment itself. Likewise, adopting a new computer system might require us to learn new skills, which could be seen as a cost to some people or as a benefit to others, if they perceived that those skills would somehow provide other benefits beyond the use of the system itself. Once again, any costs and

[7] This section benefited from conversations with Dr. Robert Briggs, research scientist at the Center for the Management of Information at the University of Arizona.

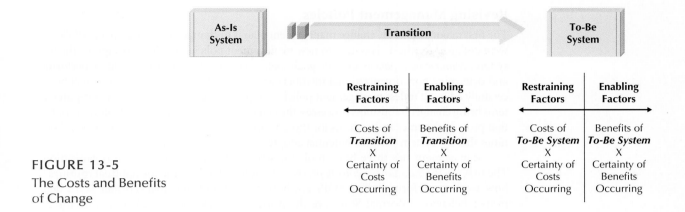

FIGURE 13-5
The Costs and Benefits of Change

benefits from the transition process must be weighted by the certainty with which they will occur (see Figure 13-5).

Taken together, these two sets of costs and benefits (and their relative certainties) affect the acceptance of change or resistance to change that project teams encounter when installing new systems in organizations. The first step in change management is to understand the factors that inhibit change—the factors that affect the perception of costs and benefits and certainty that they will be generated by the new system. It is critical to understand that the *real costs* and *real benefits* are far less important than the *perceived costs* and *perceived benefits*. People act on what they believe to be true, not on what is true. Thus, any understanding of how to motivate change must be developed from the viewpoint of the people expected to change, not from the viewpoint of those leading the change.

CONCEPTS IN ACTION

13-C Understanding Resistance to a DSS

*O*ne of the first commercial software packages I developed was a DSS to help schedule orders in a paper mill. The system was designed to help the person who scheduled orders decide when to schedule particular orders to reduce waste in the mill. This was a very challenging problem—so challenging, in fact, that it usually took the scheduler a year or two to really learn how to do the job well.

The software was tested by a variety of paper mills over the years and has always reduced the amount of waste, usually by about 25%, but sometimes by 75% when a scheduler new to the job was doing the scheduling. Although we ended up selling the package to most paper mills that tested it, we usually encountered significant resistance from the person doing the scheduling (except when the scheduler was new to the job and the package clearly saved a significant amount). At the time, I assumed that the resistance to the system was related to the amount of waste reduced: the less waste reduced, the more resistance because the payback analysis showed it took longer to pay for the software.

—Alan Dennis

Questions

1. What is another possible explanation for the different levels of resistance encountered at different mills?
2. How might this be addressed?

Revising Management Policies

The first major step in the change management plan is to change the management policies that were designed for the as-is system to new management policies designed to support the to-be system. *Management policies* provide goals, define how work processes should be performed, and determine how organizational members are rewarded. No computer system will be successfully adopted unless management policies support its adoption. Many new computer systems bring changes to business processes; they enable new ways of working. Unless the policies that provide the rules and rewards for those processes are revised to reflect the new opportunities that the system permits, potential adopters cannot easily use it.

Management has three basic tools for structuring work processes in organizations.[8] The first are the *standard operating procedures* (SOPs) that become the habitual routines for how work is performed. The SOPs are both formal and informal. Formal SOPs define proper behavior. Informal SOPs are the norms that have developed over time for how processes are actually performed. Management must ensure that the formal SOPs are revised to match the to-be system. The informal SOPs will then evolve to refine and fill in details absent in the formal SOPs.

The second aspect of management policy is defining how people assign meaning to events. What does it mean to "be successful" or "do good work"? Policies help people understand meaning by defining *measurements* and *rewards*. Measurements explicitly define meaning because they provide clear and concrete evidence about what is important to the organization. Rewards reinforce measurements because "what gets measured gets done" (an overused but accurate saying). Measurements must be carefully designed to motivate desired behavior. The IBM credit example (Your Turn 3-2) illustrates the problem when flawed measurements drive improper behavior (when the credit analysts became too busy to handle credit requests, they would find nonexistent errors so they could return them unprocessed).

A third aspect of management policy is *resource allocation*. Managers can have clear and immediate impacts on behavior by allocating resources. They can redirect funds and staff from one project to another, create an infrastructure that supports the new system, and invest in training programs. Each of these activities has both a direct and symbolic effect. The direct effect comes from the actual reallocation of resources. The symbolic effect shows that management is serious about its intentions. There is less uncertainty about management's long-term commitment to a new system when potential adopters see resources being committed to support it.

| YOUR | 13-2 Standard Operating Procedures |
| TURN | |

*I*dentify and explain three standard operating procedures for the course in which you are using this book. Discuss whether they are formal or informal.

---

[8] This section builds on the work of Anthony Giddons, *The Constitution of Society: Outline of the Theory of Structure* (Berkeley: University of California Press, 1984). A good summary of Giddons's theory that has been revised and adapted for use in understanding information systems is an article by Wanda Orlikowski and Dan Robey: "Information Technology and the Structuring of Organizations," *Information Systems Research* 2, no. 2 (1991): 143–169.

Assessing Costs and Benefits

The next step in developing a change management plan is to develop two clear and concise lists of costs and benefits provided by the new system (and the transition to it) compared with the as-is system. The first list is developed from the perspective of the organization, which should flow easily from the business case developed during the feasibility study and refined over the life of the project (see Chapter 2). This set of organizational costs and benefits should be distributed widely so that everyone expected to adopt the new system should clearly understand why the new system is valuable to the organization.

The second list of costs and benefits is developed from the viewpoints of the different potential adopters expected to change, or stakeholders in the change. For example, one set of potential adopters may be the frontline employees, another may be the first-line supervisors, and yet another might be middle management. Each of these potential adopters, or stakeholders, may have a different set of costs and benefits associated with the change—costs and benefits that can differ widely from those of the organization. In some situations, unions may be key stakeholders that can make or break successful change.

Many systems analysts naturally assume that frontline employees are the ones whose set of costs and benefits are the most likely to diverge from those of the organization and thus are the ones who most resist change. However, they usually bear the brunt of problems with the current system. When problems occur, they often experience them firsthand. Middle managers and first-line supervisors are the most likely to have a divergent set of costs and benefits and, therefore, resist change because new computer systems often change how much power they have. For example, a new computer system may improve the organization's control over a work process (a benefit to the organization) but reduce the decision-making power of middle management (a clear cost to middle managers).

An analysis of the costs and benefits for each set of potential adopters, or stakeholders, will help pinpoint those who will likely support the change and those who might resist the change. The challenge at this point is to try to change the balance of the costs and benefits for those expected to resist the change so that they support it (or at least do not actively resist it). This analysis could uncover some serious problems that have the potential to block the successful adoption of the system. It may be necessary to reexamine the management policies and make significant changes to ensure that the balance of costs and benefits is such that important potential adopters are motivated to adopt the system.

Figure 13-6 summarizes some of the factors that are important to successful change. The first and most important reason is a compelling personal reason to change. All change is made by individuals, not organizations. If there are compelling reasons for the key groups of individual stakeholders to want the change, then the change is more likely to be successful. Factors such as increased salary, reduced unpleasantness, and—depending on the individuals—opportunities for promotion and personal development can be important motivators. However, if the change makes current skills less valuable, individuals might resist the change because they have invested a lot of time and energy in acquiring those skills, and anything that diminishes those skills may be perceived as diminishing the individual (because important skills bring respect and power).

There must also be a compelling reason for the organization to need the change; otherwise, individuals become skeptical that the change is important and are less certain it will, in fact, occur. Probably the hardest organization to change is an organization that has been successful because individuals come to believe that what worked in the past will continue to work. By contrast, in an organization that is on the brink of bankruptcy, it is easier to convince individuals that change is needed. Commitment and support from credible business sponsors and top management are also important in increasing the certainty that the change will occur.

Factor		Examples	Effects	Actions to Take
Benefits of to-be system	Compelling personal reason(s) for change	Increased pay, fewer unpleasant aspects, opportunity for promotion, most existing skills remain valuable	If the new system provides clear personal benefits to those who must adopt it, they are more likely to embrace the change.	Perform a cost–benefit analysis from the viewpoint of the stakeholders, make changes where needed, and actively promote the benefits.
Certainty of benefits	Compelling organizational reason(s) for change	Risk of bankruptcy, acquisition, government regulation	If adopters do not understand why the organization is implementing the change, they are less certain that the change will occur.	Perform a cost–benefit analysis from the viewpoint of the organization and launch a vigorous information campaign to explain the results to everyone.
	Demonstrated top management support	Active involvement, frequent mentions in speeches	If top management is not seen to actively support the change, there is less certainty that the change will occur.	Encourage top management to participate in the information campaign.
	Committed and involved business sponsor	Active involvement, frequent visits to users and project team, championing	If the business sponsor (the functional manager who initiated the project) is not seen to actively support the change, there is less certainty that the change will occur.	Encourage the business sponsor to participate in the information campaign and play an active role in the change management plan.
	Credible top management and business sponsor	Management and sponsor who do what they say instead of being members of the "management fad of the month" club.	If the business sponsor and top management have credibility in the eyes of the adopters, the certainty of the claimed benefits is higher.	Ensure that the business sponsor and/or top management has credibility so that such involvement will help; if there is no credibility, involvement will have little effect.
Costs of transition	Low personal costs of change	Few new skills needed	The cost of the change is not borne equally by all stakeholders; the costs are likely to be higher for some.	Perform a cost–benefit analysis from the viewpoint of the stakeholders, make changes where needed, and actively promote the low costs.
Certainty of costs	Clear plan for change	Clear dates and instructions for change, clear expectations	If there is a clear migration plan, it will likely lower the perceived costs of transition.	Publicize the migration plan.
	Credible change agent	Previous experience with change, does what he/she promises to do	If the change agent has credibility in the eyes of the adopters, the certainty of the claimed costs is higher.	If the change agent is not credible, then change will be difficult.
	Clear mandate for change agent from sponsor	Open support for change agent when disagreements occur	If the change agent has a clear mandate from the business sponsor, the certainty of the claimed costs is higher.	The business sponsor must actively demonstrate support for the change agent.

FIGURE 13-6   Major Factors in Successful Change

The likelihood of successful change is increased when the cost of the transition to individuals who must change is low. The need for significantly different new skills or disruptions in operations and work habits can create resistance. A clear migration plan developed by a credible change agent who has support from the business sponsor is an important factor in increasing the certainty about the costs of the transition process.

Motivating Adoption

The single most important factor in motivating a change is providing clear and convincing evidence of the need for change. Simply put, everyone who is expected to adopt the change must be convinced that the benefits from the to-be system outweigh the costs of changing.

There are two basic strategies to motivating adoption: informational and political. Both strategies are often used simultaneously. With an *informational strategy*, the goal is to convince potential adopters that the change is for the better. This strategy works when the cost–benefit set of the target adopters has more benefits than costs. In other words, there really are clear reasons for the potential adopters to welcome the change.

Using this approach, the project team provides clear and convincing evidence of the costs and benefits of moving to the to-be system. The project team writes memos and develops presentations that outline the costs and benefits of adopting the system from the perspective of the organization and from the perspective of the target group of potential adopters. This information is disseminated widely throughout the target group, much like an advertising or public relations campaign. It must emphasize the benefits and increase the certainty in the minds of potential adopters that these benefits will actually be achieved. In our experience, it is always easier to sell painkillers than vitamins; that is, it is easier to convince potential adopters that a new system will remove a major problem (or other source of pain) than that it will provide new benefits (e.g., increase sales). Therefore, informational campaigns are more likely to be successful if they stress reducing or eliminating problems rather than focusing on providing new opportunities.

The other strategy for motivating change is a *political strategy*. With a political strategy, organizational power, not information, is used to motivate change. This approach is often used when the cost–benefit set of the target adopters has more costs than benefits. In other words, although the change might benefit the organization, there are no reasons for the potential adopters to welcome the change.

The political strategy is usually beyond the control of the project team. It requires someone in the organization who holds legitimate power over the target group to influence the group to adopt the change. This may be done in a coercive manner (e.g., adopt the system or you're fired) or in a negotiated manner, in which the target group gains benefits in other ways that are linked to the adoption of the system (e.g., linking system adoption to increased training opportunities). Management policies can play a key role in a political strategy by linking salary to certain behaviors desired with the new system.

In general, for any change that has true organizational benefits, about 20 to 30 percent of potential adopters will be *ready adopters*. They recognize the benefits, quickly adopt the system, and become proponents of the system. Another 20 to 30 percent are *resistant adopters*. They simply refuse to accept the change and they fight it, either because the new system has more costs than benefits for them personally or because they place such a high cost on the transition process itself that no amount of benefits from the new system can outweigh the change costs. The remaining 40 to 60 percent are *reluctant adopters*. They tend to be apathetic and will go with the flow to either support or resist the system, depending on how the project evolves and how their coworkers react to the system. Figure 13-7 illustrates the actors who are involved in the change management process.

The goal of change management is to actively support and encourage the ready adopters and help them win over the reluctant adopters. There is usually little that can be done about the resistant adopters because their set of costs and benefits may be divergent from those of the organization. Unless there are simple steps that can be taken to rebalance their costs and benefits or the organization chooses to adopt a strongly political strategy, it is often best to ignore this small minority of resistant adopters and focus on the larger majority of ready and reluctant adopters.

Sponsor	Change Agent	Potential Adopters
The sponsor wants the change to occur.	The change agent leads the change effort.	Potential adopters are the people who must change.
		20–30 percent are ready adopters.
		20–30 percent are resistant adopters.
		40–60 percent are reluctant adopters.

FIGURE 13-7
Actors in the Change Management Process

Enabling Adoption: Training

Potential adopters might want to adopt the change, but unless they are capable of adopting it, they won't. Careful *training* enables adoption by providing the skills needed to adopt the change. Training is probably the most self-evident part of any change management initiative. How can an organization expect its staff members to adopt a new system if they are not trained? However, we have found that training is one of the most commonly overlooked parts of the process. Many organizations and project managers simply expect potential adopters to find the system easy to learn. Because the system is presumed to be so simple, it is taken for granted that potential adopters should be able to learn with little effort. Unfortunately, this is usually an overly optimistic assumption.

Every new system requires new skills, either because the basic work processes have changed (sometimes radically, in the case of BPR; see Chapter 3) or because the computer system used to support the processes is different. The more radical the changes to the business processes, the more important it is to ensure the organization has the new skills required to operate the new business processes and supporting information systems. In general, there are three ways to get these new skills. One is to hire new employees who have the needed skills that the existing staff does not. Another is to outsource the processes to an organization that has the skills that the existing staff does not. Both these approaches are controversial and are usually considered only in the case of BPR when the new skills needed are likely to be the most different from the set of skills of the current staff. In most cases, organizations choose the third alternative: training existing staff in the new business processes and the to-be system. Every training plan must consider what to train and how to deliver the training.

What to Train   What training should you provide to the system users? It's obvious: how to use the system. The training should cover all the capabilities of the new system so users understand what each module does, right? Wrong. Training for business systems should focus on helping the users to accomplish their jobs, not on how to use the system. The system is simply a means to an end, not the end in itself. This focus on performing the job (i.e., the business processes), not using the system, has two important implications. First, the training must focus on the activities around the system as well as on the system itself. The training must help the users understand how the computer fits into the bigger picture of their jobs. The use of the system must be put in context of the manual business processes as well as of those that are computerized, and it must also cover the new management policies that were implemented along with the new computer system.

Second, the training should focus on what the user needs to do, not what the system can do. This is a subtle—but very important—distinction. Most systems provide far more capabilities than the users will need to use (e.g., when was the last time you wrote a macro in Microsoft Word?). Rather than attempting to teach the users all the features of the system, training should instead focus on the much smaller set of activities that users perform on a regular basis, and ensure that users are truly expert in those. When the focus is on the 20 percent of functions that the users will use 80 percent of the time (instead of attempting to cover all functions), users become confident about their ability to use the system. Training should mention the other little-used functions but only so that users are aware of their existence and know how to learn about them when their use becomes necessary.

One source of guidance for designing training materials is the use cases. The use cases outline the common activities that users perform and thus can be helpful in understanding the business processes and system functions that are likely to be most important to the users.

CONCEPTS IN ACTION

13-D Too Much Paper (Part 2)

Some clerks at the South Dakota Department of Worker's Compensation (see Concepts in Action 13-A) were afraid that the digital solution might not work. What if they could not find an electronic file on the computer? What if a hard drive crashed or files were accidentally deleted? What if they could not retrieve the electronic file?

Question

In terms of organizational feasibility and adoption, what might an analyst do to convince these clerks to adopt the new technology?

How to Train   There are many ways to deliver training. The most commonly used approach is *classroom training,* in which many users are trained at the same time by the same instructor. This has the advantage of training many users at one time with only one instructor and creates a shared experience among the users.

It is also possible to provide *one-on-one training,* in which one trainer works closely with one user at a time. This is obviously more expensive, but the trainer can design the training program to meet the needs of individual users and can better ensure that the users really do understand the material. This approach is typically used only when the users are very important or when there are very few users.

Another approach that is becoming more common is to use some form of *computer-based training (CBT),* in which the training program is delivered via computer, either on CD or over the web. CBT programs can include text slides, audio, and even video and animation. CBT is typically more costly to develop but is cheaper to deliver because no instructor is needed to actually provide the training.

Figure 13-8 summarizes four important factors to consider in selecting a training method: cost to develop, cost to deliver, impact, and reach. CBT is typically more expensive to develop than one-on-one or classroom training, but it is less expensive to deliver. One-on-one training has the most impact on the user because it can be customized to the user's precise needs, knowledge, and abilities, whereas CBT has the least impact. However, CBT has the greatest reach—the ability to train the most users over the widest distance in the shortest time—because it is much simpler to distribute than classroom and one-on-one training, simply because no instructors are needed.

	One-on-One Training	Classroom Training	Computer-Based Training
Cost to develop	Low to Medium	Medium	High
Cost to deliver	High	Medium	Low
Impact	High	Medium to High	Low to Medium
Reach	Low	Medium	High

FIGURE 13-8
Selecting a
Training Method

Figure 13-8 suggests a clear pattern for most organizations. If there are only a few users to train, one-on-one training is the most effective. If there are many users to train, many organizations turn to CBT. We believe that the use of CBT will increase in the future. Quite often, large organizations use a combination of all three methods. Regardless of which approach is used, it is important to leave the users with a set of easily accessible materials that can be referred to long after the training has ended (usually a quick reference guide and a set of manuals, whether on paper or in electronic form).

POST-IMPLEMENTATION ACTIVITIES[9]

The goal of post-implementation activities is the *institutionalization* of the use of the new system—that is, to make it the normal, accepted, routine way of performing the business processes. Post-implementation activities attempt to refreeze the organization after the successful transition to the new system. Although the work of the project team naturally winds down after implementation, the business sponsor, and sometimes the project manager, are actively involved in refreezing. These two—and, ideally, many other stakeholders—actively promote the new system and monitor its adoption and usage. They usually provide a steady flow of information about the system and encourage users to contact them to discuss issues.

In this section, we examine three key post-implementation activities: *system support* (providing assistance in the use of the system), *system maintenance* (continuing to refine and improve the system), and *project assessment* (analyzing the project to understand what activities were done well—and should be repeated—and what activities need improvement in future projects).

YOUR TURN

13-3 Developing a Training Plan

Suppose you are leading the conversion from one word processor to another in your organization. Develop an outline of topics that would be included in the training. Develop a plan for training delivery.

System Support

Once the project team has installed the system and performed the change management activities, the system is officially turned over to the *operations group*. This group is responsible for operating the system, whereas the project team was responsible for developing the system. Members of the operations group are usually closely involved in the installation activities because they are the ones who must ensure that the system actually works. After the system is installed, the project team leaves but the operations group remains.

Providing system support means helping the users to use the system. Usually, this means providing answers to questions and helping users understand how to perform a certain function; this type of support can be thought of as *on-demand training*.

Online support is the most common form of on-demand training. This includes the documentation and help screens built into the system, as well as separate websites that provide answers to *frequently asked questions (FAQs)*, which enable users to find answers without contacting a person. Obviously, the goal of most systems is to provide sufficiently good online

[9] The material in this section is related to the Enhanced Unified Process's Production Phase and the Operations and Support workflow (see Figure 1-18).

FIGURE 13-9
Elements of a
Problem Report

> - Time and date of the report
> - Name, e-mail address, and telephone number of the support person taking the report
> - Name, e-mail address, and telephone number of the person who reported the problem
> - Software and/or hardware causing problem
> - Location of the problem
> - Description of the problem
> - Action taken
> - Disposition (problem fixed or forwarded to system maintenance)

support so that the user doesn't need to contact a person, because providing online support is much less expensive than providing a person to answer questions.

Most organizations provide a *help desk* that provides a place for a user to talk with a person who can answer questions (usually over the phone but sometimes in person). The help desk supports all systems, not just one specific system, so it receives calls about a wide variety of software and hardware. The help desk is operated by *level-1 support* staff, who have very broad computer skills and are able to respond to a wide range of requests, from network problems and hardware problems to problems with commercial software and problems with the business application software developed in-house.

The goal of most help desks is to have the level-1 support staff resolve 80 percent of the help requests they receive on the first call. If the issue cannot be resolved by level-1 support staff, a *problem report* (Figure 13-9) is completed (often using a special computer system designed to track problem reports) and passed to a *level-2 support* staff member.

The level-2 support staff members are people who know the application system well and can provide expert advice. For a new system, they are usually selected during the implementation phase and become familiar with the system as it is being tested. Sometimes, the level-2 support staff members participate in training during the change management process to become more knowledgeable about the system, the new business processes, and the users themselves.

The level-2 support staff works with users to resolve problems. Most problems are successfully resolved by the level-2 staff. However, sometimes, particularly in the first few months after the system is installed, the problem turns out to be a bug in the software that must be fixed. In this case, the problem report becomes a *change request* that is passed to the system maintenance group (see the next section).

CONCEPTS IN ACTION

13-E Too Much Paper (Part 3)

The South Dakota Department of Worker's Compensation had legal hurdles to implementing a digital solution to handle workers' compensation claims (see Concepts in Action 13-A and 13-D). One hurdle was that the previous paper method had physical signatures from employees indicating that they had received treatment or from the doctor indicating that medical treatment was performed.

Question

What legal aspect might arise from only having digital signatures or only electronic or paper copies of documents instead of physical documents?

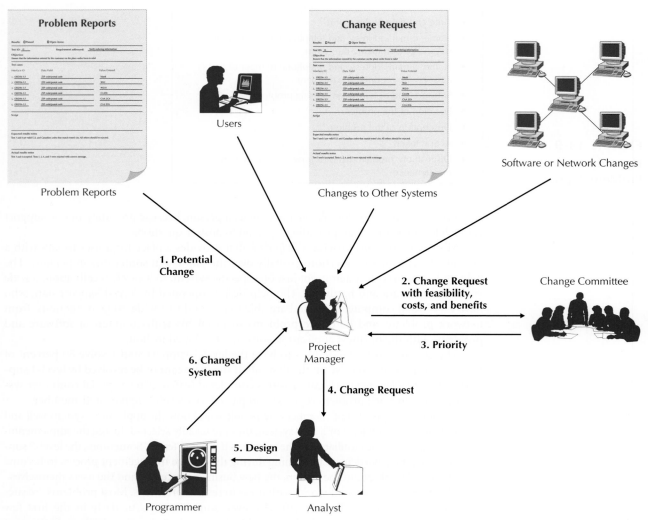

FIGURE 13-10   Processing a Change Request

System Maintenance

System maintenance is the process of refining the system to make sure it continues to meet business needs. Substantially more money and effort is devoted to system maintenance than to the initial development of the system, simply because a system continues to change and evolve as it is used. Most beginning systems analysts and programmers work first on maintenance projects; usually only after they have gained some experience are they assigned to new development projects.

Every system is "owned" by a project manager in the IS group (Figure 13-10). This individual is responsible for coordinating the system's maintenance effort for that system. Whenever a potential change to the system is identified, a change request is prepared and forwarded to the project manager. The change request is a smaller version of the *system request* discussed in Chapter 2. It describes the change requested and explains why the change is important.

Changes can be small or large. Change requests that are likely to require a significant effort are typically handled in the same manner as system requests: they follow the same process as the project described in this book, starting with project identification in Chapter 2 and following through installation in this chapter. Minor changes typically follow a smaller version of this same process. There is an initial assessment of feasibility and of costs and benefits, and the change request is prioritized. Then a systems analyst (or a programmer/analyst) performs the analysis, which might include interviewing users, and prepares an initial design before programming begins. The new (or revised) program is then extensively tested before the system is converted from the old system to the revised one.

Change requests typically come from five sources. The most common source is problem reports from the operations group that identify bugs in the system that must be fixed. These are usually given immediate priority because a bug can cause significant problems. Even a minor bug can cause major problems by upsetting users and reducing their acceptance of and confidence in the system.

The second most common source of change requests is enhancement to the system from users. As users work with the system, they often identify minor changes in the design that can make the system easier to use or identify additional functions that are needed. Such enhancements are important in satisfying the users and are often key in ensuring that the system changes as the business requirements change. Enhancements are often given second priority after bug fixes.

A third source of change requests is other system development projects. For example, if the doctor in the appointment problem decided that he or she would like to have a web-based appointment system that would allow patients to directly interact with the current appointment system, it is likely that other systems, such as billing, would have to be modified to ensure that the two systems would work together. These changes required by the need to integrate two systems are generally rare, but are becoming more common as system integration efforts become more common.

A fourth source of change requests is those that occur when underlying software or networks change. For example, new versions of Windows often require an application to change the way the system interacts with Windows or enables application systems to take advantage of new features that improve efficiency. Although users might never see these changes (because most changes are inside the system and do not affect its user interface or functionality), these changes can be among the most challenging to implement because analysts and programmers must learn about the new system characteristics, understand how application systems use (or can use) those characteristics, and then make the needed programming changes.

The fifth source of change requests is senior management. These change requests are often driven by major changes in the organization's strategy or operations. These significant change requests are typically treated as separate projects, but the project manager responsible for the initial system is often placed in charge of the new project.

Project Assessment

The goal of project assessment is to understand what was successful about the system and the project activities (and, therefore, should be continued in the next system or project) and what needs to be improved. Project assessment is not routine in most organizations, except for military organizations, which are accustomed to preparing after-action reports. Nonetheless, assessment can be an important component in organizational learning because it helps organizations and people understand how to improve their work. It is particularly important for junior staff members because it helps promote faster learning. There are two primary parts to project assessment—project team review and system review.

Project Team Review   A *project team review* focuses on the way the project team carried out its activities. Each project member prepares a short, two- to three-page document that reports and analyzes his or her performance. The focus is on performance improvement, not penalties for mistakes made. By explicitly identifying mistakes and understanding their causes, project team members will, it is hoped, be better prepared for the next time they encounter a similar situation—and less likely to repeat the same mistakes. Likewise, by identifying excellent performance, team members will be able to understand why their actions worked well and how to repeat them in future projects.

PRACTICAL   13-1 Beating Buggy Software

TIP

*H*ow do you avoid bugs in the commercial software you buy? Here are six tips:

1. *Know your software:* Find out if the few programs you use day in and day out have known bugs and patches, and track the websites that offer the latest information on them.
2. *Back up your data:* This dictum should be tattooed on every monitor. Stop reading right now and copy the data you can't afford to lose onto a second hard disk or web server. We'll wait.
3. *Don't upgrade—yet:* It's tempting to upgrade to the latest and greatest version of your favorite software, but why chance it? Wait a few months, check out other users' experiences with the upgrade on Usenet news-groups or the vendor's own discussion forum, and then go for it. But only if you must.

4. *Upgrade slowly:* If you decide to upgrade, allow your-self at least a month to test the upgrade on a separate system before you install it on all the computers in your home or office.
5. *Forget the betas:* Installing beta software on your pri-mary computer is a game of Russian roulette. If you really have to play with beta software, get a second computer.
6. *Complain:* The more you complain about bugs and demand remedies, the more costly it is for vendors to ship buggy products. It's like voting—the more people participate, the better the results.

Source: "Software Bugs Run Rampant," *PC World* 17, no. 1 (January 1999): 46.

The project manager, who meets with the team members to help them understand how to improve their performance, assesses the documents prepared by each team member. The project manager then prepares a summary document that outlines the lessons learned from the project. This summary identifies what actions should be taken in future projects to improve performance but is careful not to identify team members who made mistakes. The summary is widely circulated among all project managers to help them understand how to manage their projects better. Often, it is also circulated among regular staff members who did not work on the project so that they, too, can learn from other projects.

System Review   The focus of the *system review* is to understand the extent to which the proposed costs and benefits from the new system identified during feasibility analysis were actually recognized from the implemented system. Project team review is usually con-ducted immediately after the system is installed while key events are still fresh in team members' minds, but system review is often undertaken several months after the system is installed because it often takes a while before the system can be properly assessed.

System review starts with the system request and feasibility analysis prepared at the start of the project. The detailed analyses prepared for the expected business value (both tangible and intangible) as well as the economic feasibility analysis are reexamined and a new analysis is prepared after the system has been installed. The objective is to compare the anticipated business value against the actual realized business value from the system. This helps the organization assess whether the system actually provided the value it was planned to provide. Whether or not the system provides the expected value, future projects can benefit from an improved understanding of the true costs and benefits.

A formal system review also has important behavior implications for project initiation. Because everyone involved with the project knows that all statements about business value and the financial estimates prepared during project initiation will be evaluated at the end of the project, they have an incentive to be conservative in their assessments. No one wants to be the project sponsor or project manager for a project that goes radically over budget or fails to deliver promised benefits.

APPLYING THE CONCEPTS AT CD SELECTIONS

In this installment of the CD Selections case, we see how the new system is transitioned from the development team and put into production by the user community. To ensure a smooth transition, Alec and Margaret oversaw the necessary user training, including employees from CD Selections help desk department, and the creation of the necessary, relevant documentation. Looking back over the development of the system, Alec and Margaret evaluate the processes used and the individual development team members to identify lessons learned throughout the process. Finally, they set up a process to maintain the system.

SUMMARY

Cultural Issues and Information Technology Adoption

Given the global business environment, cultural issues become even more important when deploying information systems today. The cultural dimensions that need to be taken into consideration include Hall's speed of messages, context, and time dimensions along with Hofstede's power distance, uncertainty avoidance, individualism versus collectivism, masculinity versus femininity, and long- versus short-term orientation dimensions. Furthermore, these dimensions tend to interact with one another. From an information systems deployment perspective, the most important thing to remember is to take into consideration the local culture before deploying the new system.

Conversion

Conversion, the technical process by which the new system replaces the old system, has three major steps: install hardware, install software, and convert data. Conversion style, in which users are switched between the old and new systems, can be via either direct conversion (in which users stop using the old system and immediately begin using the new system) or parallel conversion (in which both systems are operated simultaneously to ensure the new system is operating correctly). Conversion location—what parts of the organization are converted and when—can be via a pilot conversion in one location; via a phased conversion, in which locations are converted in stages over time; or via simultaneous conversion, in which all locations are converted at the same time. The system can be converted module by module or as a whole at one time. Parallel and pilot conversions are less risky because they have a greater chance of detecting bugs before the bugs have widespread effect, but parallel conversion can be expensive.

Change Management

Change management is the process of helping people to adopt and adapt to the new system and its work processes. People resist change for very rational reasons, usually because they perceive the costs to themselves of the new system (and the transition to it) to outweigh the benefits. The first step in the change management plan is to change the management policies, devise measurements and rewards that support the new system, and allocate resources to support it. The second step is to develop a concise list of costs and benefits to the organization and to all relevant stakeholders. This points out who is likely to support and who is likely to resist the change. The third step is to motivate adoption both by providing information and by using political strategies—using power to induce potential adopters to adopt the new system. Finally, training is essential to enable successful adoption. Training should focus on the primary functions the users will perform and look beyond the system itself to help users integrate the system into their work processes.

Post-implementation Activities

System support is performed by the operations group, which provides online and help desk support to the users. System support has both a level-1 support staff, who answer the phone and handles most of the questions, and level-2 support staff, who follow up on challenging problems and sometimes generates change requests for bug fixes. System maintenance responds to change requests to fix bugs and improve the business value of the system. The goal of project assessment is to understand what was successful about the system and the project activities and what needs to be improved. Project team review focuses on the way the project team carried out its activities and usually results in documentation of key lessons learned. System review focuses on understanding the extent to which the proposed costs and benefits from the new system were actually recognized from the implemented system.

KEY TERMS

Change agent, 555
Change management, 547
Change request, 565
Classroom training, 563
Collectivism, 548
Computer-based training (CBT), 563
Context, 548
Conversion, 549
Conversion location, 550

LIBRARY, UNIVERSITY OF CHESTER

QUESTIONS

1. What are the three basic steps in managing organizational change?
2. What are the cultural issues of which developers should be aware?
3. What are the major components of a migration plan?
4. Compare and contrast direct conversion and parallel conversion.
5. Compare and contrast pilot conversion, phased conversion, and simultaneous conversion.
6. Compare and contrast modular conversion and whole-system conversion.
7. Explain the trade-offs among selecting between the types of conversion in questions 4, 5, and 6.
8. What are the three key roles in any change management initiative?
9. Why do people resist change? Explain the basic model for understanding why people accept or resist change.
10. What are the three major elements of management policies that must be considered when implementing a new system?
11. Compare and contrast an information change management strategy with a political change management strategy. Is one better than the other?
12. Explain the three categories of adopters you are likely to encounter in any change management initiative.
13. How should you decide what items to include in your training plan?
14. Compare and contrast three basic approaches to training.
15. What is the role of the operations group in system development?
16. Compare and contrast two major ways of providing system support.
17. How is a problem report different from a change request?
18. What are the major sources of change requests?
19. Why is project assessment important?
20. How is project team review different from system review?
21. What do you think are three common mistakes that novice analysts make in migrating from the as-is to the to-be system?
22. Some experts argue that change management is more important than any other part of system development. Do you agree or not? Explain.
23. In our experience, change management planning often receives less attention than conversion planning. Why do you think this happens?

EXERCISES

A. Suppose you are installing a new accounting package in your small business. What conversion strategy would you use? Develop a conversion plan (i.e., technical aspects only).

B. Suppose you are installing a new room reservation system for your university that tracks which courses are assigned to which rooms. Assume that all the rooms in each building are "owned" by one college or department and only one person in that college or department has permission to assign them. What conversion strategy would you use? Develop a conversion plan (i.e., technical aspects only).

C. Suppose you are installing a new payroll system in a very large multinational corporation. What conversion strategy would you use? Develop a conversion plan (i.e., technical aspects only).

D. Consider a major change you have experienced in your life (e.g., taking a new job, starting a new school). Prepare a cost–benefit analysis of the change in terms of both the change and the transition to the change.

E. Suppose you are the project manager for a new library system for your university. The system will improve the way students, faculty, and staff can search for books by enabling them to search over the web, rather than using only the current text-based system available on the computer terminals in the library. Prepare a cost–benefit analysis of the change in terms of both the change and the transition to the change for the major stakeholders.

F. Prepare a plan to motivate the adoption of the system in exercise E.

G. Prepare a training plan that includes both what you would train and how the training would be delivered for the system in exercise E.

H. Suppose you are leading the installation of a new DSS to help admissions officers manage the admissions process at your university. Develop a change management plan (i.e., organizational aspects only).

I. Suppose you are the project leader for the development of a new web-based course registration system for your university that replaces an old system in which students had to go to the coliseum at certain times and stand in line to get permission slips for each course they wanted to take. Develop a migration plan (including both technical conversion and change management).

J. Suppose you are the project leader for the development of a new airline reservation system that will be used by the airline's in-house reservation agents. The system will replace the current command-driven system designed in the 1970s that uses terminals. The new system uses PCs with a web-based interface. Develop a migration plan (including both conversion and change management) for your telephone operators.

K. Develop a migration plan (including both conversion and change management) for the independent travel agencies who use the airline reservation system described in exercise J.

L. For the A Real Estate Inc problem in Chapters 4 through 12:
   1. Prepare a plan to motivate adoption of the system.
   2. Prepare a training plan that includes both what you would train and how the training would be delivered.
   3. Prepare a change management plan.
   4. Develop a migration plan.

M. For the A Video Store problem in Chapters 4 through 12:
   1. Prepare a plan to motivate adoption of the system.
   2. Prepare a training plan that includes both what you would train and how the training would be delivered.
   3. Prepare a change management plan.
   4. Develop a migration plan.

N. For the gym problem in Chapters 4 through 12:
   1. Prepare a plan to motivate adoption of the system.
   2. Prepare a training plan that includes both what you would train and how the training would be delivered.
   3. Prepare a change management plan.
   4. Develop a migration plan.

O. For the Picnics R Us problem in Chapters 4 through 12:
   1. Prepare a plan to motivate adoption of the system.
   2. Prepare a training plan that includes both what you would train and how the training would be delivered.
   3. Prepare a change management plan.
   4. Develop a migration plan.

P. For Of-the-Month Club problem in Chapters 4 through 12:
   1. Prepare a plan to motivate adoption of the system.
   2. Prepare a training plan that includes both what you would train and how the training would be delivered.
   3. Prepare a change management plan.
   4. Develop a migration plan.

MINICASES

1. The development team of New York City Global Public Transportation System 2030 (NYC-GPTS2030) (see minicase 1, Chapter 11) is facing the important decision of selecting the appropriate conversion strategy for the installation of the new monitoring, management, and user services system that will automate the activity of the city's entire Public Transportation System (PTS). Since this is such a complex decision with so many variables and aspects involved, some of them not well-known or completely understood, the members of the development team ended up in an impasse due to divergent opinions on what strategy to adopt along the three dimensions of the conversion strategy: style, location, and modules. Although the team members agree on some aspects of the conversion strategy, a comprehensive consensus that would allow the project to move forward seems to be difficult to achieve. Following the layout of the main conversion strategies from the text, comment and highlight the pros and cons of each of the possible conversion decisions along the dimensions specified above.

2. San Babila Taxi operates a fleet of taxis in Milan, Italy. The company provides the usual metered taxi services as well as short-term and long-term private car services for companies in Milan. With the increase in the number of cars, drivers, and contracts, the San Babila has experienced difficulty in keeping accurate records of its business activity and the movement of its fleet of cars. It recently purchased a new information system for the business and instructed all its drivers to use the new system. Private car drivers will have to swipe an ID badge through a reader at the start and end of each shift, and key in particulars of the car used and the client(s) served during the shift. Taxi drivers will also have to swipe their ID badges, record particulars of their cars, and, using an in-taxi computer system, record each and every fare they pick up. The in-taxi computer system will also be used for dispatching taxis to customers' requests by phone or through the company's website.

   The San Babila office staff were eagerly awaiting the installation of the new system. They agreed the system would reduce problems and errors in billing and would make their work easier. The drivers were less enthusiastic, being unaccustomed to the use of computers and to having their activities closely monitored.

   a. Discuss the factors that may inhibit the acceptance of this new system by the drivers.
   b. Discuss how an informational strategy could be used to motivate adoption of the new system.
   c. Discuss how a political strategy could be used to motivate adoption of the new system.

3. Professor Takato, president of FRDI, is happy to see his long anticipated CMS readily installed with all the features and modules, as initially described in minicase 4 in Chapter 7, in place. Given the list of the main modules of the CMS application:
   a. General Presentation and Marketing – presents the conference, conference topics, venue, traveling and accommodation information, program committees, important dates, and other.
   b. Communication Automation – supports generation and submission of possibly personalized e-mails or other types of messages to various categories of recipients: potential participants drawn from a database of previous conference participants to advertise the upcoming conference, submitting authors (e.g., for confirmation of paper acceptance), reviewers (e.g., for sending list of papers to review), and other.
   c. Submission System – supports submission of papers.
   d. Review System – supports management of the paper review process.
   e. Registration System – supports conference registration and payment of conference fees.
   f. Post-conference Survey System – supports the collection and evaluation of post-conference online feedback from participants.
      Do the following:
      1. Identify the main user categories of the CMS application.
      2. Help Professor Takato develop a strategy for the post-implementation activities related to his CMS project, namely, system support and system maintenance.

4. For the Professional and Scientific Staff Management problem described in Chapters 4, and 6 through 11:
   a. Prepare a plan to motivate adoption of the system.
   b. Prepare a training plan that includes both what you would train and how the training would be delivered.
   c. Prepare a change management plan.
   d. Develop a migration plan.

INDEX

P